Probability and Mathematical Statistics (Continued)

MUIRHEAD • Aspects of Multivariate Statistical
PARZEN • Modern Probability Theory and Its Applications
PURI and SEN • Nonparametric Methods in General Linear Models
PURI and SEN • Nonparametric Methods in Multivariate Analysis
RANDLES and WOLFE • Introduction to the Theory of Nonparametric Statistics
RAO • Linear Statistical Inference and Its Applications, *Second Edition*
RAO and SEDRANSK • W.G. Cochran's Impact on Statistics
ROHATGI • An Introduction to Probability Theory and Mathematical Statistics
ROHATGI • Statistical Inference
ROSS • Stochastic Processes
RUBINSTEIN • Simulation and The Monte Carlo Method
SCHEFFE • The Analysis of Variance
SEBER • Linear Regression Analysis
SEBER • Multivariate Observations
SEN • Sequential Nonparametrics: Invariance Principles and Statistical Inference
SERFLING • Approximation Theorems of Mathematical Statistics
TJUR • Probability Based on Radon Measures
WILLIAMS • Diffusions, Markov Processes, and Martingales, Volume I: Foundations
ZACKS • Theory of Statistical Inference

Applied Probability and Statistics

ABRAHAM and LEDOLTER • Statistical Methods for Forecasting
AGRESTI • Analysis of Ordinal Categorical Data
AICKIN • Linear Statistical Analysis of Discrete Data
ANDERSON, AUQUIER, HAUCK, OAKES, VANDAELE, and WEISBERG • Statistical Methods for Comparative Studies
ARTHANARI and DODGE • Mathematical Programming in Statistics
BAILEY • The Elements of Stochastic Processes with Applications to the Natural Sciences
BAILEY • Mathematics, Statistics and Systems for Health
BARNETT • Interpreting Multivariate Data
BARNETT and LEWIS • Outliers in Statistical Data, *Second Edition*
BARTHOLOMEW • Stochastic Models for Social Processes, *Third Edition*
BARTHOLOMEW and FORBES • Statistical Techniques for Manpower Planning
BECK and ARNOLD • Parameter Estimation in Engineering and Science
BELSLEY, KUH, and WELSCH • Regression Diagnostics: Identifying Influential Data and Sources of Collinearity
BHAT • Elements of Applied Stochastic Processes, *Second Edition*
BLOOMFIELD • Fourier Analysis of Time Series: An Introduction
BOX • R. A. Fisher, The Life of a Scientist
BOX and DRAPER • Evolutionary Operation: A Statistical Method for Process Improvement
BOX, HUNTER, and HUNTER • Statistics for Experimenters: An Introduction to Design, Data Analysis, and Model Building
BROWN and HOLLANDER • Statistics: A Biomedical Introduction
BUNKE and BUNKE • Statistical Inference in Linear Models, Volume I
CHAMBERS • Computational Methods for Data Analysis
CHATTERJEE and PRICE • Regression Analysis by Example
CHOW • Econometric Analysis by Control Methods
CLARKE and DISNEY • Probability and Random Processes: A First Course with Applications, *Second Edition*
COCHRAN • Sampling Techniques, *Third Edition*
COCHRAN and COX • Experimental Designs, *Second Edition*
CONOVER • Practical Nonparametric Statistics, *Second Edition*

CONOVER and IMAN • Introduction to Modern Business Statistics
CORNELL • Experiments with Mixtures: Designs, Models and The Analysis of Mixture Data
COX • Planning of Experiments
DANIEL • Biostatistics: A Foundation for Analysis in the Health Sciences, *Third Edition*
DANIEL • Applications of Statistics to Industrial Experimentation
DANIEL and WOOD • Fitting Equations to Data: Computer Analysis of Multifactor Data, *Second Edition*
DAVID • Order Statistics, *Second Edition*
DAVISON • Multidimensional Scaling
DEMING • Sample Design in Business Research
DILLON and GOLDSTEIN • Multivariate Analysis: Methods and Applications
DODGE • Analysis of Experiments with Missing Data
DODGE and ROMIG • Sampling Inspection Tables, *Second Edition*
DOWDY and WEARDEN • Statistics for Research
DRAPER and SMITH • Applied Regression Analysis, *Second Edition*
DUNN • Basic Statistics: A Primer for the Biomedical Sciences, *Second Edition*
DUNN and CLARK • Applied Statistics: Analysis of Variance and Regression
ELANDT-JOHNSON and JOHNSON • Survival Models and Data Analysis
FLEISS • Statistical Methods for Rates and Proportions, *Second Edition*
FLEISS • The Design and Analysis of Clinical Experiments
FOX • Linear Statistical Models and Related Methods
FRANKEN, KÖNIG, ARNDT, and SCHMIDT • Queues and Point Processes
GALAMBOS • The Asymptotic Theory of Extreme Order Statistics
GIBBONS, OLKIN, and SOBEL • Selecting and Ordering Populations: A New Statistical Methodology
GNANADESIKAN • Methods for Statistical Data Analysis of Multivariate Observations
GOLDBERGER • Econometric Theory
GOLDSTEIN and DILLON • Discrete Discriminant Analysis
GREENBERG and WEBSTER • Advanced Econometrics: A Bridge to the Literature
GROSS and CLARK • Survival Distributions: Reliability Applications in the Biomedical Sciences
GROSS and HARRIS • Fundamentals of Queueing Theory, *Second Edition*
GUPTA and PANCHAPAKESAN • Multiple Decision Procedures: Theory and Methodology of Selecting and Ranking Populations
GUTTMAN, WILKS, and HUNTER • Introductory Engineering Statistics, *Third Edition*
HAHN and SHAPIRO • Statistical Models in Engineering
HALD • Statistical Tables and Formulas
HALD • Statistical Theory with Engineering Applications
HAND • Discrimination and Classification
HILDEBRAND, LAING, and ROSENTHAL • Prediction Analysis of Cross Classifications
HOAGLIN, MOSTELLER and TUKEY • Exploring Data Tables, Trends and Shapes
HOAGLIN, MOSTELLER, and TUKEY • Understanding Robust and Exploratory Data Analysis
HOEL • Elementary Statistics, *Fourth Edition*
HOEL and JESSEN • Basic Statistics for Business and Economics, *Third Edition*
HOGG and KLUGMAN • Loss Distributions

(*continued on back*)

Analysis of Experiments
with Missing Data

Analysis of Experiments with Missing Data

Yadolah Dodge

Professor of Statistics and Operations Research
University of Neuchâtel, Switzerland

JOHN WILEY & SONS

New York • Chichester • Brisbane • Toronto • Singapore

Library of Congress Cataloging in Publication Data:

Dodge, Yadolah, 1944–
Analysis of experiments with missing data.

(Wiley series in probability and mathematical statistics. Applied probability and statistics, ISSN 0271-6356)
Includes bibliographies and index.
1. Experimental design. 2. Analysis of variance.
I. Title. II. Series.
QA279.D63 1985 001.6'34 85-5296
ISBN 0-471-88736-6

Printed in the United States of America

10 9 8 7 6 5 4 3 2 1

To

David S. Birkes

Khadijeh Khatami

Justus F. Seely

Preface

The purpose of this book is to present recently developed theories and techniques for analyzing a data set resulting from designed experiments with missing data.

In factorial experiments one often has a situation where it is not possible to accomodate all treatment combinations in the experiment. The device of fractional replication pioneered by Finny (1945) has proved to be very useful in such situations. In these cases the design matrix is of maximal rank and the experiment can be analyzed in a straightforward fashion.

However, it is possible that some of the observations are missing, cannot be collected, or in some other way are not obtainable. For example, crops are destroyed in some plots, an experimenter fails to record some data, gross errors occur in recording, a patient withdraws from a treatment program, one or more animals die in the course of an experiment, or some subjects do not take all treatments as planned. Missing data of this nature occur quite frequently.

In such situations the designed experiment is no longer balanced and loses its symmetry, and in some situations, when enough observations are missing, not all of the usual parametric functions are estimable. Such a situation, unless discovered, will lead to incorrect assignment of degrees of freedom for certain sums of squares, to singular matrices that are to be inverted in a computer program, and, most importantly, will make contrasts involved in the determination of certain sums of squares to be not known precisely. Therefore, analyzing data of this nature without appropriate knowledge may lead to inappropriate inferences about the hypothesis and, consequently, to wrong conclusions about the results of the experiment.

The most common approach to this problem when there are few empty cells is to apply the analysis for a complete experiment using calculated values in place of the missing observations. Here one calculated value is placed into each empty cell. The values are chosen so that the residual sum of squares is minimized.

Allen and Wishart (1930) seem to have been the first to have considered the problem of missing observations by providing two formats for estimating the value of a single missing observation for a randomized block and a Latin squares design, respectively. Their work was later pursued by Yates (1933, 1936), Yates and Hale (1939), Cornish (1940a,b, 1941, 1944), Anderson (1946), Tocher (1952), Healy and Westmacott (1956), Hartley (1956), Wilkinson (1958a,b, 1960), Joech (1966), Preece (1971), Scolve (1972), Shrearer (1973), Rubin (1972), Haseman and Gaylor (1973), and Jarrett (1978).

A second approach to the problem of missing observations in classification designs is to employ the incidence matrix **N** as opposed to the design matrix **X**. Utilizing the incidence matrix **N** to obtain the estimability information seems to have been first considered by Bose (1949) for an additive two-way model.

For an additive two-way classification model (block by treatment, with arbitrary incidence) Bose introduced the notion of connectedness, and via this concept answered the question of whether every treatment contrast is estimable. Following Bose's work, Weeks and Williams (1964) and Srivastava and Anderson (1970) treated the additive n-way classification model.

Recently, Birkes, Dodge, and Seely (1976) presented general and complete results for estimability considerations in an additive two-way classification model which are easily programmed for computers. They introduced an algorithm, the R process, which determines what cell expectations are estimable. Furthermore, they gave a method for determining a basis for the estimable functions involving only one effect. They also provided results on estimability for an additive three-way classification model with arbitrary incidence matrices.

A third approach to the problem of missing observations in designed experiments is the exact least squares approach of writing a mathematical model for all observations that are present. The least squares normal equations are then constructed for the estimation of parameters in the model. Because the equations corresponding to the missing observations are in fact "missing," the system of normal equations loses its symmetry as compared to when all observations are present, and consequently to obtain solutions to normal equations, generalized inverses may be needed.

Computational aspects of generalized inverses have not been dealt with very extensively in the literature and at any event it seems that such computations are subject to rounding-off errors, which might even lead to erroneous conclusions. Dodge and Majumdar (1979) presented a method of analyzing a data set following an n-way classification model with completely arbitrary pattern via an algorithm which finds a least squares g inverse for the design matrix using the incidence and the design matrix simultaneously.

This book gives complete results in analyzing experiments with arbitrary patterns to both theoretical and applied statisticians. The treatment is theoretical but many examples along with computer programs are provided to clarify the concepts. Chapter 1 contains definitions and terminologies on designed experiments and linear models, and also explains the notations used in this book. Chapter 2 discusses the comparison of treatments in the one-way classification experiments and introduces several important statistical concepts and principles that will be used in other chapters. Experiments involving two factors and the two-way classification models are considered in Chapter 3. The notion of correspondences between parametrizations is also discussed in this chapter. The complete 2^n factorial experiments are the subject of Chapter 4.

The main purpose of Chapter 5 is to present general and rather complete results for estimability considerations in an additive two-way classification model which are easily programmed for electronic computers. Methods are given, based upon the completion of the R process, for determining a basis for the estimable functions involving only one effect; for determining ranks of matrices pertinent to considerations for degrees of freedom; and for determining which portions of the design are connected. The results are somewhat novel in that only integer operations and the incidence matrix $\mathbf{N}$ are utilized. In fact, one needs only the facility to distinguish between zeros and nonzeros in the incidence matrix to obtain all the results. A procedure is outlined in Chapter 5 for obtaining the usual sums of squares and for obtaining a solution for the normal equations as well as the covariance matrix associated with the solution obtained. Attention is paid to ensuring that certain matrices are invertible and to discussing various restrictions which may, if desired, be imposed upon the solution to the normal equations.

Chapters 6 and 7 address the three- and n-way classifications with missing observations, respectively. The approach in these two chapters is along the same line as that of Chapter 5. In Chapter 8, the construction of least squares generalized inverse for classification models is considered. The algorithms presented in this chapter give the g inverse with no rounding-off errors, and it is programmed for electronic computers.

Minimally connected designs are the subject of Chapter 9. Selective tables of minimally connected 2^n for $n \leqslant 9$ and $a \times b$ designs are provided at the end of Chapter 9. The computer program that is presented at the end of Chapter 9 is capable of providing more minimally connected $a \times b$ designs if desired. The material in Chapters 5 through 9 has never appeared before in classical textbooks dealing with experimental designs.

This book is written for readers with a background knowledge of mathematics and statistics at the undergraduate level. For those who are already familiar with matrix algebra and elementary analysis of variance Chapters 1 through 4 may be reviewed quickly. However, readers without the minimal mathematical background indicated above should read Chapter 1 of Rao's *Linear Statistical Inference and its Applications*, or Appendices I and II of Scheffé's *Analysis of Variance*, or Chapter 1 of Searle's *Linear Models*. Apart from matrix algebra, an attempt has been made to make this book self-contained.

The main emphasis of this book is presented in Chapters 5 through 9, with computer programs given at the end of each of these chapters. They are written in standard FORTRAN in an interactive way and can be implemented directly on any system that accepts standard FORTRAN. These programs have been tested on a variety of problems. At the end of each program one or two examples, which have already been described in the main body of the respective chapter, are given with complete input–output data. In this way it is easy to see how the output can be interpreted. Although writing these programs is a difficult task, I believe that any statistics book dealing directly with computation must be accompanied by computer programs in order to eliminate the gap between theory and application.

There is enough material in the book for a one-term course on analysis of experiments with missing data. In addition, the contents of this book may be useful as supplementary material in such courses as analysis of variance, linear models, computational statistics, and statistical methods.

A part of this book was written while I was a visiting professor at Oregon State University from 1979 to 1980. At Oregon State, I had the opportunity to work with Dr. Paul McClellan, who helped me with Chapters 1 through 4. Although those chapters have been drastically revised in this book, without his valuable help on these and other chapters, this book would never have gotten off the ground. Comments by Professor T. Calinski of the Academy of Agriculture in Poznan, Poland, by Professor D. Birkes of Oregon State University, by Dr. Dibyen Majumdar of the University of Illinois, by Professor K. Afsarinejad of Linköping University in Sweden, by Professor P. Nüsch of Ecole Polytechnique Fédérale de Lausanne, and, last but not least, by my former student Mr. K. Kazempour have been extremely useful in the preparation of the final manuscript. My assistants M. G. Zagami, A. Weber and S. Weber helped me a great deal with the numerical computations of the examples and computer programs. Mrs. H. Badan undertook the heavy burden of typing the manuscript. I wish to express my deep gratitude to all of them.

Finally, I wish to express my sincere appreciation to Professor Justus F. Seely who suggested the initial problem in 1971, and to Professor David

Birkes and Professor Tadeusz Calinski, whose helpful suggestions and criticisms contributed greatly to this manuscript. Without the guidance and encouragement of these three people, this book could not have been completed.

YADOLAH DODGE

Neuchâtel, Switzerland
April 1985

Contents

Analysis of Experiments
with Missing Data

CHAPTER 1

Prologue

Me: *Which way must I follow?*

The Sage: *If you are a real pilgrim, you will achieve the journey whichever way you go.*

SOHRAVARDI: *Persian Philosopher* (1155–1191)

1.1 INTRODUCTION

An *experiment* is an operation carried out under controlled conditions in order to discover an unknown effect or law, to test or to establish a hypothesis, or to illustrate a known law.

The experimenter wishes to clarify the relationship between the controllable conditions of the experiment and the experimental outcome or *response*. The experimenter bases the analysis upon observations which are not only affected by the controlled conditions, but also by uncontrolled conditions and measurement errors.

Why are experiments necessary? Experimentation is the basis of the scientific approach to knowledge. Through it, we discover previously unknown laws and test hypotheses. Experimentation justifies or refutes our beliefs.

Each controllable condition is called a *factor*. Each factor is allowed to take on a finite number of values or types, called the *levels* of the factor. A *treatment* is a particular choice of level for each factor. The smallest division of the experimental material such that any two parts may receive different treatments in the actual experiment is called an *experimental unit*. Experimental units are also called *plots*, *runs*, or *points*.

The *factor space*, or *domain*, is the set of all possible treatments. A *design* specifies for each treatment $ij \ldots w$ in the domain how many, say $n_{ij\ldots w}$, experimental units are to be assigned to the treatment $ij \ldots w$ and how the assignment is to be made. Here $ij \ldots w$ denotes the treatment having the first factor at level i, the second factor at level $j, \ldots$, and the last factor at level w. Sometimes there are restraints on how the experimental units can be allocated to the treatments. The collection of all allocations that meet these restraints is called the *design setting*.

For fixed assumptions about the relationship between treatment and response, the variability associated with any estimates that are made depends only on the design. *Optimal designs* are those which in some sense minimize the errors associated with making the set of estimates the experiment is aimed at. Since we cannot expect to simultaneously minimize all errors, design optimality depends upon the particular choice of an optimality criterion.

The traditional approach to experimental design is to start with examples and then to define the terminology and present the methods of analysis by analogy. Our approach is to be free of a connection to any specific field. We will stay in the framework of statistics, and linear models in particular, trying to clarify the relationship between models and designs. However, examples will be included to illustrate the statistical concepts.

1.2 MODELS AND PARAMETRIZATION

Throughout, we assume that a random variable, Y, called the *response variable*, is related to several *controllable variables* $x_1, \ldots, x_k$ by a relation of the form

$$E(Y) = f(\mathbf{x}) \quad \text{or} \quad Y = f(\mathbf{x}) + e_{\mathbf{x}}, \qquad E(e_{\mathbf{x}}) = 0, \quad \text{where } \mathbf{x} = (x_1, \ldots, x_k).$$

That is, the response of an experimental unit with treatment x is a random variable Y whose expectation is functionally related to $\mathbf{x}$. The function f may be called the *response surface*, or the *regression function*. Further, we assume that f is partially known; $f(\mathbf{x}) = f(\mathbf{x}; \boldsymbol{\theta})$, where $\boldsymbol{\theta} \in \Theta$ is a vector of unknown parameters whose specification would completely determine f. The set Θ is called the *parameter space*.

Choosing a model, $Y = f(\mathbf{x}; \boldsymbol{\theta}) + e_{\mathbf{x}}$, means accepting the hypothesis that f has a certain functional form and that the set of errors $e_{\mathbf{x}}$ for distinct runs of an experiment has a given covariance structure. A *linear model* postulates that f is linear in the components of $\boldsymbol{\theta}$:

$$f(\mathbf{x}; \boldsymbol{\theta}) = \sum_{j=1}^{p} f_j(\mathbf{x})\theta_j, \qquad \boldsymbol{\theta} \in \Theta,$$

where the function f_j may be nonlinear in $\mathbf{x}$, and Θ is a subspace of R^p. When the errors are uncorrelated and have the same unknown variance we call this the *standard linear model.* Models that are not linear are called *nonlinear models.* We will use linear models exclusively.

A *factorial domain* consists of all possible combinations of levels of two or more factors. Experiments using factorial domains are called *factorial* or *classification* experiments. Factorial experiments have several advantages. They are more efficient than one factor at a time experiments. Factorial experiments allow the effects of a factor to be estimated at several levels of the other factors, yielding conclusions that are valid over a wide range of experimental conditions. Finally, they are necessary in order to avoid misleading conclusions when interactions between factors are present.

A *2^n factorial domain* is a factorial domain with n factors, each having two levels.

We shall emphasize the use of linear models and 2^n factorial domains in the following chapters. 2^n factorials are particularly useful in the early stages of experimental work, when there are likely to be many factors to be investigated. These experiments use the smallest number of treatments with which n multilevel factors can be studied over a factorial domain. However, we must assume that the response is approximately linear over the range of the factor levels chosen. In a factorial experiment we use n to denote the number of factors and N to denote the number of observations.

EXAMPLE 1.1 One example of a 2^2 factorial experiment can be illustrated in Table 1.1. The margins indicate the level of each factor and the array specifies how many experimental units are assigned to each treatment (factor level combination). Here there are total of $N=3$ units. Table 1.1(a) specifies which observations are made, and Table 1.1(b) shows the observations (scores, yields, etc.) obtained from the experiments. Here Y_{ij} is the observation on treatment (i, j) (ith row and jth column). A way to specify the above factorial in coded form is (11, 12, 21). Each number represents the treatment

Table 1.1 ***(a) n_{ij} Values and (b) Observations***

	Factor A				Factor A		
	Level	1	2		Level	1	2
Factor B	1	1	1	Factor B	1	Y_{11}	Y_{12}
	2	1	0		2	Y_{21}	—
	(a)				(b)		

given an experimental unit. Each digit specifies the level of the corresponding factor.

We have defined experiments, domains, designs, and models. What are the relationships among these? Designs and models are defined on domains. Both a design and a model may be specified independently on a particular domain. In order to make a set of estimates based upon a particular experiment, the model may need to have a particular structure. In this sense, then, the experiment can suggest a model that is said to be associated with the experiment. However, note that *the association between experiment and model is not one to one.*

EXAMPLE 1.1 (Continued) Suppose we wish to estimate the response function for each of the four treatments. An appropriate standard linear model may be written in the following form:

$$Y_{ij}=\mu+\alpha_i+\beta_j+e_{ij}, \qquad \operatorname{Cov}(e_{ij}, e_{i'j'})=\begin{cases}\sigma^2 & \text{if } i=i' \text{ and } j=j' \\ 0 & \text{otherwise}\end{cases} \qquad i, j=1, 2.$$

Here the subscripts specify the levels of the two factors for each treatment. μ, α_i, and β_j are unknown parameters and the errors e_{ij} are uncorrelated and have mean 0 and unknown variance σ^2.

The designed experiment and model can be summarized in matrix form by

$$\mathbf{Y}=\mathbf{X}\boldsymbol{\theta}+\mathbf{e}, \qquad \boldsymbol{\Delta}'\boldsymbol{\theta}=\mathbf{0}, \qquad \operatorname{Cov}(\mathbf{Y})=\mathbf{V}.$$

Here $\mathbf{Y}$, $\mathbf{X}$, $\boldsymbol{\theta}$, $\mathbf{e}$, $\boldsymbol{\Delta}$, and $\mathbf{V}$ have the dimensions $N\times 1$, $N\times p$, $p\times 1$, $N\times 1$, $p\times s$, and $N\times N$, respectively. $\boldsymbol{\theta}$ contains the unknown parameters in the linear model. Since Θ is a subspace of R^p for linear models, there exists a $p\times s$ matrix $\boldsymbol{\Delta}$ such that $\Theta=\{\boldsymbol{\theta}: \boldsymbol{\Delta}'\boldsymbol{\theta}=\mathbf{0}\}$. Each row of $\mathbf{Y}$, $\mathbf{X}$, and $\mathbf{e}$ contains, respectively, the value of the response variable Y, the model coefficients $f_1(\mathbf{x}), \ldots, f_p(\mathbf{x})$, and the error $e_{\mathbf{x}}$ for a point in the domain. The matrix $\mathbf{X}$ is called the *design matrix*, the *regression matrix*, or the *model matrix*. Also associated with a factorial experiment is the *incidence matrix*, $\mathbf{N}=(n_{ij\ldots w})$, which specifies which observations are made. The incidence matrix is defined to be an n-dimensional matrix consisting of the $n_{ij\ldots w}$'s, where $n_{ij\ldots w}$ is the number of observations made with the first factor at its ith level, the second factor at its jth level, and so on. A position in this n-dimensional matrix is called a *cell*. A cell is said to be *occupied* if its entry is nonzero.

The experiment and model can also be specified by

$$E(\mathbf{Y})=\mathbf{X}\boldsymbol{\theta}, \qquad \boldsymbol{\Delta}'\boldsymbol{\theta}=\mathbf{0}, \qquad \operatorname{Cov}(\mathbf{Y})=\mathbf{V}. \tag{1.1}$$

The set $\Omega=\{E(\mathbf{Y}): \boldsymbol{\Delta}'\boldsymbol{\theta}=\mathbf{0}\}$ is a subset of R^N called the *expectation space*. For a particular experiment, the expectation space is determined by the chosen model. However, *infinitely* many models can give rise to the same expectation space. Given Ω, any representation $E(\mathbf{Y})=\mathbf{X}\boldsymbol{\theta}$, $\boldsymbol{\Delta}'\boldsymbol{\theta}=\mathbf{0}$, where $\mathbf{X}$ and $\boldsymbol{\Delta}$ are known matrices, will be called *parametrization* for Ω if $\Omega=\{\mathbf{X}\boldsymbol{\theta}\colon \boldsymbol{\Delta}'\boldsymbol{\theta}=\mathbf{0}\}$. The dimension of Ω, denoted by m, is called the *rank* of the model or the *regression degrees of freedom.* Note that $m\leqslant\min(N, p)$. The expectation space is also called the *regression space.* When $\boldsymbol{\Delta}=\mathbf{0}$, the parametrization is said to be *unconstrained.* Note that by the very definition of parametrization the regression spaces corresponding to two different parametrizations for $E(\mathbf{Y})$ are identical.

EXAMPLE 1.1 (Continued) The incidence matrix which specifies which observations are made is in fact Table 1.1a with $n_{11}=1$, $n_{12}=1$, $n_{21}=1$, and $n_{22}=0$. So

$\mathbf{N}=$

	β_1	β_2
α_1	1	1
α_2	1	0

.

We can write the preceding standard linear model for this experiment as

$$E(Y_{ij})=\mu+\alpha_i+\beta_j, \qquad \operatorname{Cov}(Y_{ij}, Y_{i'j'})=\begin{cases}\sigma^2 & i=i', j=j'\\ 0 & \text{otherwise}\end{cases},$$

where $i, j=1, 2$ and μ, α_i's, β_j's are unknown constants. Now set

$$\mathbf{Y}=\begin{bmatrix}Y_{11}\\ Y_{12}\\ Y_{21}\end{bmatrix}, \qquad \mathbf{X}=\begin{bmatrix}1 & 1 & 0 & 1 & 0\\ 1 & 1 & 0 & 0 & 1\\ 1 & 0 & 1 & 1 & 0\end{bmatrix}, \qquad \boldsymbol{\theta}=\begin{bmatrix}\mu\\ \alpha_1\\ \alpha_2\\ \beta_1\\ \beta_2\end{bmatrix}.$$

Then $\mathbf{Y}$ is a response variable vector such that $E(\mathbf{Y})=\mathbf{X}\boldsymbol{\theta}$, where the parameter vector $\boldsymbol{\theta}$ is completely unknown. Here $\boldsymbol{\Delta}=\mathbf{0}$, $\mathbf{V}=\mathbf{I}$, and $\Omega=R^3$. The matrix $\mathbf{X}$ is the design matrix. Now consider the same situation except suppose that the parameters are known to satisfy the conditions

$$\alpha_1+\alpha_2=0, \qquad \beta_1+\beta_2=0.$$

Let $\mathbf{Y}$ and $\mathbf{X}$ be defined as before and let

$$\boldsymbol{\Delta}'=\begin{bmatrix}0 & 1 & 1 & 0 & 0\\ 0 & 0 & 0 & 1 & 1\end{bmatrix}.$$

Then the assumptions regarding the expectation of $\mathbf{Y}$ can be stated as

$$E(\mathbf{Y}) = \mathbf{X}\boldsymbol{\theta}, \quad \boldsymbol{\Delta}'\boldsymbol{\theta} = \mathbf{0}, \quad \text{and} \quad \Omega = \{\mathbf{X}\boldsymbol{\theta}: \boldsymbol{\Delta}'\boldsymbol{\theta} = \mathbf{0}\}.$$

1.3 ERROR TERM AND ESTIMATION

We have seen that a particular experiment can suggest the response function for a linear model. The other term in the model, the error, is also important.

Error is the deviation of the response from the assumed model. There are three kinds of error. First, the assumed model may be wrong. The resulting error, which may be called *misfit error*, introduces bias in the predicted response. The second kind of error is called *error-in-variables*. We have assumed above that the points $\mathbf{x}$ in the domain are known exactly, but this may not be true in real experiments. This error causes bias in parameter estimates. We will not discuss error-in-variables further in this book. *Experimental error* is the third kind of error. This error is the variation in the response due to mistakes, uncontrolled conditions, sampling errors, and response measurement error. It may arise from the experimental units, the application of treatments, and in the measurement of responses. Randomization is used to reduce any bias in the experimental error due to inadvertent systematic effects of uncontrolled conditions.

The error term in the linear model is additive with zero mean. Thus the model assumes that the response is the sum of the response function (determined by the treatments and unknown parameter values) and an unsystematic contribution due to overall experimental error. The linear model error is assumed to have a covariance matrix $\mathbf{V}$, where $\mathbf{V}$ lies in some known set $\mathscr{V}$ of positive definite matrices. Throughout this book it is assumed that the covariance structure is of the form $\text{Cov}(\mathbf{Y}) = \sigma^2\mathbf{I}$.

If the purpose of the experiment is to estimate the unknown parameter $\boldsymbol{\theta}$ in the model, by minimizing some given function of the errors, no further assumption need to be placed on the error term. Thus estimation can be approached in a distribution-free manner. The estimators may have properties (such as unbiasedness) independent of any distributional assumptions made later.

Rather than to estimate the parameter vector $\boldsymbol{\theta}$ itself, we may wish to estimate a particular function of it. A parametric function is said to be *estimable* within some class of estimators if the class contains an unbiased estimator for that function.

Estimation is usually restricted to estimable parametric functions. However, one may wish to assign values to nonestimable parametric functions using *hyperestimators*.

The purpose of the experiment may be to test hypotheses concerning the parameters of a model. In this case, one usually makes additional distributional assumptions.

Traditionally, the error term e has been assumed to have a multivariate normal distribution, denoted $N(\mathbf{0}, \mathbf{V})$. This assumption is popular for two reasons. It will often appear to be an appropriate distribution to model the errors, although there is no way to guarantee its applicability. Perhaps more importantly, it results in a tractable distributional analysis, providing test statistics with known and tabled distributions, for example, t, F, and χ^2.

One might prefer to assume that the errors in a particular model are from a contaminated normal distribution, or from a very different distribution, such as the Laplace distribution. But then how does one test, for example, the hypothesis of equal variances for two Laplacian populations?

1.4 REPLICATION, RANDOMIZATION, AND BLOCKING

Replication in an experiment is the repetition of a treatment. This technique is used to control experimental error by averaging out the uncontrolled conditions and distributing the error over many units. Replication increases the precision of estimates. Also, it is needed to estimate the error variance unless a simplified model is used.

Randomization is an objective procedure by which treatments are randomly assigned to experimental units subject to any restrictions imposed by the design. It is used to protect against systematic error caused by subjective treatment assignments. It also helps to avoid treatment arrangements that combine with patterns in the uncontrolled conditions to produce systematic error which persists over long or repeated experiments. Randomization can be used to conceal the treatment assignments from the people involved in an experiment to protect against personal biases distorting the data. Finally, randomization may be explicit in the method used to estimate experimental error. When the experimental units are randomly assigned to the treatments we have the *completely randomized design.*

It is important to use experimental units that are as alike as possible to reduce uncontrolled variation and hence to improve the precision of the experiment. As the number of experimental units in an experiment increase, this becomes harder to achieve. *Blocking* is a method of reducing the effect of variations in the experimental material on the error of the treatment comparisons. The units are grouped into sets, called *blocks*, where all units in a block are as alike as possible. Some or all of the treatments are then randomly assigned to the units within each block, subject to any other restrictions imposed by the design. The effects of variations in the units between blocks are then eliminated from comparisons among treatments appearing together in the blocks. When no restriction besides blocking is used, we have a *randomized block design*. If every treatment appears at least once in each block, we have a *randomized complete block design*. Otherwise, it is a *randomized incomplete block design*. Several systems of grouping can be used simultaneously to reduce several sources of systematic error.

For example, in the *Latin square design* the treatments are grouped in two different ways (double grouping). The effect of double grouping is to equally eliminate from the errors all differences among rows and columns. To eliminate three sources of variability, one requires the *Graeco-Latin square design*. A Graeco-Latin square design allows the study of treatments with three different blocking variables. The reader familiar with the subject realizes that the randomized complete block design is a *two-way classification design*, the Latin square is a *three-way classification* design, and the Graeco-Latin square design is a *four-way classification* design.

The arrangement of factorial experiments in a completely randomized design (when it is possible) assures us that we are not including a new error term into the model due to the *restriction on randomization* in the design that can exist for randomized block design, Latin square designs, or other higher grouping designs. We will use completely randomized designs in Chapters 2–7 unless explicitly stated otherwise.

We remark this point once again that there are many possible models for the analysis of data following any designed experiment. Usually a nonlinear model is most realistic, but often one uses a linear model as an approximation. Even in a linear model one must choose which factors and which interactions to include.

1.5 FACTORIAL EXPERIMENTS WITH MISSING DATA

In many designed experiments, the experimenter has a precise idea about how the experiment should be analyzed. Randomized block, Latin square, Graeco-Latin square, balanced incomplete block, and fractionally replicated designs are some examples of such designs.

However, it may happen by accident or some other reason that some of the observations are missing, cannot be collected, or in some other way are not obtainable. For example, crops are destroyed in some plots, the experimenter fails to record some data, gross errors occur in recording, a patient withdraws from the treatment, one or more animals die in the course of the experiment, or some subjects do not take all treatments as planned. Missing data of this nature occur quite frequently.

In such situations the designed experiment is no longer balanced and loses its symmetry, and in some situations, when enough observations are missing, not all of the usual parametric functions are estimable. Such a situation, unless discovered, will lead to incorrect assignment of degrees of freedom for certain sums of squares, to singular matrices that are to be inverted in a computer program, and most importantly, will make contrasts involved in the determination of certain sums of squares to be not known precisely. Therefore, analyzing data of this nature without appropriate knowledge may lead to inappropriate inferences about the hypothesis, and consequently to wrong conclusions about the results of the experiment.

One possible solution to the problem is to *repeat* the experiment under similar conditions and obtain the new values for the missing observations. However, such a solution, though ideal, may not be feasible economically (with regards to time and money) and physically.

The most common approach to this problem when there are few empty cells is to apply the analysis for a complete experiment using *calculated values* in place of the missing observations. Here one calculated value is placed into each empty cell. The values are chosen so that the residual sum of squares is minimized. The residual degrees of freedom are reduced by one for each empty cell in the original data and the usual F tests are performed. Missing data formulas are available in many experimental design texts for special situations. But even in those cases the analysis is *approximate*. (See bibliographical notes at the end of Chapter 5 for a review of this approach.)

A second approach to the problem of missing observations in designed experiments is the *exact least squares* approach by writing a mathematical model for all observations that are present. The least squares normal equations are then constructed for the estimation of parameters in the model. Because the equations corresponding to the missing observations are in fact "missing," the system of normal equations loses its symmetry as compared to when all observations are present, and consequently to obtain solutions to normal equations, *generalized inverses* may be needed. In Chapter 8 we give a method which finds least squares generalized inverses for classification models without any rounding off errors.

A third approach to the problem is to employ the *incidence matrix* as opposed to the design matrix. In dealing with a general linear model, one usually investigates estimability directly in terms of the design matrix $\mathbf{X}$. For the case of a classification model the same information contained in the design matrix $\mathbf{X}$ can also be derived from the incidence matrix $\mathbf{N}$. We consider this approach in Chapters 5–7.

Although in some experiments the missing observation(s) affect the distribution of the remaining ones, it is implicitly assumed throughout this book that the distribution of the observed random variables of the model is the same with or without the missing observations.

In this book we address ourselves solely to the analysis of data sets with missing observations following a *planned experimental design.*

Whether the missing values occur in the process of experimentation (by accident, being spoilt, not obtained or rejected) or have not been employed beforehand (because of impossibility of giving treatment, or being economically not feasible) we refer to them as "missing observations" or "missing data."

1.6 SOME LINEAR ALGEBRA

The collection of all $p \times 1$ column vectors with real entries is called the p-dimensional Euclidean space and is denoted by R^p. The usual rules of scalar multiplication and vector addition are always assumed for vectors in R^p, and under these operations R^p is a *vector space* of dimension p.

A rectangular array with real numbers as elements is called a matrix. Matrices and vectors are denoted by boldface letters, and a prime on a matrix is always used to denote the transpose of the matrix. The matrix operations of additions, multiplications, and so on, are defined in the usual way.

The $s \times s$ diagonal matrix with diagonal elements $d_1, \ldots, d_s$ and zeros elsewhere is denoted by $\text{diag}(d_1, \ldots, d_s)$, and when the diagonal elements are all unity the matrix is called an identity matrix and is denoted by $\mathbf{I}_s$. The s-dimensional column vector with all its elements equal to 1 is denoted by $\mathbf{1}_s$.

A $p \times p$ matrix $\mathbf{A}$ is said to be symmetric whenever $\mathbf{A} = \mathbf{A}'$. If $\mathbf{A}$ is symmetric and if $\mathbf{x}'\mathbf{A}\mathbf{x} \geqslant 0$ for all $\mathbf{x} \in R^p$, then $\mathbf{A}$ is said to be a *nonnegative definite* matrix. If additionally $\mathbf{x}'\mathbf{A}\mathbf{x} = 0$ only when $\mathbf{x} = \mathbf{0}$, then $\mathbf{A}$ is said to be a *positive definite* matrix.

Suppose $\mathscr{V}_1$ and $\mathscr{V}_2$ are vector spaces. A function $\mathbf{A}$ from $\mathscr{V}_1$ into $\mathscr{V}_2$ satisfying $\mathbf{A}(a\mathbf{v} + b\mathbf{u}) = a\mathbf{A}\mathbf{v} + b\mathbf{A}\mathbf{u}$ for all $\mathbf{u}$, $\mathbf{v} \in \mathscr{V}_1$ and for all a, $b \in R^1$ is said to be a linear transformation from $\mathscr{V}_1$ to $\mathscr{V}_2$. A linear transformation from $\mathscr{V}_1$ to $\mathscr{V}_2$ brings about two subspaces. If $\mathbf{A}$ denotes the linear transformation, then the two subspaces are denoted by $R(\mathbf{A}) = \{\mathbf{A}\mathbf{v}\colon \mathbf{v} \in \mathscr{V}_1\}$ called the *range* of $\mathbf{A}$, and $N(\mathbf{A}) = \{\mathbf{v}\colon \mathbf{A}\mathbf{v} = \mathbf{0}\}$ called the *null space* of $\mathbf{A}$.

The dimension of $R(\mathbf{A})$ is denoted by $r(\mathbf{A})$ and is called the *rank* of $\mathbf{A}$, and the dimension of $N(\mathbf{A})$ is denoted by $n(\mathbf{A})$ and is called the *nullity* of $\mathbf{A}$. If $\mathbf{A}$ is a linear transformation from $\mathscr{V}_1$ to $\mathscr{V}_2$ then $\dim \mathscr{V}_1 = r(\mathbf{A}) + n(\mathbf{A})$. Suppose $\mathbf{A}$ is an $n \times m$ matrix. If $\mathbf{A}$ is regarded as mapping a vector $\mathbf{x}$, $\mathbf{x} \in R^m$ to a vector $\mathbf{y}$, $\mathbf{y} = \mathbf{A}\mathbf{x} \in R^n$ ($\mathbf{A}\mathbf{x}$ denotes the usual matrix multiplication), then $\mathbf{A}$ is a linear transformation. Throughout, the matrices are considered as linear transformations between Euclidean spaces.

Suppose $\mathscr{V}$ is a vector space. Let $\mathscr{V}_1$ and $\mathscr{V}_2$ be subspaces of $\mathscr{V}$. The sum $\mathscr{V}_1 + \mathscr{V}_2$ is defined according to $\mathscr{V}_1 + \mathscr{V}_2 = \{\mathbf{v}_1 + \mathbf{v}_2\colon \mathbf{v}_1 \in \mathscr{V}_1, \mathbf{v}_2 \in \mathscr{V}_2\}$.

The intersection $\mathscr{V}_1 \cap \mathscr{V}_2$ (the set of all vectors which belong to both spaces) and the sum $\mathscr{V}_1 + \mathscr{V}_2$ are both subspaces of $\mathscr{V}$.

A vector space $\mathscr{V}$ is said to be the *direct sum* of two vector spaces $\mathscr{V}_1$ and $\mathscr{V}_2$ if every vector $\mathbf{v} \in \mathscr{V}$ can be expressed *uniquely* in the form $\mathbf{v} = \mathbf{v}_1 + \mathbf{v}_2$ where $\mathbf{v}_1 \in \mathscr{V}_1$ and $\mathbf{v}_2 \in \mathscr{V}_2$. This is expressed notationally as $\mathscr{V} = \mathscr{V}_1 \oplus \mathscr{V}_2$. Notice that in this case the two subspaces $\mathscr{V}_1$ and $\mathscr{V}_2$ have only the zero vector in common, that is, $\mathscr{V}_1 \cap \mathscr{V}_2 = \{\mathbf{0}\}$.

Two vectors $\mathbf{x}, \mathbf{y} \in \mathscr{V}$ are said to be *orthogonal*, that is, $\mathbf{x} \perp \mathbf{y}$ if $\mathbf{x}'\mathbf{y} = 0$. If $\mathscr{V}_1$ and $\mathscr{V}_2$ are subspaces of $\mathscr{V}$ such that for all $\mathbf{x} \in \mathscr{V}_1$ and $\mathbf{y} \in \mathscr{V}_2$, $\mathbf{x} \perp \mathbf{y}$, then $\mathscr{V}_1$ and $\mathscr{V}_2$ are said to be *orthogonal subspaces*. This is denoted by $\mathscr{V}_1 \perp \mathscr{V}_2$.

If $\mathscr{V}_1$ and $\mathscr{V}_2$ are orthogonal subspaces of $\mathscr{V}$, then their sum $\mathscr{V}_1 + \mathscr{V}_2$ is said to be an *orthogonal direct sum*, and is denoted by $\mathscr{V}_1 \perp \mathscr{V}_2$. When two subspaces are orthogonal, their intersection is $\{\mathbf{0}\}$. Thus an orthogonal direct sum is a direct sum, but the converse statement is not true. If $\mathscr{V}$ is a vector space, then $\mathscr{V}^\perp$, the set of all vectors perpendicular to $\mathscr{V}$, called the *orthogonal complement* of $\mathscr{V}$, is also a vector space. The concept of two subspaces being orthogonal can be extended to three or more subspaces.

A set of vectors $\mathbf{v}_1, \ldots, \mathbf{v}_p$ is said to *span* a vector space $\mathscr{V}$ if every vector $\mathbf{v} \in \mathscr{V}$ can be expressed as a linear combination of these vectors, that is, if there exist constants $a_1, \ldots, a_p$ such that $\mathbf{v} = \sum_{i=1}^{p} a_i \mathbf{v}_i$. The vectors $\mathbf{v}_1, \ldots, \mathbf{v}_p$ are *linearly independent* if $\sum_{i=1}^{p} a_i \mathbf{v}_i = \mathbf{0}$ implies that $a_1 = \cdots = a_p = 0$. If the vectors $\mathbf{v}_1, \ldots, \mathbf{v}_p$ *span* $\mathscr{V}$ (denoted by sp $\mathscr{V}$) and are linearly independent, then they are said to form a *basis* for $\mathscr{V}$.

If $\mathbf{X}$ is an $s \times t$ matrix with columns $\mathbf{x}_1, \ldots, \mathbf{x}_t$, then $R(\mathbf{X}) = \{\mathbf{X}\mathbf{l} : \mathbf{l} \in R^t\} = \text{sp}\{\mathbf{x}_1, \ldots, \mathbf{x}_t\}$. That is, the range of a matrix is the space spanned by its column vectors. The null space of the matrix $\mathbf{X}$ is the space of all vectors $\mathbf{f}$ such that $\mathbf{Xf} = \mathbf{0}$. Therefore, to every matrix $\mathbf{X}$ there correspond two important vector spaces known as the range space and the null space of $\mathbf{X}$. These two spaces are related by $N(\mathbf{X}) = R(\mathbf{X}')^\perp$. That is, the orthogonal complement of the range space of $\mathbf{X}$ transposed is the null space of $\mathbf{X}$.

The dimension of $R(\mathbf{X})$ is denoted by $r(\mathbf{X})$ and is called the *rank* of $\mathbf{X}$ (= maximal number of linearly independent columns of the matrix $\mathbf{X}$), and the dimension of $N(\mathbf{X})$ is denoted by $n(\mathbf{X})$ and is known as the *nullity* of $\mathbf{X}$. The nullity of $N(\mathbf{X})$ is obtained by $r(\mathbf{X}) + n(\mathbf{X}) =$ number of columns of $\mathbf{X}$. Thus if $\mathbf{X}$ is an $s \times t$ matrix of rank p, then $\dim[R(\mathbf{X})] = r(\mathbf{X}) = p$ and $\dim[N(\mathbf{X})] = n(\mathbf{X}) = t - p$.

Suppose $\mathbf{A}$ and $\mathbf{B}$ are matrices whose row dimensions are the same but whose column dimensions are arbitrary. Then $(\mathbf{A}, \mathbf{B})$ is a partitioned matrix with

$$R(\mathbf{A}, \mathbf{B}) = R(\mathbf{A}) + R(\mathbf{B})$$

and

$$r(\mathbf{A}, \mathbf{B}) = r(\mathbf{A}) + r(\mathbf{B}) - \dim[R(\mathbf{A}) \cap R(\mathbf{B})]$$

That is, the range of the partitioned matrix $(\mathbf{A}, \mathbf{B})$ is the sum of the individual ranges, whereas the rank operation is additive if the subspaces $R(\mathbf{A})$ and $R(\mathbf{B})$ are disjoint.

Suppose $\mathscr{V}$ is a direct sum of subspaces $\mathscr{V}_1$ and $\mathscr{V}_2$ so that each $\mathbf{v} \in \mathscr{V}$ can be expressed uniquely in the form $\mathbf{v} = \mathbf{v}_1 + \mathbf{v}_2$, for $\mathbf{v}_1 \in \mathscr{V}_1$ and $\mathbf{v}_2 \in \mathscr{V}_2$. The function $\mathbf{P}$ from $\mathscr{V}$ to $\mathscr{V}$ defined as $\mathbf{P}\mathbf{v} = \mathbf{v}_1$ is called the *projection operator* on $\mathscr{V}_1$ along $\mathscr{V}_2$. If $\mathbf{P}$ is the projection operator on $\mathscr{V}_1$ along $\mathscr{V}_2$, then it is easy to prove that $\mathbf{I} - \mathbf{P}$ is the projection operator on $\mathscr{V}_2$ along $\mathscr{V}_1$. If $\mathscr{V} = \mathscr{V}_1 \oplus \mathscr{V}_1^{\perp}$, then the projection operator on $\mathscr{V}_1$ along $\mathscr{V}_1^{\perp}$ is called the *orthogonal projection* on $\mathscr{V}_1$.

A dot as in $n_.$ indicates summation over the suppressed subscript, that is $n_. = \sum_i n_i$ and $Y_. = \sum_i Y_{i\cdot}$. If $Y \in R^n$, then $\bar{Y}_. = Y_./n$. Other notation are introduced in the text as needed.

BIBLIOGRAPHICAL NOTES

1.1–1.2 The materials in these sections can be found in Ash and Hedayat (1978) and Raktoe, Hedayat, and Federer (1981). The latter authors provided a systematic treatise on the subject of factorial designs. It is an excellent book for both statisticians as well as mathematicians interested in the subject of factorials and looking for understanding the basic concepts.

1.3–1.4 Herzberg and Cox (1959) have given bibliographies on experimental designs. Federer and Balaam (1972) provided an exclusive bibliography on designs (the arrangement of treatment in an experiment) and treatment design (the selection of treatments to be used in an experiment) for the period prior to 1968. Federer and Federer (1973) gave a partial bibliography on statistical designs for the period of 1968–1971. Recently, Hedayat, and Afsarinejad (1975, 1978) gave an exclusive reference on repeated measurement (cross-over) designs. Ash and Hedayat (1978) with an introduction to optimal designs provided an extensive bibliography on the theory of optimal designs. Federer (1980, 1981) in a three-part article has given the bibliography on experiment design from 1972–1978. With an introduction to some recent results in experiment design and some preliminary concepts and definitions he presented a list of references from over 40 books and journals. For recent developments in design of experiment see Atkinson (1982). His excellent survey is based on the literature of 1975–1980.

1.5 Hoyle (1971) gives an introduction to spoilt data (missing, extra, and mixed-up observations) with an extensive bibliography on the subject. Afifi and Elashoff (1966) in two-part articles considered the problem of missing observations in multivariate statistics. The complete bibliography on "missing data" is given at the end of Chapter 5.

1.2 For parametrizations in linear models see Graybill (1976) and Seely (1979). Seely (1979) introduced the notion of correspondences and via the concept of correspondences showed how the information about parametric functions under one parametrization can be related to parametric functions under another parametrization. A brief description of correspondences will be given in Chapter 3. See also Searle (1971) and Rao (1973).

1.4 See Kempthorne (1977) for the necessity of randomizations for the validity of error assumptions.

REFERENCES

Afifi, A. A. and Elashoff, R. M. (1966). Missing observations in multivariate statistics, I. Review of the literature. *J. Am. Statis. Assoc.* **61**, 595–604.

Ash, A. and Hedayat, A. (1978). An introduction to design optimality with an overview of the literature. *Comm. Statist.* **A7**, 1295–1325.

Atkinson, A. C. (1982). Developments in the design of experiments. *Int. Statist. Rev.* **50**, 161–177.

Federer, W. T. (1964). Literature review of experimental design through 1949. Mathematics Research Center Technical Summary Report 405, University of Wisconsin, February.

Federer, W. T. and Balaam, L. N. (1972). *Biography on Experiment and Treatment Design: Pre-1968*. Edinburgh: Oliver and Boyd (for the International Statistical Institute).

Federer, W. T. and Federer, A. J. (1973). A study of design publications 1968 through 1971. *Am. Statist.* **27**, 160–163.

Federer, W. T. (1980). Some recent results in experiment design with a bibliography, I. *Int. Statist. Rev.* **48**, 337–368.

Federer, W. T. (1981). Some recent results in experiment design with a bibliography, II: A–K. *Int. Statist. Rev.* **49**, 95–109.

Federer, W. T. (1981). Some recent results in experiment design with a bibliography, III: L–Z. *Int. Statist. Rev.* **49**, 185–197.

Hedayat, A. and Afsarinejad, K. (1975). Repeated measurements designs, I. In *A Survey of Statistical Design and Linear Models*, J. N. Srirastava, Ed., pp. 229–242. Amsterdam: North Holland.

Hedayat, A. and Afsarinejad, K. (1978). Repeated measurements design, II. *Ann. Statist.* **6**, 619–628.

Herzberg, A. M. and Cox, D. R. (1959). Recent work in the design of experiments: A bibliography and a review. *J. Roy. Statist. Soc.* **A 132**, 29–67.

Hoyle, M. H. (1971). Spoilt data—An introduction bibliography. *J. Roy. Statist. Soc.* **A 134**, 429–439.

Raktoe, B. L., Hedayat, A., and Federer, W. T. (1981). *Factorial Designs*. New York: Wiley.

Kempthorne, O. (1977). Why randomized? *J. Statist. Planning and Inference* **1**, 1–25.

Rao, C. R. (1973). *Linear Statistical Inference and Its Applications*, 2nd ed. New York: Wiley.

Searle, S. R. (1971). *Linear Models*. New York: Wiley.

Seely, J. (1979). Parametrizations and correspondences in linear models. Technical Report No. 72, Oregon State University, Corvallis.

CHAPTER 2

One Factor Experiments: Comparing Treatments

Would you that spangle of Existence spend
About the Secret-quick about it, Friend!
A Hair perhaps divides the False from True
And upon what, prithee, does life depend?

KHAYYAM NAISHAPURI-RUBÁIYÁT:
Persian Astronomer and Poet (12th Century)

2.1 INTRODUCTION

This chapter discusses the comparison of treatments. The comparison of treatments is important it itself. Also, the discussion will introduce several important statistical concepts and principles that will be used later in more complex comparisons.

Why would we wish to compare treatments? Two purposes of experimentation are: to discover unknown laws and to illustrate known laws. This is done by investigating the relationship between the experimental treatments and responses. One approach to investigating this relationship is to estimate the response function for each treatment. Another approach is to compare the responses for different treatments.

We may wish to compare only two treatments. In this case, the natural comparison would be the difference in their observed responses. But the responses are not determined solely by the treatments. They also depend upon the uncontrolled conditions and measurement errors. So although there may be a difference in the observed responses for the two treatments, there may be no difference in their response function values. How large must the observed response difference be to be "significant"—to imply with some measure of confidence that the response function values are different? Also, a difference in the response function values may not be reflected in a significant response difference. How likely is this? The discipline of statistics, based upon mathematics and probability theory, addresses these questions.

There are two goals one may wish to pursue in the context of comparing treatments using the differences in their observed responses. One may wish to *estimate* the differences in the response function values or to *test* for a difference in the values. The discipline of *statistical inference* provides methods to pursue both goals.

2.2 STATISTICAL INFERENCE

Statistical inference is concerned with the following situation. A random variable Y takes values in some known set $\mathfrak{Y}$ according to a *probability law or distribution* P. The set $\mathfrak{Y}$ is called the *sample space* or the *population*. The probability distribution P is assumed to lie in some family $\mathfrak{P}$ of distributions defined on the sample space $\mathfrak{Y}$. The family $\mathfrak{P}$ is called a *statistical model* for Y. Although P is assumed to lie in $\mathfrak{P}$, we do not know which distribution in $\mathfrak{P}$ is the true distribution P. Statistical inference is a collection of methods for inferring properties about P based upon an observed value, *realization*, or *sample*, of the random variable Y. Two types of statistical inference are *estimation* and *hypothesis testing*.

Usually $\mathfrak{P}$ is parametrized. That is, $\mathfrak{P}=\{P_\theta\colon \theta \in \Theta\}$ for some known set Θ called the *parameter set*. Since $P \in \mathfrak{P}$, the parameter set Θ must contain some θ such that $P=P_\theta$. If θ is unique, it is called the *true parameter*. A statistic $S(Y)$ is a function of the random variable Y. *Statistical estimation* uses a statistic $T(Y)$ called an *estimator*, to estimate or approximate the true parameter θ or some function $g(\theta)$ of it. The value $T(y)$ of an estimator for a particular realization of Y is called an *estimate*. If an estimator provides a single value as an estimate, it is called a *point estimator*, and the estimate is called a *point estimate* or just an *estimate*. If an estimator provides an interval as an estimate, it is called an *interval estimator*, and the estimate is called a *confidence interval*. Point estimates serve to approximate the value of interest. Confidence intervals serve to bracket, or to contain, the value of interest to some predetermined level of confidence.

There are three ways to judge whether an estimator is a good one. It may possess a desirable property, such as unbiasedness. An estimator $T(Y)$ of a parametric function $g(\theta)$ is said to be *unbiased* if the expected value of $T(Y)$ under the distribution $P=P_\theta$ is $g(\theta)$ for all $\theta \in \Theta$ (all $P_\theta \in \mathfrak{P}$). The estimator may optimize some numerical criterion, such as minimum variance. Or it may be obtained by an appealing method. Some approaches that have been proposed to obtain good estimators are the method of moments, the method of minimum chi-square, the method of least squares, the method of maximum likelihood, and minimax estimation. These approaches to point estimation require distributional assumptions on the responses to varying degrees. For example, the method of least squares requires no distributional assumptions, except for some moment assumptions, while the method of maximum likelihood requires full knowledge of the form of the distribution. *Robust estimation* is an approach that attempts to preserve the performance of estimators in the presence of departures from underlying assumptions. *Nonparametric estimation* imposes very mild assumptions on the distributions to derive estimators.

Point estimators can be used to predict responses for future experiments or used in the construction of tests of hypotheses. Interval estimators are also closely related to tests of hypotheses.

A *statistical hypothesis* is a statement that the true distribution P lies in some subset $\mathfrak{P}_0$ of the family $\mathfrak{P}$, and is denoted by $H: P \in \mathfrak{P}_0$. When $\mathfrak{P}$ is parametrized, the hypothesis may state that the true parameter θ lies in some subset Θ_0 of Θ, or $H: \theta \in \Theta_0$. If the hypothesis completely specifies the distribution P, that is, $\mathfrak{P}_0 = \{P_0\}$, it is called a *simple hypothesis*. Otherwise it is called a *composite hypothesis*.

Suppose we wish to test the validity of an hypothesis. The hypothesis to be tested is called the *null hypothesis* and is denoted by H_0. Its *alternative* is the set complement of $\mathfrak{P}_0$ in $\mathfrak{P}$ and is denoted by H_A. A *statistical test of an hypothesis* is a rule $\delta(Y)$ that specifies, for each possible outcome y of the random variable Y, whether the hypothesis H_0 is to be judged to be false. If $\delta(y)=1$, H_0 is judged to be false and we are said to *reject* the null hypothesis. If $\delta(y)=0$, H_0 is judged to be true and we *accept* the null hypothesis. (Some authors prefer to say in this case that we fail to reject the null hypothesis.) If $\delta(y)=0$ or 1 for all possible realizations y, then δ is said to be a *nonrandomized test*. If $\delta(y) \in (0, 1)$ for some y, then H_0 is to be rejected with probability $\delta(y)$. Such a test is called a *randomized test*.

Statistical tests typically consist of a *test statistic* $T(Y)$ and a partitioning of the range of the test statistic into a *critical*, or *rejection region* and an *acceptance region*. If $T(y)$ is in the rejection region the hypothesis H_0 is rejected. If $T(y)$ is in the acceptance region H_0 is accepted. Randomized tests typically have value 1 in the rejection region, 0 in the acceptance region, and probabilities on the boundary, if any, between these regions.

Since a statistical test is a function of Y, it is a random variable whose distribution depends upon the unknown distribution P of Y. The *size* of a test $\delta(Y)$ of a simple null hypothesis H_0: $P = P_0$ is the expected value $E\delta(Y)$ under the distribution P_0. For a nonrandomized test this is the probability of rejecting H_0 when H_0 is true. The size of a test of a composite null hypothesis H_0: $P \in \mathfrak{P}_0$ is the supremum of $E\delta(Y)$ over the distributions in $\mathfrak{P}_0$. We say that a test has a *significance level* α if the size of the test is less than or equal to α.

The *power* of a test $\delta(Y)$ for a simple alternative hypothesis H_A: $P=P_A$ is the expected value $E\delta(Y)$ under the distribution P_A. For nonrandomized tests this is the probability of rejecting H_0 when H_0 is false. When H_A is a composite hypothesis the power of a test is a function of which particular alternative distribution is considered. This function is called the *power function* of the test.

We can make one of two types of errors when testing an hypothesis. *Type I error* occurs when we reject the null hypothesis when it is true. *Type II error* occurs when we accept the null hypothesis when it is false. We can attempt to construct statistical tests that control the probability of making either type of error.

One approach to constructing a test begins by restricting attention to some class $\mathfrak{D}$ of tests. The significance level of the test is selected and then a test δ in $\mathfrak{D}$ is chosen which gives the best power function relative to some criterion, if such a test exists.

An example is to construct a test of a simple null hypothesis against a simple alternative hypothesis by considering the class of all tests with significance levels at most some value, say $\alpha(0 \leqslant \alpha \leqslant 1)$, and then finding that test or those tests that have maximum power against the simple alternative. Such a test is called a *most powerful* (MP) *test of size* α. The Neyman–Pearson lemma guarantees that a MP test always exists for testing a simple null hypothesis against a simple alternative when $0 < \alpha \leqslant 1$.

To test a composite null hypothesis H_0 against a composite alternative H_A one can again restrict attention to the class of tests with level α and then attempt to find a test that has the greatest power among all tests within this class for each alternative distribution in H_A. If such a test exists, it is called a *uniformly most powerful* (UMP) *test of size* α for testing H_0 against H_A.

A UMP test will not always exist for general H_0 and H_A. But in this case a uniformly most powerful test within a smaller class of tests may exist. A test of H_0 against H_A is said to be *unbiased* if the power of the test for every distribution in H_A is no less than the significance level of the test for every distribution in H_0. This says that it is at least as likely to reject H_0 when H_A is true as when H_0 is true. A *uniformly most powerful unbiased* (UMPU) *test of size* α for testing H_0 against H_A is the UMP test within the class of unbiased tests of level α.

Another approach is the likelihood ratio test procedure, which will be discussed later and will be the basis for most of the tests discussed in this book.

Constructing tests of hypotheses requires some knowledge of the distributions of the random variable Y when H_0 and when H_A are true. The common distributional assumption in the context of designed experiments is that the responses are normally distributed with the response function as its mean and the covariance matrix $\sigma^2\mathbf{I}$, where $\sigma^2 > 0$ is unknown. This is a popular choice, simply because it leads to tractable analysis-test statistics with known and tabulated distributions for common hypotheses.

2.3 TREATMENT COMPARISON

The experimental designs defined in Chapter 1 are for comparative experiments involving two or more treatments. The object of any such design is to obtain information on the treatments relative to one another. Therefore, constraints such as single blocking or double blocking of the experimental materials or conditions are done in order to compare treatments under less variable conditions. When there are enough homogeneous experimental units (material) available to the experimenter we can randomly assign treatments to experimental units, subject to the condition that an equal number of experimental units is assigned to each treatment. Such an arrangement of treatments to experimental units is called completely randomized design, or *one-way classification design.*

A treatment is a particular combination of factor levels. However, when considering two distinct treatments we can consider each to be one of two possible levels of a composite factor. In this way, we can treat the case of two treatments as a single factor with two levels. An extension of comparing two treatments is that of comparing t treatments of a factor. For example, let us suppose we are interested in investigating the effect of fertilizers on the yield of a plant. Since many fertilizers (factors) are involved, we may select one of them, say phosphate (in a given form). Having selected the fertilizer and its form, next we need to determine the range (upper and lower limits) over which this particular form of phosphate will be studied. This range will be determined by the experimenter with some knowledge about the plant life and the objectives of the experimenter. After the range has been fixed, we must determine the number of levels for this factor. Choosing the number of levels depends on the amount of information we need from the response (yield) in the given range, and the experimenter knowledge of the form of the response function.

Assuming that the response is linear, and we have chosen t levels, then we have the following linear model:

$$Y_{ij}=\mu+\tau_i+e_{ij}, \qquad i=1, 2, \ldots, t; \qquad j=1, 2, \ldots, n_i. \tag{2.1}$$

Y_{ij} is the observed response for the jth experimental unit receiving treatment i. μ denotes the part of the response that does not depend upon the treatment and τ_i denotes the effect on the response for treatment i, $i=1, 2, \ldots, t$. The e_{ij} term represents the experimental error for the jth unit receiving treatment

i. The e_{ij} are uncorrelated random variables with mean zero and the same variance σ^2. The parameters μ, σ^2, and the τ_i are unknown constants.

There are three important properties of this model. First, the treatment effect adds to the part of the response that does not depend upon the treatment. For some models, it may be necessary to transform experimental data in some way to satisfy this property. Secondly, the treatment effects are constant; they do not vary from one experimental unit to another. A model satisfying these properties is called an *additive model*. In Chapter 3, we will also use the phrase "additive model" to mean a model without interaction terms. In that context, the phrase will mean that each factor effect is additive and does not depend upon the level of the other factors. The third important property is that the response for one experimental unit is unaffected by the particular assignment of treatments to the other units. That is, there is no interference between experimental units.

Note that the model error term and the design's method of randomization in the design are one to one. The model error term implies that the units were assigned to the treatments completely at random and the completely randomized design implies that the errors are uncorrelated.

The model (2.1) is said to be a one-way classification model. In this chapter the one-way classification model is presented along with some basic facts about such models. To avoid technicalities let us assume that each n_i (number of observations in each group) is nonzero and the total number of observations $N=\sum_{i=1}^{t} n_i$ is greater than t (number of levels or groups). We denote the $N\times 1$ vector of ones by $\mathbf{1}_N$, and we use the standard dot notation indicating summation over the suppressed subscript, and we use the bar to denote an average in conjunction with the dot notation.

In presenting the analysis of a single-factor experiment, we consider first the case of two treatments. We refer to this case throughout this chapter as Case 1. The comparison of two treatments also serves as an example to clarify some of our discussions in this chapter. The results are later generalized to the case of t treatments (Case 2).

Case 1. Comparing Two Treatments

Having two treatments, suppose treatment 1 is assigned to n_1 experimental units completely at random (with no restrictions on randomizations) and similarly treatment 2 to n_2 experimental units. The data obtained from such an experiment can be displayed in a one-way classification table as in Table 2.1. This designed experiment and model has the parametrization

$$E(Y_{ij})=\mu+\tau_i, \qquad \operatorname{Cov}(e_{ij}, e_{i'j'})=\begin{cases}\sigma^2 & \text{if } i=i' \text{ and } j=j' \\ 0 & \text{otherwise}\end{cases}$$

Table 2.1 Response Values of Two Treatments Applied to $N = n_1 + n_2$ Experimental Units with No Restriction on Randomizations

Treatments 1	Treatments 2
Y_{11}	Y_{21}
Y_{12}	Y_{22}
$\vdots$	$\vdots$
Y_{1n_1}	Y_{2n_2}

or in matrix form as

$$E(\mathbf{Y}) = \mathbf{X}\boldsymbol{\theta}, \qquad \text{Cov}(\mathbf{Y}) = \sigma^2\mathbf{I}, \qquad \boldsymbol{\theta} \in \Theta \tag{2.2}$$

where $\mathbf{Y} = (Y_{11}, Y_{12}, \ldots, Y_{1n_1}, Y_{21}, Y_{22}, \ldots, Y_{2n_2})'$, $\boldsymbol{\theta} = (\mu, \tau_1, \tau_2)'$, and

$$\mathbf{X} = \begin{bmatrix} 1 & 1 & 0 \\ 1 & 1 & 0 \\ \vdots & \vdots & \vdots \\ 1 & 1 & 0 \\ 1 & 0 & 1 \\ 1 & 0 & 1 \\ \vdots & \vdots & \vdots \\ 1 & 0 & 1 \end{bmatrix} \begin{matrix} \left.\vphantom{\begin{matrix}1\\1\\\vdots\\1\end{matrix}}\right\} n_1 \text{ rows} \\ \left.\vphantom{\begin{matrix}1\\1\\\vdots\\1\end{matrix}}\right\} n_2 \text{ rows} \end{matrix} .$$

Using the terminology of Chapter 1, $\mathbf{X}$ is the design matrix. The parameter space $\Theta = R^3$ since μ, τ_1, and τ_2 are unknown. The expectation space is $R(\mathbf{X})$.

To begin with, we wish to estimate the components of the unknown vector $\boldsymbol{\theta}$, or at least to estimate some parametric function $g(\boldsymbol{\theta})$. One might approach this problem by defining a class of estimators that possess some desired properties and then find an estimator within that class that optimizes some numerical criterion. An example of this approach is to consider the class of linear estimators of the form $\mathbf{t}'\mathbf{Y}$, where $\mathbf{t} \in R^N$, which are unbiased for $g(\boldsymbol{\theta})$. We could then seek an estimator within that class which had minimum variance. This approach is known as the best linear unbiased estimation or BLUE.

Another approach is to estimate $\boldsymbol{\theta}$ by a value $\hat{\boldsymbol{\theta}} \in \Theta$ that in some sense minimizes the deviations between the responses Y_{ij} and the fitted response function values $\hat{Y}_{ij} = \hat{\mu} + \hat{\tau}_i$. The deviations $Y_{ij} - \hat{Y}_{ij}$ may be negative as well as positive. Since both types of deviations are undesirable, we want to minimize some function of the absolute values of the deviations, rather than the deviations themselves.

The classical approach to estimation problems in linear models is the method of least squares. Section 2.4 is devoted to a brief investigation of the method of least squares.

2.4 LEAST SQUARES ESTIMATION

Our aim is to estimate the unknown parameters μ, τ_1, and τ_2 by $\hat{\mu}$, $\hat{\tau}_1$, and $\hat{\tau}_2$ by the method of least squares, which minimizes the sum of squared deviations. The deviations are defined in the following way. Let $\hat{Y}_{ij}$ denote the estimate of the $E(Y_{ij})$ obtained by substituting $\hat{\boldsymbol{\theta}}$ for the unknown $\boldsymbol{\theta}$, that is,

$$\hat{Y}_{ij} = \hat{\mu} + \hat{\tau}_i$$

then $Y_{ij} - \hat{Y}_{ij}$ is the jth deviation, and we want to minimize the sum of squares

$$\sum_{i=1}^{2} \sum_{j=1}^{n_i} (Y_{ij} - \hat{Y}_{ij})^2.$$

In matrix notation the function that is minimized is

$$\|\mathbf{Y} - \mathbf{X}\hat{\boldsymbol{\theta}}\|^2$$

where $\| . \|$ denotes the Euclidean or L_2 norm, that is,

$$\mathbf{a} \in R^N \Rightarrow \|\mathbf{a}\| = \left[\sum_{i=1}^{N} a_i^2 \right]^{1/2}.$$

$\hat{\boldsymbol{\theta}}$ is said to be least squares solution for $\boldsymbol{\theta}$ if it minimizes the sum of squared deviations over all possible choices of $\boldsymbol{\theta}$ in the parameter space Θ:

$$\|\mathbf{Y} - \mathbf{X}\hat{\boldsymbol{\theta}}\|^2 = \min_{\boldsymbol{\theta} \in \Theta} \|\mathbf{Y} - \mathbf{X}\boldsymbol{\theta}\|^2.$$

Why has the sum of squared deviations been traditionally used? In the first place, its use provides closed form solutions with attractive properties. In the second place, under the additional assumption that the responses have a multivariate normal distribution, the distributions of the least squares estimators and of test statistics based on them are known and tabulated.

There are, of course, other functions one might choose. One might wish to minimize the sum of the absolute deviations

$$\sum_{i=1}^{2} \sum_{j=1}^{n_i} |Y_{ij} - \hat{\mu} - \hat{\tau}_i| = \|\mathbf{Y} - \mathbf{X}\hat{\boldsymbol{\theta}}\|_1$$

using the L_1 norm. Or one may choose another norm as a measure of the deviations of the responses from their fitted values. However, we will emphasize the method of least squares or its generalization in this and later chapters.

We now consider how to calculate least squares solution $\hat{\boldsymbol{\theta}}$ for the following parametrization:

$$E(\mathbf{Y})=\mathbf{X}\boldsymbol{\theta}, \quad \boldsymbol{\Delta}'\boldsymbol{\theta}=\mathbf{0}, \quad R(\boldsymbol{\Delta})\cap R(\mathbf{X}')=\{\mathbf{0}\}, \quad \mathrm{Cov}(\mathbf{Y})=\sigma^2\mathbf{I}. \tag{2.3}$$

This is parametrization (1.1) with the restriction that the constraints satisfy $R(\boldsymbol{\Delta})\cap R(\mathbf{X}')=\{\mathbf{0}\}$. This means that the rows of $\boldsymbol{\Delta}'$ are linearly independent of the rows of $\mathbf{X}$. We want to find some $\hat{\boldsymbol{\theta}}$ minimizing

$$\|\mathbf{Y}-\mathbf{X}\boldsymbol{\theta}\|^2=(\mathbf{Y}-\mathbf{X}\boldsymbol{\theta})'(\mathbf{Y}-\mathbf{X}\boldsymbol{\theta})=\mathbf{Y}'\mathbf{Y}-2\boldsymbol{\theta}'\mathbf{X}'\mathbf{Y}+\boldsymbol{\theta}'\mathbf{X}'\mathbf{X}\boldsymbol{\theta} \tag{2.4}$$

subject to the constraint $\boldsymbol{\Delta}'\boldsymbol{\theta}=\mathbf{0}$. This problem can be solved by the method of *Lagrangian multipliers*.

The Lagrangian $L(\boldsymbol{\theta}, \boldsymbol{\lambda})$ of this problem is

$$L(\boldsymbol{\theta}, \boldsymbol{\lambda})=(\mathbf{Y}-\mathbf{X}\boldsymbol{\theta})'(\mathbf{Y}-\mathbf{X}\boldsymbol{\theta})+\boldsymbol{\lambda}'\boldsymbol{\Delta}'\boldsymbol{\theta}.$$

Equating $\partial L/\partial\boldsymbol{\theta}$ and $\partial L/\partial\boldsymbol{\lambda}$ to zero we obtain

$$-2\mathbf{X}'\mathbf{Y}'+2\mathbf{X}'\mathbf{X}\boldsymbol{\theta}+\boldsymbol{\Delta}\boldsymbol{\lambda}=\mathbf{0}$$

$$\boldsymbol{\Delta}'\boldsymbol{\theta}=\mathbf{0}.$$

The first equation implies that $\boldsymbol{\Delta}\boldsymbol{\lambda}$ is a linear combination of the columns of $\mathbf{X}'$, or $\boldsymbol{\Delta}\boldsymbol{\lambda}\in R(\mathbf{X}')$. But we have assumed that $R(\boldsymbol{\Delta})\cap R(\mathbf{X}')=\{\mathbf{0}\}$. Therefore $\boldsymbol{\Delta}\boldsymbol{\lambda}=\mathbf{0}$ and we obtain the following necessary condition for $\hat{\boldsymbol{\theta}}$ to be least squares for $\boldsymbol{\theta}$:

$$\mathbf{X}'\mathbf{X}\hat{\boldsymbol{\theta}}=\mathbf{X}'\mathbf{Y} \quad \text{and} \quad \boldsymbol{\Delta}'\hat{\boldsymbol{\theta}}=\mathbf{0}. \tag{2.5}$$

Furthermore, the *Hessian matrix* of the Lagrangian is

$$\begin{bmatrix} \dfrac{\partial^2 L}{\partial\boldsymbol{\theta}^2} & \dfrac{\partial^2 L}{\partial\boldsymbol{\theta}\,\partial\boldsymbol{\lambda}} \\ \dfrac{\partial^2 L}{\partial\boldsymbol{\lambda}\,\partial\boldsymbol{\theta}} & \dfrac{\partial^2 L}{\partial\boldsymbol{\lambda}^2} \end{bmatrix}=\begin{bmatrix} \mathbf{X}'\mathbf{X} & \mathbf{0} \\ \mathbf{0} & \mathbf{0} \end{bmatrix},$$

which is non-negative definite. Therefore any solution $\hat{\boldsymbol{\theta}}$ to (2.5) minimizes (2.4) subject to $\boldsymbol{\Delta}'\boldsymbol{\theta}=\mathbf{0}$ and is therefore a least squares for $\boldsymbol{\theta}$. Equations (2.5) represent a system of linear equations called *augmented normal equations.* When $\boldsymbol{\Delta}=\mathbf{0}$, the system of equations is called the *normal equations.* We have proved the following.

THEOREM 2.1 For parametrization (2.3), $\hat{\boldsymbol{\theta}}$ is least squares for $\boldsymbol{\theta}$ if and only if it satisfies (2.5).

Consider the unconstrained parametrization, that is, (2.3) with $\mathbf{\Delta}=\mathbf{0}$. If $r(\mathbf{X})=p$, then $\mathbf{X}'\mathbf{X}$ is nonsingular and there exists a unique solution, which will be called least squares estimator $\hat{\boldsymbol{\theta}}=(\mathbf{X}'\mathbf{X})^{-1}\mathbf{X}'\mathbf{Y}$. The normal equations are consistent $[R(\mathbf{X}'\mathbf{X})=R(\mathbf{X}')]$, so when $\mathbf{X}'\mathbf{X}$ is singular there are infinitely many least squares solutions for $\boldsymbol{\theta}$. A particular solution can be calculated by $\hat{\boldsymbol{\theta}}=(\mathbf{X}'\mathbf{X})^{-}\mathbf{X}'\mathbf{Y}$, where $\mathbf{A}^{-}$ denotes any generalized inverse (g inverse) of the matrix $\mathbf{A}$. The topic of Chapter 8 is the construction of g inverses for solving normal equations arising from classification models.

Another way to obtain a particular solution to the normal equations is to include constraints on the solution components $\mathbf{\Delta}'\hat{\boldsymbol{\theta}}=\mathbf{0}$. The solution must then satisfy the augmented normal equations (2.5). For a solution to always exist, the rows of $\mathbf{\Delta}'$ must be linearly independent of the rows of $\mathbf{X}$. For the solution to be unique, $\mathbf{\Delta}$ must satisfy $r(\mathbf{X}', \mathbf{\Delta})=p$. If $r(\mathbf{X})=p$ in the unconstrained parametrization, or if $r(\mathbf{X}', \mathbf{\Delta})=p$ in the constrained parametrization, the least squares estimator $\hat{\boldsymbol{\theta}}$ for $\boldsymbol{\theta}$ is unique and the parametrization is said to have *full rank*.

Suppose we wish to estimate $\boldsymbol{\theta}$ using the method of least squares, where $\boldsymbol{\theta}$ is subject to consistent linear equality restrictions. That is,

$$\begin{aligned}&\text{Minimize } \|\mathbf{Y}-\mathbf{X}\boldsymbol{\theta}\|^2\\ &\quad \boldsymbol{\theta}\in\Theta\\ &\text{subject to}\quad \mathbf{\Delta}'\boldsymbol{\theta}=\mathbf{C}\end{aligned}$$

where $\mathbf{\Delta}$ is a $p\times s$ matrix of rank s and $\mathbf{C}$ is a known column vector.

The Lagrangian function is

$$\mathbf{L}=(\mathbf{Y}-\mathbf{X}\boldsymbol{\theta})'(\mathbf{Y}-\mathbf{X}\boldsymbol{\theta})+(\boldsymbol{\theta}'\mathbf{\Delta}-\mathbf{C}')\boldsymbol{\lambda}.$$

Now, equating $\partial\mathbf{L}/\partial\boldsymbol{\theta}=\mathbf{0}$ and $\partial\mathbf{L}/\partial\boldsymbol{\lambda}=\mathbf{0}$, we obtain

$$\begin{aligned}-2\mathbf{X}'\mathbf{Y}+2\mathbf{X}'\mathbf{X}\boldsymbol{\theta}+\mathbf{\Delta}\boldsymbol{\lambda}&=\mathbf{0}\\ \mathbf{\Delta}'\boldsymbol{\theta}&=\mathbf{C}.\end{aligned}\tag{2.5a}$$

Let the solution of (2.5a) be $\hat{\boldsymbol{\theta}}_R$ and $\hat{\boldsymbol{\lambda}}_R$, the subscript R denoting the restricted least squares. Then assuming $r(\mathbf{X})=p$ we have

$$\begin{aligned}\hat{\boldsymbol{\theta}}_R&=(\mathbf{X}'\mathbf{X})^{-1}\mathbf{X}'\mathbf{Y}-\tfrac{1}{2}(\mathbf{X}'\mathbf{X})^{-1}\mathbf{\Delta}\boldsymbol{\lambda}_R\\ &=\hat{\boldsymbol{\theta}}-\tfrac{1}{2}(\mathbf{X}'\mathbf{X})^{-1}\mathbf{\Delta}\hat{\boldsymbol{\lambda}}_R.\end{aligned}\tag{2.5b}$$

As $\hat{\boldsymbol{\theta}}_R$ satisfies $\mathbf{\Delta}'\boldsymbol{\theta}=\mathbf{C}$,

$$\begin{aligned}\mathbf{C}&=\mathbf{\Delta}'\hat{\boldsymbol{\theta}}_R\\ &=\mathbf{\Delta}'\hat{\boldsymbol{\theta}}-\tfrac{1}{2}\mathbf{\Delta}'(\mathbf{X}'\mathbf{X})^{-1}\mathbf{\Delta}\hat{\boldsymbol{\lambda}}_R.\end{aligned}$$

We have assumed that $\mathbf{\Delta}$ is of rank s, and that $(\mathbf{X}'\mathbf{X})^{-1}$ is positive definite as

$(\mathbf{X}'\mathbf{X})$ is positive definite. And so $\boldsymbol{\Delta}'(\mathbf{X}'\mathbf{X})^{-1}\boldsymbol{\Delta}$ is positive definite. Hence

$$-\frac{1}{2\hat{\lambda}_R}=[\boldsymbol{\Delta}'(\mathbf{X}'\mathbf{X})^{-1}\boldsymbol{\Delta}]^{-1}(\mathbf{C}-\boldsymbol{\Delta}'\hat{\boldsymbol{\theta}}). \tag{2.5c}$$

Thus $\hat{\boldsymbol{\theta}}_R$ can be obtained from (2.5b) using (2.5c), as

$$\hat{\boldsymbol{\theta}}_R=\hat{\boldsymbol{\theta}}+(\mathbf{X}'\mathbf{X})^{-1}\boldsymbol{\Delta}[\boldsymbol{\Delta}'(\mathbf{X}'\mathbf{X})^{-1}\boldsymbol{\Delta}]^{-1}(\mathbf{C}-\boldsymbol{\Delta}'\hat{\boldsymbol{\theta}}). \tag{2.5d}$$

The estimate $\hat{\boldsymbol{\theta}}_R$ obtained in this manner is a least squares estimator with consistent linear restrictions or *restricted* least squares estimate.

REMARK The statement $\hat{\boldsymbol{\theta}}$ is least squares solution for $\boldsymbol{\theta}$ might be read as implying that $\hat{\boldsymbol{\theta}}$ is an estimator for $\boldsymbol{\theta}$. This is definitely not the case when the design matrix is less than full rank. The nonuniqueness of $\hat{\boldsymbol{\theta}}$ in such a situation excludes it from being a candidate for estimating $\boldsymbol{\theta}$. Even though $\hat{\boldsymbol{\theta}}$ is not necessarily unique, it is possible that $\boldsymbol{\Lambda}'\hat{\boldsymbol{\theta}}$ is unique for particular choices of the matrix $\boldsymbol{\Lambda}'$.

For our parametrization of Case 1, $\hat{\boldsymbol{\theta}}$ must satisfy the normal equations:

$$\begin{bmatrix} n_1+n_2 & n_1 & n_2 \\ n_1 & n_1 & 0 \\ n_2 & 0 & n_2 \end{bmatrix}\begin{bmatrix}\hat{\mu}\\ \hat{\tau}_1\\ \hat{\tau}_2\end{bmatrix}=\begin{bmatrix}Y_{..}\\ Y_{1.}\\ Y_{2.}\end{bmatrix}.$$

Here

$$Y_{i.}=\sum_{j=1}^{n_i}Y_{ij} \quad\text{and}\quad Y_{..}=\sum_{i=1}^{2}\sum_{j=1}^{n_i}Y_{ij}=Y_{1.}+Y_{2.}.$$

Since $\mathbf{X}'\mathbf{X}$ is singular there are infinitely many solutions to this equation.

The response for treatment i is $\mu+\tau_i$. If μ is to denote the average treatment response, then

$$\mu=(\mu+\tau_1+\mu+\tau_2)/2=\mu+(\tau_1+\tau_2)/2$$

and we must have $\tau_1+\tau_2=0$. If this constraint is included with the normal equations, the system will have the unique solution

$$\hat{\mu}=(\bar{Y}_{1.}+\bar{Y}_{2.})/2,$$
$$\hat{\tau}_1=(\bar{Y}_{1.}-\bar{Y}_{2.})/2,$$
$$\hat{\tau}_2=(\bar{Y}_{2.}-\bar{Y}_{1.})/2,$$

where $\bar{Y}_{i.}=Y_{i.}/n_i$, $i=1, 2$. The constraint $\hat{\tau}_1+\hat{\tau}_2=0$ has been called the *usual constraint* by some authors.

Some remarks should be made about the use of the "usual constraint"

$\hat{\tau}_1+\hat{\tau}_2=0$. First, it is not necessary to impose constraints on the solution $\hat{\boldsymbol{\theta}}$ to obtain a particular least squares solution. We could have used any g inverse $(\mathbf{X}'\mathbf{X})^-$ to obtain one from $\hat{\boldsymbol{\theta}}=(\mathbf{X}'\mathbf{X})^-\mathbf{X}'\mathbf{Y}$. Secondly, if constraints are to determine uniquely a solution to the normal equations, not just any constraints may be used. The constraints $\boldsymbol{\Delta}'\hat{\boldsymbol{\theta}}=\mathbf{0}$ must satisfy $R(\mathbf{X}')\cap R(\boldsymbol{\Delta}) =\{\mathbf{0}\}$ and $r(\mathbf{X}', \boldsymbol{\Delta})=p$. Thirdly, the usual constraint $\hat{\tau}_1+\hat{\tau}_2=0$ may not be the simplest constraint that might be used. For example, we might use $\hat{\tau}_2=0$ instead. Finally, we can use constraints to determine a particular solution whether or not they are included in the model (in the specification of Θ).

As it stands, our parametrization for Case 1 does not have full rank. However, if the constraint $\tau_1+\tau_2=0$ is included in the parameterization, then the parametrization does have full rank.

$g(\hat{\boldsymbol{\theta}})$ is called *least squares for the parametric function* $g(\boldsymbol{\theta})$ whenever $\hat{\boldsymbol{\theta}}$ is least squares for $\boldsymbol{\theta}$. It is common to consider parametric functions of the form $\boldsymbol{\lambda}'\boldsymbol{\theta}$, where $\boldsymbol{\lambda}\in R^P$ is a known vector. $\boldsymbol{\lambda}'\boldsymbol{\theta}$ is then a linear combination of the components of $\boldsymbol{\theta}$. One example is the parametric function $\boldsymbol{\lambda}'\boldsymbol{\theta}=\tau_1-\tau_2$, where $\boldsymbol{\lambda}=(0, 1, -1)'$.

A *parametric vector* is any vector whose elements are parametric functions. We will consider only *linear parametric vectors* $\boldsymbol{\Lambda}'\boldsymbol{\theta}$, where $\boldsymbol{\Lambda}'$ is an $s\times p$ matrix. $\boldsymbol{\Lambda}'\boldsymbol{\theta}$ is a least squares estimator for $\boldsymbol{\Lambda}'\boldsymbol{\theta}$ whenever $\hat{\boldsymbol{\theta}}$ is least squares for $\boldsymbol{\theta}$. Note that $\boldsymbol{\Lambda}'\hat{\boldsymbol{\theta}}$ is a vector-valued estimator.

The following theorem characterizes those linear parametric vectors that have unique least squares estimators.

THEOREM 2.2 In the context of parametrization (2.3) $\boldsymbol{\Lambda}'\hat{\boldsymbol{\theta}}$ is unique for all $\hat{\boldsymbol{\theta}}$ least squares for $\boldsymbol{\theta}$ if and only if $R(\boldsymbol{\Lambda})\subset R(\mathbf{X}', \boldsymbol{\Delta})$.

Proof Suppose we have uniqueness and let $\hat{\boldsymbol{\theta}}$ be a fixed least squares estimator for $\boldsymbol{\theta}$. Let $\mathbf{f}$ denote any vector such that $\mathbf{f}\in N(\mathbf{X})\cap N(\boldsymbol{\Delta}')$. Set $\tilde{\boldsymbol{\theta}}=\hat{\boldsymbol{\theta}}+\mathbf{f}$. Then

$$\mathbf{X}'\mathbf{X}\tilde{\boldsymbol{\theta}}=\mathbf{X}'\mathbf{X}\hat{\boldsymbol{\theta}}+\mathbf{X}'\mathbf{X}\mathbf{f}=\mathbf{X}'\mathbf{X}\hat{\boldsymbol{\theta}}=\mathbf{X}'\mathbf{Y}$$

and

$$\boldsymbol{\Delta}'\tilde{\boldsymbol{\theta}}=\boldsymbol{\Delta}'\hat{\boldsymbol{\theta}}+\boldsymbol{\Delta}'\mathbf{f}=\boldsymbol{\Delta}'\hat{\boldsymbol{\theta}}=\mathbf{0}.$$

So $\tilde{\boldsymbol{\theta}}$ is also a least squares estimator for $\boldsymbol{\theta}$. By uniqueness, we then have

$$\boldsymbol{\Lambda}'\hat{\boldsymbol{\theta}}=\boldsymbol{\Lambda}'\tilde{\boldsymbol{\theta}}=\boldsymbol{\Lambda}'\hat{\boldsymbol{\theta}}+\boldsymbol{\Lambda}'\mathbf{f}\Rightarrow\mathbf{f}\in N(\boldsymbol{\Lambda}').$$

So

$$N(\mathbf{X})\cap N(\boldsymbol{\Delta}')\subset N(\boldsymbol{\Lambda}')\Rightarrow R(\boldsymbol{\Lambda}')\subset R(\mathbf{X}')+R(\boldsymbol{\Delta})=R(\mathbf{X}', \boldsymbol{\Delta}).$$

Now suppose $\mathbf{\Lambda}$ is such that $R(\mathbf{\Lambda}) \subset R(\mathbf{X}', \mathbf{\Delta})$ and let $\hat{\boldsymbol{\theta}}$ and $\tilde{\boldsymbol{\theta}}$ be any two least squares estimators for $\boldsymbol{\theta}$. Note that $\mathbf{X}\hat{\boldsymbol{\theta}} = \mathbf{X}\tilde{\boldsymbol{\theta}}$. This is true because

$$\mathbf{X}'\mathbf{X}\hat{\boldsymbol{\theta}} = \mathbf{X}'\mathbf{X}\tilde{\boldsymbol{\theta}} = \mathbf{X}'\mathbf{Y} \Rightarrow \mathbf{X}'\mathbf{X}(\hat{\boldsymbol{\theta}} - \tilde{\boldsymbol{\theta}}) = \mathbf{0}$$
$$\Rightarrow \mathbf{X}(\hat{\boldsymbol{\theta}} - \tilde{\boldsymbol{\theta}}) \in N(\mathbf{X}') = R(\mathbf{X})^{\perp},$$

yet $\mathbf{X}(\hat{\boldsymbol{\theta}} - \tilde{\boldsymbol{\theta}}) \in R(\mathbf{X})$, so $\mathbf{X}(\hat{\boldsymbol{\theta}} - \tilde{\boldsymbol{\theta}}) = \mathbf{0}$. Since $R(\mathbf{\Lambda}) \subset R(\mathbf{X}', \mathbf{\Delta})$, $\mathbf{\Lambda} = \mathbf{X}'\mathbf{A} + \mathbf{\Delta}\mathbf{B}$ for some matrices $\mathbf{A}$ and $\mathbf{B}$. Then

$$\mathbf{\Lambda}'\hat{\boldsymbol{\theta}} = \mathbf{A}'\mathbf{X}\hat{\boldsymbol{\theta}} + \mathbf{B}'\mathbf{\Delta}'\hat{\boldsymbol{\theta}} = \mathbf{A}'\mathbf{X}\hat{\boldsymbol{\theta}} = \mathbf{A}'\mathbf{X}\tilde{\boldsymbol{\theta}} = \mathbf{A}'\mathbf{X}\tilde{\boldsymbol{\theta}} + \mathbf{B}'\mathbf{\Delta}'\tilde{\boldsymbol{\theta}} = \mathbf{\Lambda}'\tilde{\boldsymbol{\theta}}.$$

Since $\hat{\boldsymbol{\theta}}$ and $\tilde{\boldsymbol{\theta}}$ are arbitrary least squares estimators, $\mathbf{\Lambda}'\hat{\boldsymbol{\theta}}$ is unique for all least squares $\hat{\boldsymbol{\theta}}$. □

The parametric function $g(\boldsymbol{\theta})$ is called *identifiable* if $g(\bar{\boldsymbol{\theta}}) = g(\bar{\bar{\boldsymbol{\theta}}})$ whenever $\mathbf{X}\bar{\boldsymbol{\theta}} = \mathbf{X}\bar{\bar{\boldsymbol{\theta}}}$, $\mathbf{\Delta}'\bar{\boldsymbol{\theta}} = \mathbf{\Delta}'\bar{\bar{\boldsymbol{\theta}}} = \mathbf{0}$. A parametric function $g(\boldsymbol{\theta})$ is identifiable if and only if $g(\hat{\boldsymbol{\theta}})$ is unique for all least squares $\hat{\boldsymbol{\theta}}$. A parametric vector is said to be identifiable if its components are identifiable parametric functions. We have just seen that a linear parametric vector $\mathbf{\Lambda}'\boldsymbol{\theta}$ for parametrization (2.3) is identifiable if and only if $R(\mathbf{\Lambda}) \subset R(\mathbf{X}', \mathbf{\Delta})$. A linear parametric function $\boldsymbol{\lambda}'\boldsymbol{\theta}$ is identifiable if and only if $\boldsymbol{\lambda} \in R(\mathbf{X}', \mathbf{\Delta})$.

Least squares estimators $\mathbf{\Lambda}'\hat{\boldsymbol{\theta}}$ for identifiable parametric vectors are unbiased:

THEOREM 2.3 Let $\mathbf{\Lambda}'\boldsymbol{\theta}$ be an identifiable parametric vector for parametrization (2.3). Then $\mathbf{\Lambda}'\hat{\boldsymbol{\theta}}$ is unbiased for $\mathbf{\Lambda}'\boldsymbol{\theta}$ for every least squares estimator $\hat{\boldsymbol{\theta}}$.

Proof Since $R(\mathbf{\Lambda}) \subset R(\mathbf{X}', \mathbf{\Delta}) = R(\mathbf{X}'\mathbf{X}, \mathbf{\Delta})$, there exist matrices $\mathbf{A}$ and $\mathbf{B}$ such that $\mathbf{\Lambda} = \mathbf{X}'\mathbf{X}\mathbf{A} + \mathbf{\Delta}\mathbf{B}$. Let $\hat{\boldsymbol{\theta}}$ be least squares for $\boldsymbol{\theta}$. Then

$$E(\mathbf{\Lambda}'\hat{\boldsymbol{\theta}}) = E[\mathbf{A}'\mathbf{X}'\mathbf{X}\hat{\boldsymbol{\theta}} + \mathbf{B}'\mathbf{\Delta}'\hat{\boldsymbol{\theta}}] = E[\mathbf{A}'\mathbf{X}'\mathbf{X}\hat{\boldsymbol{\theta}}] = E[\mathbf{A}'\mathbf{X}'\mathbf{Y}] = \mathbf{A}'\mathbf{X}'\mathbf{X}\boldsymbol{\theta}$$
$$= \mathbf{A}'\mathbf{X}'\mathbf{X}\boldsymbol{\theta} + \mathbf{B}'\mathbf{\Delta}'\boldsymbol{\theta} = \mathbf{\Lambda}'\boldsymbol{\theta},$$

because $E(\mathbf{X}'\mathbf{X}\hat{\boldsymbol{\theta}}) = E(\mathbf{X}'\mathbf{Y}) = \mathbf{X}'\mathbf{X}\hat{\boldsymbol{\theta}}$ and $\mathbf{\Delta}'\boldsymbol{\theta} = \mathbf{0}$. □

In summary, in the context of parametrization (2.3) we have

$$R(\mathbf{\Lambda}) \subset R(\mathbf{X}', \mathbf{\Delta}) \Leftrightarrow \mathbf{\Lambda}'\boldsymbol{\theta} \text{ is identifiable,}$$
$$\Rightarrow \mathbf{\Lambda}'\hat{\boldsymbol{\theta}} \text{ is unbiased for } \mathbf{\Lambda}'\boldsymbol{\theta}$$
$$\text{for all least squares } \hat{\boldsymbol{\theta}}.$$

Consider again our parametrization of Case 1. The components μ, τ_1, and τ_2 of $\boldsymbol{\theta}$ are not identifiable parametric functions, because none of $(1,0,0)(0,1,0)$, or (0, 0, 1) are linear combinations of the rows of $\mathbf{X}$. However, $\tau_1 - \tau_2$ is identifiable because (0, 1, -1) is a linear combination of the rows of $\mathbf{X}$. It represents the increase in response due to using treatment 1 over using treatment 2. Using our choices for $\hat{\tau}_1$ and $\hat{\tau}_2$, we have $\hat{\tau}_1 - \hat{\tau}_2 = \bar{Y}_{1.} - \bar{Y}_{2.}$. This is just the difference in mean response between the two treatments, and we would obtain the same least squares estimators for $\tau_1 - \tau_2$ no matter which $\hat{\boldsymbol{\theta}}$ we used. The variance of this estimator is $\sigma^2(1/n_1 + 1/n_2)$.

So we can obtain least squares estimators for parametric functions $g(\boldsymbol{\theta})$. How are these estimators to be used? Some people take the estimates and use them as they are to make decisions. Others prefer to use further properties of the estimators to construct interval estimates or to test hypotheses.

Estimating identifiable parametric functions of the form $\boldsymbol{\lambda}'\boldsymbol{\theta}$ for linear models by the method of least squares is same as the method of *best linear unbiased estimation.*

2.5 BEST LINEAR UNBIASED ESTIMATION

Best linear unbiased estimation, adapted to the needs of this text, assumes a linear model $\mathbf{Y} = \mathbf{X}\boldsymbol{\theta} + \boldsymbol{\varepsilon}$ with a parameter space Θ that can be represented as $\Theta = \{\boldsymbol{\theta} \in R^P\colon \boldsymbol{\Delta}'\boldsymbol{\theta} = \mathbf{0}\}$ for some known matrix $\boldsymbol{\Delta}$ and with a covariance matrix $\sigma^2\mathbf{I}$ for the observed responses where $\sigma^2 > 0$ is unknown. We use the parametrization

$$E(\mathbf{Y}) = \mathbf{X}\boldsymbol{\theta}, \quad \boldsymbol{\Delta}'\boldsymbol{\theta} = \mathbf{0}, \quad \mathrm{Cov}(\mathbf{Y}) = \sigma^2\mathbf{I}. \tag{2.6}$$

We wish to estimate parametric functions of the form $\boldsymbol{\lambda}'\boldsymbol{\theta}$ using estimators of the form $\mathbf{t}'\mathbf{Y}$ where $\mathbf{t} \in R^N$. In this context, $\boldsymbol{\lambda}'\boldsymbol{\theta}$ is *estimable* if there exists an estimator $\mathbf{t}'\mathbf{Y}$ that is unbiased for $\boldsymbol{\lambda}'\boldsymbol{\theta}$: $E(\mathbf{t}'\mathbf{Y}) = \boldsymbol{\lambda}'\boldsymbol{\theta}$ for all $\boldsymbol{\theta} \in \Theta$. More generally, a parametric vector $\boldsymbol{\Lambda}'\boldsymbol{\theta}$ is said to be estimable if there exists a matrix $\mathbf{A}$ such that $\mathbf{A}'\mathbf{Y}$ is unbiased for $\boldsymbol{\Lambda}'\boldsymbol{\theta}$.

THEOREM 2.4 A parametric vector $\boldsymbol{\Lambda}'\hat{\boldsymbol{\theta}}$ is estimable if and only if $R(\boldsymbol{\Lambda}) \subset R(\mathbf{X}', \boldsymbol{\Delta})$.

Proof Suppose $R(\boldsymbol{\Lambda}) \subset R(\mathbf{X}', \boldsymbol{\Delta})$. Then there exist matrices $\mathbf{A}$ and $\mathbf{B}$ such that $\boldsymbol{\Lambda} = \mathbf{X}'\mathbf{A} + \boldsymbol{\Delta}\mathbf{B}$. Then

$$E[\mathbf{A}'\mathbf{Y}] = \mathbf{A}'\mathbf{X}\boldsymbol{\theta} = \mathbf{A}'\mathbf{X}\boldsymbol{\theta} + \mathbf{B}'\boldsymbol{\Delta}'\boldsymbol{\theta} = \boldsymbol{\Lambda}'\boldsymbol{\theta},$$

so $\boldsymbol{\Lambda}'\boldsymbol{\theta}$ is estimable. Suppose now that $\boldsymbol{\Lambda}'\boldsymbol{\theta}$ is estimable. Then there exists some matrix $\mathbf{A}$ such that $E(\mathbf{A}'\mathbf{Y}) = \mathbf{A}'\mathbf{X}\boldsymbol{\theta} = \boldsymbol{\Lambda}'\boldsymbol{\theta}$ for all $\boldsymbol{\theta} \in \Theta = N(\boldsymbol{\Delta}')$. Then $(\boldsymbol{\Lambda} - \mathbf{X}'\mathbf{A})'\boldsymbol{\theta} = \boldsymbol{\Lambda}'\boldsymbol{\theta} - \mathbf{A}'\mathbf{X}\boldsymbol{\theta} = \mathbf{0}$ for all $\boldsymbol{\theta} \in N(\boldsymbol{\Delta}')$ so

$$R(\boldsymbol{\Lambda} - \mathbf{X}'\mathbf{A}) \subset N(\boldsymbol{\Delta}')^{\perp} = R(\boldsymbol{\Delta}),$$

and, hence,

$$\mathbf{\Lambda} = \mathbf{X}'\mathbf{A} + \mathbf{\Delta B} \quad \text{for some matrix } \mathbf{B}.$$

Therefore $R(\mathbf{\Lambda}) \subset R(\mathbf{X}', \mathbf{\Delta})$. □

Combining Theorems 2.2 and 2.4 we get the following theorem.

THEOREM 2.4a A parametric vector $\mathbf{\Lambda}'\boldsymbol{\theta}$ is estimable if and only if it is identifiable.

Let $\bar{\Theta}$ denote the set of estimable parametrization functions of the form $\boldsymbol{\lambda}'\boldsymbol{\theta}$. Then

$$\bar{\Theta} = \{\boldsymbol{\lambda}'\boldsymbol{\theta} \colon \boldsymbol{\lambda} \in R(\mathbf{X}', \mathbf{\Delta})\}.$$

$\bar{\Theta}$ is a vector space of estimators, and a linear combination of estimable linear parametric functions is also an estimable linear parametric function.

A *best linear unbiased estimator* (BLUE) is any estimator of the form $\mathbf{t}'\mathbf{Y}$ such that

$$\text{Var}(\mathbf{t}'\mathbf{Y}|\boldsymbol{\theta}, \sigma^2) \leqslant \text{Var}(\mathbf{a}'\mathbf{Y}|\boldsymbol{\theta}, \sigma^2) \quad \text{for all } \boldsymbol{\theta} \in \Theta, \qquad \sigma^2 > 0$$

for every linear estimator $\mathbf{a}'\mathbf{Y}$ having the same expectation as $\mathbf{t}'\mathbf{Y}$. Thus a BLUE is a uniformly minimum variance unbiased estimator (UMVUE) within the class of linear estimators. $\mathbf{t}'\mathbf{Y}$ is said to be BLUE for $\boldsymbol{\lambda}'\boldsymbol{\theta}$ if $\mathbf{t}'\mathbf{Y}$ is BLUE and is unbiased for $\boldsymbol{\lambda}'\boldsymbol{\theta}$.

Recall that the expectation space is denoted by $\Omega = \{\mathbf{X}\boldsymbol{\theta} \colon \mathbf{\Delta}'\boldsymbol{\theta} = \mathbf{0}\}$ and that it is a vector space. From Zyskind's theorem (see bibliographical notes) we have the following corollary:

COROLLARY 2.1 For parametrization (2.6), $\mathbf{t}'\mathbf{Y}$ is a BLUE if and only if $\mathbf{t} \in \Omega$.

We have the following encouraging theorem:

THEOREM 2.5 For parametrization (2.6), there exists a unique BLUE for every estimable linear parametric function.

Proof Let $\boldsymbol{\lambda}'\boldsymbol{\theta}$ be estimable. Then $\boldsymbol{\lambda} = \mathbf{X}'\mathbf{X}\mathbf{a} + \mathbf{\Delta d}$ for some vectors $\mathbf{a}$ and $\mathbf{d}$ (Theorem 2.4), and $\mathbf{a}'\mathbf{X}'\mathbf{Y}$ is an unbiased estimator for $\boldsymbol{\lambda}'\boldsymbol{\theta}$:

$$E(\mathbf{a}'\mathbf{X}'\mathbf{Y}) = \mathbf{a}'\mathbf{X}'\mathbf{X}\boldsymbol{\theta} = (\mathbf{a}'\mathbf{X}'\mathbf{X} + \mathbf{d}'\mathbf{\Delta}')\boldsymbol{\theta} = \boldsymbol{\lambda}'\boldsymbol{\theta}.$$

So there exists an unbiased estimator $\mathbf{t}'\mathbf{Y}$ for every estimable $\boldsymbol{\lambda}'\boldsymbol{\theta}$. Since $R^N = \Omega \oplus \Omega^{\perp}$ and $\mathbf{t} \in R^N$, $\mathbf{t} = \mathbf{b} + \mathbf{f}$ for $\mathbf{b} \in \Omega$ and $\mathbf{f} \in \Omega^{\perp}$. $\mathbf{b}'\mathbf{Y}$ is then a BLUE for

$\boldsymbol{\lambda}'\boldsymbol{\theta}$:

$$E(\mathbf{b}'\mathbf{Y}) = E(\mathbf{t}'\mathbf{Y} - \mathbf{f}'\mathbf{Y}) = \boldsymbol{\lambda}'\boldsymbol{\theta} - \mathbf{f}'\mathbf{X}\boldsymbol{\theta} = \boldsymbol{\lambda}'\boldsymbol{\theta}.$$

Suppose now that $\mathbf{b}'\mathbf{Y}$ and $\mathbf{c}'\mathbf{Y}$ are both BLUEs for $\boldsymbol{\lambda}'\boldsymbol{\theta}$. Then $\mathbf{b}'\mathbf{X}\boldsymbol{\theta} = \mathbf{c}'\mathbf{X}\boldsymbol{\theta}$ for all $\boldsymbol{\theta} \in \Theta$ or

$$(\mathbf{b} - \mathbf{c})'\mathbf{X}\boldsymbol{\theta} = \mathbf{0} \qquad \text{for all } \boldsymbol{\theta} \in \Theta \Rightarrow \mathbf{b} - \mathbf{c} \in \Omega^{\perp}.$$

But $\mathbf{b}, \mathbf{c} \in \Omega \Rightarrow \mathbf{b} - \mathbf{c} \in \Omega$ and $\mathbf{b} - \mathbf{c} \in \Omega \cap \Omega^{\perp} = \{\mathbf{0}\}$. So $\mathbf{b} = \mathbf{c}$ and BLUEs are unique. □

How can BLUEs be found? One way is to look for random vectors that are *Gauss–Markov*. $\hat{\boldsymbol{\theta}}$ is said to be a *Gauss–Markov* for* $\boldsymbol{\theta}$ provided that $\boldsymbol{\lambda}'\hat{\boldsymbol{\theta}}$ is the BLUE for $\boldsymbol{\lambda}'\boldsymbol{\theta}$ whenever $\boldsymbol{\lambda}'\boldsymbol{\theta}$ is an estimable parametric function.

We have the following result:

THEOREM 2.6 Let $\mathbf{W}$ be any matrix satisfying $R(\mathbf{W}) = \Omega$, and let $\mathbf{P}$ denote the orthogonal projection on Ω. Then the following statements are equivalent:

(a) $\hat{\boldsymbol{\theta}}$ is a Gauss–Markov estimator for $\boldsymbol{\theta}$.
(b) $\mathbf{W}'\mathbf{X}\hat{\boldsymbol{\theta}} = \mathbf{W}'\mathbf{Y}$ and $\boldsymbol{\Delta}'\hat{\boldsymbol{\theta}} = \mathbf{0}$.
(c) $\mathbf{X}\hat{\boldsymbol{\theta}} = \mathbf{P}\mathbf{Y}$ and $\boldsymbol{\Delta}'\hat{\boldsymbol{\theta}} = \mathbf{0}$.

Proof Note that $\mathbf{P} = \mathbf{W}(\mathbf{W}'\mathbf{W}(^{-}\mathbf{W}'$. Applying $\mathbf{W}(\mathbf{W}'\mathbf{W})^{-}$ to (b) yields (c), because $\mathbf{X}\hat{\boldsymbol{\theta}} \in \Omega$ and hence $\mathbf{P}\mathbf{X}\hat{\boldsymbol{\theta}} = \mathbf{X}\hat{\boldsymbol{\theta}}$. Conversely, applying $\mathbf{W}'$ to (c) yields (b) because $\mathbf{P}\mathbf{W} = \mathbf{W}$ and hence $\mathbf{W}'\mathbf{P} = \mathbf{W}'$. Thus (b) and (c) are equivalent. Assume (a). For arbitrary $\mathbf{a}$ and $\mathbf{b}$, let $\boldsymbol{\lambda} = \mathbf{X}'\mathbf{a} + \boldsymbol{\Delta}\mathbf{b}$. By Theorem 2.4, $\boldsymbol{\lambda}'\boldsymbol{\theta}$ is estimable, so $\boldsymbol{\lambda}'\hat{\boldsymbol{\theta}} = \mathbf{a}'\mathbf{X}\hat{\boldsymbol{\theta}} + \mathbf{b}'\boldsymbol{\Delta}'\hat{\boldsymbol{\theta}}$ is a BLUE for $\boldsymbol{\lambda}'\boldsymbol{\theta}$. But $E(\mathbf{a}'\mathbf{P}\mathbf{Y}) = \mathbf{a}'\mathbf{P}\mathbf{X}\boldsymbol{\theta} = \mathbf{a}'\mathbf{X}\boldsymbol{\theta} + \mathbf{b}'\boldsymbol{\Delta}\boldsymbol{\theta} = \boldsymbol{\lambda}'\boldsymbol{\theta}$ for all $\boldsymbol{\theta} \in \Omega$. Also $\mathbf{P}\mathbf{a} \in \Omega$, so $\mathbf{a}'\mathbf{P}\mathbf{Y}$ is a BLUE for $\boldsymbol{\lambda}'\boldsymbol{\theta}$ by Corollary 2.1. By Theorem 2.5, BLUEs are unique, so $\mathbf{a}'\mathbf{X}\hat{\boldsymbol{\theta}} + \mathbf{b}'\boldsymbol{\Delta}'\hat{\boldsymbol{\theta}} = \mathbf{a}'\mathbf{P}\mathbf{Y}$. This is true for all $\mathbf{a}$ and $\mathbf{b}$, so $\mathbf{X}\hat{\boldsymbol{\theta}} = \mathbf{P}\mathbf{Y}$ and $\boldsymbol{\Delta}'\hat{\boldsymbol{\theta}} = \mathbf{0}$. Conversely, assume (c). Suppose $\boldsymbol{\lambda}'\hat{\boldsymbol{\theta}}$ is estimable. Then $\boldsymbol{\lambda} = \mathbf{X}'\mathbf{a} + \boldsymbol{\Delta}\mathbf{b}$ for some $\mathbf{a}$ and $\mathbf{b}$. Now $\boldsymbol{\lambda}'\hat{\boldsymbol{\theta}} = \mathbf{a}'\mathbf{X}\hat{\boldsymbol{\theta}} + \mathbf{b}'\boldsymbol{\Delta}'\hat{\boldsymbol{\theta}} = \mathbf{a}'\mathbf{P}\mathbf{Y}$, $E(\boldsymbol{\lambda}'\hat{\boldsymbol{\theta}}) = E(\mathbf{a}'\mathbf{P}\mathbf{Y}) = \boldsymbol{\lambda}'\boldsymbol{\theta}$, and $\mathbf{P}\mathbf{a} \in \Omega$, so $\boldsymbol{\lambda}'\hat{\boldsymbol{\theta}}$ is BLUE for $\boldsymbol{\lambda}'\boldsymbol{\theta}$. Thus (a) and (c) are equivalent. □

When $R(\boldsymbol{\Delta}) \cap R(\mathbf{X}') = \{\mathbf{0}\}$, we may choose $\mathbf{W} = \mathbf{X}$ and statement (b) then becomes the augmented normal equations (2.5).

Because the covariance matrix is $\sigma^2\mathbf{I}$, $\hat{\boldsymbol{\theta}}$ being Gauss–Markov for $\boldsymbol{\theta}$ is the same as $\hat{\boldsymbol{\theta}}$ being (constrained) least squares for $\boldsymbol{\theta}$. For any estimable (identifiable) parametric function $\boldsymbol{\lambda}'\boldsymbol{\theta}$, $\boldsymbol{\lambda}'\hat{\boldsymbol{\theta}}$ being BLUE for $\boldsymbol{\lambda}'\boldsymbol{\theta}$ is the same as $\boldsymbol{\lambda}'\hat{\boldsymbol{\theta}}$

*Some authors restrict themselves to using random vectors due to measurability considerations, but we will not concern ourselves with this and call $\hat{\theta}$ an estimator.

being least squares for $\boldsymbol{\lambda}'\boldsymbol{\theta}$. This, then, is how least squares estimation is a special case of best linear unbiased estimation.

When $\hat{\boldsymbol{\theta}}$ is least squares for $\boldsymbol{\theta}$, $\hat{R}=\|\mathbf{Y}-\mathbf{X}\hat{\boldsymbol{\theta}}\|^2$ is called the *residual sum of squares*. Then

$$E(\hat{R})=(N-m)\sigma^2, \quad \text{where } m=\dim \Omega.$$

Except where explicitly stated otherwise, we will use the method of best linear unbiased estimation to estimate linear parametric functions.

2.6 MAXIMUM LIKELIHOOD ESTIMATION

Gauss–Markov estimators, and consequently least squares estimators, are *maximum likelihood estimators* when the random vector $\mathbf{Y}$ has a multivariate normal distribution. Suppose a random vector $\mathbf{Y}$ in R^N has a multivariate density function $f_{\mathbf{Y}}(\mathbf{y}|\boldsymbol{\theta})$, where $\mathbf{y} \in R^N$ and $\boldsymbol{\theta} \in \Theta$ is a vector of parameters. The *likelihood function* of the random vector $\mathbf{Y}$ is the density function evaluated at $\mathbf{Y}=\mathbf{y}$ and considered as a function of $\boldsymbol{\theta}$ defined on Θ. It is denoted by $L_{\mathbf{Y}}(\boldsymbol{\theta})$, and $L_{\mathbf{Y}}(\boldsymbol{\theta})=f_{\mathbf{Y}}(\mathbf{y}|\boldsymbol{\theta})$, $\boldsymbol{\theta} \in \Theta$. A maximum likelihood estimator for $\boldsymbol{\theta}$, if it exists, is any vector $\hat{\boldsymbol{\theta}} \in \Theta$ satisfying

$$L_{\mathbf{Y}}(\hat{\boldsymbol{\theta}})=\sup_{\boldsymbol{\theta} \in \Theta} L_{\mathbf{Y}}(\boldsymbol{\theta}).$$

A random vector $\mathbf{Y}$ in R^N is said to have a *multivariate normal distribution* with mean vector $\boldsymbol{\mu} \in R^N$ and covariance matrix $\sigma^2\mathbf{I}$, denoted $\mathbf{Y} \sim N_N(\boldsymbol{\mu}, \sigma^2\mathbf{I})$, if it has the density function

$$f_{\mathbf{Y}}(\mathbf{y}|\boldsymbol{\mu}, \sigma^2)=(2\pi\sigma^2)^{-N/2} \exp\left(-\frac{1}{2\sigma^2}\|\mathbf{y}-\boldsymbol{\mu}\|^2\right), \qquad \mathbf{y} \in R^N.$$

Suppose $\mathbf{Y}$ has parametrization (2.6). Since $\mathbf{X}$ and $\boldsymbol{\Delta}$ are known matrices, the likelihood function of $\mathbf{Y}$, expressed in terms of the unknown parameters $\boldsymbol{\theta}$ and σ^2, is

$$L_{\mathbf{Y}}(\boldsymbol{\theta}, \sigma^2)=(2\pi\sigma^2)^{-N/2} \exp\left(-\frac{1}{2\sigma^2}\|\mathbf{y}-\mathbf{X}\boldsymbol{\theta}\|^2\right),$$

for $\boldsymbol{\theta} \in \Theta=\{\boldsymbol{\theta}\colon \boldsymbol{\Delta}'\boldsymbol{\theta}=\mathbf{0}\}$ and $\sigma^2>0$. When $N>m=\dim \Omega$, with probability 1 (wp 1), the likelihood function is bounded. It is then maximized by $(\hat{\boldsymbol{\theta}}, \hat{\sigma}^2)$, where $\hat{\boldsymbol{\theta}}$ is any Gauss–Markov estimator for $\boldsymbol{\theta}$ and $\hat{\sigma}^2=\hat{R}/N$, where $\hat{R}$ is the residual sum of squares under parametrization (2.6). Since $\boldsymbol{\Delta}'\hat{\boldsymbol{\theta}}=\mathbf{0}$ and $\hat{R}>0$ (wp 1), we see that $(\hat{\boldsymbol{\theta}}, \hat{\sigma}^2)$ is maximum likelihood for $(\boldsymbol{\theta}, \sigma^2)$ (wp 1). When $\hat{R}=0$, the likelihood function can be made arbitrarily large, and therefore no maximum likelihood estimator for $(\boldsymbol{\theta}, \sigma^2)$ exists in this case.

2.7 THE LIKELIHOOD RATIO TEST AND THE ANALYSIS OF VARIANCE

Now consider the problem of testing hypotheses concerning the expectation space Ω or the components of the parameter $\boldsymbol{\theta}$. To proceed, we must make assumptions on the probability distribution of the responses. Up to this point, we have assumed parametrization (2.6). We now add that the vector $\mathbf{Y}$ has a multivariate normal distribution:

$$\mathbf{Y} \sim N_N(\mathbf{X}\boldsymbol{\theta}, \sigma^2\mathbf{I}), \qquad \boldsymbol{\Delta}'\boldsymbol{\theta}=\mathbf{0}, \qquad \sigma^2>0.$$

It is known that UMPU tests exist for testing certain hypotheses about the parameters of a normal distribution. However, not all hypotheses have UMPU tests. Even when they do exist, exact UMPU tests of given size for particular hypotheses can be difficult to construct.

Another method of constructing tests of hypotheses is the *likelihood ratio test procedure*. If the likelihood function of $\mathbf{Y}$ is larger for distributions in H_0 than for distributions in H_A, we would tend to accept H_0 rather than H_A. If the reverse were true, we would tend to reject H_0 in favor of H_A. For testing H_0 against H_A, the *likelihood ratio* is the ratio of the maximum (actually the supremum) of the likelihood function of $\mathbf{Y}$ over all distributions in H_0 to the maximum overall distributions in H_0 and H_A whenever the ratio is defined. The likelihood ratio is denoted $\lambda(\mathbf{Y})$ and $0 \leqslant \lambda(\mathbf{Y}) \leqslant 1$. Under certain distributional assumptions and for testing particular types of hypotheses, the likelihood ratio $\lambda(\mathbf{Y})$ or some function of it has a known distribution when H_0 is true. The likelihood ratio test procedure rejects H_0 in favor of H_A when $\lambda(\mathbf{Y})$ is less than some critical value that is determined by the distribution of $\lambda(\mathbf{Y})$ under H_0 and the desired level of the test. The same procedure is used under more general distributional assumptions by using large-sample theory to obtain the asymptotic distribution of $-2\ln[\lambda(\mathbf{Y})]$ when H_0 is true.

The likelihood ratio test procedure determines the critical value λ_c of $\lambda(\mathbf{Y})$ for a given level of significance by using the known distribution of $\lambda(\mathbf{Y})$ when H_0 is true. Given an observation $\mathbf{y}$, the procedure then rejects H_0 in favor of H_A if $\lambda(\mathbf{y})<\lambda_c$. So the procedure rejects H_0; the observation is sufficiently unlikely when H_0 is true. Is this a reasonable approach to testing an hypothesis? If so, when?

The probability of Type I error is at most the specified level of the test. So when we reject H_0 in favor of H_A, we are assured that the probability of our making a mistake is bounded by the significance level of the test.

However, the probability of Type II error is not controlled in this procedure. The likelihood ratio $\lambda(\mathbf{Y})$ may depend on only one distribution in H_A for a particular $\mathbf{y}$, and λ_c is determined by assuming H_0 to be true. It is possible that the probability that $\lambda(\mathbf{y})<\lambda_c$ is lower for some distributions in

H_A than for any distribution in H_0. The likelihood ratio test will then be biased. It is therefore important to consider the distribution of the test statistic when H_A is true.

Let us apply this method to test the hypothesis H_0: $\tau_1 = \tau_2$ versus H_A: $\tau_1 \neq \tau_2$ for the two-treatment experiment (Case 1), with normally distributed responses.

The hypothesis H_0 means that the treatment expectations are equal: $\mu + \tau_1 = \mu + \tau_2$. When the constraint $\tau_1 + \tau_2 = 0$ is included, we see that H_0 is equivalent to H_0: $\tau_1 = \tau_2 = 0$. The family of distributions represented by H_0 is

$$\mathfrak{P}_0 = \{N_N(\mu \mathbf{1}_N, \sigma^2 \mathbf{I}) : \mu \in R^1, \qquad \sigma^2 > 0\}.$$

The likelihood function for $\mathfrak{P}_0$ is

$$L_0(\mu, \sigma^2) = (2\pi\sigma^2)^{-N/2} \exp\left(-\sum_{i=1}^{2} \sum_{j=1}^{n_i} \frac{(Y_{ij} - \mu)^2}{2\sigma^2}\right).$$

L_0 is maximized over $\mathfrak{P}_0$ by the maximum likelihood estimates

$$\tilde{\mu} = \bar{Y}_{..} = \sum_{i=1}^{2} \sum_{j=1}^{n_i} \frac{Y_{ij}}{N}$$

and

$$\tilde{\sigma}^2 = \frac{\hat{R}_0}{N} = \sum_{i=1}^{2} \sum_{j=1}^{n_i} \frac{(Y_{ij} - \bar{Y}_{..})^2}{N}.$$

$\hat{R}_0$ is the residual sum of squares under the hypothesis.

The family of distributions represented by $H_0 \cup H_A$ is

$$\mathfrak{P} = \{N_N(\mathbf{X}\boldsymbol{\theta}, \sigma^2 \mathbf{I}) : \boldsymbol{\theta} \in R^3, \qquad \sigma^2 > 0\}.$$

The likelihood function for $\mathfrak{P}$ is

$$L(\boldsymbol{\theta}, \sigma^2) = (2\pi\sigma^2)^{-N/2} \exp\left(\frac{-\|\mathbf{Y} - \mathbf{X}\boldsymbol{\theta}\|^2}{2\sigma^2}\right)$$

$$= (2\pi\sigma^2)^{-N/2} \exp\left(-\sum_{i=1}^{2} \sum_{j=1}^{n_i} \frac{(Y_{ij} - \mu - \tau_i)^2}{2\sigma^2}\right).$$

The likelihood function is, of course, maximized over $H_0 \cup H_A$ (L is maximized over $\mathfrak{P}$) by the maximum likelihood estimates

$$\hat{\mu} + \hat{\tau}_i = \bar{Y}_{i.} = \sum_{j=1}^{n_i} \frac{Y_{ij}}{n_i}, \qquad i = 1, 2$$

and

$$\hat{\sigma}^2 = \frac{\hat{R}}{N} = \sum_{i=1}^{2} \sum_{j=1}^{n_i} \frac{(Y_{ij} - \bar{Y}_{i.})^2}{N}.$$

The likelihood ratio is then $\lambda(\mathbf{Y})=(\hat{R}/\hat{R}_0)^{N/2}$, when $\hat{R}_0>0$.

The likelihood ratio test procedure says to reject $H_0:\tau_1=\tau_2$ when $\lambda(\mathbf{Y})<K$ where K is some constant chosen to yield the desired significance level for this test.

However, rather than using $\lambda(\mathbf{Y})$ directly, we use an equivalent test statistic. Define

$$\hat{F}=\hat{F}(\mathbf{Y})=\frac{(\hat{R}_0-\hat{R})}{\hat{R}/(N-2)}.$$

Since $\hat{R}/\hat{R}_0=[1+\hat{F}/(N-2)]^{-1}$, and $0\leqslant\lambda(\mathbf{Y})\leqslant 1$, rejecting H_0 when $\lambda(\mathbf{Y})$ is small is equivalent to rejecting H_0 when $\hat{F}$ is large. Furthermore, with our normality assumption the distribution of $\hat{F}$ is the central F distribution with 1 and $N-2$ degrees of freedom when the hypothesis H_0 is true. So, given a significance level α, we compute the $(1-\alpha)$ 100th percentile of the F distribution with 1 and $N-2$ degrees of freedom. We then reject $H_0:\tau_1=\tau_2$ in favor of H_A: $\tau_1\neq\tau_2$ at the level of significance α if, for our sample $\mathbf{Y}$, $\hat{F}(\mathbf{Y})$ is greater than this percentile. $\hat{F}(\mathbf{Y})$ has a noncentral F distribution when H_A is true, so this test is unbiased.

$\hat{F}(\mathbf{Y})$ can be computed in the following manner. Compute the (mean corrected) total sum of squares, SSTO, treatment sum of squares, SSTR, and the residual (or error) sum of squares, SSE, as

$$\text{SSTO}=\hat{R}_0=\sum_{i=1}^{2}\sum_{j=1}^{n_i}(Y_{ij}^2-\bar{Y}_{..})^2=\sum_{i=1}^{2}\sum_{j=1}^{n_i}Y_{ij}^2-N\bar{Y}_{..}^2,$$

$$\text{SSTR}=\hat{R}_0-\hat{R}=\sum_{i=1}^{2}n_i(\bar{Y}_{i.}-\bar{Y}_{..})^2=\sum_{i=1}^{2}n_i\bar{Y}_{i.}^2-N\bar{Y}_{..}^2,$$

and

$$\text{SSE}=\hat{R}=\sum_{i=1}^{2}\sum_{j=1}^{n_i}(Y_{ij}-\bar{Y}_{i.})^2=\text{SSTO}-\text{SSTR}.$$

Then

$$\hat{F}=\frac{\text{SSTR}}{\text{SSE}/(N-2)}=\frac{\text{MSTR}}{\text{MSE}},$$

in the case of two treatments.

This is the one-way analysis of variance procedure, which can be summarized in a table, as in Table 2.2.

Each Y_{ij} can be expressed as a sum of three components:

$$Y_{ij}=(Y_{ij}-\bar{Y}_{i.})+(\bar{Y}_{i.}-\bar{Y}_{..})+\bar{Y}_{..}.$$

The first component is the deviation of the response from the average of

Table 2.2 Analysis of Variance of One-Way Classification

Source of Variations	Degrees of Freedom	Sums of Squares	Mean Squares	F Ratio
Treatments	1	SSTR	MSTR	$\hat{F}$
Residual	$N-2$	SSE	MSE	
Total	$N-1$	SSTO		

the responses for the same treatment. The second is the deviation of that average treatment response from the overall response mean. The last is the overall response mean. After some algebra, we can obtain

$$\sum_{i=1}^{2}\sum_{j=1}^{n_i}(Y_{ij}-\bar{Y}_{..})^2=\sum_{i=1}^{2}\sum_{j=1}^{n_i}(Y_{ij}-\bar{Y}_{i.})^2+\sum_{i=1}^{2}n_i(\bar{Y}_{i.}-\bar{Y}_{..})^2. \tag{2.7}$$

When devided by $N-1$, the left-hand side of (2.7) is the sample variance of the observed responses. This equation partitions the *total* (mean-adjusted) *sum of squares* into two parts. The first, called the *within sum of squares,* is due to the variations in the observed responses for the same treatment. The second, called the *between sum of squares,* is due to the variations in the average treatment responses from the overall mean. The *analysis of variance (ANOVA) is a data analysis technique that compares the relative sizes of the within and between sums of squares to infer whether* $\tau_1=\tau_2$ or $\tau_1\neq\tau_2$. It will arise naturally in our following discussion of the method of least squares.

The likelihood ratio test procedure can be applied more generally to test linear hypotheses for linear models under normality assumptions. We begin by assuming that $\mathbf{Y}$ is a random vector with a distribution in the family

$$\mathfrak{P}=\{N_N(\boldsymbol{\mu},\sigma^2\mathbf{I}):\boldsymbol{\mu}\in\Omega,\sigma^2>0\},$$

where Ω is a subspace of R^N.

Suppose that Ω_0 is a subspace such that $\Omega_0\subset\Omega$. Let $\Omega_A=\Omega-\Omega_0$, the set complement of Ω_0 in Ω. Then the hypotheses H_0: $\boldsymbol{\mu}\in\Omega_0$, $\sigma^2>0$, and H_A: $\boldsymbol{\mu}\in\Omega_A$, $\sigma^2>0$ are jointly called a linear hypothesis (on Ω). It is the usual practice to only indicate the restrictions in the hypothesis statements, for example, H_0: $\boldsymbol{\mu}\in\Omega_0$ and H_A: $\boldsymbol{\mu}\in\Omega_A$.

The linear hypotheses discussed in the context of experimental designs will arise from hypotheses on the parameter $\boldsymbol{\theta}$ of some parametrization

$$E(\mathbf{Y})=\mathbf{X}\boldsymbol{\theta},\quad \boldsymbol{\theta}\in\Theta,\qquad \text{Cov}(\mathbf{Y})=\sigma^2\mathbf{I}$$

where Θ is a known subspace of R^P. Suppose Θ_0 is a subspace such that $\Theta_0\subset\Theta$. Then the hypotheses H_0: $\boldsymbol{\theta}\in\Theta_0$ and H_A: $\boldsymbol{\theta}\in\Theta_A=\Theta-\Theta_0$ is called a *linear hypothesis on* $\boldsymbol{\theta}$ if the sets $\Omega_0 = \{\mathbf{X}\boldsymbol{\theta}: \boldsymbol{\theta} \in \Theta_0\}$ and $\Omega_A = \{\mathbf{X}\boldsymbol{\theta}: \boldsymbol{\theta} \in \Theta_A\}$ are disjoint. We will often only write $H_0: \boldsymbol{\theta} \in \Theta_0$ to represent this linear hypothesis.

THEOREM 2.7 $H_0: \boldsymbol{\Lambda}'\boldsymbol{\theta} = \mathbf{0}$ is a linear hypothesis on $\boldsymbol{\theta}$ if and only if $\boldsymbol{\Lambda}'\boldsymbol{\theta}$ is an estimable parametric vector.

Proof Suppose $H_0: \boldsymbol{\Lambda}'\boldsymbol{\theta} = \mathbf{0}$ is a linear hypothesis on $\boldsymbol{\theta}$. Then the sets $\Omega_0 = \{\mathbf{X}\boldsymbol{\theta}: \boldsymbol{\Delta}'\boldsymbol{\theta} = \mathbf{0}, \boldsymbol{\Lambda}'\boldsymbol{\theta} = \mathbf{0}\}$ and $\Omega_A = \{\mathbf{X}\boldsymbol{\theta}: \boldsymbol{\Delta}'\boldsymbol{\theta} = \mathbf{0}, \boldsymbol{\Lambda}'\boldsymbol{\theta} \neq \mathbf{0}\}$ are disjoint. Letting $\boldsymbol{\theta} = \mathbf{0}$, we see that $\mathbf{0} \in \Omega_0$. Let $\bar{\boldsymbol{\theta}}$ be any vector such that $\mathbf{X}\bar{\boldsymbol{\theta}} = \mathbf{0}$, $\boldsymbol{\Delta}'\bar{\boldsymbol{\theta}} = \mathbf{0}$. Then $\boldsymbol{\Lambda}'\bar{\boldsymbol{\theta}} = \mathbf{0}$, because otherwise $\mathbf{0} \in \Omega_0 \cap \Omega_A$. Therefore, $\boldsymbol{\theta} \in N(\mathbf{X}) \cap N(\boldsymbol{\Delta}')$ implies $\bar{\boldsymbol{\theta}} \in N(\boldsymbol{\Lambda}')$, which means that $N(\mathbf{X}) \cap N(\boldsymbol{\Delta}') \subset N(\boldsymbol{\Lambda}')$. Taking orthogonal complements, we then find that $R(\boldsymbol{\Lambda}) \subset R(\mathbf{X}', \boldsymbol{\Delta})$ and $\boldsymbol{\Lambda}'\boldsymbol{\theta}$ is estimable.

Now suppose $\boldsymbol{\Lambda}'\boldsymbol{\theta}$ is estimable. Then there exist matrices $\mathbf{A}$ and $\mathbf{B}$ such that $\boldsymbol{\Lambda} = \mathbf{X}'\mathbf{A} + \boldsymbol{\Delta}\mathbf{B}$. Suppose $\mathbf{x} \in \Omega_0 \cap \Omega_A$. Then there exist $\boldsymbol{\theta}_1$, $\boldsymbol{\theta}_2$ such that

$$\mathbf{X}\boldsymbol{\theta}_1 = \mathbf{x} = \mathbf{X}\boldsymbol{\theta}_2, \quad \boldsymbol{\Delta}'\boldsymbol{\theta}_1 = \boldsymbol{\Delta}'\boldsymbol{\theta}_2 = \mathbf{0}, \quad \text{and} \quad \boldsymbol{\Lambda}'\boldsymbol{\theta}_1 = \mathbf{0}, \quad \text{while } \boldsymbol{\Delta}'\boldsymbol{\theta}_2 \neq \mathbf{0}.$$

But $\boldsymbol{\Lambda}'\boldsymbol{\theta}_1 = \mathbf{A}'\mathbf{X}\boldsymbol{\theta}_1 + \mathbf{B}'\boldsymbol{\Delta}'\boldsymbol{\theta}_1 = \mathbf{A}'\mathbf{X}\boldsymbol{\theta}_2 + \mathbf{B}'\boldsymbol{\Delta}'\boldsymbol{\theta}_2 = \boldsymbol{\Lambda}'\boldsymbol{\theta}_2$, a contradiction. So Ω_0 and Ω_A must be disjoint, and $H_0: \boldsymbol{\Lambda}'\boldsymbol{\theta} = \mathbf{0}$ is a linear hypothesis on Θ. □

We define the random variables

$$\hat{R} = \inf_{\boldsymbol{\mu} \in \Omega} \|\mathbf{Y} - \boldsymbol{\mu}\|^2 \quad \text{and} \quad \hat{R}_0 = \inf_{\boldsymbol{\mu} \in \Omega_0} \|\mathbf{Y} - \boldsymbol{\mu}\|^2,$$

the *residual sum of squares under the model* and the *residual sum of squares under the hypothesis*, respectively. Clearly $\hat{R} \leqslant \hat{R}_0$. The likelihood ratio test statistic for testing H_0 against H_A is given by

$$\lambda(\mathbf{Y}) = \frac{\sup_{\boldsymbol{\mu} \in \Omega_0, \sigma^2 > 0} f_Y(\mathbf{Y}|\boldsymbol{\mu}, \sigma^2\mathbf{I})}{\sup_{\boldsymbol{\mu} \in \Omega, \sigma^2 > 0} f_Y(\mathbf{Y}|\boldsymbol{\mu}, \sigma^2\mathbf{I})} = \left(\frac{\hat{R}}{\hat{R}_0}\right)^{N/2},$$

which is well defined when $\hat{R}_0 > 0$ (this happens with probability one). The likelihood ratio test procedure is to reject H_0 in favor of H_A for small values of $\lambda(\mathbf{Y})$.

We again consider an alternative form of the test. Let $m = \dim \Omega$ and $m_0 = \dim \Omega_0$, the dimensions of the subspaces Ω and Ω_0. If $m_0 = m$, then $\Omega_0 = \Omega$ and there is nothing to test. So suppose $m_0 < m$. Define

$$\hat{F} = \frac{(\hat{R}_0 - \hat{R})/(m - m_0)}{\hat{R}/(N - m)}.$$

Then a test procedure that is equivalent to the likelihood ratio test procedure is to reject H_0 in favor of H_A when $\hat{F}$ is large.

Let $\chi^2(n, \phi^2)$ denote the noncentral chi-square distribution with n degrees of freedom and noncentrality parameter ϕ^2. When $\phi^2 = 0$, we denote the distribution by $\chi^2(n)$ and call it the central chi-square distribution with n degrees of freedom. As before, let Ω denote a known subspace of R^N. We have the following:

THEOREM 2.8 Suppose $\mathbf{Y} \sim N_N(\boldsymbol{\mu}, \sigma^2\mathbf{I})$ where $\boldsymbol{\mu} \in \Omega$ and $\sigma^2 > 0$. Let $H_0: \boldsymbol{\mu} \in \Omega_0$ be a linear hypothesis with $m_0 < m$. Then:

(a) $\hat{R}/\sigma^2 \sim \chi^2(N-m)$.
(b) $(\hat{R}_0 - \hat{R})/\sigma^2 \sim \chi^2(m - m_0, \phi^2)$ for some ϕ^2.
(c) $\phi^2 = 0$ in (b) if and only if $\boldsymbol{\mu} \in \Omega_0$.
(d) $\hat{R}_0 - \hat{R}$ and $\hat{R}$ are independent.

So when testing a linear hypothesis $H_0: \boldsymbol{\mu} \in \Omega_0$ when $m > m_0$, the test statistic $\hat{F}$ has the central F distribution with $(m - m_0)$ and $(N - m)$ degrees of freedom under H_0 and the noncentral F distribution with the same degrees of freedom under H_A. Given a particular significance level α, the critical value for the likelihood ratio test can be constructed. Also the power of the test for each distribution in H_A can be calculated.

The statistics $\hat{R}$ and $\hat{R}_0$ can be computed from least squares $\hat{\boldsymbol{\theta}}$ and $\hat{\boldsymbol{\theta}}_0$ under models $E(\mathbf{Y}) \in \Omega$ and $E(\mathbf{Y}) \in \Omega_0$, respectively:

$$\hat{R} = \|\mathbf{Y} - \mathbf{X}\hat{\boldsymbol{\theta}}\|^2 \quad \text{and} \quad \hat{R}_0 = \|\mathbf{Y} - \mathbf{X}\hat{\boldsymbol{\theta}}_0\|^2.$$

This method is called *general linear regression.* The total sum of squares $\|\mathbf{Y}\|^2$ can be partitioned using the regression results and displayed in Table 2.2a.

The regression under Ω_0 line is often omitted and its sum of squares subtracted from the total sum of squares. This was done in Table 2.2 for the test of $H_0: \tau_1 = \tau_2$ in Case 1. There $\Omega_0 = R(\mathbf{1}_N)$ and $\|\mathbf{Y}\hat{\boldsymbol{\theta}}_0\|^2 = N\bar{Y}_{..}^2$. The total sum of

Table 2.2a Partitioning Sum of Squares

Source	Degrees of Freedom		Sum of Squares	
Regression under Ω	m		$\|\mathbf{X}\hat{\boldsymbol{\theta}}\|^2$	
Under Ω_0		m_0		$\|\mathbf{X}\hat{\boldsymbol{\theta}}_0\|^2$
Deviation from Ω_0		$m - m_0$		$\hat{R}_0 - \hat{R}$
Residuals under Ω	$N - m$		$\hat{R}$	
Total	N		$\|\mathbf{Y}\|^2$	

squares (not corrected for the mean)

$$\text{SSTO} = \mathbf{Y}'\mathbf{Y} = \sum_{i=1}^{t} \sum_{j=1}^{n_i} Y_{ij}^2,$$

can be partitioned into the residual sum of squares, given by

$$\begin{aligned}\text{SSE} &= (\mathbf{Y} - \mathbf{X}\hat{\boldsymbol{\theta}})'(\mathbf{Y} - \mathbf{X}\hat{\boldsymbol{\theta}}) \\ &= \mathbf{Y}'\mathbf{Y} - \hat{\boldsymbol{\theta}}'\mathbf{X}'\mathbf{Y}\end{aligned}$$

where $\hat{\boldsymbol{\theta}}$ is any solution satisfying the normal equations $\mathbf{X}'\mathbf{X}\hat{\boldsymbol{\theta}} = \mathbf{X}'\mathbf{Y}$, and

$$\text{SSR} = \text{SSTO} - \text{SSE} = \hat{\boldsymbol{\theta}}'\mathbf{X}'\mathbf{Y}$$

which represents that portion of the total sum of squares attributable to having fitted the regression model

$$\text{MI}: E(\mathbf{Y}) = \mathbf{X}\boldsymbol{\theta}; \qquad \boldsymbol{\theta} \in R^p, \qquad \boldsymbol{\theta} = (\mu, \tau_1, \ldots, \tau_t)', \qquad \Omega = R(\mathbf{X})$$

and is called the sum of squares due to regression. SSR is also called *reduction in sum of squares.*

Now suppose that the model has no treatments in it, that is,

$$\text{MII}: E(\mathbf{Y}) = \mathbf{1}_N\mu, \qquad \mu \in \Omega_0, \qquad \Omega_0 = R(\mathbf{1}_N)$$

which means all $\tau_i = 0$. Then $\hat{\mu} = \bar{Y}_{..}$, and SSR (regression under Ω_0) which is usually called correction for the mean, becomes

$$\text{SSM} = \mathbf{Y}'N^{-1}\mathbf{1}'\mathbf{1}\mathbf{Y} = N\bar{Y}_{..}^2.$$

Notice that SSM is the regression sum of squares for the submodel MII: $E(\mathbf{Y}) \in \Omega_0$, and can naturally be viewed as a reduction in the total sum of squares $\mathbf{Y}'\mathbf{Y}$ that is accounted for by fitting the submodel MII while ignoring the remaining portion of the regression space Ω. In short, we can say that in the general linear regression approach to the analysis of variance we are comparing the adequacy of different models for the same set of data. Such a comparison can therefore be made by comparing different values of SSR that result from fitting different models. For convenience, a reduction in sum of squares will be written as SS($-$), where ($-$) indicates the model fitted. Therefore, SS(μ, τ) is the reduction in sum of squares for fitting model MI, and SS(μ) is the reduction in sum of squares when fitting model MII.

We can now summarize the partitioning of total sum of squares using the results of the general linear regression in Table 2.2b.

The quantity SS(μ, τ) $-$ SS(μ) is the reduction in sum of squares due to fitting two different models, namely MI and MII, and can be viewed as the

Table 2.2b Partitioning Sum of Squares

Source	Degrees of Freedom	Sum of Squares
Regression under Ω	m	$SS(\mu, \tau)$
Under Ω_0	m_0	$SS(\mu)$
Deviation from Ω_0	$m-m_0$	$SS(\mu, \tau)-SS(\mu)$
Residual under Ω	$N-m$	$\mathbf{Y}'\mathbf{Y}-SS(\mu, \tau)$
Total	N	$\|\mathbf{Y}\|^2$

additional reduction in sum of squares due to fitting τ after μ. We call

$$SS(\mu, \tau)-SS(\mu)$$

the sum of squares for τ *adjusted* for μ, and we denote it by $SS(\tau|\mu)$.

Table 2.2b suggests the common association of the dimensions $m=\dim \Omega$ and $m_0=\dim \Omega_0$ with the *degrees of freedom* for the regressions under Ω and Ω_0, respectively. $m-m_0$ is the difference in the dimensions of Ω and Ω_0 and is called the degrees of freedom for testing H_0: $\mu \in \Omega_0$ against H_A: $\mu \in \Omega - \Omega_0$.

This table also illustrates that regression analysis is the method underlying the analysis of variance. Indeed, the analysis of variance approaches to testing hypotheses is merely a convenient computational method for obtaining the above information derived by exploiting the characteristics of particular models.

Case 2. Comparing t Treatments

We now apply the methods introduced in this chapter to compare t treatments. Again, each treatment could be considered a different level of a composite factor with t levels.

We will use the model

$$Y_{ij}=\mu+\tau_1+e_{ij}, \qquad i=1, 2, \ldots, t; \quad j=1, 2, \ldots, n_i,$$

where, without any loss in generality, $n_i>0$ for all i. As before, μ and the τ_i are unknown constants. μ denotes the part of the response that does not depend upon the treatments and τ_i denotes the effect on the response for treatment i, $i=1, 2, \ldots, t$. The e_{ij} term represents the experimental error for the jth experimental unit receiving treatment i. The e_{ij} are independent random variables with zero mean and the same variance σ^2. When constructing hypothesis tests we will also assume that the errors e_{ij} are normally distributed.

Models defined on factorial domains are called *classifications*. When there are n factors, we have an *n-way classification*. The model we are using here is a one-way classification.

In matrix form, the parametrization for this model and experiment is

$$E(\mathbf{Y}) = \mathbf{X}\boldsymbol{\theta}, \qquad \boldsymbol{\Delta}'\boldsymbol{\theta} = \mathbf{0}, \qquad \mathrm{Cov}(\mathbf{Y}) = \sigma^2\mathbf{I}, \tag{2.8}$$

where

$$\mathbf{Y} = (Y_{11}, \ldots, Y_{1n_1}, Y_{21}, \ldots, Y_{2n_2}, \ldots, Y_{t1}, \ldots, Y_{tn_t})',$$

$$\boldsymbol{\theta} = (\mu, \tau_1, \tau_2, \ldots, \tau_t)',$$

$$\boldsymbol{\Delta} = \mathbf{0} \quad \text{so} \quad \Theta = R^p, \qquad p = t+1,$$

$$\mathbf{V} = \mathbf{I}_N, \text{ the } N \times N \text{ identity matrix, } N = \sum_{i=1}^{t} n_i, \text{ and}$$

$$\mathbf{X} = \left[\begin{array}{ccccccc} 1 & 1 & 0 & 0 & \cdots & 0 & 0 \\ & & \cdots & & & & \\ 1 & 1 & 0 & 0 & \cdots & 0 & 0 \\ 1 & 0 & 1 & 0 & \cdots & 0 & 0 \\ & & \cdots & & & & \\ 1 & 0 & 1 & 0 & \cdots & 0 & 0 \\ & & & \vdots & & & \\ 1 & 0 & 0 & 0 & \cdots & 0 & 1 \\ & & \cdots & & & & \\ 1 & 0 & 0 & 0 & \cdots & 0 & 1 \end{array}\right] \begin{array}{l} \left.\begin{array}{l} \\ \\ \\ \end{array}\right\} n_1 \text{ rows} \\ \left.\begin{array}{l} \\ \\ \\ \end{array}\right\} n_2 \text{ rows.} \\ \quad \vdots \\ \left.\begin{array}{l} \\ \\ \\ \end{array}\right\} n_t \text{ rows} \end{array}$$

$$\underbrace{\hphantom{1\ 0\ 0\ 0\ \cdots\ 0\ 1}}_{p = t+1 \text{ columns}}$$

The set of estimable linear parametric functions is

$$\bar{\Theta} = \{\boldsymbol{\lambda}'\boldsymbol{\theta} : \boldsymbol{\lambda} \in R(\mathbf{X}', \boldsymbol{\Delta})\}.$$

We saw before that $\bar{\Theta}$ is a vector space over the field of real numbers. The elements of $\bar{\Theta}$ are functions $\boldsymbol{\lambda}'\boldsymbol{\theta}$ from Θ into R^1. Elements $\boldsymbol{\lambda}_1'\boldsymbol{\theta}, \boldsymbol{\lambda}_2'\boldsymbol{\theta}, \ldots, \boldsymbol{\lambda}_q'\boldsymbol{\theta}$ of $\bar{\Theta}$ are said to be *linearly* independent if $a_1 = a_2 = \cdots = a_q = 0$ are the only real numbers for which

$$\sum_{i=1}^{q} a_i \boldsymbol{\lambda}_i'\boldsymbol{\theta} = 0 \quad \text{for all } \boldsymbol{\theta} \in \Theta.$$

If $\{\boldsymbol{\lambda}_1'\boldsymbol{\theta}, \ldots, \boldsymbol{\lambda}_q'\boldsymbol{\theta}\}$ is a set of elements of $\bar{\Theta}$ such that every $\boldsymbol{\lambda}'\boldsymbol{\theta} \in \bar{\Theta}$ can be expressed as a linear combination of these elements, then the set is called a *spanning set* for $\bar{\Theta}$. A spanning set for $\bar{\Theta}$ that consists of linearly independent elements is called a *basis* for $\bar{\Theta}$. The number of elements in any basis for $\bar{\Theta}$

is the same and is called the *dimension* of $\bar{\Theta}$, denoted dim $\bar{\Theta}$. We have

$$\dim \bar{\Theta} = r(\mathbf{X}', \mathbf{\Delta}) - r(\mathbf{\Delta}) = \dim \Omega,$$

where Ω is the expectation space for the parametrization.

For a particular parametrization, $\bar{\Theta}$ could be described as the set of all linear combinations of the elements of some basis for $\bar{\Theta}$. We earlier saw that the set of all BLUEs was of the form $\{\mathbf{t}'\mathbf{Y}: \mathbf{t} \in \Omega\}$ where Ω is a vector space in R^N. The set of all BLUEs is also a vector space, so any linear combination of BLUEs is also a BLUE for its expectation. We can find the BLUE of the elements on a basis for $\bar{\Theta}$. Then every estimable parametric function $\boldsymbol{\lambda}'\boldsymbol{\theta}$ can be expressed as a linear combination of these basis elements, and its BLUE $\boldsymbol{\lambda}'\hat{\boldsymbol{\theta}}$ can be found as the same linear combination of the corresponding BLUEs for the basis.

For our parametrization of Case 2 experiment, we have $\Omega = R(\mathbf{X})$ and $\bar{\Theta} = \{\boldsymbol{\lambda}'\boldsymbol{\theta}: \boldsymbol{\lambda} \in R(\mathbf{X}')\}$. One basis for $\bar{\Theta}$ is the set $\boldsymbol{\lambda}'\boldsymbol{\theta}$ with each $\boldsymbol{\lambda}'$ a distinct row of $\mathbf{X}$, that is, the set of distinct treatment expectations $\{\mu+\tau_1, \mu+\tau_2, \ldots, \mu+\tau_t\}$. The estimable linear parametric functions are just those that can be expressed as linear combinations of these expectations.

Another basis for $\bar{\Theta}$ is $\{\mu+\tau_1, \tau_1-\tau_2, \ldots, \tau_1-\tau_t\}$. The parametric functions $\tau_1-\tau_2, \ldots, \tau_1-\tau_t$ are examples of estimable linear parametric functions involving only the τ parameters. We shall characterize *the set T of estimable linear parametric functions involving only the τ parameters.*

THEOREM 2.9 A parametric function of the form $\sum_{i=1}^t \lambda_i\tau_i$ is estimable if and only if $\sum_{i=1}^t \lambda_i = 0$.

Proof Let $\boldsymbol{\lambda} = (0, \lambda_1, \lambda_2, \ldots, \lambda_t)'$. Suppose $\boldsymbol{\lambda}'\boldsymbol{\theta} = \sum_{i=1}^t \lambda_i\tau_i$ is estimable. Then it must be expressible as a linear combination of the elements in the basis $\{\mu+\tau_1, \ldots, \mu+\tau_t\}$ for $\bar{\Theta}$. That is,

$$\sum_{i=1}^t \lambda_i\tau_i = \sum_{i=1}^t a_i(\mu+\tau_i) = \left(\sum_{i=1}^t a_i\right)\mu + \sum_{i=1}^t a_i\tau_i$$

for some $\mathbf{a} = (a_1, a_2, \ldots, a_t)'$. Then we must have $a_i = \lambda_i$ and

$$\sum_{i=1}^t \lambda_i = \sum_{i=1}^t a_i = 0.$$

Suppose now that $\sum_{i=1}^t \lambda_i = 0$. Then

$$\boldsymbol{\lambda}'\boldsymbol{\theta} = \sum_{i=1}^t \lambda_i\tau_i = \sum_{i=1}^t \lambda_i(\mu+\tau_i) \qquad \text{where } \mu+\tau_i \in \bar{\Theta},$$

so $\boldsymbol{\lambda}'\boldsymbol{\theta}$ is estimable. $\square$

A τ *contrast* is any parametric function of the form $\sum_{i=1}^{t} \lambda_i \tau_i$ where $\sum_{i=1}^{t} \lambda_i = 0$. Theorem 2.9 says that the set T consists exactly of the τ contrasts. Furthermore, T is a vector space.

THEOREM 2.10 The set T of estimable linear parametric functions involving only the τ parameters is a vector space with dimension $f_\tau = t-1$.

Proof Since T consists exactly of the τ contrasts, and because every linear combination of τ contrasts is also a τ contrast, T is a vector space. The estimable parametric function $\mu + \tau_1$ is linearly independent of the elements of T, which is a subspace of $\bar{\Theta}$. Therefore dim $T \leqslant t-1$. On the other hand, the set of linearly independent elements $\{\tau_1 - \tau_2, \tau_1 - \tau_3, \ldots, \tau_1 - \tau_t\} \subset T$, so dim $T \geqslant t-1$, and we conclude that dim $T = t-1$. □

We can summarize these results by saying that the vector space T of τ contrasts consists of all estimable linear parametric functions involving only the τ parameters and has dimension $t-1$. Furthermore, a basis for T is $\{\tau_1 - \tau_2, \tau_1 - \tau_3, \ldots, \tau_1 - \tau_t\}$.

We now consider how to obtain BLUEs for estimable linear parametric functions. Since $\mathbf{\Delta} = \mathbf{0}$, we have $\Omega = R(\mathbf{X})$. So $\mathbf{W} = \mathbf{X}$ satisfies $R(\mathbf{W}) = \Omega$ and by Theorem 2.6 we see that $\hat{\boldsymbol{\theta}}$ is Gauss–Markov for $\boldsymbol{\theta}$ if and only if it satisfies $\mathbf{X}'\mathbf{X}\hat{\boldsymbol{\theta}} = \mathbf{X}'\mathbf{Y}$, or

$$\begin{bmatrix} N & n_1 & n_2 & \cdots & n_t \\ n_1 & n_2 & 0 & \cdots & 0 \\ n_2 & 0 & n_2 & \cdots & 0 \\ & & & \cdots & \\ n_t & 0 & 0 & \cdots & n_t \end{bmatrix} \begin{bmatrix} \hat{\mu} \\ \hat{\tau}_1 \\ \hat{\tau}_2 \\ \vdots \\ \hat{\tau}_t \end{bmatrix} = \begin{bmatrix} Y_{..} \\ Y_{1.} \\ Y_{2.} \\ \vdots \\ Y_{t.} \end{bmatrix}.$$

This is an underdetermined system of equations. By adding the constraint $\hat{\tau}_1 + \hat{\tau}_2 + \cdots + \hat{\tau}_t = 0$, we obtain the particular solution:

$$\hat{\mu} = \sum_{i=1}^{t} \frac{\bar{Y}_{i.}}{t}, \qquad \text{where } \bar{Y}_{i.} = \sum_{j=1}^{n_i} \frac{Y_{ij}}{n_i},$$

$$\hat{\tau}_i = \bar{Y}_{i.} - \hat{\mu} = \bar{Y}_{i.} - \sum_{i=1}^{t} \frac{\bar{Y}_{i.}}{t} \qquad \text{for } i = 1, 2, \ldots, t.$$

Then $\hat{\mu} + \hat{\tau}_i = \bar{Y}_{i.}$ and $\hat{\tau}_i - \hat{\tau}_j = \bar{Y}_{i.} - \bar{Y}_{j.}$, for $i, j = 1, 2, \ldots, t$. The BLUEs for the basis $\{\mu + \tau_1, \ldots, \mu + \tau_t\}$ of $\bar{\Theta}$ are the observed treatment response means $\{\bar{Y}_{1.}, \ldots, \bar{Y}_{t.}\}$. The BLUEs for $\{\mu + \tau_1, \tau_1 - \tau_2, \ldots, \tau_1 - \tau_t\}$ are $\{\bar{Y}_{1.}, \bar{Y}_{1.} - \bar{Y}_{2.}, \ldots, \bar{Y}_{1.} - \bar{Y}_{t.}\}$. Note that any constraint $\mathbf{\Delta}'\hat{\boldsymbol{\theta}} = \mathbf{0}$ satisfying $R(\mathbf{X}') \cap R(\mathbf{\Delta}) = \{\mathbf{0}\}$ would have provided the same BLUEs.

The BLUE of any estimable linear parametric function can be found as the appropriate linear combination of the elements of either basis above. Furthermore, the variance of any BLUE can be found from the covariance matrix of the BLUEs of any basis for $\bar{\Theta}$. For example, we have

$$\text{Cov}\begin{bmatrix} \hat{\mu}+\hat{\tau}_1 \\ \hat{\mu}+\hat{\tau}_2 \\ \vdots \\ \hat{\mu}+\hat{\tau}_t \end{bmatrix} = \sigma^2 \begin{bmatrix} \frac{1}{n_1} & 0 & \cdots & 0 \\ 0 & \frac{1}{n_2} & \cdots & 0 \\ & & \vdots & \\ 0 & 0 & \cdots & \frac{1}{n_t} \end{bmatrix}$$

The residual sum of squares is $\hat{R}=\sum_{i=1}^{t}\sum_{j=1}^{n_i}(Y_{ij}-\bar{Y}_{i.})^2$. Since $m=\dim\Omega=t$, $\hat{R}/(N-t)$ is an unbiased estimator for σ^2. The covariance matrix above can then be estimated by replacing σ^2 by this estimate.

We can construct likelihood ratio tests of linear hypotheses under the assumption that $\mathbf{Y}$ has a multivariate normal distribution. Any hypothesis of the form H_0: $\mathbf{\Lambda}'\boldsymbol{\theta}=\mathbf{0}$ where $\mathbf{\Lambda}'\boldsymbol{\theta}$ is an estimable parametric vector is a linear hypothesis.

The most commonly tested hypothesis for Case 2 is that the treatment expectations are all equal:

$$H_0: \mu+\tau_1=\mu+\tau_2=\cdots=\mu+\tau_t \quad \text{versus } H_A\text{: not all expectations are equal.}$$

This is equivalent to

$$H_0: \tau_1=\tau_2=\cdots=\tau_t \quad \text{versus } H_A: \tau_i\neq\tau_j \quad \text{for some } i, j.$$

We can also express the null and alternative hypotheses as

$$H_0: \mathbf{\Lambda}'\boldsymbol{\theta}=\mathbf{0} \quad \text{versus} \quad H_A: \mathbf{\Lambda}'\boldsymbol{\theta}\neq\mathbf{0}$$

where

$$\mathbf{\Lambda}'\boldsymbol{\theta}=\begin{bmatrix} \tau_1-\tau_2 \\ \tau_1-\tau_3 \\ \cdots \\ \tau_1-\tau_t \end{bmatrix} = \begin{bmatrix} 0 & 1 & -1 & 0 & \cdots & 0 \\ 0 & 1 & 0 & -1 & \cdots & 0 \\ & & & & \cdots & \\ 0 & 1 & 0 & 0 & \cdots & -1 \end{bmatrix}\begin{bmatrix} \mu \\ \tau_1 \\ \vdots \\ \tau_t \end{bmatrix}.$$

The components of $\mathbf{\Lambda}'\boldsymbol{\theta}$ are the elements of a basis for T, the vector space of τ contrasts. The null hypothesis is then equivalent to the statement that all τ contrasts are zero.

The $\hat{F}$ statistic for testing this hypothesis can be computed using an

Table 2.3 ANOVA *for Comparing t Treatments*

Source	Degrees of Freedom	Sum of Squares	Mean Squares	$\hat{F}$
Between	$t-1$	$\text{SSTR}=\sum_{i=1}^{t} n_i\bar{Y}_{i.}^2-N\bar{Y}_{..}^2$	$\text{MSTR}=\dfrac{\text{SSTR}}{t-1}$	$\dfrac{\text{MSTR}}{\text{MSE}}$
within	$N-t$	$\text{SSE}=\sum_{ij}(Y_{ij}-\bar{Y}_{i.})^2$	$\text{MSE}=\dfrac{\text{SSE}}{N-t}$	
Total	$N-1$	$\sum_{i=1}^{t}\sum_{j=1}^{n_i} Y_{ij}^2-N\bar{Y}_{..}^2$		

analysis of variance table, with

$$\text{SSTO}=\hat{R}_0=\sum_{i=1}^{t}\sum_{j=1}^{n_i}(Y_{ij}-\bar{Y}_{..})^2=\sum_{i=1}^{t}\sum_{j=1}^{n_i}Y_{ij}^2-N\bar{Y}_{..}^2,$$

$$\text{SSTR}=\hat{R}_0-\hat{R}=\sum_{i=1}^{t}n_i(\bar{Y}_{i.}-\bar{Y}_{..})^2=\sum_{i=1}^{t}n_i\bar{Y}_{i.}^2-N\bar{Y}_{..}^2,$$

and

$$\text{SSE}=\hat{R}=\text{SSTO}-\text{SSTR}.$$

The degrees of freedom associated with the hypothesis is $m-m_0$, the dimension associated with the hypothesis. For our hypothesis, $m-m_0=t-1=f_\tau$, the dimension of T. The degrees of freedom for the residuals (under Ω) is $N-m=N-t$. The total degrees of freedom (adjusted for the hypothesis) is $N-m_0=N-1$. $\hat{F}$ is computed as the ratio

$$\hat{F}=\frac{(\hat{R}_0-\hat{R})/(m-m_0)}{\hat{R}/(N-m)}=\frac{\text{SSTR}/(t-1)}{\text{SSE}/(N-t)}=\frac{\text{MSTR}}{\text{MSE}}$$

and the null hypothesis is rejected at level α if $\hat{F}$ exceeds the upper $100\alpha\%$ point of the F distribution with $t-1$ and $N-t$ degrees of freedom.

The above information is conveniently summarized in the ANOVA of Table 2.3. In this table we have used the usual terminology of between, within, and total as opposed to deviations from the hypothesis, residual, and residual under the hypothesis.

NUMERICAL EXAMPLE Suppose the following observations resulted from a one-way classification experiment with $t=4$ groups and $n_1=3$, $n_2=n_3=4$, and $n_4=6$.

1	2	3	4
1.6	2.5	3.5	4.1
2.3	3.0	4.2	3.9
1.2	2.8	4.3	4.6
	2.9	4.5	5.2
			5.9
			6.0

The most common way of parametrizing the one-way classification model is

$$\text{MI: } E(Y_{ij}) = \mu + \tau_i, \qquad i = 1, \ldots, 4$$
$$j = 1, \ldots, n_i$$

or in matrix form

$$E(\mathbf{Y}) = \mathbf{X}\boldsymbol{\theta}, \qquad \boldsymbol{\theta} \in R^5,$$

where

$$\boldsymbol{\theta} = (\mu, \tau_1, \tau_2, \tau_3, \tau_4)', \qquad \mathbf{X} = (\mathbf{1}_N, \mathbf{A}).$$

Notice that $\mathbf{1}_N \in R(\mathbf{A})$, which implies the above presentation is indeed a parametrization for $E(\mathbf{Y})$.

In such an experiment, an experimenter is initially interested in determining if there are differences among four treatments, that is,

$$H_0: \tau_1 = \tau_2 = \tau_3 = \tau_4 \quad \text{versus} \quad H_A: \tau_i \neq \tau_j \qquad \text{for some } i \neq j.$$

The $\hat{F}$ statistic, which is a measure of differences for the null hypothesis can be computed using an initial ANOVA table with

$$\hat{R}_0 = \sum_{i=1}^{4} \sum_{j=1}^{n_i} (Y_{ij} - \bar{Y}_{..})^2 = \sum_{i=1}^{4} \sum_{j=1}^{n_i} Y_{ij}^2 - N\bar{Y}_{..}^2$$

$$N = n_1 + n_2 + n_3 + n_4 = 3 + 4 + 4 + 6 = 17$$

$$N\bar{Y}_{..}^2 = N \frac{\left(\sum_{i=1}^{4} \sum_{j=1}^{n_i} Y_{ij}\right)^2}{N^2} = \frac{\left(\sum_{i=1}^{4} \sum_{j=1}^{n_i} Y_{ij}\right)^2}{N}$$

$$= \frac{(1.6 + 2.3 + \cdots + 6.0)^2}{17} = \frac{(62.5)^2}{17} = \frac{3906.25}{17} = 229.78$$

$$\hat{R}_0 = (1.6)^2 + (2.3)^2 + \cdots + (6.0)^2 - 229.78 = 260.45 - 229.78 = 30.67.$$

$$\hat{R}_0 - \hat{R} = \sum_{i=1}^{4} n_i(\bar{Y}_{i.} - \bar{Y}_{..})^2 = \sum_{i=1}^{4} n_i \bar{Y}_{i.}^2 - N\bar{Y}_{..}^2,$$

we have

$$\bar{Y}_{1.}=1.70, \qquad \bar{Y}_{2.}=2.80, \qquad \bar{Y}_{3.}=4.125, \qquad \bar{Y}_{4.}=4.95.$$

so that

$$\hat{R}_0-\hat{R}=3(1.7)^2+4(2.8)^2+4(4.13)^2+6(4.95)^2-229.78$$
$$=8.67+31.36+68.06+147.02-229.78=255.11-229.78=25.33.$$

And finally

$$\hat{R}=\text{SSTO}-\text{SSTR}=30.67-25.33=5.34.$$

We now have the relevant information for testing the hypothesis of treatments equality. This information is conveniently summarized in the initial ANOVA of Table 2.4.

The upper 5% point of the central F distribution with 3 and 13 degrees of freedom is 3.41, so the null hypothesis $H_0: \tau_1=\tau_2=\tau_3=\tau_4$ is rejected.

In addition to the ANOVA table, other analysis are often desired for the one-way classification model. The estimates and their standard errors, confidence intervals, and other hypothesis tests are examples of such analysis. Here we find the estimates to be

$$\hat{\mu}=\sum_{i=1}^{4}\frac{\bar{Y}_{i.}}{4}=\frac{1.7+2.8+4.125+4.950}{4}=3.39$$
$$\hat{\tau}_1=\bar{Y}_{1.}-\hat{\mu}=1.7-3.39=-1.69$$

and similarly

$$\hat{\tau}_2=-0.59,\ \hat{\tau}_3=0.73, \text{ and } \hat{\tau}_4=1.55.$$

There are of course many other possibilities for our estimates to be a Gauss–Markov estimates. The above solution is the standard one given in text books and it has the property that $\sum_{i=1}^{4}\hat{\tau}_i=0$. The BLUEs for $\{\mu+\tau_1, \tau_1-\tau_2, \tau_1-\tau_3, \tau_1-\tau_4\}$ are $\{1.7, -1.1, -2.425, -3.25\}$ and the BLUEs for the basis $\{\mu+\tau_1, \mu+\tau_2, \mu+\tau_3, \mu+\tau_4\}$ are the observed treatment response means $\{1.7, 2.8, 4.125, 4.95\}$.

Table 2.4 **ANOVA**

Source	Degrees of Freedom	Sum of Squares	Mean Squares	$\hat{F}$
Between	3	25.33	8.44	20.54
Within	13	5.34	0.41	
Total	16	30.67		

The hypothesis stated above is in terms of τ. We can state the same hypothesis as a linear hypothesis on $E(\mathbf{Y})$. That is,

$$H_0: E(\mathbf{Y}) = \mathbf{1}_N \mu, \qquad \mu \in R^1.$$

The preferred statement for the null hypothesis usually depends on the problem at hand and on the preferences of the person formulating the hypothesis.

If the null hypothesis H_0: $\tau_1 = \tau_2 = \cdots = \tau_t$ is rejected, the alternative H_A: $\tau_i \neq \tau_j$ for some i, j is accepted. When the null hypothesis is rejected, it is reasonable to seek out the reason for its rejection and to relate this to treatment comparisons.

As we saw before, the null hypothesis is equivalent to the statement that all τ contrasts are zero. H_0 is rejected when $\hat{F} > F_{t-1,N-t}(\alpha)$, the upper $100\alpha\%$ point of the central F distribution with $t-1$ and $N-t$ degrees of freedom. This is equivalent to there existing some τ contrast that is significantly different from zero at the same significance level α. That is, H_0 is rejected at level α if and only if there exists some τ contrast $\boldsymbol{\lambda}'\boldsymbol{\theta}$ such that

$$\hat{T} = \frac{|\boldsymbol{\lambda}'\hat{\boldsymbol{\theta}}|}{s(\boldsymbol{\lambda}'\hat{\boldsymbol{\theta}})} > T_{N-t}(\alpha/2),$$

where $s(\boldsymbol{\lambda}'\hat{\boldsymbol{\theta}})$ denotes the estimated standard deviation of $\boldsymbol{\lambda}'\hat{\boldsymbol{\theta}}$ and $T_{N-t}(\alpha/2)$ denotes the upper $100\alpha/2\%$ point of the t distribution with $N-t$ degrees of freedom.

The equivalence of the F test with a T test for some τ contrast is easily seen when $t=2$ (Case 1). Then

$$\hat{F} = \frac{\sum_{i=1}^{2} n_i \bar{Y}_{i.}^2 - N\bar{Y}_{..}^2}{\sum_{i=1}^{2}\sum_{j=1}^{n_i} (Y_{ij} - \bar{Y}_{i.})^2/(N-2)} = \frac{(\bar{Y}_{1.} - \bar{Y}_{2.})^2}{S^2}$$

where

$$S^2 = \left(\frac{1}{n_1} + \frac{1}{n_2}\right) \sum_{i=1}^{2}\sum_{j=1}^{n_i} \frac{(Y_{ij} - \bar{Y}_{i.})^2}{n_1 + n_2 - 2}$$

is the pooled sample variance of $\bar{Y}_{1.} - \bar{Y}_{2.}$. Therefore $\hat{F} = \hat{T}^2$, where $\hat{T}$ is the pooled t statistic for testing H_0: $\tau_1 = \tau_2$ or, equivalently H_0: $\tau_1 - \tau_2 = 0$, and

$$\hat{F} > F_{1,N-2}(\alpha) \quad \text{if and only if } |\hat{T}| > T_{N-2}(\alpha/2).$$

So rejecting H_0 at a particular level α implies that *some* τ contrast is significantly different from zero at level α. For $t > 2$ this does not mean, however, that some treatment comparison $\tau_i - \tau_j$ is significantly different from zero.

Nevertheless, when the null hypothesis is rejected we accept the hypothesis that the treatment effects τ_i are not all equal, and it may be useful to test all $t(t-1)/2$ treatment comparisons $\tau_i - \tau_j$, $i<j$, to see if any are, indeed, significant. The purpose of the next section is to discuss the problem of multiple comparisons and to present one method of making these comparisons.

2.8 MULTIPLE COMPARISONS

The pairwise multiple comparison problem can be formulated as follows: We have t independent random samples taken from t populations whose densities differ only in location

$$f_i(y) = f(y - \tau_i) \quad \text{for } i = 1, 2, \ldots, t. \tag{2.9}$$

The $t(t-1)/2$ possible pairwise comparison hypotheses can be expressed as

$$H_{ij}^0\colon \tau_i = \tau_j \quad \text{versus} \quad H_{ij}^-: \tau_i < \tau_j \quad \text{or} \quad H_{ij}^+: \tau_i > \tau_j \tag{2.10}$$

for $1 \leqslant i < j \leqslant t$. We want to make the correct decision for each of these comparisons.

For each comparison, we will do one of the following:

1. Accept H_{ij}^0, H_{ij}^-, or H_{ij}^+ when it is true.
2. Reject H_{ij}^0 when it is true.
3. Accept H_{ij}^0 when it is false.
4. Accept H_{ij}^+ when H_{ij}^- is true or accept H_{ij}^- when H_{ij}^+ is true.

In case 1 we make a correct decision. In case 2 we make a *Type I error* and in case 3 we make a *Type II error*. In case 4 we are said to make a *Type III error*. These decisions and errors are called *comparisonwise* decisions and errors.

We can consider all the comparisons together and say that we commit an *experimentwise* (or *family*) Type I, II, or III *error* if we commit a Type I, II, or III error in at least one of the individual comparisons.

Multiple comparisons may be made by using tests of significance or by using confidence intervals. These are not equivalent approaches and each have advantages and disadvantages. Miller points out some of these in *Simultaneous Statistical Inference* (1966):

> . . . For some problems a non-null effect is worthy of consideration only if the magnitude of the effect is sufficient to produce a scientific, technologic, philosophic, or social change. Mere statistical significance is not enough to warrant its notice; the effect must also be biologically, physically, socially, etc., significant as well. To answer the question of the apparent magnitude of the effect, the

experimenter needs point and interval estimates; tests of significance will not suffice.

In other situations it may be that any difference from nullity, no matter how small or large, is of importance. If the existence of any non-null effect will demolish an existing theory or give a clue to understanding some previously unexplained phenomenon, then a test of significance is what is required. Confidence intervals here would be a luxury and afterthought, and nothing, in particular power, should be sacrificed to gain them.

In either the confidence interval approach or the significance test approach it is customary to control the probability of Type I error. However, if only the probability of comparisonwise Type I error is controlled, say at level α, the probability of experimentwise Type I error can grow to as large as $1-(1-\alpha)^c$ for c comparisons. Multiple comparison procedures typically control the probability of experimentwise Type I error under certain assumptions and for certain classes of comparisons.

There is no universally agreed upon multiple comparison procedure, even if we restrict attention to confidence intervals or to significance tests. Some procedures assume that the distributions are normal, and others are nonparametric. Some procedures are designed for situations where only a prespecified number of certain types (for example, pairwise) comparisons are to be made, while another will allow infinitely many contrasts to be tested. We present a simple, yet robust and relatively powerful procedure for making all possible pairwise comparisons while still keeping the probability *of experimentwise Type I error below a specified level.* We also give the nonparametric analogs of this "normal theory" procedure.

The method is known as the least significant difference (LSD) with a preliminary F test (FLSD). It assumes that the distributions are normal. The name LSD comes from the critical t percentile used in step 3 below. The procedure is as follows:

1 Select a significance level α. This will be the bound on the probability of experimentwise Type I error.

2 Let $\{Y_{ij}: j=1, 2, \ldots, n_i\}$ denote the sample from population i, $i=1, 2, \ldots, t$. Perform the $\hat{F}$ test of H_0: $\tau_1=\tau_2=\cdots=\tau_t$ by rejecting H_0 if

$$\hat{F}=\frac{\sum_{i=1}^{t} n_i(\bar{Y}_{i.}-\bar{Y}_{..})^2/(t-1)}{\sum_{i=1}^{t}\sum_{j=1}^{n_i}(Y_{ij}-\bar{Y}_{i.})^2/(N-t)}>F_{t-1,N-t}(\alpha)$$

where $N=n_.$ and $F_{t-1,N-t}(\alpha)$ is the upper $100\alpha\%$ point of the central F distribution with $t-1$ and $N-t$ degrees of freedom. Otherwise accept H_0 and stop. This bounds the probability of experimentwise Type I error at α.

3 For each of the $t(t-1)/2$ pairwise comparisons, compute the statistic

$$\hat{T}_{ij} = \frac{\bar{Y}_{i.} - \bar{Y}_{j.}}{S(1/n_i + 1/n_j)},$$

where

$$S^2 = \sum_{i=1}^{t} \sum_{j=1}^{n_i} \frac{(Y_{ij} - \bar{Y}_{i.})^2}{N-t}.$$

Let $T_{N-t}(\alpha/2)$ denote the upper $100\alpha/2\%$ point of the t distribution with $N-t$ degrees of freedom. Then

$$\text{if} \quad \hat{T}_{ij} < -T_{N-t}(\alpha/2), \quad \text{accept } H_{ij}^{-},$$

$$\text{if} \quad \hat{T}_{ij} > T_{N-t}(\alpha/2), \quad \text{accept } H_{ij}^{+}$$

and otherwise accept H_{ij}^{0}.

It can happen that the F test rejects H_0 in the step 2, but none of the pairwise differences at the third step are significant.

There are several other parametric (normal theory based) pairwise multiple comparison procedures, for example: the least significant difference without the preliminary F test, Boneferroni inequality, Tukey studentized range, Duncan multiple range, Newman–Keuls multiple range, and the Scheffé F projection method.

Other techniques whose test statistics do not depend on the form of the underlying distribution of the sample, are *nonparametric* procedures. They generally use rank sum statistics based on two different ranking methods, namely, *t sample ranking*, and *pairwise sample ranking*. These techniques are more applicable than their normal theory based analogs.

In t sample ranking, all observations in t samples are ranked in a single joint ranking, and in pairwise sample ranking the observations are ranked in only two combined samples i, j for comparison of τ_i and τ_j.

When all samples are combined to give the t sample ranking of the tn (equal sample size) observations, the nonparametric tests of (2.10) are based on the differences of mean ranks in sample i and j

$$\bar{R}_i - \bar{R}_j \tag{2.11}$$

where R_i denotes the rank of the ith sample in the combine ranking. Under the null hypothesis

$$\tau_1 = \tau_2 = \cdots = \tau_t,$$

the difference of mean ranks in (2.11) has the expectation zero and standard deviation

$$\sigma_d \sqrt{t(t+1)/6}.$$

The critical values for t sample ranking are based on a large-sample multivariate normal approximation for the distribution of mean ranks. If ties occur we can adjust the standard deviation. The critical value for the least significant "difference" is thus

$$z(\alpha)\sigma_d,$$

where $P[|Z| \geqslant z(\alpha)] = \alpha$ for the standard normal variate Z. Thus H_{ij}^0 is rejected in favor of $H_{ij}^-(H_{ij}^+)$ when

$$\bar{R}_j - \bar{R}_i \geqslant z(\alpha)\sigma_d \text{ or } (\bar{R}_i - \bar{R}_j) \geqslant z(\alpha)\sigma_d.$$

Here α gives the approximate comparisonwise error rate when H_{ij}^0 holds. This test is analogous to the parametric method LSD without the preliminary one-way analysis of variance F test. To obtain the nonparametric analog to the FLSD in t sample ranking, use the Kruskal–Wallis H test as the preliminary test. We may call this HLSD. The null hypothesis $H_0: \tau_1 = \tau_2 = \cdots = \tau_t$ is rejected by H test if

$$H = \frac{12}{t(tn+1)} \sum_{i=1}^{t} \bar{R}_i^2 - 3(tn+1) \geqslant \chi^2(\alpha, t-1).$$

Only if H_{ij}^0 is rejected are then the LSD tests performed. The LSD procedures thus control the approximate experimentwise error rate when H_{ij}^0 holds.

For the pairwise ranking where two samples i and j are jointly ranked, let W_{ij} denote the sum of ranks in the sample i. The least significant "difference" critical value is given by the upper $\alpha/2$ quantile of the Wilcoxon rank sum statistic for sample sizes n and n

$$\omega(\alpha, n, n)$$

that is, reject H_{ij}^0 in favor of $H_{ij}^-(H_{ij}^+)$ if

$$W_{ji} \text{ or } W_{ij} \geqslant \omega(\alpha, n, n)$$

where

$$P(\max(W_{ij}, W_{ji}) \geqslant \omega(\alpha, n, n)) = \alpha$$

when H_{ij}^0 holds. The normal approximation yields

$$\omega(\alpha, n, n) = n(n + \tfrac{1}{2}) + z(\alpha)\sqrt{n^2(2n+1)/12}.$$

For the preliminary test to the least significant difference (HLSD) the Kruskal–Wallis H test is made using the t sample rank means $\bar{R}_i$. Only for the case of rejection with this preliminary test are then the LSD (ordinary two-sample Wilcoxon) tests performed. It is not necessary for the Wilcoxon test that the sample sizes be equal. The last two tests for the pairwise ranking procedures are analogs to the LSD and FLSD of the parametric procedures.

The Kruskal–Wallis test briefly mentioned above is actually the nonparametric analog to the F test in the one-way analysis of variance. If in a one-way analysis of variance we use the ranks of the observations rather than their actual values, the F test and the Kruskal–Wallis H statistics will be equivalent, and this equivalency can be expressed as

$$F = \left[\frac{k-1}{N-k}\left(\frac{N-1}{H} - 1\right)\right]^{-1}.$$

BIBLIOGRAPHICAL NOTES

2.1–2.2 For an exclusive treatise on testing of statistical hypotheses see Lehmann (1959). For an introductory approach to the topic see Hogg and Craig (1978), Brunk (1975), Lindgren (1968), and Rao (1973). A decision theoretic approach is given by Ferguson (1967). The historical development of testing statistical hypotheses is given on page 120 of Lehmann (1959). For the mathematical programming approach to the problem of testing statistical hypotheses along with references to the subject from programming point of view, see Arthanari and Dodge (1981).

2.3 Experiments with single factor is a subject that can be found in almost every experimental design books. See, for example, Fisher (1935), Federer (1955), John (1971), Cochran and Cox (1957), Scheffé (1959), and Anderson and McLean (1974). Worked out examples can be found in all the above books. The comparison of two samples (two levels from a single factor) is also given in all statistical methods books. For example, see Snedecor and Cochran (1980). For the notion of additivity and other fundamental concepts in experimental designs see Cox (1958).

2.4 For an interesting historical development of the principle of least squares from its inception by Legendre–Gauss see Seal (1967) and Plackett (1972). The method was invented independently by Gauss (1809) and Legendre (1806). Harter in a series of six articles (1974a, 1974b, 1975a, 1975b, 1975c, 1976) has accomplished the enormous job of collecting the available literature on regression methods both for L_1 and L_2 norms. See Gentle (1977) for literature available on L_1 norm. See also Arthanari and Dodge for other references on L_1, L_2, and L_∞ norms along with the mathematical programming approach to the problem. The presentation in this section follows that of Seely's lecture notes. His approach to the problem of general linear hypotheses is unique and provides an excellent and systematic treatise on the subject. See also Searle (1971) for an excellent presentation on solving linear equations and an introduction to linear models, constructing the design matrices (he called incidence matrix), and other interesting results on estimability.

2.5 Certain improvements on the theory of least squares have been made after the work of Markov (1900) by Aitken (1935), Bose (1950–1951), Neyman and David (1938), Parzen (1961), Rao (1945b, 1945c, 1946, 1962), Goldman and Zelen (1964), Rao and Mitra (1968), and Zyskind and Martin (1969). In Rao (1973) a complete chapter is devoted to the theory of least squares and the analysis of variance. The idea of an estimable parametric function in linear model context was introduced by Bose (1944).

2.6 The principle of the likelihood function as well as the solution method using logarithmic derivatives were known and used by Gauss (1880) in his original development of theory of least squares estimation. As a general method of estimation it was first introduced by Fisher (1912) and was further developed by Fisher in a series of works (1922, 1925). See also Deutsch (1965), Graybill (1961), Rao (1973), Rao (1971), Godambe and Sprott (1971), Rao (1957, 1958), and Kiefer and Wolfowitz (1956).

2.7 The likelihood-ratio principle was formulated by Neyman and Pearson (1928). See Kolodziejczyk (1935). The symbol F was introduced by Snedecor 1934 in the honor of R. A. Fisher. See Scheffé (1953) for an interesting discussion on the subject of likelihood-ratio test. For a discussion on general linear hypotheses see Searle (1971), Seely (1979), and Seber (1977). An excellent discussion on the multiple regression applied to analysis of variance is provided by Draper and Smith (1966). Also see Chapter 9 of Seber (1977) for similar discussion, Chapter 6 of Searle (1971), and Chapter 3 of Scheffé (1959). For the proof of Theorem 2.8 see Seely (1979) lecture notes Chapter 9, pages 6–7.

2.8 The fundamental ideas of multiple comparisons and simultaneous confidence intervals were laid down by Duncan (1947), Fisher (1935), Scheffé (1953), and Tukey (1949). Scheffé (1953) provided the link between multiple comparisons and the general linear hypotheses. An excellent and complete treatise was published by Miller (1966). The revised edition (1981) includes his recent article (1977), which covers the recent development in multiple comparisons up to 1976. It is "the" book available on the subject. Calinski and Corsten (1982) recently introduced two clustering methods to be embedded in a consistent manner into appropriate simultaneous test procedures. Their purpose is to group the treatments into possibly small numbers of distinct but internally homogeneous clusters. The performance of several parametric and nonparametric multiple comparison procedures were compared in a simulation study by Dodge and Thomas (1980).

2.5 See Seely (1977) on estimability and linear hypotheses. In that work Seely via a proposition provides an easily justified reason as to why attention should be confined to estimable parametric vectors when formulating linear hypotheses. See Reiersol (1963) for equivalent condition via identifiability. When $\mathrm{Cov}(\mathbf{Y})=\sigma^2\mathbf{V}$ with $\mathbf{V}$ a known non-negative definite matrix, Zyskind theorem says that $\mathbf{t}'\mathbf{Y}$ is a BLUE if and only if $\mathbf{Vt}\in\Omega$, where $\Omega=\{\mathbf{X}\boldsymbol{\theta}:\boldsymbol{\Delta}'\boldsymbol{\theta}=\mathbf{0}\}$. See Zyskind (1967) for details.

REFERENCES

Aitken, A. C. (1935). On least squares and linear combination of observations. *Proc. Roy. Soc. Edin. A* **55**, 42–48.

Anderson, V. L. and McLean, R. A. (1974). *Design of Experiments*, New York: Dekker.

Arthanari, T. S. and Dodge, Y. (1981). *Mathematical Programming in Statistics.* New York: Wiley.

Bose, R. C. (1944). The fundamental problem of linear estimation. Proc. 31th Indian Science Congress, Part III, 2–3.

Bose, R. C. (1950–51). Least square aspects of the analysis of variance. Mimeographed series, No. 9. North Carolina University.

Brunk, H. D. (1975). *An Introduction to Mathematical Statistics*, 3rd ed. New York: Wiley.

Caliński, T. and Corsten, L. C. A. (1982). Clustering means in ANOVA by simultaneous testing. Technical Report, Academy of Agriculture, Poznan, Poland.

Cochran, W. G. and Cox, G. M. (1957). *Experimental Designs*, 2nd ed. New York: Wiley.

Cox, D. R. (1958). *Planning of Experiments.* New York: Wiley.

Deutsch, R. (1965). *Estimation Theory.* Englewood Cliff, N.J.: Prentice-Hall.

Dodge, Y. and Thomas, D. R. (1980). On the performance of non-parametric and normal theory multiple comparison procedures. *Sankhyā, Series B, Parts 1 and 2* **42**, 11–27.

Draper, N. R. and Smith, H. (1966). *Applied Regression Analysis.* New York: Wiley.

Duncan, D. B. (1947). Significance tests for differences between ranked variates drawn from normal populations. Unpublished doctoral thesis, Iowa State College, Ames, Iowa.

Federer, W. T. (1955). *Experimental Design: Theory and Applications.* New York: Macmillan.

Ferguson, T. S. (1967). *Mathematical Statistics.* New York: Academic Press.

Fisher, R. A. (1912). On an absolute criterion for fitting frequency curves. *Mess. of Math.* **41**, 155.

Fisher, R. A. (1922). On the mathematical foundations of theoretical statistics. *Philos. Trans. Royal Soc. London*, 222, 309.

Fisher, R. A. (1925). Theory of statistical estimation. *Proc. Cambridge Philos. Soc.* **22**, 700.

Fisher, R. A. (1935). *The Design of Experiments*, 1st ed. Edinburgh: Oliver and Boyd (1966, 8th ed.).

Gauss, K. F. (1809). *Werke* **4**, 1–93, Göttingen.

Gauss, C. F. (1880). *Werke*, **4**, Göttingen.

Gentle, T. E. (1977). Least absolute values estimation: An introduction. *Commun. Statist.* **B6**, 313–328.

Godambe, V. P. and Sprott, D. A. (1971). *Foundations of Statistical Inference*. Canada: Holt, Reinhart, and Winston.

Goldman, A. J. and Zelen, M. (1964). Weak generalized inverses and minimum variance linear unbiased estimation. *J. Res. Nat. Bur. Stand.* **68B**, 151–172.

Graybill, F. A. (1961). *An Introduction to Linear Statistical Models*, Vol. 1. New York: McGraw-Hill.

Harter, H. L. (1974a). The method of least squares and some alternatives I. *Int. Statist. Rev.* **42**, 147–174.

Harter, H. L. (1974b). The method of least squares and some alternatives II. *Int. Statist. Rev.* **42**, 235–264.

Harter, H. L. (1975a). The method of least squares and some alternatives III. *Int. Statist. Rev.* **43**, 1–44.

Harter, H. L. (1975b). The method of least squares and some alternatives IV. *Int. Statist. Rev.* **43**, 125–130 and 273–278.

Harter, H. L. (1975c). The method of least squares and some alternatives V. *Int. Statist. Rev.* **43**, 269–272.

Harter, H. L. (1976). The method of least squares and some alternatives VI. *Int. Statist. Rev.* **44**, 113–159.

Hogg, R. V. and Craig, A. T. (1978). *Introduction to Mathematical Statistics*, 4th ed. London: Macmillan.

John, P. M. (1971). *Statistical Design and Analysis of Experiments*. New York: Macmillan.

Kiefer, J. and Wolfowitz, J. (1956). Consistency of the maximum likelihood estimator in the presence of infinitely many incidental parameters. *Annals of Mathematical Statistics*, **27**, 887–906.

Kolodziejczyk, S. (1935). On an important class of statistical hypotheses. *Biometrika*, **27**, 161–190.

Legendre, A. L. (1806). Nouvelles méthodes pour la détermination des orbites des comètes, Paris.

Lehmann, E. L. (1959). *Testing Statistical Hypotheses*. New York: Wiley.

Lindgren, B. W. (1968). *Statistical Theory*, 2nd ed. New York: Macmillan.

Markov, A. A. (1900). *Wahrscheinlichkeitsrechnung*. Leipzig: Tebner.

Miller, R. G. Jr. (1966). *Simultaneous Statistical Inference*. New York: McGraw-Hill. (2nd ed. 1981, Springer-Verlag, New York).

Neyman, J. and David, F. N. (1938). Extension of the Markoff theorem on least squares. *Statis. Res. Mem.* **2**, 105–116.

Neyman, J. and Pearson, E. S. (1928). On the use and interpretation of certain test criteria for purposes of statistical inference. *Biometrika* **20A**, 175–240, 263–294.

Parzen, E. (1961). An approach to time series analysis. *Ann. Math. Statist.* **32**, 951–989.

Plackett, R. L. (1972). Studies in the history of probability and statistics. XXIX. The discovery of least squares. *Biometrika* **59**, 239–252.

Rao, C. R. (1945b). Generalization of Markoff's theorem and tests of linear hypotheses. *Sankhyā* **7**, 9–16.

Rao, C. R. (1945c). Markoff's theorem with linear restrictions on parameters. *Sankhyā* **7**, 16–19.

Rao, C. R. (1946). On the linear combination of observations and the general theory of least squares. *Sankhyā* **7**, 237–256.

Rao, C. R. (1957). Maximum likelihood estimation for multinomial distribution. *Sankhyā* **18**, 139–148.

Rao, C. R. (1958). Maximum likelihood estimation for the multinomial distribution with an infinite number of cells. *Sankhyā* **20**, 211–218.

Rao, C. R. (1962). A note on a generalized inverse of a matrix with applications to problems in mathematical statistics. *J. Roy. Statist. Soc. B* **24**, 152–158.

Rao, C. R. (1971). *Some Aspects of Statistical Inference in Problems of Sampling from Finite Populations. Foundations of Statistical Inference*, Canada: Holt, Rinehart, and Winston, pp. 177–202.

Rao, C. R. (1973). *Linear Statistical Inference and Its applications*, 2nd ed. New York: Wiley.

Rao, C. R. and Mitra, S. K. (1968f). Some results in estimation and tests of linear hypotheses under the Gauss-Markoff model. *Sankhyā A* **30**, 281–290.

Reiersol, O. (1963). Identifiability, estimability, phenorestricting specifications, and zero Lagrange multiplies in the analysis of variance. *Skandanavia Aktuarietidskr* **46**, 131–142.

Scheffé, H. (1953). A method for judging all contrast in the analysis of variance. *Biometrika* **40**, 87–104.

Scheffé, H. (1959). *The Analysis of Variance*. New York: Wiley.

Seal, H. L. (1967). The historical development of the Gauss linear model. *Biometrika* **54**, 1–24.

Searle, S. R. (1971). *Linear Models*. New York: Wiley.

Seber, G. A. F. (1977). *Linear Regression Analysis*. New York: Wiley.

Seely, J. (1977). Estimability and linear hypotheses. *Am. Statistician* **31**, 121–123.

Seely, J. (1979). Lecture notes on general linear hypothesis. Oregon State University.

Snedecor, G. W. and Cochran, W. G. (1980). *Statistical Methods*, 7th ed. Ames, Iowa: The Iowa State University Press.

Tukey, J. W. (1949). Comparing individual means in the analysis of variance. *Biometric* **5**, 99–114.

Zyskind, G. and Martin, F. B. (1969). On best linear estimation and a general Gauss–Markoff theorem in linear models with arbitrary non-negative covariance structure. *SIAM, J. Appl. Math.* **17**, 1190–1202.

Zyskind, G. (1967). On canonical forms, nonnegative covariance matrices and best and simple least squares estimators in linear models. *Ann. Math. Statist.* **38**, 1092–1109.

CHAPTER 3

Experiments Involving Two Factors

... And so the talk goes on
Between the believers and those who are not.

MOULAVI RUMI: *Persian Soufi Poet (1207–1273)*

3.1 INTRODUCTION

In Chapter 2 we discussed a single-factor experiment in which an experimenter is initially interested in determining if there are differences among treatments. The experimental design used was completely randomized design, in which the treatments were allocated to experimental units completely at random.

In planning a controlled experiment, the experimenter should have some knowledge of the experimental materials. Such a knowledge can be used to increase the accuracy of an experiment. If an experiment for comparing v treatments is to be conducted and the number of experimental units available to the experimenter is $N=vb$, and if we divide N experimental units into b homogeneous groups (strips of land, say), and allocate the v treatments to experimental units at random in each group and independently from group to group, then we have a two-way classification design, since any observation is classified by the treatment it receives and the group to which it belongs. The name given to the group varies with the kind of application. For example, in agricultural field experiments, groups are called blocks, and the experimental plan is called a block design. In such designs if every treatment appears at least once in each block, the plan is called a *complete randomized block design.* Figure 3.1 shows the arrangement of four treatments denoted by A, B, C, and D in five blocks.

Note that there is a *restriction on randomization* in the design, because the experimental units are not selected in the same manner as in a completely randomized design. The experimental units are grouped by some kind of blocking variable and treatments are assigned randomly to experimental units within each block. The blocks may either be fixed or random, depending

I	A	B	C	D
II	D	C	A	B
III	C	B	A	D
IV	A	C	B	D
V	D	C	A	B

Figure 3.1 Randomized complete Block design with $b=5$ and $v=4$.

on the method of selection of the blocks from the population of all blocks, and each block contains a group of experimental units which are restricted to that particular block.

In such a design, treatment differences are of major interest to the experimenter, and block contributions are to be eliminated. That is, the interest is still centered on one factor (treatments), but a restriction is placed on the randomization.

A different kind of two-way arrangement is a *replicated factorial* design in which the main effects of two factors and their interactions are all of equal interest. Suppose that a levels of one fixed set of treatments (factor A) and b levels of another set of treatments (factor B) are run in all possible ab combinations on ab experimental units assigned completely at random. The design of such an experiment is again a completely randomized design. There are no restrictions on randomizations (no blocking) in the design of this experiment.

This arrangement may be called a two-way factorial completely randomized design or simply two-way factorial. Other names, such as a two-way factorial experiment, a two-way factorial arrangement, or a two-way factorial design are also common. The model associated with such designs is called the two-way classification model (such a term is also used for randomized block designs).

REMARK The mathematical model suggested by the experiment when there is a restriction on randomization will naturally be different from that with no restriction. This will also affect the way that the data should be analyzed, especially the F tests, and hence the conclusions. Throughout this book we assume that there is no restriction in the design of the experiment. We only consider factorial experiments in a completely randomized design.

Why should one wish to investigate two factors simultaneously, rather than one at a time? These experiments are of importance for the following reasons:

1 A two-way factorial experiment is more efficient with regard to reresources than is the combination of two single-factor experiments. The former takes less time and requires fewer experimental units for a given level of precision.

2 A two-way factorial experiment allows the effect on the response due to increasing the level of each factor to be estimated at each level of the other factor. This yields conclusions that are valid over a wider range of experimental conditions than successive single-factor designs allow.

3 Finally, the simultaneous investigation of two factors is necessary when interaction between the factors is present, that is, when a factor effect depends up the level of the other factor.

A class of experiments that are of great importance are two-level factorial experiments. In these experiments each factor (variable) occurs at only two levels. We begin our discussion by first considering $2\times 2=2^2$ factorial experiment.

3.2 2^2 FACTORIAL EXPERIMENT

The 2^2 factorial experiment investigates the relationship between two controllable conditions (factors), each with two levels, and the experimental response. A treatment is a particular combination of levels of two factors. The domain consists of all four possible combinations of the levels and the design includes at least one experimental unit per treatment.

A classification model suggested by the 2^2 factorial design is

$$Y_{ijk}=\mu+\alpha_i+\beta_j+(\alpha\beta)_{ij}+e_{ijk},\qquad \mathrm{Var}(e_{ijk})=\sigma^2,$$
$$i,j=1,2 \quad\text{and}\quad k=1,2,\ldots,n_{ij}. \tag{3.1}$$

Here Y_{ijk} denotes the kth observation on treatment (i,j), and μ denotes the part of the response that does not depend upon the treatment. The sum $\alpha_i+\beta_j+(\alpha\beta)_{ij}$ represents the contribution to the response due to treatment (i,j), the combination with factor 1 at level i and factor 2 at level j. The e_{ijk} term represents the experimental error for experimental unit k given treatment (i,j), and n_{ij} is the number of units assigned to this treatment. The e_{ijk} are independent random variables with zero mean and the same variance.

The treatment contribution $\alpha_i+\beta_j+(\alpha\beta)_{ij}$ to the response is the sum of the contributions α_i of factor 1 at level i alone, β_j of factor 2 at level j alone, and $(\alpha\beta)_{ij}$ of the combination of factors 1 and 2 at levels i and j, respectively, acting jointly. μ and the α_i, β_j, and $(\alpha\beta)_{ij}$ are constants. Hence the treatment contributions are constant and add to the part of the response that does not depend upon the treatment. The model is therefore an additive one, in the sense of Chapter 2.

The contribution of factor 1 at level i to the response is $\alpha_i+(\alpha\beta)_{ij}$ and adds to the part of the response that does not depend upon that factor. However, this contribution depends upon the level of factor 2, so it is not constant. The same is true for the contribution of factor 2 to the response.

In the context of factorial experiments, it is common practice to call a model additive if each factor contribution is constant and adds to the part of the response that does not depend upon the levels of the other factors. An example of a model that is additive in this sense for the 2^2 factorial experiment is that model obtained from the preceding one by dropping the $(\alpha\beta)_{ij}$ term. The resulting model is often called the additive two-way classification. The $(\alpha\beta)_{ij}$ term is called the interaction term and the original model is then called the two-way classification with interaction.

Another consideration that is important in the analysis and interpretation of the 2^2 factorial experiment is whether the factor levels are the only ones of interest (represent all treatments of interest) or whether the factor levels are a sample of those of interest. Section 3.3 discusses briefly such considerations.

3.3 FIXED, RANDOM, AND MIXED MODELS

As explained above, the treatment contribution to the response is constant. That is, the contribution to the response that is due to the treatment is the same for all experimental units receiving the same treatment. However, the particular choice of treatments used in an experiment may not be the only treatments of interest. In this case, the choice of treatments and therefore the treatment contributions to the response may vary from one experiment to another.

A treatment is a particular combination of levels of the factors. If the particular levels of a factor in an experiment are the levels of that factor that are of interest, then those levels would be fixed from one repetition of the experiment to another and we would call that factor *fixed*. Model parameters that refer only to fixed factors are *fixed effects*. Inferences concerning fixed factors are then confined to estimable linear combinations of the effects of those factors.

If, on the other hand, the levels of a factor in an experiment are merely a random sample of levels from a larger population about which we wish to draw inferences, then the factor levels would vary from one repetition of the experiment to another and we would call the factor *random*. The parameters that refer to at least one random factor are called *random effects*. Random effects are then random variables and by including a constant term in the model they can, without loss of generality, be assumed to have zero mean and some unknown variance. The selection of levels for each random factor is assumed to be independent of the selection of levels for the other random factors and of the experimental error. Hence the variance of the experimental response is the sum of the experimental error variance and the random effect variances, and each variance is called a *variance component.* Inferences concerning random factors are confined to their variance components.

If aside from the error term the model consists only of fixed effects, it is called a *fixed effects model.* If aside from a constant term common to all treatments the model consists only of random effects, it is called a *random effects model*, or a *variance component model.* If neither case holds, the model is called a *mixed model.*

Unless explicitly stated otherwise, we shall restrict our attention for the remainder of this book to fixed effects models.

3.4 SPECIFYING FACTORIAL ARRANGEMENTS

Here we introduce a few more definitions and some notation that will be useful in discussing factorial experiments. Consider the 2^n factorial experiment, with n factors, each with two levels. The domain consists of $T=2^n$ distinct treatments. Let $r_1, r_2, \ldots, r_T$ denote the number of units the design associates with each point in the domain. r_j is called the *replication number* of the jth treatment. Let $N=\sum_{j=1}^{T} r_j$ be the total number of experimental units in the experiment and let t denote the number of nonzero r_j. Then we say we have a 2^n *factorial arrangement with parameters* $t; N; r_1, \ldots, r_T$, denoted by FA($n; t; N; r_1, \ldots, r_T$). For example, the 2^2 factorial introduced in Section 3.2 is FA($2; 4; N; n_{11}, n_{12}, n_{21}, n_{22}$), where $N=n_{..}$ and $n_{ij}>0$ for $i, j=1, 2$.

The pattern of nonzero r_j will be seen to be a very important property of an arrangement. Because of this, arrangements are classified accordingly. A factorial arrangement is said to be *complete* if $r_j>0$ for all j, otherwise it is called incomplete. This means that there is at least one unit assigned to every treatment combination in the domain. If $r_j=r_i$ for all i, j, we have equal replication. Otherwise, we have unequal replications. A complete factorial arrangement is said to be *minimal* if $r_j=1$ for all j and it is denoted by MFA(n). Here $t=N=T=2^n$. A factorial arrangement is said to be *fractional* if some but not all $r_j>0$, and it is denoted by FFA($n; t; N; r_1, \ldots, r_T$). We will use the general notation to denote an arbitrary 2^n factorial arrangement and use the MFA and FFA notation only when these specific cases are considered. In this chapter then, we consider the complete 2^2 factorial arrangement FA($2; 4; N; r_1, r_2, r_3, r_4$) where $r_1=n_{11}, r_2=n_{12}, r_3=n_{21}, r_4=n_{22}$, and $N=n_{..}$.

3.5 A PARAMETRIZATION AND ANALYSIS OF THE 2^2 FACTORIAL

Consider again the 2^2 factorial experiment and the model given in (3.1). The parametrization for this experiment and model in matrix form is

$$E(\mathbf{Y}) = \mathbf{X}\boldsymbol{\theta}, \qquad \boldsymbol{\Delta}'\boldsymbol{\theta} = \mathbf{0}, \qquad \text{Cov}(\mathbf{Y}) = \sigma^2\mathbf{I}, \tag{3.2}$$

where

$$\mathbf{Y} = \begin{bmatrix} Y_{111} \\ \vdots \\ Y_{11n_{11}} \\ Y_{121} \\ \vdots \\ Y_{12n_{12}} \\ Y_{211} \\ \vdots \\ Y_{21n_{21}} \\ Y_{221} \\ \vdots \\ Y_{22n_{22}} \end{bmatrix}, \qquad \mathbf{X} = \begin{bmatrix} 1 & 1 & 0 & 1 & 0 & 1 & 0 & 0 & 0 \\ & & & & \vdots & & & & \\ 1 & 1 & 0 & 1 & 0 & 1 & 0 & 0 & 0 \\ 1 & 1 & 0 & 0 & 1 & 0 & 1 & 0 & 0 \\ & & & & \vdots & & & & \\ 1 & 1 & 0 & 0 & 1 & 0 & 1 & 0 & 0 \\ 1 & 0 & 1 & 1 & 0 & 0 & 0 & 1 & 0 \\ & & & & \vdots & & & & \\ 1 & 0 & 1 & 1 & 0 & 0 & 0 & 1 & 0 \\ 1 & 0 & 1 & 0 & 1 & 0 & 0 & 0 & 1 \\ & & & & \vdots & & & & \\ 1 & 0 & 1 & 0 & 1 & 0 & 0 & 0 & 1 \end{bmatrix} \begin{matrix} \left.\vphantom{\begin{matrix}1\\1\\1\end{matrix}}\right\} n_{11} \text{ rows} \\ \left.\vphantom{\begin{matrix}1\\1\\1\end{matrix}}\right\} n_{12} \text{ rows} \\ \left.\vphantom{\begin{matrix}1\\1\\1\end{matrix}}\right\} n_{21} \text{ rows} \\ \left.\vphantom{\begin{matrix}1\\1\\1\end{matrix}}\right\} n_{22} \text{ rows} \end{matrix}$$

$$\boldsymbol{\theta} = [\mu, \alpha_1, \alpha_2, \beta_1, \beta_2, (\alpha\beta)_{11}, (\alpha\beta)_{12}, (\alpha\beta)_{21}, (\alpha\beta)_{22}]'$$

and

$$\boldsymbol{\Delta} = \mathbf{0}, \ (\Theta = R^9).$$

Here $\Omega = R(\mathbf{X})$.

For this parametrization, the set of estimable linear parametric functions $\boldsymbol{\lambda}'\boldsymbol{\theta}$ is $\bar{\Theta} = \{\boldsymbol{\lambda}'\boldsymbol{\theta}\colon \boldsymbol{\lambda} \in R(\mathbf{X}', \boldsymbol{\Delta})\} = \{\boldsymbol{\lambda}'\boldsymbol{\theta}\colon \boldsymbol{\lambda} \in R(\mathbf{X}')\}$ and $\dim \bar{\Theta} = r(\mathbf{X}) = \dim \Omega$. One basis for $\bar{\Theta}$ is the set of $\boldsymbol{\lambda}'\boldsymbol{\theta}$ with each $\boldsymbol{\lambda}'$ a distinct row of $\mathbf{X}$, that is, the set of distinct treatment expectations $\{\mu+\alpha_1+\beta_1+(\alpha\beta)_{11}, \ \mu+\alpha_1+\beta_2+(\alpha\beta)_{12}, \ \mu+\alpha_2+\beta_1+(\alpha\beta)_{21}, \ \mu+\alpha_2+\beta_2+(\alpha\beta)_{22}\}$. The estimable linear parametric functions are just those that can be expressed as linear combinations of these expectations.

Another basis is the set

$$\{\mu+\bar{\alpha}_{.}+\bar{\beta}_{.}+\overline{(\alpha\beta)}_{..}, \ \alpha_1-\alpha_2+\overline{(\alpha\beta)}_{1.}-\overline{(\alpha\beta)}_{2.}, \ \beta_1-\beta_2+\overline{(\alpha\beta)}_{.1}-\overline{(\alpha\beta)}_{.2}, \ ((\alpha\beta)_{11} \\ -(\alpha\beta)_{12}-(\alpha\beta)_{21}+(\alpha\beta)_{22})/2\}$$

where, for example, $\bar{\alpha}_{.} = (\alpha_1+\alpha_2)/2$, $\overline{(\alpha\beta)}_{1.} = [(\alpha\beta)_{11}+(\alpha\beta)_{12}]/2$, and $\overline{(\alpha\beta)}_{..} = [(\alpha\beta)_{11}+(\alpha\beta)_{12}+(\alpha\beta)_{21}+(\alpha\beta)_{22}]/4$. The first basis element is the response function averaged over all points in the domain. The second element is the difference in the response function for the levels of factor 1 averaged over the levels of factor 2 and it is called the *main effect* for factor 1. The third element is the main effect for factor 2. The last element is the *two-way interaction* between factors 1 and 2. It is one-half of the difference between the response function increase due to increasing the level of one factor for each level of the other factor. There are, of course, many other possible bases to choose from.

Any $\hat{\boldsymbol{\theta}}$ that is a least squares solution for $\boldsymbol{\theta}$ must satisfy

$$\begin{bmatrix} n_{..} & n_{1.} & n_{2.} & n_{.1} & n_{.2} & n_{11} & n_{12} & n_{21} & n_{22} \\ n_{1.} & n_{1.} & 0 & n_{11} & n_{12} & n_{11} & n_{12} & 0 & 0 \\ n_{2.} & 0 & n_{2.} & n_{21} & n_{22} & 0 & 0 & n_{21} & n_{22} \\ n_{.1} & n_{11} & n_{21} & n_{.1} & 0 & n_{11} & 0 & n_{21} & 0 \\ n_{.2} & n_{12} & n_{22} & 0 & n_{.2} & 0 & n_{12} & 0 & n_{22} \\ n_{11} & n_{11} & 0 & n_{11} & 0 & n_{11} & 0 & 0 & 0 \\ n_{12} & n_{12} & 0 & 0 & n_{12} & 0 & n_{12} & 0 & 0 \\ n_{21} & 0 & n_{21} & n_{21} & 0 & 0 & 0 & n_{21} & 0 \\ n_{22} & 0 & n_{22} & 0 & n_{22} & 0 & 0 & 0 & n_{22} \end{bmatrix} \begin{bmatrix} \hat{\mu} \\ \hat{\alpha}_1 \\ \hat{\alpha}_2 \\ \hat{\beta}_1 \\ \hat{\beta}_2 \\ \widehat{(\alpha\beta)}_{11} \\ \widehat{(\alpha\beta)}_{12} \\ \widehat{(\alpha\beta)}_{21} \\ \widehat{(\alpha\beta)}_{22} \end{bmatrix} = \begin{bmatrix} Y_{...} \\ Y_{1..} \\ Y_{2..} \\ Y_{.1.} \\ Y_{.2.} \\ Y_{11.} \\ Y_{12.} \\ Y_{21.} \\ Y_{22.} \end{bmatrix}.$$

This is a redundant set of equations. Dropping equations 3, 5, 7, 8, and 9 and then using constraints

$$\hat{\alpha}_1+\hat{\alpha}_2=\hat{\beta}_1+\hat{\beta}_2=\widehat{(\alpha\beta)}_{11}+\widehat{(\alpha\beta)}_{12}=\widehat{(\alpha\beta)}_{21}+\widehat{(\alpha\beta)}_{22}=\widehat{(\alpha\beta)}_{11}+\widehat{(\alpha\beta)}_{21}=0, \quad (3.3)$$

we obtain the system

$$\begin{bmatrix} n_{..} & n_{1.} & n_{2.} & n_{.1} & n_{.2} & n_{11} & n_{12} & n_{21} & n_{22} \\ n_{1.} & n_{1.} & 0 & n_{11} & n_{12} & n_{11} & n_{12} & 0 & 0 \\ n_{.1} & n_{11} & n_{21} & n_{.1} & 0 & n_{11} & 0 & n_{21} & 0 \\ n_{11} & n_{11} & 0 & n_{11} & 0 & n_{11} & 0 & 0 & 0 \\ 0 & 1 & 1 & 0 & 0 & 0 & 0 & 0 & 0 \\ 0 & 0 & 0 & 1 & 1 & 0 & 0 & 0 & 0 \\ 0 & 0 & 0 & 0 & 0 & 1 & 1 & 0 & 0 \\ 0 & 0 & 0 & 0 & 0 & 0 & 0 & 1 & 1 \\ 0 & 0 & 0 & 0 & 0 & 1 & 0 & 1 & 0 \end{bmatrix} \begin{bmatrix} \hat{\mu} \\ \hat{\alpha}_1 \\ \hat{\alpha}_2 \\ \hat{\beta}_1 \\ \hat{\beta}_2 \\ \widehat{(\alpha\beta)}_{11} \\ \widehat{(\alpha\beta)}_{12} \\ \widehat{(\alpha\beta)}_{21} \\ \widehat{(\alpha\beta)}_{22} \end{bmatrix} = \begin{bmatrix} Y_{...} \\ Y_{1..} \\ Y_{.1.} \\ Y_{11.} \\ 0 \\ 0 \\ 0 \\ 0 \\ 0 \end{bmatrix}.$$

The solution of this system is

$$\hat{\mu}=(\bar{Y}_{11.}+\bar{Y}_{12.}+\bar{Y}_{21.}+\bar{Y}_{22.})/4,$$

$$\hat{\alpha}_1=-\hat{\alpha}_2=(\bar{Y}_{11.}+\bar{Y}_{12.}-\bar{Y}_{21.}-\bar{Y}_{22.})/4,$$

$$\hat{\beta}_1 = -\hat{\beta}_2 = (\bar{Y}_{11.} - \bar{Y}_{12.} + \bar{Y}_{21.} - \bar{Y}_{22.})/4,$$

$$\widehat{(\alpha\beta)}_{11} = -\widehat{(\alpha\beta)}_{12} = -\widehat{(\alpha\beta)}_{21} = \widehat{(\alpha\beta)}_{22} = (\bar{Y}_{11.} - \bar{Y}_{12.} - \bar{Y}_{21.} + \bar{Y}_{22.})/4,$$

and it satisfies

$$\hat{\mu} + \hat{\alpha}_i + \hat{\beta}_j + \widehat{(\alpha\beta)}_{ij} = \bar{Y}_{ij.} \qquad \text{for } i, j = 1, 2.$$

So the BLUE for each treatment expectations is just the average response for the treatment. Since these four parametric functions span $\bar{\Theta}$, the BLUE of any estimable parametric function is a linear combination of the average treatment responses.

The covariance matrix of our solution $\hat{\boldsymbol{\theta}}$ is

$\text{Cov}(\hat{\boldsymbol{\theta}}) =$

$$\frac{\sigma^2}{16}\begin{bmatrix} K & K_A & -K_A & K_B & -K_B & K_{AB} & -K_{AB} & -K_{AB} & K_{AB} \\ K_A & K & -K & K_{AB} & -K_{AB} & K_B & -K_B & -K_B & K_B \\ -K_A & -K & K & -K_{AB} & K_{AB} & -K_B & K_B & K_B & -K_B \\ K_B & K_{AB} & -K_{AB} & K & -K & K_A & -K_A & -K_A & K_A \\ -K_B & -K_{AB} & K_{AB} & -K & K & -K_A & -K_A & K_A & K_A \\ K_{AB} & K_B & -K_B & K_A & -K_A & K & -K & -K & K \\ -K_{AB} & -K_B & K_B & -K_A & -K_A & -K & K & K & -K \\ -K_{AB} & -K_B & K_B & -K_A & K_A & -K & K & K & -K \\ K_{AB} & K_B & -K_B & K_A & K_A & K & -K & -K & K \end{bmatrix}$$

where

$$K = \frac{1}{n_{11}} + \frac{1}{n_{12}} + \frac{1}{n_{21}} + \frac{1}{n_{22}},$$

$$K_A = \frac{1}{n_{11}} + \frac{1}{n_{12}} - \frac{1}{n_{21}} - \frac{1}{n_{22}},$$

$$K_B = \frac{1}{n_{11}} - \frac{1}{n_{12}} + \frac{1}{n_{21}} - \frac{1}{n_{22}},$$

and

$$K_{AB} = \frac{1}{n_{11}} - \frac{1}{n_{12}} - \frac{1}{n_{21}} + \frac{1}{n_{22}}.$$

The covariance between any two BLUEs $\boldsymbol{\lambda}_1'\hat{\boldsymbol{\theta}}$ and $\boldsymbol{\lambda}_2'\hat{\boldsymbol{\theta}}$ is

$$\text{Cov}(\boldsymbol{\lambda}_1'\hat{\boldsymbol{\theta}}, \boldsymbol{\lambda}_2'\hat{\boldsymbol{\theta}}) = \boldsymbol{\lambda}_1' \,\text{Cov}(\hat{\boldsymbol{\theta}})\boldsymbol{\lambda}_2.$$

3.6 SOME ALTERNATIVE PARAMETRIZATIONS

We introduced the constraints (3.3) in order to find a solution $\hat{\boldsymbol{\theta}}$ by solving a nonsingular system of equations. These constraints can be expressed in matrix form as $\boldsymbol{\Delta}_c'\hat{\boldsymbol{\theta}}=\mathbf{0}$, where

$$\boldsymbol{\Delta}_c'=\begin{bmatrix} 0 & 1 & 1 & 0 & 0 & 0 & 0 & 0 & 0 \\ 0 & 0 & 0 & 1 & 1 & 0 & 0 & 0 & 0 \\ 0 & 0 & 0 & 0 & 0 & 1 & 1 & 0 & 0 \\ 0 & 0 & 0 & 0 & 0 & 0 & 0 & 1 & 1 \\ 0 & 0 & 0 & 0 & 0 & 1 & 0 & 1 & 0 \end{bmatrix}.$$

Our parametrization (3.2) with $\boldsymbol{\Delta}=\mathbf{0}$ has the expectation space $\Omega=R(\mathbf{X})$. But $\{\mathbf{X}\boldsymbol{\theta}^c: \boldsymbol{\Delta}_c'\boldsymbol{\theta}^c=0\}=R(\mathbf{X})=\Omega$, as, because of $r(\mathbf{X}')+r(\boldsymbol{\Delta}_c)=9(=P$, say,) and $R(\mathbf{X}')\cap R(\boldsymbol{\Delta}_c)=\{\mathbf{0}\}$, $\boldsymbol{\theta}^c=(\mathbf{X}'\mathbf{X}+\boldsymbol{\Delta}_c\boldsymbol{\Delta}_c')^{-1}\mathbf{X}'\mathbf{X}\boldsymbol{\theta}$, giving $\mathbf{X}\boldsymbol{\theta}^c=\mathbf{X}\boldsymbol{\theta}$. So

$$E(\mathbf{Y})=\mathbf{X}\boldsymbol{\theta}^c, \qquad \boldsymbol{\Delta}_c'\boldsymbol{\theta}^c=\mathbf{0}, \qquad \operatorname{Cov}(\mathbf{Y})=\sigma^2\mathbf{I} \tag{3.4}$$

is also a parametrization for Ω.

Constraints $\boldsymbol{\Delta}'\boldsymbol{\theta}=\mathbf{0}$ for which $R(\mathbf{X}')\cap R(\boldsymbol{\Delta})=\{\mathbf{0}\}$ will be called *nonpre-estimable* (some authors prefer the term nonestimable) with respect to the unconstrained parametrization $E(\mathbf{Y})=\mathbf{X}\boldsymbol{\theta}$. We see that if nonpreestimable constraints are added to the unconstrained parametrization the result will also be a parametrization for the same expectation space Ω.

For parametrization (3.4), the space of estimable parametric functions $\boldsymbol{\lambda}'\boldsymbol{\theta}^c$ is

$$\bar{\Theta}_c=\{\boldsymbol{\lambda}'\boldsymbol{\theta}^c: \boldsymbol{\lambda}\in R(\mathbf{X}', \boldsymbol{\Delta}_c)=R^9\},$$

so the parametric vector $\boldsymbol{\theta}^c$ is estimable. The BLUE for $\boldsymbol{\theta}^c$ is the least squares $\hat{\boldsymbol{\theta}}$ calculated before under parametrization (3.2) using the constraints (3.3). Under this constrained parametrization, the average of the response function over the domain is just μ^c. The main effects for factors 1 and 2 are, respectively, $2\alpha_1^c$ and $2\beta_1^c$. The two-way interaction is $2(\alpha\beta)_{11}^c$. This parametrization represents the average response, the main effects, and the two-way interaction as simple parametric functions.

Another unconstrained parametrization for Ω is

$$E(Y_{ijk})=\phi_{ij}, \qquad \operatorname{Var}(Y_{ijk})=\sigma^2, \qquad i,j=1,2$$

or in matrix form

$$E(\mathbf{Y})=\mathbf{U}\boldsymbol{\phi}, \qquad \boldsymbol{\phi}\in R^4, \qquad \operatorname{Cov}(\mathbf{Y})=\sigma^2\mathbf{I} \tag{3.5}$$

where **Y** is as before, $\boldsymbol{\phi}=(\phi_{11}, \phi_{12}, \phi_{21}, \phi_{22})'$, and

$$\mathbf{U}=\begin{bmatrix} 1 & 0 & 0 & 0 \\ & \cdots & & \\ 1 & 0 & 0 & 0 \\ 0 & 1 & 0 & 0 \\ & \cdots & & \\ 0 & 1 & 0 & 0 \\ 0 & 0 & 1 & 0 \\ & \cdots & & \\ 0 & 0 & 1 & 0 \\ 0 & 0 & 0 & 1 \\ & \cdots & & \\ 0 & 0 & 0 & 1 \end{bmatrix} \begin{matrix} \left.\begin{matrix} \\ \\ \\ \end{matrix}\right\} n_{11} \text{ rows} \\ \left.\begin{matrix} \\ \\ \\ \end{matrix}\right\} n_{12} \text{ rows} \\ \left.\begin{matrix} \\ \\ \\ \end{matrix}\right\} n_{21} \text{ rows} \\ \left.\begin{matrix} \\ \\ \\ \end{matrix}\right\} n_{22} \text{ rows} \end{matrix}.$$

The parameter ϕ_{ij} represents the expected response for treatment (i, j). The space of estimable linear parametric functions is

$$\bar{\Phi}=\{\boldsymbol{\pi}'\boldsymbol{\phi}: \boldsymbol{\pi} \in R(\mathbf{U}')=R^4\}.$$

So $\boldsymbol{\phi}$ is an estimable parametric vector. The BLUE $\hat{\boldsymbol{\phi}}$ for $\boldsymbol{\phi}$ is the solution of

$$\begin{bmatrix} n_{11} & 0 & 0 & 0 \\ 0 & n_{12} & 0 & 0 \\ 0 & 0 & n_{21} & 0 \\ 0 & 0 & 0 & n_{22} \end{bmatrix} \begin{bmatrix} \hat{\phi}_{11} \\ \hat{\phi}_{12} \\ \hat{\phi}_{21} \\ \hat{\phi}_{22} \end{bmatrix} = \begin{bmatrix} Y_{11.} \\ Y_{12.} \\ Y_{21.} \\ Y_{22.} \end{bmatrix}.$$

So $\hat{\phi}_{ij}=\bar{Y}_{ij.}$. Note that these are the same BLUEs for the expected treatment responses as those found under parametrizations (3.2) and (3.4). The covariance matrix for $\hat{\boldsymbol{\phi}}$ is easily found:

$$\operatorname{Cov}(\hat{\boldsymbol{\phi}})=\sigma^2 \begin{bmatrix} \frac{1}{n_{11}} & 0 & 0 & 0 \\ 0 & \frac{1}{n_{12}} & 0 & 0 \\ 0 & 0 & \frac{1}{n_{21}} & 0 \\ 0 & 0 & 0 & \frac{1}{n_{22}} \end{bmatrix}.$$

Computationally, this is a very convenient parametrization. The average of the response function over the domain, the main effects for factors 1 and 2, and the two-way interactions are

$$(\phi_{11}+\phi_{12}+\phi_{21}+\phi_{22})/4,$$

$$(\phi_{11}+\phi_{12}-\phi_{21}-\phi_{22})/2,$$

$$(\phi_{11}-\phi_{12}+\phi_{21}-\phi_{22})/2,$$

and

$$(\phi_{11}-\phi_{12}-\phi_{21}+\phi_{22})/2.$$

There are, of course, infinitely many other parametrizations to choose from. Given any two parametrizations for Ω. How can we relate information about parametric functions under one parametrization to parametric functions under the other? This can be done through the concept of correspondence.

3.7 CORRESPONDENCES BETWEEN PARAMETRIZATIONS

Suppose we have two parametrizations for Ω:

$$P_1\colon E(\mathbf{Y})=\mathbf{X}\boldsymbol{\theta}, \qquad \boldsymbol{\Delta}'\boldsymbol{\theta}=\mathbf{0},$$

$$P_2\colon E(\mathbf{Y})=\mathbf{U}\boldsymbol{\phi}, \qquad \boldsymbol{\Gamma}'\boldsymbol{\phi}=\mathbf{0},$$

and suppose that $\boldsymbol{\Lambda}'\boldsymbol{\theta}$ and $\boldsymbol{\pi}'\boldsymbol{\phi}$ are parametric vectors. We say that $\boldsymbol{\Lambda}'\boldsymbol{\theta}$ and $\boldsymbol{\pi}'\boldsymbol{\phi}$ *correspond* and we write $\boldsymbol{\Lambda}'\boldsymbol{\theta}\doteq\boldsymbol{\pi}'\boldsymbol{\phi}$ provided that

$$\boldsymbol{\Delta}'\bar{\boldsymbol{\theta}}=\mathbf{0}, \qquad \boldsymbol{\Gamma}'\bar{\boldsymbol{\phi}}=\mathbf{0}, \qquad \text{and} \qquad \mathbf{X}\bar{\boldsymbol{\theta}}=\mathbf{U}\bar{\boldsymbol{\phi}} \quad \text{imply} \quad \boldsymbol{\Lambda}'\bar{\boldsymbol{\theta}}=\boldsymbol{\pi}'\bar{\boldsymbol{\phi}}.$$

The notation $\doteq$ is meant to remind us that although $\boldsymbol{\Lambda}'\boldsymbol{\theta}=\boldsymbol{\pi}'\boldsymbol{\phi}$ makes no sense in general (the parametric vectors are defined on different parameter spaces), $\boldsymbol{\Lambda}'\bar{\boldsymbol{\theta}}=\boldsymbol{\pi}'\bar{\boldsymbol{\phi}}$ whenever valid $\bar{\boldsymbol{\theta}}$ and $\bar{\boldsymbol{\phi}}$ describe the same mean vector $E(\mathbf{Y})$.

The following result is very useful for finding BLUEs.

THEOREM 3.1 Suppose that $\boldsymbol{\Lambda}'\boldsymbol{\theta}\doteq\boldsymbol{\pi}'\boldsymbol{\phi}$ and that $\hat{\boldsymbol{\theta}}$ and $\hat{\boldsymbol{\phi}}$ are least squares for $\boldsymbol{\theta}$ and $\boldsymbol{\phi}$, respectively. Then $\boldsymbol{\Lambda}'\boldsymbol{\theta}$ and $\boldsymbol{\pi}'\boldsymbol{\phi}$ are estimable parametric vectors and $\boldsymbol{\Lambda}'\hat{\boldsymbol{\theta}}=\boldsymbol{\pi}'\hat{\boldsymbol{\phi}}$.

Proof Suppose $\mathbf{X}\bar{\boldsymbol{\theta}}=\mathbf{X}\bar{\bar{\boldsymbol{\theta}}}$, $\boldsymbol{\Delta}'\bar{\boldsymbol{\theta}}=\boldsymbol{\Delta}'\bar{\bar{\boldsymbol{\theta}}}=\mathbf{0}$. Choose $\bar{\boldsymbol{\phi}}$ such that $\mathbf{X}\bar{\boldsymbol{\theta}}=\mathbf{U}\bar{\boldsymbol{\phi}}$, $\boldsymbol{\Gamma}'\bar{\boldsymbol{\phi}}=\mathbf{0}$. Then $\boldsymbol{\Lambda}'\bar{\boldsymbol{\theta}}=\boldsymbol{\pi}'\bar{\boldsymbol{\phi}}$. In the same way $\boldsymbol{\Lambda}'\bar{\bar{\boldsymbol{\theta}}}=\boldsymbol{\pi}'\bar{\boldsymbol{\phi}}$. This shows $\boldsymbol{\Lambda}'\boldsymbol{\theta}$ is identifiable. By Theorem 2.4a, $\boldsymbol{\Lambda}'\boldsymbol{\theta}$ is estimable. Similarly, $\boldsymbol{\pi}'\boldsymbol{\phi}$ is an estimable parametric vector. Since $\hat{\boldsymbol{\theta}}$ and $\hat{\boldsymbol{\phi}}$ are least squares for $\boldsymbol{\theta}$ and $\boldsymbol{\phi}$, respectively, $\boldsymbol{\Delta}'\hat{\boldsymbol{\theta}}=\mathbf{0}$, $\boldsymbol{\Gamma}'\hat{\boldsymbol{\phi}}=\boldsymbol{\theta}$, and $\mathbf{X}\hat{\boldsymbol{\theta}}=\mathbf{P}\mathbf{Y}=\mathbf{U}\hat{\boldsymbol{\phi}}$ by Theorem 2.6c. Then $\boldsymbol{\Lambda}'\hat{\boldsymbol{\theta}}=\boldsymbol{\pi}'\hat{\boldsymbol{\phi}}$. □

Suppose we want to determine the BLUE $\boldsymbol{\Lambda}'\hat{\boldsymbol{\theta}}$ of some estimable parametric vector $\boldsymbol{\Lambda}'\boldsymbol{\theta}$. If we know that $\boldsymbol{\Lambda}'\boldsymbol{\theta}\doteq\boldsymbol{\pi}'\boldsymbol{\phi}$ where $\boldsymbol{\pi}'\boldsymbol{\phi}$ is defined with respect to a computationally more convenient parametrization, then $\boldsymbol{\pi}'\hat{\boldsymbol{\phi}}$ is the desired BLUE for $\boldsymbol{\Lambda}'\boldsymbol{\theta}$.

As an example, consider parametrizations (3.2) and (3.5). Suppose we want the BLUEs of the treatment expectations $\mu+\alpha_i+\beta_j+(\alpha\beta)_{ij}$ under parametrizations (3.2). Treatment expectations for different parametrizations correspond (we are assuming in this chapter that $n_{ij}>0$); so $\mu+\alpha_i+\beta_j+(\alpha\beta)_{ij}\doteq\phi_{ij}$. But $\hat{\phi}_{ij}=\bar{Y}_{ij.}$, so $\hat{\mu}+\hat{\alpha}_i+\hat{\beta}_j+\widehat{(\alpha\beta)}_{ij}=\bar{Y}_{ij.}$, which is what we saw before under (3.2).

Reparametrization can also be useful in constructing tests of linear hypotheses. Recall that relative to the parametrization $\Omega=\{\mathbf{X}\boldsymbol{\theta}\colon \boldsymbol{\theta}\in\Theta\}$ where Θ is a subspace of R^p, $H_0\colon \boldsymbol{\theta}\in\Theta_0$ is called a linear hypothesis on $\boldsymbol{\theta}$ if Θ_0 is a subspace of Θ and if the sets $\Omega_0=\{\mathbf{X}\boldsymbol{\theta}\colon \boldsymbol{\theta}\in\Theta_0\}$ and $\Omega_A=\{\mathbf{X}\boldsymbol{\theta}\colon \boldsymbol{\theta}\in\Theta-\Theta_0\}$ are disjoint. Recall also that a linear hypothesis on $\boldsymbol{\theta}$ is equivalent to a linear hypothesis on Ω.

Consider again parametrizations P_1 and P_2. We have:

THEOREM 3.2 Suppose $\boldsymbol{\Lambda}'\boldsymbol{\theta}\doteq\boldsymbol{\pi}'\boldsymbol{\phi}$ where $\boldsymbol{\Lambda}'\boldsymbol{\theta}$ and $\boldsymbol{\pi}'\boldsymbol{\phi}$ are parametric vectors. Then:

- **(a)** $H_0\colon \boldsymbol{\Lambda}'\boldsymbol{\theta}=\mathbf{0}$ is a linear hypothesis on $\boldsymbol{\theta}$.
- **(b)** $H_0\colon \boldsymbol{\pi}'\boldsymbol{\phi}=\mathbf{0}$ is a linear hypothesis on $\boldsymbol{\phi}$.
- **(c)** The hypotheses in (a) and (b) describe the same linear hypothesis on Ω.

Proof Since corresponding parametric vectors are estimable (Theorem 3.1), both $\boldsymbol{\Lambda}'\boldsymbol{\theta}$ and $\boldsymbol{\pi}'\boldsymbol{\phi}$ are estimable. Theorem 2.7 then gives us statements (a) and (b). To prove (c), we need to show that $\Omega_0^a=\{\mathbf{X}\boldsymbol{\theta}\colon \boldsymbol{\Delta}'\boldsymbol{\theta}=\mathbf{0},\ \boldsymbol{\Lambda}'\boldsymbol{\theta}=\mathbf{0}\}$ and $\Omega_0^b=\{\mathbf{U}\boldsymbol{\phi}\colon \boldsymbol{\Gamma}'\boldsymbol{\phi}=\mathbf{0},\ \boldsymbol{\pi}'\boldsymbol{\phi}=\mathbf{0}\}$ are the same sets. Suppose $\mathbf{x}\in\Omega_0^a$. Then there exists some $\bar{\boldsymbol{\theta}}$ such that

$$\mathbf{x}=\mathbf{X}\bar{\boldsymbol{\theta}},\qquad \boldsymbol{\Delta}'\bar{\boldsymbol{\theta}}=\mathbf{0},\qquad \boldsymbol{\Lambda}'\bar{\boldsymbol{\theta}}=\mathbf{0}.$$

But $\mathbf{x}\in\Omega=\{\mathbf{U}\boldsymbol{\phi}\colon \boldsymbol{\Gamma}'\boldsymbol{\phi}=\mathbf{0}\}$, so there exists some $\bar{\boldsymbol{\phi}}$ such that

$$\mathbf{x}=\mathbf{U}\bar{\boldsymbol{\phi}},\qquad \boldsymbol{\Gamma}'\bar{\boldsymbol{\phi}}=\mathbf{0}.$$

Since $\boldsymbol{\Lambda}'\boldsymbol{\theta}\doteq\boldsymbol{\pi}'\boldsymbol{\phi}$, we have $\boldsymbol{\pi}'\bar{\boldsymbol{\phi}}=\boldsymbol{\Lambda}'\bar{\boldsymbol{\theta}}=\mathbf{0}$ by the definition of $\doteq$. So $\Omega_0^a\subset\Omega_0^b$. We can show $\Omega_0^b\subset\Omega_0^a$ in a similar way, and the sets must be the same. □

We have seen that the BLUE of an estimable parametric vector under one parametrization can be calculated by calculating the BLUE of a corresponding parametric vector under a more computationally convenient parametrization. We have seen as well that a linear hypothesis under one parametrization can be tested by testing a corresponding linear hypothesis under another parametrization.

In what follows, when considering a hypothesis $H_0\colon \boldsymbol{\Lambda}'\boldsymbol{\theta}=\mathbf{0}$, it will always be assumed that $\Lambda'\theta$ is estimable, which ensures that the hypothesis is equivalent to a linear hypothesis on Ω.

3.8 AN ALTERNATIVE FORM FOR $\hat{R}_0 - \hat{R}$

Theorem 3.2 is particularly useful after we develop an alternative expression for $\hat{R}_0 - \hat{R}$, the deviation sum of squares for a linear hypothesis. Consider the parametrization (2.6)

$$E(\mathbf{Y}) = \mathbf{X}\boldsymbol{\theta}, \qquad \boldsymbol{\Delta}'\boldsymbol{\theta} = \mathbf{0}, \qquad \mathrm{Cov}(\mathbf{Y}) = \sigma^2\mathbf{I},$$

where $\sigma^2 > 0$. Then we have:

THEOREM 3.3 Suppose H_0: $\boldsymbol{\Lambda}'\boldsymbol{\theta} = \mathbf{0}$ is a linear hypothesis on $\boldsymbol{\theta}$, $\hat{\boldsymbol{\theta}}$ is a least squares for $\boldsymbol{\theta}$, and $\mathbf{V}_\Lambda$ is a matrix such that $\mathrm{Cov}(\boldsymbol{\Lambda}'\hat{\boldsymbol{\theta}}) = \sigma^2\mathbf{V}_\Lambda$. Then

$$\hat{R}_0 - \hat{R} = (\boldsymbol{\Lambda}'\hat{\boldsymbol{\theta}})'\mathbf{V}_\Lambda^-(\boldsymbol{\Lambda}'\hat{\boldsymbol{\theta}}),$$

where $\mathbf{V}_\Lambda^-$ can be any g inverse of $\mathbf{V}_\Lambda$. Furthermore, if $\boldsymbol{\theta} \in \Theta$, then

$$E\{\hat{R}_0 - \hat{R}\} = (m - m_0)\sigma^2 + (\boldsymbol{\Lambda}'\boldsymbol{\theta})'\mathbf{V}_\Lambda^-(\boldsymbol{\Lambda}'\boldsymbol{\theta}),$$

where $m_0 = \dim \Omega_0 = \dim\{\mathbf{X}\boldsymbol{\theta}\colon \boldsymbol{\Delta}'\boldsymbol{\theta} = \mathbf{0}, \boldsymbol{\Lambda}'\boldsymbol{\theta} = \mathbf{0}\}$.

When $\mathbf{Y}$ has a multivariate normal distribution, which will be assumed henceforth any time a hypothesis testing problem is considered, Theorem 2.8 says that $(\hat{R}_0 - \hat{R})/\sigma^2$ has a noncentral chi-square distribution with $m - m_0$ degrees of freedom. We see from Theorem 3.3 that the noncentrality parameter is $(\boldsymbol{\Lambda}'\boldsymbol{\theta})'\mathbf{V}_\Lambda^-(\boldsymbol{\Lambda}'\boldsymbol{\theta})$. When H_0: $\boldsymbol{\Lambda}'\boldsymbol{\theta} = \mathbf{0}$ is true, the noncentrality parameter is zero, and $(\hat{R}_0 - \hat{R})/\sigma^2$ has a (central) chi-square distribution.

When is $\mathbf{V}_\Lambda$ positive definite, so $\mathbf{V}_\Lambda^{-1}$ can be used in the expression for $\hat{R}_0 - \hat{R}$?

THEOREM 3.4 Let $\mathbf{V}_\Lambda$ be as in Theorem 3.3. Then $r(\mathbf{V}_\Lambda) = m - m_0$ and $\mathbf{V}_\Lambda$ is positive definite if and only if the rows of $\boldsymbol{\Lambda}'$ are linearly independent and $R(\boldsymbol{\Lambda}) \cap R(\boldsymbol{\Delta}) = \{\mathbf{0}\}$.

Suppose that given any parametrization P_1 we want to test the linear hypothesis H_0: $\boldsymbol{\Lambda}'\boldsymbol{\theta} = \mathbf{0}$ on $\boldsymbol{\theta}$ and we know that $\boldsymbol{\Lambda}'\boldsymbol{\theta} \doteq \boldsymbol{\pi}'\boldsymbol{\phi}$ where P_2 is unconstrained ($\boldsymbol{\Gamma} = \mathbf{0}$), $\mathbf{U}$ has full column rank, and the rows of $\boldsymbol{\pi}'$ are linearly independent. Then

$$\hat{R}_0 - \hat{R} = (\boldsymbol{\pi}'\hat{\boldsymbol{\phi}})'\mathbf{V}_\pi^{-1}(\boldsymbol{\pi}'\hat{\boldsymbol{\phi}}),$$

where $\boldsymbol{\pi}'\hat{\boldsymbol{\phi}} = \boldsymbol{\pi}'(\mathbf{U}'\mathbf{U})^{-1}\mathbf{U}'\mathbf{Y}$ and $\mathbf{V}_\pi = \boldsymbol{\pi}'(\mathbf{U}'\mathbf{U})^{-1}\boldsymbol{\pi}$.

3.9 TESTS OF HYPOTHESES FOR THE 2^2 FACTORIAL

We now discuss some tests of hypotheses for the 2^2 factorial experiment, applying what we have learned about correspondences between parametrizations. The test most people begin with is the test of the hypothesis of no interaction. Before constructing the test, it seems worthwhile to discuss what interaction is in the context of the 2^2 factorial experiment.

Interaction in factorial experiments is said to exist when the effect on the response of a subset of the factors depends not only on the levels of those factors, but also on the levels of other factors.

In the 2^2 factorial, interaction is present if the change in the response due to increasing the level of one of the factors depends on the level of the other. This is the two-way interaction between the factors, defined earlier in the chapter.

The two-way interaction was defined with respect to parametrization (3.2) as the parametric function

$$[(\alpha\beta)_{11}-(\alpha\beta)_{12}-(\alpha\beta)_{21}+(\alpha\beta)_{22}]/2.$$

Now $\mu+\alpha_i+\beta_j+(\alpha\beta)_{ij}\doteq\phi_{ij}$ where ϕ_{ij} is from parametrization (3.5). Using some algebra we find that

$$\begin{aligned}[(\alpha\beta)_{11}-(\alpha\beta)_{12}-(\alpha\beta)_{21}+(\alpha\beta)_{22}]/2 &\doteq (\phi_{11}-\phi_{12}-\phi_{21}+\phi_{22})/2\\ &=[(\phi_{22}-\phi_{21})-(\phi_{12}-\phi_{11})]/2,\end{aligned}$$

which expresses the interaction as one-half the difference in the expected response to changing factor 2 from level one to level two between factor 1 at its second and that at its first level. This could be rearranged so that the roles of the two factors were reversed.

Figure 3.2 graphically illustrates examples with and without interaction. Note that the roles of factors 1 and 2 could be reversed. Main effects for both factors exist in both examples. Graphically the hypothesis of no interaction in the 2^2 factorial is equivalent to the hypothesis of parallel lines.

The hypothesis of no interaction can be stated with respect to parametrizations (3.2), (3.4), and (3.5), respectively, as

$$H_0\colon [(\alpha\beta)_{11}-(\alpha\beta)_{12}-(\alpha\beta)_{21}+(\alpha\beta)_{22}]/2=0,$$

$$H_0\colon 2(\alpha\beta)^c_{11}=0,$$

$$H_0\colon (\phi_{11}-\phi_{12}-\phi_{21}+\phi_{22})/2=0.$$

These statements describe the same linear hypothesis because the parametric functions correspond.

Since parametrization (3.5) is unconstrained and has full column rank, we

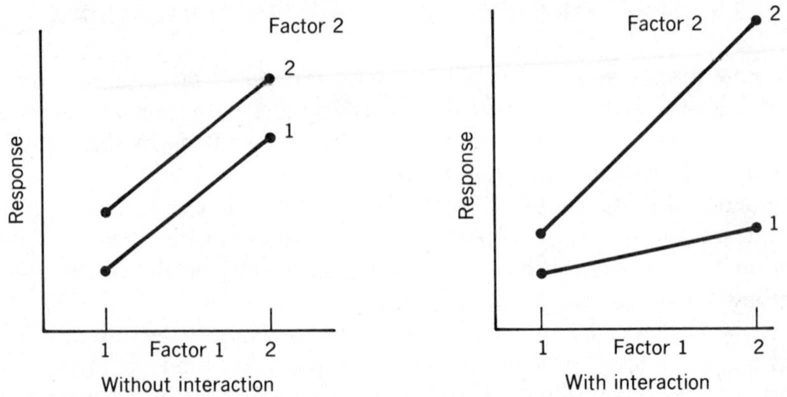

Figure 3.2 Examples of 2^2 factorials with and without interaction.

will construct the test of our hypothesis using this parametrization. We apply Theorems 3.3 and 3.4 to H_0: $\boldsymbol{\pi}'\boldsymbol{\phi}=\mathbf{0}$, where $\boldsymbol{\pi}'=(\frac{1}{2}, -\frac{1}{2}, -\frac{1}{2}, \frac{1}{2})$.

$$\boldsymbol{\pi}'\hat{\boldsymbol{\phi}}=(\bar{Y}_{11.}-\bar{Y}_{12.}-\bar{Y}_{21.}+\bar{Y}_{22.})/2,$$

$$\mathbf{V}_\pi=\text{Var}(\boldsymbol{\pi}'\hat{\boldsymbol{\phi}})/\sigma^2=(n_{11}^{-1}+n_{12}^{-1}+n_{21}^{-1}+n_{22}^{-1})/4=K/4,$$

$$\hat{R}_0-\hat{R}=(\boldsymbol{\pi}'\hat{\boldsymbol{\phi}})'\mathbf{V}_\pi^{-1}(\boldsymbol{\pi}'\hat{\boldsymbol{\phi}})=(\bar{Y}_{11.}-\bar{Y}_{12.}-\bar{Y}_{21.}+\bar{Y}_{22.})^2/K,$$

$$m-m_0=r(\mathbf{V}_\pi)=1,$$

$$E(\hat{R}_0-\hat{R})=\sigma^2+(\phi_{11}-\phi_{12}-\phi_{21}+\phi_{22})^2/K,$$

$$\hat{R}=\|\mathbf{Y}-\mathbf{U}\hat{\boldsymbol{\phi}}\|^2=\sum_{i=1}^{2}\sum_{j=1}^{2}\sum_{k=1}^{n_{ij}}(Y_{ijk}-\bar{Y}_{ij.})^2,$$

and

$$N-m=N-4.$$

The $\hat{F}$ statistic is then

$$\hat{F}=\frac{(\hat{R}_0-\hat{R})/(m-m_0)}{\hat{R}/(N-m)}=\frac{(\hat{R}_0-\hat{R})}{\hat{R}/(N-4)}.$$

Other very common tests are those for the hypotheses of no main effects. The hypothesis of no main effect for factor 1 can be stated with respect to our three parametrizations as

$$H_0\colon \alpha_1-\alpha_2+\overline{(\alpha\beta)}_{1.}-\overline{(\alpha\beta)}_{2.}=0,$$

$$H_0\colon 2\alpha_1^c=0,$$

$$H_0\colon (\phi_{11}+\phi_{12}-\phi_{21}-\phi_{22})/2=0.$$

We again apply Theorems 3.3 and 3.4 to $H_0\colon \boldsymbol{\pi}'\boldsymbol{\phi}=\mathbf{0}$, where now

$$\begin{aligned}
\boldsymbol{\pi}' &= (\tfrac{1}{2}, \tfrac{1}{2}, -\tfrac{1}{2}, -\tfrac{1}{2})\colon\\
\boldsymbol{\pi}'\hat{\boldsymbol{\phi}} &= (\bar{Y}_{11.}+\bar{Y}_{12.}-\bar{Y}_{21.}-\bar{Y}_{22.})/2,\\
\mathbf{V}_{\pi} &= \mathrm{Var}(\boldsymbol{\pi}'\hat{\boldsymbol{\phi}})/\sigma^2 = K/4,\\
\hat{R}_0-\hat{R} &= (\boldsymbol{\pi}'\hat{\boldsymbol{\phi}})'\mathbf{V}_{\pi}^{-1}(\boldsymbol{\pi}'\hat{\boldsymbol{\phi}}) = (\bar{Y}_{11.}+\bar{Y}_{12.}-\bar{Y}_{21.}-\bar{Y}_{22.})^2/K,\\
m-m_0 &= r(\mathbf{V}_{\pi}) = 1,\\
E(\hat{R}_0-\hat{R}) &= \sigma^2+(\phi_{11}+\phi_{12}-\phi_{21}-\phi_{22})^2/K,
\end{aligned}$$

and $\hat{R}$ and $N-m$ are the same as for the test of hypothesis of no interaction. The hypothesis of no main effects for factor 2 can be stated as

$$\begin{aligned}
&H_0\colon \beta_1-\beta_2+\overline{(\alpha\beta)}_{.1}-\overline{(\alpha\beta)}_{.2}=0,\\
&H_0\colon 2\beta_1^c=0,\\
&H_0\colon (\phi_{11}-\phi_{12}+\phi_{21}-\phi_{22})/2=0.
\end{aligned}$$

We obtain

$$\begin{aligned}
\hat{R}_0-\hat{R} &= (\bar{Y}_{11.}-\bar{Y}_{12.}+\bar{Y}_{21.}-\bar{Y}_{22.})^2/K\\
m-m_0 &= 1,\\
E(\hat{R}_0-\hat{R}) &= \sigma^2+(\phi_{11}-\phi_{12}+\phi_{21}-\phi_{22})^2/K,
\end{aligned}$$

and $\hat{R}$ and $N-m$ are the same as before.

We can state the hypothesis that both main effects and the interaction are all zero as $H_0\colon \boldsymbol{\pi}'\boldsymbol{\phi}=\mathbf{0}$, where

$$\boldsymbol{\pi}' = \frac{1}{2}\begin{bmatrix} 1 & 1 & -1 & -1\\ 1 & -1 & 1 & -1\\ 1 & -1 & -1 & 1 \end{bmatrix}.$$

The parametrization (3.5) is unconstrained and the rows of $\boldsymbol{\pi}'$ are linearly independent, so $\mathbf{V}_{\pi}$ is positive definite, and $\hat{R}_0-\hat{R}=(\boldsymbol{\pi}'\hat{\boldsymbol{\phi}})'\mathbf{V}_{\pi}^{-1}(\boldsymbol{\pi}'\hat{\boldsymbol{\phi}})$. Here

$$\boldsymbol{\pi}'\hat{\boldsymbol{\phi}} = \frac{1}{2}\begin{bmatrix} \bar{Y}_{11.} & + & \bar{Y}_{12.} & - & \bar{Y}_{21.} & - & \bar{Y}_{22.}\\ \bar{Y}_{11.} & - & \bar{Y}_{12.} & + & \bar{Y}_{21.} & - & \bar{Y}_{22.}\\ \bar{Y}_{11.} & - & \bar{Y}_{12.} & - & \bar{Y}_{21.} & + & \bar{Y}_{22.} \end{bmatrix},$$

$$\mathbf{V}_{\pi} = \boldsymbol{\pi}'(\mathbf{U}'\mathbf{U})^{-1}\boldsymbol{\pi} = \frac{1}{4}\begin{bmatrix} K & K_{AB} & K_B\\ K_{AB} & K & K_A\\ K_B & K_A & K \end{bmatrix},$$

and $m-m_0=r(\mathbf{V}_\pi)=3$. We see that the deviation sum of squares $\hat{R}_0-\hat{R}$ for the hypothesis $H_0\colon \boldsymbol{\pi}'\boldsymbol{\theta}=\mathbf{0}$ is equal to the sum of the deviation sums of squares for the individual hypotheses derived above if and only if $\mathbf{V}_\pi=\frac{1}{4}\operatorname{diag}(K, K, K)$, which is true if and only if $K_A=K_B=K_{AB}=0$, or

$$\frac{1}{n_{11}}+\frac{1}{n_{12}}-\frac{1}{n_{21}}-\frac{1}{n_{22}}=0,$$

$$\frac{1}{n_{11}}-\frac{1}{n_{12}}+\frac{1}{n_{21}}-\frac{1}{n_{22}}=0,$$

and

$$\frac{1}{n_{11}}-\frac{1}{n_{12}}-\frac{1}{n_{21}}+\frac{1}{n_{22}}=0.$$

This is true if and only if $n_{11}=n_{12}=n_{21}=n_{22}=r$ (say), that is, if there is an equal number of observations for each treatment. This is called a 2^2 factorial arrangement with equal replications, denoted by FA(2, 4, 4r, r, r, r, r). Furthermore, we see that this is also the necessary and sufficient condition that the BLUEs of the individual parametric functions are uncorrelated, and, consequently that under the assumption of normality the individual deviation sums of squares are independent random variables.

In the equal replication case we have the simplifications

$$\boldsymbol{\pi}'\hat{\boldsymbol{\phi}}=\begin{bmatrix}(Y_{1..} & - & Y_{2..})/2r & & & & \\ (Y_{.1.} & - & Y_{.2.})/2r & & & & \\ (Y_{11.} & - & Y_{12.} & - & Y_{21.} & + & Y_{22.})/2r\end{bmatrix}$$

and

$$\hat{R}_0-\hat{R}=(Y_{1..}-Y_{2..})^2/4r+(Y_{.1.}-Y_{.2.})^2/4r+(Y_{11.}-Y_{12.}-Y_{21.}+Y_{22.})^2/4r,$$

where each term in $\hat{R}_0-\hat{R}$ is the deviation sum of squares for an individual hypothesis associated with a row of $\boldsymbol{\pi}'\boldsymbol{\phi}$. $\hat{R}_0-\hat{R}$ has a noncentral chi-square distribution with 3 degrees of freedom. Each term has a noncentral chi-square distribution with 1 degree of freedom and is independent of $\hat{R}$ and of the other terms in $\hat{R}_0-\hat{R}$.

To derive $\hat{R}_0-\hat{R}$ in the arbitrary complete 2^2 factorial more easily, we consider which linear hypothesis on Ω, $H_0\colon E(\mathbf{Y})\in\Omega_0$, is equivalent to $H_0\colon \boldsymbol{\pi}'\boldsymbol{\phi}=\mathbf{0}$. Now, $\Omega_0=\{\mathbf{U}\boldsymbol{\phi}\colon \boldsymbol{\pi}'\boldsymbol{\phi}=\mathbf{0},\ \boldsymbol{\phi}\in R^4\}$ and $m_0=r(\mathbf{U}', \boldsymbol{\pi})-r(\boldsymbol{\pi})=4-3=1$. We will consider an alternative parametrization for Ω_0. Note that the matrix $\mathbf{B}=\mathbf{1}_4=(1, 1, 1, 1)'$ satisfies $\boldsymbol{\pi}'\mathbf{B}=\mathbf{0}$ and $r(\mathbf{UB})=r(\mathbf{B})=1$. Then an alternative unconstrained and full column rank parametrization for Ω_0 is

$$E(\mathbf{Y})=\mathbf{1}_N\mu, \qquad \mu\in R^1 \quad \text{or} \quad \Omega_0=R(\mathbf{1}_N).$$

Now $\hat{\mu}=\bar{Y}_{...}$ is the least squares for μ, so $\hat{\boldsymbol{\phi}}=\mathbf{B}\hat{\mu}=\mathbf{1}_N\hat{\mu}$ is least squares for $\boldsymbol{\phi}$ under H_0: $E(\mathbf{Y})\in\Omega_0$. Then

$$\hat{R}_0=\sum_{i=1}^{2}\sum_{j=1}^{2}\sum_{k=1}^{n_{ij}}(Y_{ijk}-\bar{Y}_{...})^2,$$

$$\hat{R}=\sum_{i=1}^{2}\sum_{j=1}^{2}\sum_{k=1}^{n_{ij}}(Y_{ijk}-\bar{Y}_{ij.})^2,$$

and therefore,

$$\hat{R}_0-\hat{R}=\sum_{i=1}^{2}\sum_{j=1}^{2}n_{ij}(\bar{Y}_{ij.}-\bar{Y}_{...})^2.$$

Note that $\hat{R}_0$ depends only upon Ω_0, and not upon the particular linear hypotheses on $\boldsymbol{\theta}$ or $\boldsymbol{\phi}$, or equivalently upon Θ_0 or Φ_0, which gave rise to Ω_0. The same is therefore true of the deviation sum of squares, $\hat{R}_0-\hat{R}$. That is, it is the hypothesized expectation space that is crucial in testing a linear hypothesis, and not any particular linear hypothesis defined on a parametrization.

3.10 PARTITIONING DEVIATION SUMS OF SQUARES

We have presented a particular partition of the deviation sum of squares for the hypothesis H_0: $E(\mathbf{Y})\in R(\mathbf{1}_N)$ in the 2^2 factorial with equal replication. Suppose we now wish to test a linear hypothesis H_0: $E(\mathbf{Y})\in\Omega_0$ for an arbitrary complete factorial using a particular parametrization P_1. Can we always partition the deviation sum of squares for H_0 into independent, one-degree-of-freedom sums of squares, each representing the deviation sum of squares for a less restrictive hypothesis?

The answer is yes, although the less restrictive hypotheses may be very difficult to interpret. This is true because the expectation space Ω can be expressed as $\Omega=\Omega_0\oplus\mathbf{A}$ where $\mathbf{A}$ is a subspace of Ω and $\oplus$ is used to denote an orthogonal direct sum of subspaces. $\mathbf{A}$ can be partitioned as $\mathbf{A}=\mathbf{A}_2\oplus\cdots\oplus\mathbf{A}_{m-m_0}$, where $\dim\mathbf{A}_j=1$ for $j=1,2,\ldots,m-m_0$. Let $\hat{R}_j$ denote the residual sum of squares for the linear hypothesis H_j: $E(\mathbf{Y})\in\Omega_j=\Omega_0\oplus\mathbf{A}_1\oplus\cdots\oplus\mathbf{A}_j$ for $j=0,1,2,\ldots,m-m_0$. Note that $\Omega_{m-m_0}=\Omega$ and therefore $\hat{R}_{m-m_0}=\hat{R}$. Then

$$\hat{R}_0-\hat{R}=\sum_{j=1}^{m-m_0}(\hat{R}_{j-1}-\hat{R}_j),$$

where $\hat{R}_{j-1}-\hat{R}_j$ is the deviation sum of squares for testing H_{j-1} in the model $E(\mathbf{Y})\in\Omega_j$. The deviation sums of squares $\{\hat{R}_{j-1}-\hat{R}_j, j=1,2,\ldots,m-m_0\}$ are independent and each have 1 degree of freedom.

It may be that we wish to test the linear hypothesis H_0: $\boldsymbol{\Lambda}'\boldsymbol{\theta}=\mathbf{0}$ in some parametrization P_1 where $\boldsymbol{\Lambda}=(\boldsymbol{\lambda}_1, \boldsymbol{\lambda}_2, \ldots, \boldsymbol{\lambda}_q)$ is a $p\times q$ matrix. Under what conditions can we express the deviation sum of squares $\hat{R}_0-\hat{R}$ as the sum of the deviation sums of squares $\hat{R}_j-\hat{R}$ for the individual hypotheses H_j: $\boldsymbol{\lambda}'_j\boldsymbol{\theta}=\mathbf{0}$, $j=1, 2, \ldots, q$? Also, when will these individual deviation sums of squares be independent? The following two conditions are sufficient, but not necessary:

1 For any $\hat{\boldsymbol{\theta}}$ that is a least squares for $\boldsymbol{\theta}$, $\mathrm{Cov}(\boldsymbol{\lambda}_i'\hat{\boldsymbol{\theta}}, \boldsymbol{\lambda}_j'\hat{\boldsymbol{\theta}})=\mathbf{0}$ for $i\neq j$ (the BLUEs are independent) and
2 $R(\boldsymbol{\Lambda})\cap R(\boldsymbol{\Delta})=\{\mathbf{0}\}$.

The first condition means that the matrix $\mathbf{V}_\Lambda$ is diagonal. The second condition excludes trivial hypotheses and with condition 1 guarantees, on account of Theorem 3.4, that $\mathbf{V}_\Lambda$ is positive definite. Then

$$\hat{R}_0-\hat{R}=(\boldsymbol{\Lambda}'\hat{\boldsymbol{\theta}})'\mathbf{V}_\Lambda^{-1}(\boldsymbol{\Lambda}'\hat{\boldsymbol{\theta}})=\sum_{j=1}^{q}\frac{(\boldsymbol{\lambda}_j'\hat{\boldsymbol{\theta}})}{v_j}=\sum_{j=1}^{q}(\hat{R}_j-\hat{R}),$$

where $v_j=\mathrm{Var}(\boldsymbol{\lambda}_j'\hat{\boldsymbol{\theta}})$. Furthermore, since the $\boldsymbol{\lambda}_j'\hat{\boldsymbol{\theta}}$ are independent, so are the $\hat{R}_j-\hat{R}$. Our partitioning of the deviation sum of squares for H_0: $E(\mathbf{Y})\in R(\mathbf{1}_N)$ in the 2^2 factorial with equal replication satisfies conditions 1 and 2 and is therefore an example of this result.

3.11 THE TWO-WAY CROSSED CLASSIFICATION WITH ONE OBSERVATION PER CELL

We now consider a factorial experiment involving two factors A and B with a and b levels, respectively. The term "crossed classification" or "classification" comes from the fact that in many investigations the observations can be classified in a two-way table by letting the rows of the table correspond to the levels of factor A and the columns to the levels of B. It is customary in such situations to refer to the level combinations as cells rather than treatments. Therefore the cell (i, j) in the two-way table refers to the (i, j) treatment combination, where the factor A is at the ith level and the factor B is at the jth level. The domain consists of all ab possible combinations of the levels and the design includes at least one experimental unit per cell. An example of a hypothetical two-factor experiment with exactly one experimental unit per

Table 3.1 Observation Values for a Two-Factor Experiment with One Observation per Cell

		Factor B			
	Levels	1	2	...	b
Factor A	1	Y_{11}	Y_{12}		Y_{1b}
	2	Y_{21}	Y_{22}	...	Y_{2b}
	$\vdots$	$\vdots$	$\vdots$		$\vdots$
	a	Y_{a1}	Y_{a2}		Y_{ab}

cell is given in Table 3.1. The data of this table are described as coming from a two-way crossed classification with one observation per cell. The completely general case, the two-way classification with missing observations which includes the case of unequal numbers of observations per cell will be treated in Chapter 5.

The additive two-way classification model as defined in Section 3.2 for this experiment is of the form

$$Y_{ij}=\mu+\alpha_i+\beta_j+e_{ij}, \qquad i=1,\ldots,a; \qquad j=1,\ldots,b,$$

where Y_{ij} is the observation made on the ith level of factor A and jth level of factor B, and e_{ij} is the random error with the usual assumptions. The other terms in the model are unknown parameters; μ represents the average response, α_i is the main effect of the ith level of A, and β_j is the main effect of the jth level of factor B.

The parametrization for this experiment and model in matrix form is

$$E(\mathbf{Y})=\mathbf{X}\boldsymbol{\theta}, \qquad \boldsymbol{\Delta}'\boldsymbol{\theta}=\mathbf{0}, \qquad \text{and} \qquad \operatorname{Cov}(\mathbf{Y})=\sigma^2\mathbf{I}, \tag{3.6}$$

where

$$\mathbf{Y}=(Y_{11},\ldots,Y_{1b},Y_{21},\ldots,Y_{2b},\ldots,Y_{a1},\ldots,Y_{ab})',$$

$$\boldsymbol{\theta}=(\mu,\alpha_1,\ldots,\alpha_a,\beta_1,\ldots,\beta_b)',$$

$$\Theta=R^p, \qquad p=1+a+b,$$

and the design matrix $\mathbf{X}$ is

$$
\begin{array}{cccccccccc}
\mu & \alpha_1 & \alpha_2 & \cdots & \alpha_a & \beta_1 & \beta_2 & \cdots & \beta_b
\end{array}
$$

$$
\mathbf{X}=\left[\begin{array}{ccccccccc}
1 & 1 & 0 & \cdots & 0 & 1 & 0 & & 0 \\
1 & 1 & 0 & \cdots & 0 & 0 & 1 & \cdots & 0 \\
\vdots & \vdots & \vdots & \cdots & \vdots & \vdots & \vdots & \cdots & \vdots \\
1 & 1 & 0 & \cdots & 0 & 0 & 0 & \cdots & 1 \\
\hline
1 & 0 & 1 & \cdots & 0 & 1 & 0 & \cdots & 0 \\
1 & 0 & 1 & \cdots & 0 & 0 & 1 & \cdots & 0 \\
\vdots & \vdots & \vdots & \vdots & \vdots & \vdots & \vdots & & \\
1 & 0 & 1 & \cdots & 0 & 0 & 0 & \cdots & 1 \\
\hline
\vdots & \vdots & \vdots & \vdots & \vdots & \vdots & \vdots & \vdots & \vdots \\
\hline
1 & 0 & 0 & \cdots & 1 & 1 & 0 & \cdots & 0 \\
1 & 0 & 0 & \cdots & 1 & 0 & 1 & \cdots & 0 \\
\vdots & \vdots & \vdots & \vdots & \vdots & \vdots & \vdots & \vdots & \vdots \\
1 & 0 & 0 & \cdots & 1 & 0 & 0 & \cdots & 1
\end{array}\right]=(\mathbf{1}, \mathbf{A}, \mathbf{B}).
$$

Therefore we can write (3.6) as

$$E(\mathbf{Y})=\mathbf{1}_N\mu+\mathbf{A}\boldsymbol{\alpha}+\mathbf{B}\boldsymbol{\beta}=\mathbf{X}\boldsymbol{\theta}, \qquad \boldsymbol{\theta}\in R^p,$$

where $\boldsymbol{\alpha}=(\alpha_1,\ldots,\alpha_a)'$ and $\boldsymbol{\beta}=(\beta_1,\ldots,\beta_b)'$. Here $\Omega=R(\mathbf{X})$.

The design matrix has $N=ab$ rows and $a+b+1$ columns. The first column is $\mathbf{1}_N$, and the sum of the next a columns as well as that of the sum of the last b columns is also $\mathbf{1}_N$, that is $\mathbf{1}_N\in R(\mathbf{A})\cap R(\mathbf{B})$ which implies $R(\mathbf{X})=R(\mathbf{A},\mathbf{B})$ and $r(\mathbf{X})=a+b-1$.

Let $\mathbf{N}$ denote the $a\times b$ incidence matrix whose (i, j) component is 1 [when we have unequal numbers of observations per cell the (i, j) component of $\mathbf{N}$ is n_{ij} as we see later]. Observe that $\mathbf{N}=\mathbf{A}'\mathbf{B}=\mathbf{1}_a\mathbf{1}_b'$.

We now consider how to obtain BLUEs for estimable parametric functions, by determining a $\hat{\boldsymbol{\theta}}$ that is a least squares for $\boldsymbol{\theta}$. By Theorem 2.6 we see that $\hat{\boldsymbol{\theta}}$ is least squares for $\boldsymbol{\theta}$ if and only if it satisfies the normal equations $\mathbf{X}'\mathbf{X}\hat{\boldsymbol{\theta}}=\mathbf{X}'\mathbf{Y}$.

The coefficient matrix $\mathbf{X}'\mathbf{X}$ can be written out from the knowledge of the incidence matrix $\mathbf{N}$ as

$$\mathbf{X}'\mathbf{X}=\begin{bmatrix} N & \mathbf{1}'\mathbf{A} & \mathbf{1}'\mathbf{B} \\ \mathbf{A}'\mathbf{1} & \mathbf{A}'\mathbf{A} & \mathbf{N} \\ \mathbf{B}'\mathbf{1} & \mathbf{N}' & \mathbf{B}'\mathbf{B} \end{bmatrix}.$$

Note that $N=ab$ is the sum of entries in $\mathbf{N}$, and that $\mathbf{A}'\mathbf{A}$ and $\mathbf{B}'\mathbf{B}$ are both diagonal matrices whose diagonal entries are the row sums and column sums, respectively, of the matrix $\mathbf{N}$. That is,

$$\mathbf{A}'\mathbf{A}=b\mathbf{I}_a \quad\text{and}\quad \mathbf{B}'\mathbf{B}=a\mathbf{I}_b.$$

The right-hand side of the normal equation is $\mathbf{X}'\mathbf{Y}=(\mathbf{Y}'\mathbf{1},\ \mathbf{Y}'\mathbf{A},\ \mathbf{Y}'\mathbf{B})'$. Therefore we can write the normal equations in matrix form as

$$\begin{bmatrix} N & \mathbf{1}'\mathbf{A} & \mathbf{1}'\mathbf{B} \\ \mathbf{A}'\mathbf{1} & \mathbf{A}'\mathbf{A} & \mathbf{N} \\ \mathbf{B}'\mathbf{1} & \mathbf{N}' & \mathbf{B}'\mathbf{B} \end{bmatrix} \begin{bmatrix} \hat{\mu} \\ \hat{\boldsymbol{\alpha}} \\ \hat{\boldsymbol{\beta}} \end{bmatrix} = \begin{bmatrix} \mathbf{1}'\mathbf{Y} \\ \mathbf{A}'\mathbf{Y} \\ \mathbf{B}'\mathbf{Y} \end{bmatrix}.$$

Or, in summation form,

$$\begin{aligned} ab\hat{\mu} + b\Big(\sum_i \hat{\alpha}_i\Big) + a\Big(\sum_j \hat{\beta}_j\Big) &= Y_{..} \\ b\hat{\mu} + b\hat{\alpha}_i \qquad + \quad \sum_j \hat{\beta}_j &= Y_{i.}, \qquad i=1,\ldots,a \\ a\hat{\mu} + \quad \sum_i \hat{\alpha}_i + a\hat{\beta}_j \qquad &= Y_{.j}, \qquad j=1,\ldots,b. \end{aligned}$$

From these equations it is immediate to see that one possible solution for $\hat{\boldsymbol{\theta}}$ is given by

$$\begin{aligned} \hat{\mu} &= \bar{Y}_{..}, \\ \hat{\alpha}_i &= \bar{Y}_{i.} - \bar{Y}_{..}, \qquad i=1,\ldots,a, \\ \hat{\beta}_j &= \bar{Y}_{.j} - \bar{Y}_{..}, \qquad j=1,\ldots,b. \end{aligned}$$

The above solution is obtained by letting

$$\boldsymbol{\Delta}' = \begin{bmatrix} 0 & 1 & 1 & \cdots & 1 & 0 & 0 & \cdots & 0 \\ 0 & 0 & 0 & \cdots & 0 & 1 & 1 & \cdots & 1 \end{bmatrix}.$$

That is, we temporarily impose on the model two linearly independent constraints, the so-called usual constraints of the form

$$\sum_i \alpha_i = 0,$$

and

$$\sum_j \beta_j = 0,$$

which satisfies the nonpreestimability condition $R(\mathbf{X}') \cap R(\boldsymbol{\Delta}) = \{\mathbf{0}\}$.

There are, of course, many other possibilities for a $\hat{\boldsymbol{\theta}}$ that is least squares for $\boldsymbol{\theta}$. The above solution is a standard solution given in methods textbooks.

Now consider the two submodels

$$\text{MI: } E(\mathbf{Y}) = \mathbf{1}_N \mu + \mathbf{A}\boldsymbol{\alpha}$$

and

$$\text{MII: } E(\mathbf{Y}) = \mathbf{1}_N \mu + \mathbf{B}\boldsymbol{\beta}.$$

Model MI is a one-way classification model with a treatments. Note that a least squares solution for $\boldsymbol{\alpha}$ is

$$(\mathbf{A}'\mathbf{A})^{-1}\mathbf{A}'\mathbf{Y}=(\bar{Y}_{1.},\ldots,\bar{Y}_{a.})'.$$

Similarly MII is a one-way classification and a least squares solution for $\boldsymbol{\beta}$ is

$$(\mathbf{B}'\mathbf{B})^{-1}\mathbf{B}'\mathbf{Y}=(\bar{Y}_{.1},\ldots,\bar{Y}_{.b})'.$$

In what follows we characterize the estimable parametric functions.

THEOREM 3.5 Suppose $\boldsymbol{\pi}'\boldsymbol{\theta}=c\mu+\sum_i\lambda_i\alpha_i+\sum_j\eta_j\beta_j$ is a parametric function. Then $\boldsymbol{\pi}'\boldsymbol{\theta}$ is estimable if and only if $\sum_i\lambda_i=\sum_j\eta_j=c$.

Proof Suppose $\boldsymbol{\pi}'\boldsymbol{\theta}$ is estimable. Then $c\mu+\sum_i\lambda_i\alpha_i$ must be estimable with respect to MI, and $c\mu+\sum_j\eta_j\beta_j$ must be estimable with respect to MII. Thus, the necessity follows immediately. Now suppose $\sum_i\lambda_i=\sum_j\eta_j=c$ and let

$$\boldsymbol{l}=\left(\frac{-c}{ab},\frac{\lambda_1}{b},\ldots,\frac{\lambda_a}{b},\frac{\eta_1}{a},\ldots,\frac{\eta_b}{a}\right)'.$$

Then $\mathbf{X}'\mathbf{X}\boldsymbol{l}=\boldsymbol{\pi}$ which implies $\boldsymbol{\pi}'\boldsymbol{\theta}$ is estimable. □

The following results for the discussed experimental model under (3.6) with $\boldsymbol{\Delta}=\mathbf{0}$ are the immediate consequences of Theorem 3.5.

COROLLARY 3.1 A parametric function $\boldsymbol{\lambda}'\boldsymbol{\alpha}$ or $\boldsymbol{\eta}'\boldsymbol{\beta}$ where $\boldsymbol{\lambda}'=(\lambda_1,\ldots,\lambda_a)$ and $\boldsymbol{\eta}'=(\eta_1,\ldots,\eta_b)$, is estimable if and only if it is a contrast.

COROLLARY 3.2 Let f_α and f_β denote the dimensions of the vector spaces of estimable α- and β-contrasts, respectively. Then

- **(a)** $f_\alpha=a-1$.
- **(b)** $f_\beta=b-1$.
- **(c)** $r(\mathbf{X})=a+b-1$.
- **(d)** All cell expectations $\mu+\alpha_i+\beta_j$ are estimable.

Now suppose $\boldsymbol{\eta}'\boldsymbol{\beta}$ and $\boldsymbol{\omega}'\boldsymbol{\beta}$ are β contrasts and let $\widehat{\boldsymbol{\eta}'\boldsymbol{\beta}}$ and $\widehat{\boldsymbol{\omega}'\boldsymbol{\beta}}$ be their respective BLUEs. Then we see that $\widehat{\boldsymbol{\eta}'\boldsymbol{\beta}}=\sum_j\eta_j\bar{Y}_{.j}=\boldsymbol{\eta}'\mathbf{B}'\mathbf{Y}/a$, hence, $\text{Cov}(\widehat{\boldsymbol{\eta}'\boldsymbol{\beta}},\widehat{\boldsymbol{\omega}'\boldsymbol{\beta}})=\boldsymbol{\eta}'\mathbf{B}'\mathbf{B}\boldsymbol{\omega}\sigma^2/a^2=(\sum_j\eta_j\omega_j)\sigma^2/a$. A similar statement is true for α contrasts. In addition, if we let $\boldsymbol{\lambda}'\boldsymbol{\alpha}$ denote an α contrast and $\widehat{\boldsymbol{\lambda}'\boldsymbol{\alpha}}$ its BLUE, it follows that

$$\text{Cov}(\widehat{\boldsymbol{\lambda}'\boldsymbol{\alpha}},\widehat{\boldsymbol{\eta}'\boldsymbol{\beta}})=\mathbf{0},$$

that is, the BLUEs for α and β contrasts are independent. The residual sum of squares is

$$\hat{R}=\|\mathbf{Y}-\mathbf{X}\hat{\boldsymbol{\theta}}\|^2=\sum_{i=1}^{a}\sum_{j=1}^{b}(Y_{ij}-\bar{Y}_{i.}-\bar{Y}_{.j}+\bar{Y}_{..})^2.$$

Since $m=\dim\,\Omega=a+b-1$, $\hat{R}/(ab-a-b+1)$ is an unbiased estimator for σ^2.

We can construct likelihood ratio tests of linear hypotheses under the assumption that $\mathbf{Y}$ has a multivariate normal distribution. Any hypothesis of the form $H_0\colon \boldsymbol{\Lambda}'\boldsymbol{\alpha}=\mathbf{0}$ or $H_0\colon \boldsymbol{\Gamma}'\boldsymbol{\beta}=\mathbf{0}$, where $\boldsymbol{\Lambda}'\boldsymbol{\alpha}$ and $\boldsymbol{\Gamma}'\boldsymbol{\beta}$ are estimable parametric vectors, is a linear hypothesis. The most popular hypotheses are that the α_i are all equal:

$$H_0\colon \alpha_1=\alpha_2=\cdots=\alpha_a \quad \text{versus } H_A\colon \alpha_i\neq\alpha_j \text{ for some } i, j,$$

and that the β_j are all equal:

$$H_0\colon \beta_1=\beta_2=\cdots=\beta_b \qquad H_A\colon \beta_i\neq\beta_j \text{ for some } i, j.$$

We may also express the null and alternative hypotheses as

$$H_0\colon \boldsymbol{\Lambda}'\boldsymbol{\alpha}=\mathbf{0} \quad \text{versus} \quad H_A\colon \boldsymbol{\Lambda}'\boldsymbol{\alpha}\neq\mathbf{0}$$

where

$$\boldsymbol{\Lambda}'\boldsymbol{\alpha}=\begin{bmatrix}\alpha_1-\alpha_2\\ \alpha_1-\alpha_3\\ \vdots\\ \alpha_1-\alpha_a\end{bmatrix}=\begin{bmatrix}1 & -1 & 0 & \cdots & 0\\ 1 & 0 & -1 & \cdots & 0\\ & & \cdots & & \\ 1 & 0 & 0 & \cdots & -1\end{bmatrix}\begin{bmatrix}\alpha_1\\ \alpha_2\\ \vdots\\ \alpha_a\end{bmatrix}.$$

The components of $\boldsymbol{\Lambda}'\boldsymbol{\alpha}$ are the elements of a basis for the vector space of α contrasts. Similarly, we can construct $H_0\colon \boldsymbol{\Gamma}'\boldsymbol{\beta}=\mathbf{0}$ for the estimable β contrasts. Under the null hypothesis $H_0\colon \boldsymbol{\Lambda}'\boldsymbol{\alpha}=\mathbf{0}$, we must minimize

$$\sum_i\sum_j(Y_{ij}-\hat{\mu}-\hat{\beta}_j)^2,$$

since under H_0 the model reduces to MII. On equating to zero the partial derivatives we find $\hat{\mu}$ and $\hat{\boldsymbol{\beta}}$ to have the same values as under Ω (this is true only when the number of observations is the same for all cells). We obtain the sum of squares for main effect A, denoted by SS_A, as

$$\mathrm{SS}_A=\sum_i(\bar{Y}_{i.}-\bar{Y}_{..})^2.$$

Since H_0 states that $a-1$ linearly independent estimable functions are zero, the number of degrees of freedom for SS_A is $a-1$. Hence the $\hat{F}$ statistic for

Table 3.2 Analysis of Variance for Two-Way Classification with One Observation per Cell

Source of Variations	Degrees of Freedom	Sum of Squares	Mean Squares	$\hat{F}$
Mean	1	$\bar{Y}^2_{..}$		
Rows	$a-1$	SS_A	MS_A	MS_A/MS_E
Columns	$b-1$	SS_B	MS_B	MS_B/MS_E
Residual	$ab-a-b+1$	SS_E	MS_E	
Total	ab			

testing this hypothesis can be computed as

$$\hat{F}=\frac{SS_A/a-1}{\hat{R}/(ab-a-b+1)}=\frac{MS_A}{MS_E}$$

and the null hypothesis is rejected at the α level if $\hat{F}$ exceeds the upper $100\alpha\%$ point of the F distribution with $a-1$ and $ab-a-b+1$ degrees of freedom. In a similar way we can find SS_B to be

$$SS_B=\sum_j (\bar{Y}_{.j}-\bar{Y}_{..})^2,$$

with $b-1$ degrees of freedom for testing the null hypothesis $H_0: \boldsymbol{\eta}'\boldsymbol{\beta}=\mathbf{0}$. The $\hat{F}$ statistic for testing this hypothesis is

$$\hat{F}=\frac{SS_B/b-1}{\hat{R}/(ab-a-b+1)}=\frac{MS_B}{MS_E}.$$

The null hypothesis is rejected if $\hat{F}>F_{\alpha,b-1,ab-a-b+1}$. We can summarize these results in Table 3.2.

Usually the term due to the mean is substracted from the total to give the total sum of squares.

BIBLIOGRAPHICAL NOTES

3.1 Anderson (1970) introduced a random error component into linear models for analyzing data set following designed experiments. He called this error component a "restricted error" and it corresponds to every restriction on randomization introduced in the design. Later Anderson and McLean (1974) presented this concept in detail with variety of examples in a book form. Suppose that a levels of one fixed set of treatments and b levels of another fixed set are run in all possible *ab* combinations on *ab* experimental units completely at random. The design of such an experiment is a completely randomized design, and the model for the analysis of the data following such an experiment if no interaction is present may be written as

$$Y_{ij}=\mu+\alpha_i+\beta_j+\varepsilon_{ij},$$
$$i=1,\ldots,a$$
$$j=1,\ldots,b \tag{1}$$

Since there are no restrictions on randomizations in the design of experiment, it may be called a *two-way factorial completely randomized design.* Now, suppose that there are blocks in the experiment and *a* treatments are completely randomized onto *b* experimental units in each block such that there is a different randomization in each block. Since there is a different randomization of treatments within each block, the experimental units are not selected in the same manner as in the completely randomized design. Therefore, it seems natural that the model for this experiment is different from that of (1). That is, we now have

$$Y_{ij}=\mu+\alpha_i+\beta_j+\delta_{(j)}+\varepsilon_{(ij)} \tag{2}$$

A *restriction error* refers explicitly to the fact that experimental units are restricted together in a homogeneous group. Whenever there is a restriction on a randomization another error term may be introduced into the model. The extra term, $\delta_{(i)}$, that is placed into the model associated with the randomized complete block has the same subscript as the block, which means the two terms are completely confounded with one another. See also Anderson and McLean's (1974) paper for detail. Randomized complete block design was developed by Fisher (1926).

3.2 The factorial experiment was called the "complex experiment" prior to 1926 when Fisher (1925) named it as the factorial experiment. Yates (1935, 1937) developed the concept and analysis of factorial experiments. His classic work (1937) on the subject of factorial experiment is highly recommended to the interested reader.

3.3 Eisenhart (1947) gives the assumptions underlying the analysis of variance by introducing two different models, Model I and Model II. In Model I, it is assumed that the treatment effects are fixed and that the ε_{ij} are independent random variables with mean 0 and common variance σ^2; and in Model II, it is assumed that the treatment effects and ε_{ij} are independent random variables with mean zero. The $V(\varepsilon_{ij})=\sigma_\varepsilon^2$, and $\mathrm{Var}(\tau_i)=\sigma_\tau^2$. See also Cramp (1946, 1951) for other models.

3.4 See Raktoe, Hedayat, and Federer (1981) for complete and systematic definitions and specification of factorial designs. The treatment here follows their work.

3.5 See Cox (1984) for a broad review on various aspects of interaction. Some interesting open problems are provided at the end of his paper.

3.6–3.8 Follows Seely's (1979) lecture notes and Seely's (1979) technical report on parametrizations and correspondences. As mentioned in the bibliographical notes in Chapter 1, the concept of correspondences was introduced by Seely in order to relate two parametrizations. Apart from this important concept, other reparametrization techniques are discussed in detail in his paper.

3.9 See Tukey (1949) for testing the hypothesis about interactions.

3.10 See Cochran and Cox (1957) sections 3.42–3.43 for various subdivisions of a treatment sun of squares.

3.11 See Scheffé (1959) for a complete discussion on higher-way layouts. Theorem 3.5 and Corollaries 3.1 and 3.2 are from Seely (1979) lecture notes.

REFERENCES

Anderson, V. L. (1970). Restriction errors for linear models (an aid to develop models for designed experiments). *Biometrics* **26**, 255–268.

Anderson, V. L. and McLean, R. A. (1974). *Design of Experiments: A Realistic Approach.* New York: Dekker.

Anderson, V. L. and McLean, R. A. (1974). Restriction errors: Another dimension in teaching experimental statistics. *The American Statistician* **28**, 145–152.

Cochran, W. G. and Cox, G. M. (1957). *Experimental Designs*, 2nd ed. New York: Wiley.

Cox, D. R. (1984). Interaction. *International Statist. Review* **52**, 1–31.

Crump, S. L. (1946). The estimation of variance components in analysis of variance. *Biometrics* **2**, 7–11.

Crump, S. L. (1951). The present status of variance component analysis. *Biometrics* **7**, 1–16.

Eisenhart, C. (1947). The assumptions underlying the analysis of variance. *Biometrics* **3**, 1–21.

Fisher, R. A. (1925). *Statistical Methods for Research Workers.* London: Oliver and Boyd.

Fisher, R. A. (1926). The arrangement of field experiments. *J. Agric.* **33**, 503–513.

Raktoe, B. L., Hedayat, A., and Federer, W. T. (1981). *Factorial Designs.* New York: Wiley.

Scheffé, H. (1959). *Analysis of Variance.* New York: Wiley.

Seely, J. (1979). General linear hypotheses. Lecture notes. Department of Statistics, Oregon State University.

Seely, J. (1979). Parametrizations and correspondences in linear models. Technical Report No. 72, Department of Statistics, Oregon State University.

Tukey, J. W. (1949). One degree of freedom for non-additivity. *Biometrics* **5**, 232–242.

Yates, F. (1935). Complex experiments. *J. Roy. Statist. Soc.*, Suppl. 2, 181–247.

Yates, F. (1937). The design and analysis of factorial experiments. Imperial Bureau of Soil Science, Harpenden, London.

CHAPTER 4

Complete 2^n Factorial Experiments

Within each particle, however small it may be, there lies a sun.

MOULAVI RUMI

4.1 INTRODUCTION

The purpose of this chapter is three fold. We first introduce the complete 2^n factorial experiment and some parametrizations for the full interaction model. Estimability and correspondences between the parametrizations are discussed. Then we define main effects and interactions with respect to the various parametrizations. We introduce some algorithms for calculating their BLUEs and deviation sums of squares and for performing tests of hypotheses concerning them. Finally, we consider simpler models for the 2^n design. Of particular interest is the additive model, which will be heavily used in the following chapters when handling missing observations.

4.2 COMPLETE 2^n FACTORIAL EXPERIMENT

A complete 2^n factorial experiment, as previously defined, is any experiment defined on a factorial domain with n factors, each at two levels, such that at least one experimental unit is assigned to each point in the domain. As before, we assume that the treatments are arranged in a completely randomized design. A *full interaction model* for this experiment is any linear model defined on the 2^n factorial domain such that the response function values of all the points in the domain are functionally independent. The change in the response function due to changing the levels of a subset of the factors is then not independent of the levels of the other factors; that is, there are interactions among the factors.

We will denote by $Y_{ij\ldots wh}$ the hth observation for the treatment $(i, j, \ldots, w)$, $h=1, 2, \ldots, n_{ij\ldots w}$. Keep in mind that the alphabetic ordering $i, j, \ldots, w$ is *not* meant to imply that there are 15 factors. It is only meant to illustrate a notational pattern applicable to any number of factors. In this book we again consider only standard linear models, with $\text{Var}(Y_{ij\ldots wh})=\sigma^2$. Estimability results and correspondences between parametrizations developed here will carry over to more general covariance structures. Estimation and tests of hypotheses in these more general cases can be approached using the general method of best linear unbiased estimation and the likelihood ratio test procedure.

4.3 PARAMETRIZATION AND ANALYSIS

Consider first the *cell mean* parametrization

$$E(Y_{ij\ldots wh})=\phi_{ij\ldots w}$$

where $i, j, \ldots, w=1, 2$; $h=1, 2, \ldots, n_{ij\ldots w}$ or in matrix form

$$E(\mathbf{Y})=\mathbf{U}\boldsymbol{\phi}, \qquad \boldsymbol{\phi}\in\Phi=R^{2^n}. \tag{4.1}$$

Here $\mathbf{Y}$ is an N-dimensional vector with elements $Y_{ij\ldots wh}$ in lexicographic order and $\mathbf{U}$ is a $N\times 2^n$ matrix with ones in the first column of the first $n_{11\ldots 1}$ rows, in the second column of the next $n_{11\ldots 2}$ rows, and so on to ones in the last column of the last $n_{22\ldots 2}$ rows, and with zeros elsewhere. $\phi_{ij\ldots w}$ then represents the expected response for cell (treatment) $(i, j, \ldots, w)$. $\mathbf{U}$ has full column rank and $\boldsymbol{\phi}$ is an estimable parametric vector. The space of estimable linear parametric functions is

$$\bar{\Phi}=\{\boldsymbol{\pi}'\boldsymbol{\phi}\colon \boldsymbol{\pi}\in R(\mathbf{U}')=R^{2^n}\}, \qquad m=\dim\bar{\Phi}=2^n.$$

The BLUE of $\phi_{ij\ldots w}$ is $\hat{\phi}_{ij\ldots w}=\bar{Y}_{ij\ldots w.}$, the average response for treatment $(i, j, \ldots, w)$, the cell mean. The covariance matrix of $\hat{\boldsymbol{\phi}}$ is

$$\text{Cov}(\hat{\boldsymbol{\phi}})=\sigma^2\,\text{diag}(n_{11\ldots 1}^{-1}, n_{11\ldots 2}^{-1}, \ldots, n_{22\ldots 2}^{-1}).$$

The residual sum of squares for this model is

$$\hat{R}=\|\mathbf{Y}-\mathbf{U}\hat{\boldsymbol{\phi}}\|^2=\sum_{i=1}^{2}\sum_{j=1}^{2}\cdots\sum_{w=1}^{2}\sum_{h=1}^{n_{ij\ldots wh}}(Y_{ij\ldots wh}-\bar{Y}_{ij\ldots w.})^2.$$

If the experimental error is normally distributed and $N>2^n$, then $\hat{R}$ has a central chi-square distribution with $N-2^n$ degrees of freedom.

Another popular parametrization is the unconstrained parametrization:

$$E(Y_{ij\ldots wh})=\mu+\alpha_i+\beta_j+(\alpha\beta)_{ij}+\gamma_k+(\alpha\gamma)_{ik}+(\beta\gamma)_{jk} +(\alpha\beta\gamma)_{ijk}+\cdots+(\alpha\beta\gamma\ldots\eta)_{ijk\ldots w} \tag{4.2}$$

or in matrix form,

$$E(\mathbf{Y})=\mathbf{X}\boldsymbol{\theta}, \qquad \boldsymbol{\theta}\in\Theta=R^{3^n},$$

where

$$\boldsymbol{\theta}=[\mu, \alpha_1, \alpha_2, \beta_1, \beta_2, (\alpha\beta)_{11}, (\alpha\beta)_{12}, (\alpha\beta)_{21}, (\alpha\beta)_{22}, \ldots, (\alpha\beta\ldots\eta)_{22\ldots2}]',$$

$\mathbf{X}$ is the appropriate $N\times 3^n$ matrix of zeros and ones, and $\mathbf{Y}$ is as before. This is not a full rank parametrization and $\boldsymbol{\theta}$ is not estimable. One basis for $\bar{\Theta}$ is the set of distinct treatment expectations under (4.2), so $m=2^n$ as under the cell mean parametrization (4.1). Since cell means correspond under different parametrizations, $\hat{R}$ for (4.2) is the same as for (4.1). Actually, m and $\hat{R}$ are invariant under reparametrizations because Ω is. BLUEs and deviation sums of squares for linear hypotheses under this parametrization can be found by expressing the estimable parametric functions as linear combinations of the cell expectations and then using the correspondences

$$\mu+\alpha_i+\beta_j+(\alpha\beta)_{ij}+\cdots+(\alpha\beta\ldots\eta)_{ij\ldots w}\doteq\phi_{ij\ldots w}; \qquad i, j, \ldots, w=1, 2.$$

A full rank parametrization can be obtained from (4.2) by adding the following nonpreestimable constraints:

$$0=\alpha_{.}=\beta_{.}=(\alpha\beta)_{i.}=(\alpha\beta)_{.j}=\gamma_{.}=(\alpha\gamma)_{i.}=(\alpha\gamma)_{.k}=(\beta\gamma)_{j.}=(\beta\gamma)_{.k}=(\alpha\beta\gamma)_{ij.}$$
$$=(\alpha\beta\gamma)_{i.k}=(\alpha\beta\gamma)_{.jk}=\cdots=(\alpha\beta\gamma\ldots\eta)_{.j\ldots w}, \tag{4.3}$$

for $i, j, k, \ldots, w=1, 2$. That is, the sum over any one subscript in each term (excluding μ) in the response function is zero. These constraints can be expressed in matrix form as $\boldsymbol{\Delta}_c'\boldsymbol{\theta}=\mathbf{0}$, where $\boldsymbol{\Delta}_c'$ is the appropriate $n2^{n-1}\times 3^n$ matrix of zeros and ones. There are many redundant constraints. The effect of the constraints is to force each component of $\boldsymbol{\theta}$ to be the negative of any like term with only one subscript different from itself. Thus, knowing the value of one component of $\boldsymbol{\theta}$ uniquely determines the values of all other like terms. For example, $(\alpha\beta)_{11}=-(\alpha\beta)_{12}=(\alpha\beta)_{22}=-(\alpha\beta)_{21}$ and assigning a value to any one of these like terms will determine the values of the others. There are 2^n distinct terms in the response function and 2^n equations

$$\hat{\mu}+\hat{\alpha}_i+\hat{\beta}_j+(\widehat{\alpha\beta})_{ij}+\cdots+(\widehat{\alpha\beta\ldots\eta})_{ij\ldots w}=\hat{\phi}_{ij\ldots w}=\bar{Y}_{ij\ldots w}.$$

that any least squares $\hat{\boldsymbol{\theta}}$ must satisfy. There is a unique $\hat{\boldsymbol{\theta}}^c$ that is a least squares solution for $\boldsymbol{\theta}$ and satisfies $\boldsymbol{\Delta}_c'\hat{\boldsymbol{\theta}}^c=\mathbf{0}$. We see now that the constrained parametrization

$$E(\mathbf{Y})=\mathbf{X}\boldsymbol{\theta}^c, \qquad \boldsymbol{\Delta}_c'\boldsymbol{\theta}^c=\mathbf{0} \tag{4.4}$$

is of full rank. Hence

$$\bar{\Theta}^c=\{\boldsymbol{\lambda}'\boldsymbol{\theta}^c:\boldsymbol{\lambda}\in R(\mathbf{X}', \boldsymbol{\Delta}_c)=R^{2n}\}$$

and $\boldsymbol{\theta}^c$ is an estimable parametric vector. Now

$$\mu^c+\alpha_i^c+\beta_i^c+(\alpha\beta)_{ij}^c+\cdots+(\alpha\beta\ldots\eta)_{ij\ldots w}^c\doteq\phi_{ij\ldots w}, \tag{4.5}$$

and by applying the constraints (4.3) we obtain the correspondences

$$\begin{aligned}
\mu^c &\doteq \sum_{i=1}^{2}\sum_{j=1}^{2}\sum_{k=1}^{2}\cdots\sum_{w=1}^{2}\phi_{ijk\ldots w}/2^n,\\
\alpha_1^c &\doteq \left[\sum_{j=1}^{2}\sum_{k=1}^{2}\cdots\sum_{w=1}^{2}\phi_{1jk\ldots w}-\sum_{j=1}^{2}\sum_{k=1}^{2}\cdots\sum_{w=1}^{2}-\phi_{2jk\ldots w}\right]\Big/2^n,\\
\beta_1^c &\doteq \left[\sum_{i=1}^{2}\sum_{k=1}^{2}\cdots\sum_{w=1}^{2}\phi_{i1k\ldots w}-\sum_{i=1}^{2}\sum_{k=1}^{2}\cdots\phi_{i2k\ldots w}\right]\Big/2^n,\\
(\alpha\beta)_{11}^c &\doteq \left[\sum_{k=1}^{2}\cdots\sum_{w=1}^{2}\phi_{11k\ldots w}-\sum_{k=1}^{2}\cdots\sum_{w=1}^{2}\phi_{12k\ldots w}\right.\\
&\quad\left.+\sum_{k=1}^{2}\cdots\sum_{w=1}^{2}\phi_{22k\ldots w}-\sum_{k=1}^{2}\cdots\sum_{w=1}^{2}\phi_{21k\ldots w}\right]\Big/2^n\\
&\ \ \vdots\\
(\alpha\beta\ldots\eta)_{11\ldots1}^c &\doteq \sum_{i=1}^{2}\sum_{j=1}^{2}\cdots\sum_{w=1}^{2}(-1)^s\phi_{ij\ldots w}/2^n,
\end{aligned} \tag{4.6}$$

where $s=s(i,j,\ldots,w)$ denotes the number of indices $i, j, \ldots, w$ that are equal to 2. To find the correspondence between any term in (4.4) and the terms in (4.1), first sum the $\phi_{ij\ldots w}$ over all indices not appearing in the term for parametrization (4.4), and then add these sums with their signs determined by $(-1)^s$, where s is the number of remaining indices that disagree with the indices in the term for parametrization (4.4).

The BLUEs of the components of $\boldsymbol{\theta}^c$ can be found by replacing $\phi_{ij\ldots w}$ by $\bar{Y}_{ij\ldots w.}$ in the preceding discussion.

The components of $\boldsymbol{\theta}^c$ in parametrization (4.4) form a spanning set for the space of estimable linear parametric functions $\bar{\Theta}^c$. Furthermore, using the correspondences (4.6), the constraints (4.3), and the fact that $\boldsymbol{\phi}$ is unconstrainted in parametrization (4.1), we see that the set of estimable parametric functions $\{\mu^c, \alpha_1^c, \beta_1^c, (\alpha\beta)_{11}^c, \ldots, (\alpha\beta\ldots\eta)_{11\ldots1}^c\}$ forms a basis for $\bar{\Theta}^c$.

μ^c denotes the *mean effect*—the response function averaged over the 2^n domain. The main effect for factor 1 is defined to be the parametric function $2\alpha_1^c=\alpha_1^c-\alpha_2^c$. The main effects for the other $n-1$ factors are defined similarly. The *two-way interaction* between factors 1 and 2 is $2(\alpha\beta)_{11}^c=[(\alpha\beta)_{11}^c-(\alpha\beta)_{12}^c-(\alpha\beta)_{21}^c+(\alpha\beta)_{22}^c]/2$. There are a total of $\binom{n}{2}$ distinct two-way interactions, all defined similarly. In general, there $\binom{n}{k}$ distinct *k-way interactions*, where the k-way interaction between factors $1, 2, \ldots, k$ is $2(\alpha\beta\ldots\delta)_{11\ldots1}^c$. There is only one highest order (n-way) interaction, namely $2(\alpha\beta\ldots\eta)_{11\ldots1}^c$.

The mean effect, the main effects, and the various interactions are defined in terms of parametrization (4.1) using the correspondences (4.6). For future reference, we have

$$\mu^c \doteq \sum_{i=1}^{2} \sum_{j=1}^{2} \sum_{k=1}^{2} \cdots \sum_{w=1}^{2} \phi_{ijk\ldots w}/2^n,$$

$$2\alpha_1^c = \alpha_1^c - \alpha_2^c \doteq \left[\sum_{j=1}^{2} \sum_{k=1}^{2} \cdots \sum_{w=1}^{2} \phi_{1jk\ldots w} - \sum_{j=1}^{2} \sum_{k=1}^{2} \sum_{w=1}^{2} \phi_{2jk\ldots w} \right] \Big/ 2^{n-1},$$

$$2(\alpha\beta)_{11}^c = [(\alpha\beta)_{11}^c - (\alpha\beta)_{12}^c + (\alpha\beta)_{22}^c - (\alpha\beta)_{21}^c]/2$$

$$\doteq \left[\sum_{k=1}^{2} \cdots \sum_{w=1}^{2} \phi_{11k\ldots w} - \sum_{k=1}^{2} \cdots \sum_{w=1}^{2} \phi_{12k\ldots w} + \sum_{k=1}^{2} \cdots \sum_{w=1}^{2} \phi_{22k\ldots w} - \sum_{k=1}^{2} \cdots \sum_{w=1}^{2} \phi_{21k\ldots w} \right] \Big/ 2^{n-1}, \tag{4.7}$$

$$\vdots$$

$$2(\alpha\beta\ldots\eta)_{11\ldots1}^c = \sum_{i=1}^{2} \sum_{j=1}^{2} \sum_{k=1}^{2} \cdots \sum_{w=1}^{2} (-1)^s (\alpha\beta\ldots\eta)_{ijk\ldots w}^c/2^{n-1}$$

$$\doteq \sum_{i=1}^{2} \sum_{j=1}^{2} \sum_{k=1}^{2} \cdots \sum_{w=1}^{2} (-1)^s \phi_{ijk\ldots w}/2^{n-1}.$$

Note that the set consisting of the mean effect, the n main effects, the $\binom{n}{2}$ two-way interactions, ..., and the n-way interaction also forms a basis for $\bar{\Theta}^c$. Excluding μ^c, each function in this set is a *contrast* with respect to parametrization (4.4). That is, each is a linear combination $\Sigma_i \lambda_i \theta_i^c$, where $\Sigma_i \lambda_i = 0$.

The BLUEs for the elements in the basis are easily found by replacing $\phi_{ijk\ldots w}$ by $\bar{Y}_{ijk\ldots w.}$ in the representative equations (4.7). The deviation sum of squares for testing the hypothesis that any one of the basis elements is zero, say H_0: $2(\alpha\beta\ldots\delta)_{11\ldots1}^c = 0$, is then

$$\hat{R}_0 - \hat{R} = [2^n \widehat{(\alpha\beta\ldots\delta)_{11\ldots1}^c}]^2/K, \qquad m - m_0 = 1,$$

where

$$K = \sum_{i=1}^{2} \sum_{j=1}^{2} \cdots \sum_{w=1}^{2} n_{ij\ldots w}^{-1}.$$

The linear hypothesis that all main effects and interactions are zero can be expressed in the form H_0:$\boldsymbol{\Lambda}'\boldsymbol{\theta}^c = \mathbf{0}$, where $\boldsymbol{\Lambda}'$ is a $(2^n - 1) \times 3^n$ matrix satisfying $R(\boldsymbol{\Lambda}) \cap R(\boldsymbol{\Delta}_c) = \{\mathbf{0}\}$. An equivalent linear hypothesis is H_0: $E(\mathbf{Y}) \in R(\mathbf{1}_N)$. Let

$\bar{Y}$ denote the average response over all experimental units. Then

$$\hat{R}_0 = \sum_{i=1}^{2} \sum_{j=1}^{2} \cdots \sum_{w=1}^{2} \sum_{h=1}^{n_{ij\ldots w}} (Y_{ij\ldots wh} - \bar{Y})^2, \qquad m_0 = 1,$$

and

$$\hat{R}_0 - \hat{R} = \sum_{i=1}^{2} \sum_{j=1}^{2} \cdots \sum_{w=1}^{2} n_{ij\ldots w}(\bar{Y}_{ij\ldots w.} - \bar{Y})^2, \qquad m - m_0 = 2^n - 1.$$

In order for this deviation sum of squares to be expressible as the sum of the deviation sums of squares for the individual tests, it is again necessary and sufficient for the 2^n factorial to have equal replication, with

$$n_{ij\ldots w} = r > 0 \qquad i, j, \ldots, w = 1, 2.$$

For the complete 2^n factorial with equal replication, FA(n, 2^n, $r2^n$, r, r, ..., r),

$$\begin{aligned} \hat{\mu}^c &= \bar{Y} = [Y_{\cdot\cdot\,\ldots\,\cdot}]/r2^n, \\ \widehat{2\alpha_1^c} &= [Y_{1\cdot\,\ldots\,\cdot} - Y_{2\cdot\,\ldots\,\cdot}]/r2^{n-1}, \\ 2\widehat{(\alpha\beta)}{}_{11}^c &= [Y_{11\cdot\,\ldots\,\cdot} - Y_{12\cdot\,\ldots\,\cdot} + Y_{22\cdot\,\ldots\,\cdot} - Y_{21\cdot\,\ldots\,\cdot}]/r2^{n-1}, \\ &\vdots \\ 2\widehat{(\alpha\beta\ldots\eta)}{}_{11\ldots1}^c &= \left[\sum_{i=1}^{2}\sum_{j=1}^{2}\cdots\sum_{w=1}^{2}(-1)^s Y_{ij\ldots w\cdot}\right]\Big/ r^{2n-1}. \end{aligned} \tag{4.8}$$

The bracketed quantities in (4.8) are called the *effect totals*. The deviation sum of squares for testing H_0: $2(\alpha\beta\ldots\delta)_{11\ldots1}^c = 0$ is then

$$\hat{R}_0 - \hat{R} = \frac{[2^n \widehat{(\alpha\beta\ldots\delta)}{}_{11\ldots1}^c]^2}{2^n/r} = \frac{[\ ]^2}{r2^n}$$

where [ ] denotes the effect total for $2(\alpha\beta\ldots\delta)_{11\ldots1}^c$.

4.4 TESTING EFFECTS AND INTERACTIONS

When $N > 2^n$ in the complete 2^n factorial experiment, $\hat{R} > 0$ with probability one and we can test the hypothesis that a particular effect or interaction is zero using the ratio

$$\hat{F} = \frac{\hat{R}_0 - \hat{R}}{\hat{R}/(N - 2^n)}.$$

The calculation of $\hat{R}_0 - \hat{R}$ and $\hat{R}$ is discussed earlier in this chapter. When the experiment has equal replication the calculation of $\hat{R}_0 - \hat{R}$ and $\hat{R}$ is simplified and the deviation sum of squares $\hat{R}_0 - \hat{R}$ for testing that one effect or interaction is zero is independent of that for testing that some other effect or interaction is zero. Furthermore, the deviation sum of squares for the hypothesis that all main effects and interactions are zero is the sum of these $2^n - 1$ sums of squares.

This method of hypothesis testing fails when $N = 2^n$. There are three common ways to proceed in this case. If a prior estimate $\tilde{\sigma}^2$ of σ^2 is available, for example, from an earlier experiment, we can replace the denominator in $\hat{F}$ by $\tilde{\sigma}^2$.

The most common approach is to assume certain of the (usually higher order) interactions are zero and then pool their corresponding deviation sums of squares to obtain an estimate of σ^2. The pooled sum of squares replaces $\hat{R}$ and the number pooled replaces $N - 2^n$ in $\hat{F}$.

A third approach is a graphical method (see bibliographical notes) for deciding which of the effects and interactions are statistically significant when there is no prior estimate of error. The method can also be used for investigating the validity of the model assumptions. However, the method works best when only a few of the effects and interactions are significant.

4.5 THE ADDITIVE MODEL

The full interaction model was defined to be one such that the response function values of all points in the domain are functionally independent. A simpler model that is very popular and useful for factorials is the *additive model*. An additive model defined on a factorial domain is one for which each factor level adds a constant to the part of the response function that does not depend upon that factor, regardless of the levels of the other factors. That is, the contribution of one factor at a particular level to the response is the same for all levels of the other factors. An additive model is therefore also called a *model without interaction.*

A constrained full rank parametrization for the additive 2^n model is

$$E(Y_{ijk\ldots wh}) = \mu^a + \alpha_i^a + \beta_j^a + \gamma_k^a + \cdots + \eta_w^a,$$
$$0 = \alpha_\cdot^a = \beta_\cdot^a = \gamma_\cdot^a = \cdots = \eta_\cdot^a, \tag{4.9}$$

or in matrix form,

$$E(\mathbf{Y}) = \mathbf{W}\boldsymbol{\theta}^a, \qquad \boldsymbol{\theta}^a \in \Theta^a = R^{2n+1}, \qquad \boldsymbol{\Delta}_a'\boldsymbol{\theta}^a = \mathbf{0}, \tag{4.10}$$

where $\boldsymbol{\theta}^a = (\mu^a, \alpha_1^a, \alpha_2^a, \beta_1^a, \beta_2^a, \ldots, \eta_1^a, \eta_2^a)'$, $\mathbf{W}$ is the appropriate $N \times (2n+1)$ matrix of zeros and ones, $\boldsymbol{\Delta}_a'$ is the constraint matrix, and $\boldsymbol{\Delta}$ is as before. Since parametrization (4.10) has full rank, $\boldsymbol{\theta}^a$ is an estimable parametric vector. A basis for $\bar{\Theta}^a$ is $\{\mu^a, 2\alpha_1^a, 2\beta_1^a, \ldots, 2\eta_1^a\}$, so $m_a = \dim \bar{\Theta}^a = n+1$. μ^a is called the mean effect and $2\alpha_1^a, 2\beta_1^a, \ldots, 2\eta_1^a$ are called the main effects for the additive model. The BLUEs for the basis elements are

$$\hat{\mu}^a = \sum_{i=1}^{2}\sum_{j=1}^{2}\sum_{k=1}^{2}\cdots\sum_{w=1}^{2}\bar{Y}_{ijk\ldots w\cdot},$$

$$2\hat{\alpha}_1^a = \left[\sum_{j=1}^{2}\sum_{k=1}^{2}\cdots\sum_{w=1}^{2}\bar{Y}_{1jk\ldots w\cdot} - \sum_{j=1}^{2}\sum_{k=1}^{2}\cdots\sum_{w=1}^{2}\bar{Y}_{2jk\ldots w\cdot}\right]\Big/2^{n-1},$$

$$2\hat{\beta}_1^a = \left[\sum_{i=1}^{2}\sum_{k=1}^{2}\cdots\sum_{w=1}^{2}\bar{Y}_{i1k\ldots w\cdot} - \sum_{i=1}^{2}\sum_{k=1}^{2}\cdots\sum_{w=1}^{2}\bar{Y}_{i2k\ldots w\cdot}\right]\Big/2^{n-1},$$

$$\vdots$$

$$2\hat{\eta}_1^a = \left[\sum_{i=1}^{2}\sum_{j=1}^{2}\cdots\sum_{v=1}^{2}\bar{Y}_{ij\ldots v1\cdot} - \sum_{i=1}^{2}\sum_{j=1}^{2}\cdots\sum_{v=1}^{2}\bar{Y}_{ij\ldots v2\cdot}\right]\Big/2^{n-1}.$$

Note that these are the same BLUEs as were found for $\mu^c, 2\alpha_1^c, 2\beta_1^c, \ldots, 2\eta_1^c$ in the constrained parametrization (4.4) for the full interaction model. The deviation sum of squares for testing that one of the above effects is zero is also the same as for the full interaction model. The residual sum of squares for fitting the additive model is

$$\hat{R}_a = \|\mathbf{Y} - \mathbf{W}\hat{\boldsymbol{\theta}}^a\|^2, \quad \text{with } N - m_a = N - n - 1 \text{ degrees of freedom.}$$

The deviation sum of squares for testing the null hypothesis that all main effects are zero against the alternative that the additive model holds is $\hat{R}_0 - \hat{R}_a$ with $m_a - m_0 = n$ degrees of freedom, where $\hat{R}_0$ and m_0 were derived earlier in the chapter.

The deviation sum of squares for testing the hypothesis that the additive model holds against the alternative that the full interaction model is true is $\hat{R} - \hat{R}_a$ with the $m - m_a = 2^n - n - 1$ degrees of freedom.

This multidegree of freedom deviation sum of squares can be partitioned into the associated single degree of freedom interaction sums of squares when the complete 2^n factorial has equal replication. Furthermore, when the additive model holds, the BLUEs for the interactions estimate zero and their associated deviation sums of squares are included in $\hat{R}_a$. When $N = 2^n$, these sums of squares sum to $\hat{R}_a$. Yates' method can then be used to calculate the BLUEs and sums of squares. The mean and main effect BLUEs and sums of squares are the same and the interaction sums of squares are added to get $\hat{R}_a$ (see bibliographical notes for Yates' method).

One can, with some effort, generalize the results of this chapter from 2^n factorial experiments to general n-way factorial experiments.

BIBLIOGRAPHICAL NOTES

4.1 See Yates (1937) for statistical analysis of 2^n factorial experiments. This document contains a great deal of information on factorial experiments.

4.2 See Seely (1979) lecture notes for estimation and tests of hypotheses in more general cases.

4.3 The BLUEs and deviation sums of squares can be computed individually using Eqs. (4.8). However, when most or all of these quantities are needed, it is more convenient to compute them all simultaneously. Yates' (1937) method for computing the factorial effect totals does this without explicitly using the 2^n different equations in (4.8).

The first step in Yates' method is to construct a table with $n+2$ columns and 2^n rows. In the first column, list the treatments in the systematic order $111\ldots1$, $211\ldots1$, $121\ldots1$, $221\ldots1$, $112\ldots1$, ..., $222\ldots2$. These treatments can also be specified by (1), a, b, ab, c, ..., $ab\ldots g$, where (1) denotes the first treatment above and the absence or presence of a letter in the other terms indicates that the corresponding factor appears at level one or two, respectively. The corresponding treatment totals are listed in the second column. The succeeding columns are labeled (1), (2), ..., (n):

Treatment	Treatment Total	(1)	(2)	...	(n)
(1)	$Y_{111\ldots1.}$				
a	$Y_{211\ldots1.}$	$(k)_j=(k-1)_{2j}+(k-1)_{2j-1}$			
b	$Y_{121\ldots1.}$				
ab	$Y_{221\ldots1.}$				
c	$Y_{112\ldots1.}$	$(k)_{2^{n-1}+j}=(k-1)_{2j}-(k-1)_{2j-1}$			
$\vdots$	$\vdots$				
$ab\ldots g$	$Y_{222\ldots2.}$				

The next step is to perform the following procedure on each of the last n columns. Let $(k)_j$ denote the jth entry of column (k), for $j=1, 2, \ldots, 2^n$ and $k=0, 1, 2, \ldots, n$. Here column (0) refers to the column of treatment totals. Then for $k=1, 2, \ldots, n$, let

$$(k)_j=(k-1)_{2j}+(k-1)_{2j-1}$$

$$(k)_{2^{n-1}+j}=(k-1)_{2j}-(k-1)_{2j-1}$$

for $j=1, 2, \ldots, 2^{n-1}$. That is, the first half of column (k) consists of sums of pairs of entries in column ($k-1$), and the second half consists of differences of pairs of entries in column ($k-1$).

Column (n) will then contain the effect totals for μ^c, $2\alpha_1^c$, $2\beta_1^c$, $2(\alpha\beta)_{11}^c$, $2\gamma_1^c, \ldots, 2(\alpha\beta \ldots \eta)_{11\ldots1}^c$. The BLUEs are obtained by dividing the effect totals by $r2^n$ for the mean effect and by $r2^{n-1}$ for the main effects and interactions. The deviation sum of squares for any effect is the square of the effect total divided by $r2^n$. The residual sum of squares is equal to the total sum of squares, $\|\mathbf{Y}\|^2$, minus these 2^n deviation sums of squares. See also Hunter (1966) and Box, Hunter, and Hunter (1978).

4.4 Daniel (1959) presented a graphical method for deciding which of the effects and interactions are statistically significant when there is no prior estimate of error. In this method the effects are plotted on normal probability paper. The reader is urged to study Daniel's book (1976) for a good discussion and criticism of factorial experiments in general. Many interesting methods for diagnostic checking, including the plotting of residuals on normal probability paper, are provided in his text. This is an exceptionally excellent book on the analysis of 2^n factorial experiments. See also Birbaum (1959).

REFERENCES

Birbaum, A. (1959). On the analysis of factorial experiments without replications. *Technometrics* **1**, 343–357.

Box, G. E. P., Hunter, W. G., and Hunter, J. S. (1978). *Statistics for Experimenters.* New York: Wiley.

Daniel, C. (1959). Use of half-normal plot in interpreting two-level experiments. *Technometrics* **1**, 149.

Daniel, C. (1976). *Applications of Statistics in Industrial Experimentation.* New York: Wiley.

Hunter, J. S. (1966). Inverse Yates algorithm. *Technometrics* **8**, 171.

Seely, J. (1979). General linear hypotheses. Lecture notes. Oregon State University.

Yates, F. (1937). The design and analysis of factorial experiments. Bulletin 35, Imperial Bureau of Soil Science, Harpenden, London.

CHAPTER 5

Additive Two-Way Classification with Missing Observations: Estimability and Analysis

Whether at Naishapur or Babylon,
Whether the Cup with sweet or bitter run,
The Wine of Life keeps oozing drop by drop,
The leaves of Life keep falling one by one.

KHAYYAM NAISHAPURI-RUBÁIYÁT

5.1 INTRODUCTION

In Chapter 3 we considered the complete two-way classification experiment (randomized complete block design, and two-factor experiment) using additive two-way classification model. In that experiment we had exactly one observation in the cell i, j. In the terminology of Chapter 3, such experiments may be called *complete balanced*, *single complete replicate*, or *equal numbers* two-way classifications.

In almost all experiments, the investigator would like to reduce the number of observations required for a complete factorial or block design and still obtain the desired information. Therefore, it is desirable and sometimes necessary (due to physical restrictions) to adopt a design in which there are no observations in some of the cells. For example, in a randomized block design, it may happen that the blocks available are not large enough to accommodate one unit for each treatment, and in a two-way factorial, it may not be possible to accommodate all treatment combinations in the experiment.

In this case we have what is known in literature as *incomplete*, *unequal numbers*, *unbalanced*, *missing data*, or *messy data*. If the subset of the cells in which the observations are taken is selected according to some *specified pattern*, we have *balanced incomplete* two-way classification. An example of such a classification is a balanced incomplete block design (BIB) in which the block size, though equal, is smaller than the total number of treatments to be compared (each of the blocks does not contain a complete set of treatments). In such a case the experimenter has a precise idea about how the experiment should be analyzed. The traditional analysis of variance techniques are applicable, and they can be found in almost every standard textbook on designs of experiments.

However, it may happen by accident or some other reasons that some of the observations (cells) are missing, cannot be collected, or in any other way are not obtainable in a balanced (complete or incomplete) two-way classification. For example, crops are destroyed in some plots, there is failing to record some data, gross errors occur in recording, a patient withdraws from the treatment, one or more animals are dying in the course of experiment, or some subjects do not take all treatments as planned. Missing data of this nature occur quite frequently.

In such situations the designed experiment is no longer balanced and loses its symmetry, and in some situations, when enough observations are *arbitrarily* missing, not all of the usual parametric functions are estimable. Such a situation, unless discovered, will lead to incorrect assignment of degrees of freedom for certain sums of squares, to singular matrices that are to be inverted in a computer program, and most importantly, will make contrasts involved in the determination of certain sums of squares to be not known precisely.

One possible solution to the problem is to repeat the experiment under similar conditions and obtain new values for the missing observations. However, such a solution, though ideal, may not be feasible, if not impossible economically (with regards to time and money) and physically.

The most common approach to this problem when there are few empty cells is to apply the analysis for a complete experiment using *calculated values* in place of the missing observations. Here one calculated value is placed into each empty cell. The values are chosen so that residual sum of squares is minimized. The residual degrees of freedom is reduced by one for each empty cell in the original data and the usual F tests are performed. Missing data formula are available in many experimental design texts for special situations. But even in those cases the analysis is approximate (see bibliographical notes for this approach).

A second approach to the problem of missing observations in designed experiments is the exact least squares [regression on dummy (0, 1) variables or method of fitting constants] approach by writing a mathematical model for all observations that are present. The least squares normal equations are then constructed for estimation of parameters in the model. Because the equations corresponding to missing observations are in fact "missing," the system of normal equations loses its symmetry (as compared to when all observations are present) and consequently the solutions require computation of *generalized inverses*. We consider this approach in Chapter 8.

An alternative approach to the problem is to employ the *incidence matrix* **N** as opposed to the design matrix **X**. The main purpose of this chapter is to present general and rather complete results for estimability considerations in an additive two-way classification model which are easily programmed for electronic computers. An algorithm, the *R* process, is described for determining which cell expectations are estimable. Methods are given, based upon the completion of the *R* process: for determining a basis for the estimable functions involving only one effect; for determining ranks of matrices pertinent to considerations on degrees of freedom; and for determining which portions of the design are connected. The results are somewhat novel in that only integer operations and the incidence matrix **N** are utilized. In fact, one needs only the facility to distinguish between zeros and nonzeros in the incidence matrix to obtain all the results. A second feature about the *R* process is that it would seem to be a more efficient algorithm for determining which cells are estimable than the search procedures one is led to by directly applying Bose's definition of connectedness, as illustrated by some authors.

A secondary purpose of this chapter is to incorporate the results obtained by the general theory of partitioned linear models to illustrate a way to obtain the analysis for a set of data following an additive two-way classification model. A procedure is outlined in this chapter for obtaining the usual sums of squares and for obtaining a solution for the normal equations as well as the covariance matrix associated with the solution obtained. Attention is paid to ensuring that certain matrices are invertible and to discussing various restrictions which may, if desired, be imposed upon the solution to the normal equations.

We begin our discussion by first briefly considering the construction of BIB designs in which the positions of the occupied cells are prespecified. Later, we consider the general two-way classification (in which blocks are another factor of interest) with completely arbitrary patterns.

5.2 BALANCED INCOMPLETE BLOCK DESIGNS

Complete block designs require that each treatment be assigned to each block, where a block is a collection of experimental units. Suppose there are a treatments. In each block we require at least a experimental units. If the design has b blocks we require in all ab experimental units. But in many practical situations, for various reasons, the experimenter may not be able to choose ab experimental units grouped into b blocks for his experiment. So naturally there is a need for minimizing the number of experimental units required for the experiment, subject to restrictions on the precision of the estimates obtained from the analysis of the experimental data. Balanced incomplete block designs were introduced with this problem in mind.

Let $V=\{1, 2, \ldots, a\}$ be the set of treatments and let $a \sum k$ be the set of all distinct subsets of size k based on V. Let aCk denote the cardinality of $a \sum k$. Let b denote the total number of blocks in the design.

A balanced incomplete block design, d, with parameters a, b, r, k and λ, denoted by BIB (a, b, r, k, λ) is a collection of b elements of $a \sum k$ (not necessarily distinct), called blocks, with these properties: (1) each element of V occurs in exactly r blocks, and (2) each pair of distinct elements of V appears in exactly λ blocks.

Thus a *BIB* design, *BIB* (a, b, r, k, λ), is a combinatorial arrangement of a treatments in b blocks, containing k experimental units in each, and these a treatments occur in such a way that each treatment does not occur more than once in any block, each treatment occurs on r experimental units, and each pair of treatments occurs λ times.

It is necessary for a, b, k, r, λ to satisfy the following relations.

1 $bk=ar$. The reason is that each treatment appears r times in the design and k treatments appear in each block. Thus both of them give the total number of experimental units.

2 $\lambda(a-1)=r(k-1)$. Since we have $bk(k-1)=\lambda a(a-1)$ and from (1), $bk=ar$,

$$ar(k-1)=\lambda a(a-1) \quad \text{or} \quad \lambda(a-1)=r(k-1).$$

3 From (1) and (2), we have $\lambda(a-1)\equiv 0 \bmod(k-1)$ and $\lambda a(a-1)\equiv 0 \bmod[k(k-1)]$.

4 $b\geqslant a$. This inequality, due to Fisher, can be shown by noticing that the rank of the treatment-block incidence matrix $\mathbf{N}$ associated with an incomplete block design, where $\mathbf{N}$ has a rows and b columns, and $\mathbf{N}=((n_{ij}))$ is such that $n_{ij}=1$ if the ith treatment appears in the jth block, $n_{ij}=0$ otherwise. Here $\mathbf{N}$ has rank a. Now $a\leqslant \min(a, b)$. Hence $b\geqslant a$.

These conditions can be shown to be not sufficient for the existence of a *BIB* (a, b, r, k, λ). Therefore we have to find ways to construct such designs. Classically such designs are obtained by using results from orthogonal Latin squares, finite geometrics, and difference sets, among other possibilities.

For any pair of integers a, k $(k<a)$, a BIB design can be obtained by taking all possible combinations of k out of a treatments. For these designs,

$$b=\binom{a}{k}, \qquad r=\binom{a-1}{k-1}, \qquad \lambda=\binom{a-2}{k-2}.$$

EXAMPLE 5.1 A design with $a=4$, $b=6$, $r=3$, $k=2$, and $\lambda=1$ is shown in Table 5.1. Here we have four treatments (1, 2, 3, 4), six blocks (I, II, III, IV, V, VI), of the size $k=2$ only, which is too small to accommodate all treatments simultaneously.

Table 5.1 A BIB Design

Block					
I	II	III	IV	V	VI
1	3	1	2	1	2
2	4	3	4	4	3

Each pair of treatments occurs together within a block $\lambda=1$ time. Alternatively, we can set out Table 5.1 as Table 5.2.

Table 5.2 Alternative Presentation of Table 5.1

		Block					
		I	II	III	IV	V	VI
	1	1	0	1	0	1	0
	2	1	0	0	1	0	1
$\mathbf{N}=$ treatment	3	0	1	1	0	0	1
	4	0	1	0	1	1	0

Table 5.2 is just the incidence matrix $\mathbf{N}=(n_{ij})$ with four rows and six columns; n_{ij} is the number of times that the ith treatment occurs in the jth block. For a BIB design, n_{ij} is either one or zero.

Notice that we have not explicitly restricted b blocks, should all be distinct. Some of the blocks corresponding to certain elements of $a \sum k$ may occur more than once. BIB designs in which some blocks are repeated are called BIB designs with *repeated blocks.* Such designs are of practical significance as they allow us to restrict certain treatment combinations being excluded from experiment, for various considerations. The analysis of BIB designs is a special case of the additive two-way classification designs with arbitrary patterns, which we shall consider in the following sections.

5.3 ADDITIVE TWO-WAY CLASSIFICATION WITH ARBITRARY PATTERNS

In this chapter it is assumed that $\{Y_{ijk}\}$ is a collection of independent and normally distributed random variables each having a common unknown variance σ^2 and each having an expectation of the form

$$E(Y_{ijk}) = \mu + \alpha_i + \beta_j, \tag{5.1}$$

where $i=1,\ldots,a; j=1,\ldots,b$; and $k=1,\ldots,n_{ij}$ with the usual interpretation that when $n_{ij}=0$ no random variables with the first two subscripts i, j occur in the collection. Note that no restrictions are imposed upon the unknown parameters occurring in the above expectation. It is further assumed, and without loss in generality, that

$$\begin{aligned} n_{i.} &= \sum_j n_{ij} \neq 0 \quad \text{for } i=1,\ldots,a, \\ n_{.j} &= \sum_i n_{ij} \neq 0 \quad \text{for } j=1,\ldots,b. \end{aligned} \tag{5.2}$$

Thus, we are assuming a fixed effects two-way classification model without interaction; and when the data pattern is viewed in the form of a two-way table we suppose that there are n_{ij} observations in cell (i, j) and that each row, as well as each column, has at least one observation.

Associated with the two-way classification model just described, the incidence matrix $\mathbf{N}$ is defined in the usual fashion, that is, $\mathbf{N}$ denotes the $a \times b$ matrix consisting of the n_{ij}'s, and N will denote the total number of random variables in the collection $\{Y_{ijk}\}$. Further, let $\mathbf{Y}$ denote the $N \times 1$ random vector $(Y_{111}, \ldots, Y_{11n_{11}}, Y_{121}, \ldots)'$ and let $\mathbf{1}$, $\mathbf{A}$, $\mathbf{B}$, and $\mathbf{X} = (\mathbf{1}, \mathbf{A}, \mathbf{B})$ denote matrices defined so that

$$E(\mathbf{Y}) = \mathbf{1}\mu + \mathbf{A}\boldsymbol{\alpha} + \mathbf{B}\boldsymbol{\beta} = \mathbf{X}\boldsymbol{\theta}, \qquad \boldsymbol{\theta} \in R^P \tag{5.3}$$

where $\boldsymbol{\alpha}$ denotes the $a \times 1$ vector consisting of the α_i parameters, $\boldsymbol{\beta}$ denotes the $b \times 1$ vector consisting of the β_j parameters, and $\boldsymbol{\theta}$ denotes a $(1+a+b) \times 1$ vector consisting of μ, $\boldsymbol{\alpha}$, $\boldsymbol{\beta}$ and $P = 1+a+b$. Also, let $r(\mathbf{X}) = m$ denotes the rank of $\mathbf{X}$ and observe that $\mathbf{N} = \mathbf{A}'\mathbf{B}$. For a set S, $\#(S)$ denotes the number of elements in the set. Other notations will be introduced as needed.

EXAMPLE 5.2 Consider a 2^2 factorial experiment with the following incidence matrix **N**, and the observation matrix as follows:

Incidence Matrix

$$\mathbf{N}=\begin{array}{c|cc} & 1 & 2 \\ \hline 1 & 1 & 0 \\ 2 & 2 & 3 \end{array}$$

Observation Matrix

	1	2
1	Y_{111}	—
2	Y_{211}, Y_{212}	Y_{221}, Y_{222}, Y_{223}

Here, we have an example of an unbalanced two-way classification. An additive two-way classification model that corresponds to this experiment is given by the equation

$$E(Y_{ijk})=\mu+\alpha_i+\beta_j, \qquad i=1,2, \quad j=1,2, \quad k=1,\ldots,n_{ij}.$$

In matrix notation the model for the observations is written as

$$E\begin{bmatrix} Y_{111} \\ Y_{211} \\ Y_{212} \\ Y_{221} \\ Y_{222} \\ Y_{223} \end{bmatrix}=\begin{bmatrix} 1 & 1 & 0 & 1 & 0 \\ 1 & 0 & 1 & 1 & 0 \\ 1 & 0 & 1 & 1 & 0 \\ 1 & 0 & 1 & 0 & 1 \\ 1 & 0 & 1 & 0 & 1 \\ 1 & 0 & 1 & 0 & 1 \end{bmatrix}\begin{bmatrix} \mu \\ \alpha_1 \\ \alpha_2 \\ \beta_1 \\ \beta_2 \end{bmatrix}=\mathbf{X}\boldsymbol{\theta}.$$

Here

$$\mathbf{A}=\begin{bmatrix} 1 & 0 \\ 0 & 1 \\ 0 & 1 \\ 0 & 1 \\ 0 & 1 \\ 0 & 1 \end{bmatrix}, \quad \mathbf{B}=\begin{bmatrix} 1 & 0 \\ 1 & 0 \\ 1 & 0 \\ 0 & 1 \\ 0 & 1 \\ 0 & 1 \end{bmatrix}, \quad \text{and} \quad \mathbf{N}=\mathbf{A}'\mathbf{B}=\begin{bmatrix} 1 & 0 \\ 2 & 3 \end{bmatrix}.$$

In the incidence matrix, we see that three cells are occupied and one cell is missing. In terms of the cell expectation the entries of the incidence matrix can be written as

	1	2
1	$\mu+\alpha_1+\beta_1$	0
2	$\mu+\alpha_2+\beta_1$ $\mu+\alpha_2+\beta_1$	$\mu+\alpha_2+\beta_2$ $\mu+\alpha_2+\beta_2$ $\mu+\alpha_2+\beta_2$

For an additive two-way classification model with an arbitrary $a\times b$ incidence matrix **N**, level i of factor 1 and level j of factor 2 are said to be *associated* if $n_{ij}>0$. Two levels of factor 1, two levels of factor 2, or a level of factor 1 and a level of factor 2 are said to be *connected* if it is possible to pass from one to the other by means of a chain consisting alternatively of levels of factor 1 and levels of factor 2 such that any two adjacent members of the chain are associated. A design (or a portion of a design) is said to be a *connected design* (or a connected portion of a design) if every level in the design (or portion of a design) is connected to every other.

The design matrix **X** associated with an incidence matrix **N** has maximal rank, $r(\mathbf{X})=a+b-1$, if and only if the design is connected. We shall see that the concept of connectedness is important in characterizing estimability and in estimation for the additive two-way classification model with missing observations.

5.4 ESTIMABILITY

A linear combination of the parameters $\mu, \alpha_1, \ldots, \alpha_a, \beta_1, \ldots, \beta_b$ is a linear parametric function. Such a function is estimable if it can be expressed as a linear combination of the expectations $\mu+\alpha_i+\beta_j$ of those cells for which observations are available.

A procedure is described below for determining exactly which cell expectations are estimable. That is, even though a particular n_{ij} may be zero, it is possible that the cell expectation $\mu+\alpha_i+\beta_j$ is estimable. Additionally, a procedure is given for obtaining a basis for the vector space $\bar{\Theta}$ of estimable linear parametric functions. Both these procedures use only the incidence matrix **N** are combinatorial in nature, so that if programmed for a computer only integer operations are required.

The *R process* is a procedure applied to the incidence matrix **N** to obtain a matrix **M**, called the final matrix, which determines what cell expectations are estimable. The R process is defined by the following steps:

$$\left.\begin{array}{ll} 1 & \text{Let } \mathbf{M} \text{ be the } a\times b \text{ zero matrix.} \\ 2 & \text{For each pair } i, j, \text{ if } n_{ij}\neq 0 \text{ set } m_{ij}=1. \\ 3 & \text{For each pair } i, j, \text{ if there exist } k, \text{ and } l \text{ such that } m_{il}=m_{kl}=m_{kj}=1, \text{ then set } m_{ij}=1. \\ & \text{(Pictorially, we add the fourth corner whenever three corners of a rectangle appear in the matrix.)} \\ 4 & \text{Continue step 3, using the new nonzero } m_{ij}\text{'s as corners of new rectangles, until no more entries can be changed.} \end{array}\right\} \tag{5.4}$$

Observe that the final matrix $\mathbf{M}$ is a matrix of the same dimensions as the incidence matrix $\mathbf{N}$. Also, note that if all of the n_{ij}'s are nonzero, then the final matrix may immediately be formed, that is, $\mathbf{M}$ is an $a \times b$ matrix consisting of ones only.

It is convenient at this point to partition the row and column indices of the final matrix $\mathbf{M}$. Let J_1 denote the column indices of $\mathbf{M}$ for which there is a one in the first row. J_1 is nonempty because of our assumptions (5.2). Remembering that we are dealing with the final matrix, observe that the columns of $\mathbf{M}$ corresponding to the indices in J_1 have ones in precisely the same rows and that no other column has a one in any of these rows. Let $j \in J_1$ and let I_1 denote the row indices of $\mathbf{M}$ for which there is a one in column j. Note that $1 \in I_1$ and that any $j \in J_1$ will lead to the same set of indices for the set I_1. Choose (if possible) an $i' \notin I_1$ and let J_2 denote the column indices of $\mathbf{M}$ for which there is a one in row i'. Let I_2 denote the row indices of $\mathbf{M}$ for which there is a one in column j where j may be selected arbitrarily from J_2. Continue in this fashion until all row indices of $\mathbf{M}$ are exhausted. This will lead to disjoint unions

$$\{1, \ldots, a\} = \bigcup_{k=1}^{s} I_k \quad \text{and} \quad \{1, \ldots, b\} = \bigcup_{k=1}^{s} J_k$$

such that $m_{ij}=1$ if and only if there is some k $(1 \leqslant k \leqslant s)$ such that $i \in I_k$ and $j \in J_k$. The pairs $(I_1, J_1), \ldots, (I_s, J_s)$ describe the s *connected portions* of the design. These s connected portions are easily visualized in terms of the final matrix $\mathbf{M}$ after rearranging the rows and columns so that the rows indexed by I_1 and the columns indexed by J_1 occur first, then the rows corresponding to I_2 and the columns corresponding to J_2 occur next, and so on. Such a rearrangement of rows and columns leads to the following matrix:

$$\mathbf{M}=\begin{bmatrix} \mathbf{M}_1 & \mathbf{0} & \cdots & \mathbf{0} \\ \mathbf{0} & \mathbf{M}_2 & \cdots & \mathbf{0} \\ \vdots & \vdots & & \vdots \\ \mathbf{0} & \mathbf{0} & \cdots & \mathbf{M}_s \end{bmatrix} \tag{5.5}$$

where $\mathbf{M}_k$ is an $a_k \times b_k$ matrix of ones with $a_k = \#(I_k)$ and $b_k = \#(J_k)$ for each $k=1, \ldots, s$.

THEOREM 5.1 The parametric function $\mu+\alpha_i+\beta_j$ is estimable if and only if $m_{ij}=1$.

Proof Suppose $m_{ij}=1$ and consider the R process. Of course, at step 2, $m_{ij}=1$ means $n_{ij}\neq 0$ so that $\mu+\alpha_i+\beta_j$ is clearly estimable. Whenever one sets $m_{ij}=1$ in some iteration of step 3 it is because there exist k and h such that $m_{ih}=m_{kh}=m_{kj}=1$. Applying an induction argument on the number of iterations of step 3, we can assume $\mu+\alpha_i+\beta_h$, $\mu+\alpha_k+\beta_h$, and $\mu+\alpha_k+\beta_j$ are estimable; hence,

$$\mu+\alpha_i+\beta_j=(\mu+\alpha_i+\beta_h)-(\mu+\alpha_k+\beta_h)+(\mu+\alpha_k+\beta_j)$$

is estimable.

Conversely, suppose $\mu+\alpha_i+\beta_j$ is estimable. Let $k(1\leqslant k\leqslant s)$ be such that $i\in I_k$. Also note that $\{\mu+\alpha_u+\beta_t\colon m_{ut}=1\}$ is a spanning set for the estimable linear parametric functions because it contains the set $\{\mu+\alpha_i+\beta_j\colon n_{ij}>0\}$ which, as seen above, spans it. Then we may write $\mu+\alpha_i+\beta_j$ as $\sum_{h=1}^{a}\sum_{l=1}^{b} c_{hl} \times(\mu+\alpha_h+\beta_l)$ where $c_{hl}\neq 0$ only if $m_{hl}=1$. In this representation the coefficient of α_i is $1=\sum_{l=1}^{b} c_{il}$. But $\sum_{l=1}^{b} c_{il}=\sum_{t=1}^{s}\sum_{l\in J_t} c_{il}=\sum_{l\in J_k} c_{il}$ and for $h\in I_k$ $(h\neq i)$, $0=\sum_{l=1}^{b} c_{hl}=\sum_{t=1}^{s}\sum_{l\in J_t} c_{hl}=\sum_{l\in J_k} c_{hl}$. Thus,

$$1=\sum_{h\in I_k}\left(\sum_{l\in J_k} c_{hl}\right)=\sum_{l\in J_k}\left(\sum_{h\in I_k} c_{hl}\right)=\sum_{l\in J_k}\left(\sum_{t=1}^{s}\sum_{h\in I_t} c_{hl}\right)=\sum_{l\in J_k}\left(\sum_{h=1}^{a} c_{hl}\right).$$

Since $\sum_{h=1}^{a} c_{hl}$ is the coefficient of β_l in the above representation for $\mu+\alpha_i+\beta_j$ and since this coefficient is 1 when $l=j$ and 0 otherwise, we must have $j\in J_k$. Hence, $i\in I_k$ and $j\in J_k$ so that $m_{ij}=1$. □

REMARK An immediate consequence of Theorem 5.1 is that it provides another characterization when $\mathbf{X}$ is of maximal rank. That is, we can say that $\mathbf{X}$ is of maximal rank if and only if $\mathbf{M}$ has *no zero entries*. Therefore, the concept of connectedness reduces down to saying that the final matrix $\mathbf{M}$ has no zero entries.

EXAMPLE 5.3 Consider Example 5.1 under the usual additive two-way model with an incidence matrix $\mathbf{N}$ whose nonzero entries occur in the cells occupied by ones in the following matrix:

N =

	I	II	III	IV	V	VI
1	1	0	1	0	1	0
2	1	0	0	1	0	1
3	0	1	1	0	0	1
4	0	1	0	1	1	0

By applying the *R* process we obtain the final matrix **M** as

	I	II	III	IV	V	VI
1	1	1*	1	1*	1	1*
2	1	1*	1*	1	1*	1
3	1*	1	1	1*	1*	1
4	1*	1	1*	1	1	1*

The 1* is in the final matrix are in those cells that become occupied after applying the *R* process. Notice that **M** has *no nonzero* entries, and as a result, we know that **X** is of maximal rank, that is, connected, as we knew beforehand.

Let us now consider the same example to show an interesting consequence of Theorem 5.1.

EXAMPLE 5.4 Consider Example 5.1 again with the cells (1, 1), (2, 6), (3, 2), and (4, 5) being empty. That is, suppose we have

N =

	I	II	III	IV	V	VI
1	0	0	1	0	1	0
2	1	0	0	1	0	0
3	0	0	1	0	0	1
4	0	1	0	1	0	0

After applying the *R* process, we have the following final matrix **M**

M =

	I	II	III	IV	V	VI
1	0	0	1	0	1	1*
2	1	1*	0	1	0	0
3	0	0	1	0	1*	1
4	1*	1	0	1	0	0

Notice that **M** has some *zero* entries, and as a result **X** is *not* of maximal rank. It is interesting to note that, even if one eliminates the cells (1, 1), (2, 6), and (3, 2), the design matrix **X** remains of maximal rank. This shows that not all the occupied cells are required for a two-way classification with an additive model to be connected. Such a fact can be utilized for characterization of two-way classification designs with additive model. See Chapter 9 for details.

Using the final matrix **M** and Theorem 5.1 we can easily determine whether a parametric function of the form $\mu+\alpha_i+\beta_j$ is estimable. We now wish to construct a basis for $\bar{\Theta}$, the vector space of estimable linear parametric functions.

We begin by presenting a method for constructing a basis for the vector space $\bar{\mathscr{A}}$ of estimable linear parametric functions involving only the α parameters. An α *contrast* is any linear parametric function of the form $\sum_{i=1}^{a} \lambda_i \alpha_i$ satisfying $\sum_{i=1}^{a} \lambda_i = 0$. The following theorem shows that we need only consider estimable α contrasts in describing $\bar{\mathscr{A}}$.

THEOREM 5.2 The set of estimable linear parametric functions involving only the α parameters is a vector space and is the same as the vector space of estimable α contrasts.

Proof The set of estimable linear parametric functions involving only the α parameters and the set of estimable α contrasts are clearly vector spaces over the field of real numbers. Furthermore, the second set is a subset of the first. All we need to show to complete the proof is that every estimable linear parametric function involving only the α parameters is also an α contrast. Suppose $\lambda(\theta)$ is such a function. Then it must be of the form

$$\lambda(\theta) = \sum_{i=1}^{a} \lambda_i \alpha_i \qquad \text{for some } \lambda_1, \lambda_2, \ldots, \lambda_a.$$

Since $\lambda(\theta)$ is estimable, it can be expressed as a linear combination of the expectations $\mu+\alpha_i+\beta_j$ of the cells with $n_{ij}>0$. That is,

$$\sum_{i=1}^{a} \lambda_i \alpha_i = \sum_{i=1}^{a} \sum_{j=1}^{b} c_{ij}(\mu+\alpha_i+\beta_j) = c_{..}\mu + \sum_{i=1}^{a} c_{i.}\alpha_i + \sum_{j=1}^{b} c_{.j}\beta_j,$$

where $c_{ij}=0$ if $n_{ij}=0$. Because there are no restrictions on the parameters we must have $\lambda_i = c_{i.}$ and $c_{..}=0$, so $\sum_{i=1}^{a} \lambda_i = c_{..} = 0$. □

When considering estimability of linear parametric functions for an experiment with incidence matrix **N**, one may equivalently consider an experiment with incidence matrix **M**. It is convenient, however, to view **M** in its rearranged form exhibited in (5.5). The design matrix associated with the incidence matrix in (5.5) is of the form

$$\begin{bmatrix} 1 & \mathbf{A}_1 & \mathbf{B}_1 & \mathbf{0} & \mathbf{0} & \cdots & \mathbf{0} & \mathbf{0} \\ 1 & \mathbf{0} & \mathbf{0} & \mathbf{A}_2 & \mathbf{B}_2 & \cdots & \mathbf{0} & \mathbf{0} \\ \vdots & \vdots & \vdots & \vdots & \vdots & & \vdots & \vdots \\ 1 & \mathbf{0} & \mathbf{0} & \mathbf{0} & \mathbf{0} & \cdots & \mathbf{A}_s & \mathbf{B}_s \end{bmatrix} \tag{5.6}$$

where $(\mathbf{1}, \mathbf{A}_k, \mathbf{B}_k)$ is the design matrix associated with the incidence matrix $\mathbf{M}_k$ for $1 \leqslant k \leqslant s$.

Let $\bar{\mathscr{A}}$ denote the vector space of estimable α contrasts in the subexperiment represented by the model matrix $\mathbf{X}_k = (\mathbf{1}, \mathbf{A}_k, \mathbf{B}_k)$, for $k = 1, 2, \ldots, s$. Since each subexperiment is connected,

$$r(\mathbf{1}, \mathbf{A}_k, \mathbf{B}_k) = a_k + b_k - 1,$$

$$r(\mathbf{X}) = \sum_{k=1}^{s} r(\mathbf{1}, \mathbf{A}_k, \mathbf{B}_k) = \sum_{k=1}^{s} (a_k + b_k - 1) = a + b - s,$$

and

$$\dim \bar{\mathscr{A}}_k = a_k - 1.$$

Now $\bar{\mathscr{A}} = \bar{\mathscr{A}}_1 \oplus \bar{\mathscr{A}}_2 \oplus \cdots \oplus \bar{\mathscr{A}}_s$, so

$$f_\alpha = \dim \bar{\mathscr{A}} = \sum_{k=1}^{s} \dim \bar{\mathscr{A}}_k = \sum_{k=1}^{s} (a_k - 1) = a - s = r(\mathbf{X}) - b.$$

We have established the following Theorem:

THEOREM 5.3 The dimension of the vector space of estimable α contrasts is $f_\alpha = a - s = r(\mathbf{X}) - b$.

THEOREM 5.4 A parametric function of the form $\sum_{i=1}^{a} \lambda_i \alpha_i$ is estimable if and only if $\sum_{i \in I_k} \lambda_i = 0$ for $k = 1, 2, \ldots, s$.

Proof

$$\bar{\mathscr{A}}_k = \left\{ \sum_{i \in I_k} v_i \alpha_i : \sum_{i \in I_k} v_i = 0 \right\}, \qquad k = 1, 2, \ldots, s.$$

Since $\bar{\mathscr{A}} = \bar{\mathscr{A}}_1 \oplus \bar{\mathscr{A}}_2 \oplus \cdots \oplus \bar{\mathscr{A}}_s$, $\sum_{i=1}^{a} \lambda_i \alpha_i \in \bar{\mathscr{A}}$ if and only if

$$\sum_{i=1}^{a} \lambda_i \alpha_i = \sum_{i=1}^{a} v_i \alpha_i \quad \text{where} \quad \sum_{i \in I_k} v_i = 0 \qquad \text{for } k = 1, 2, \ldots, s.$$

Since the α_i are unconstrained, $\lambda_i = v_i$ for all i, and the theorem is proved. □

By considering the kth submatrix $\mathbf{M}_k$ of the incidence matrix in (5.5), we see that there are $a_k - 1$ linearly independent estimable differences $\alpha_i - \alpha_h$

with $i, h \in I_k$. Since the sets $I_1, I_2, \ldots, I_s$ are disjoint, we can obtain

$$\sum_{k=1}^{s} (a_k - 1) = a - s$$

linearly independent estimable differences, each of which is also an estimable α contrast. Therefore, $\bar{\mathscr{A}}$ is also the vector space of estimable α differences, and a basis for $\bar{\mathscr{A}}$ may be constructed from estimable α differences.

Although Theorem 5.4 may be used to determine which α differences are estimable, an alternative and more direct procedure for determining which α differences are estimable is given in the next theorem.

THEOREM 5.5 An α difference $\alpha_i - \alpha_j (1 \leqslant i, j \leqslant a)$ is estimable if and only if the (i, j) element of $\mathbf{MM}'$ is nonzero.

Proof Let $\mathbf{S} = \mathbf{MM}'$; then if its element $s_{ij} \neq 0$ there is some k such that $m_{ik} = 1 = m_{jk}$ so that $\mu + \alpha_i + \beta_k$ and $\mu + \alpha_j + \beta_k$ are estimable. By taking the difference of these two estimable cell expectations it is seen that $\alpha_i - \alpha_j$ is estimable. Conversely, if $\alpha_i - \alpha_j$ is estimable there must be some k such that $\mu + \alpha_i + \beta_k$ is estimable (since $\alpha_i - \alpha_j$ is a linear combination of estimable cell expectations) and so $m_{ik} = 1$. But also $(\mu + \alpha_i + \beta_k) - (\alpha_i - \alpha_j) = \mu + \alpha_j + \beta_k$ is estimable, so $m_{jk} = 1$, so $m_{ik} m_{jk} = 1$ and $s_{ij} \neq 0$. □

To actually obtain a basis for the estimable α contrasts or to investigate which α differences are estimable, it is convenient to summarize Theorem 5.5 by means of the *counter triangle* $\mathbf{C}_\alpha$. We define $\mathbf{C}_\alpha$ to be the subdiagonal portion of $\mathbf{MM}'$ with all nonzero entries replaced by ones. That is, we define

$$c_{ij}^{(\alpha)} = \begin{cases} 1 & \text{if } \sum_{k=1}^{a} m_{ik} m_{jk} \neq 0 \\ 0 & \text{if } \sum_{k=1}^{a} m_{ik} m_{jk} = 0 \end{cases}, \qquad 1 \leqslant j < i \leqslant a.$$

Associated with the counter triangle $\mathbf{C}_\alpha$ we define I_α to be the set

$$I_\alpha = \{i : c_{ij}^{(\alpha)} = 1 \quad \text{for some } j,\ 1 \leqslant j \leqslant i\}.$$

I_α then contains the triangle's row indices corresponding to nonzero rows. Note that if $i \in I_k$ and i is not the smallest number in I_k, then there is some $h \in I_k\ (h < i)$ such that $\alpha_i - \alpha_h$ is estimable. Thus, it follows that $\#(I_\alpha) \geqslant \sum_{k=1}^{s} \times (a_k - 1) = a - s$. Now for each $i \in I_\alpha$ select one integer $j(i)$, $1 \leqslant j(i) < i$, such that $c_{ij(i)}^{(\alpha)} = 1$.

THEOREM 5.6 The set of α differences $\{\alpha_i - \alpha_{j(i)} : i \in I_\alpha\}$ constitutes a basis for the vector space $\bar{\mathscr{A}}$ of estimable α contrasts.

Proof Clearly, for $i \in I_\alpha$ the difference $\alpha_i - \alpha_{j(i)}$ is estimable. Now suppose $\sum_{i \in I_\alpha} d_i(\alpha_i - \alpha_{j(i)}) = 0$ with some $d_i \neq 0$. Let k be the largest element of I_α such that $d_k \neq 0$. Since $j(i) < i < k$ for all $i \in I_\alpha$ such that $i \neq k$ and $d_i \neq 0$, and since $j(k) < k$, the coefficient of α_k in $\sum_{i \in I_\alpha} d_i(\alpha_i - \alpha_{j(i)})$ is d_k. But then $d_k = 0$. This contradiction shows that the $\alpha_i - \alpha_{j(i)}$, $i \in I_\alpha$, are linearly independent. But they are contained in $\bar{\mathscr{A}}$, $\dim \bar{\mathscr{A}} = f_\alpha = a - s$, and $\#(I_\alpha) \geqslant a - s$. Therefore, they form a basis for $\bar{\mathscr{A}}$. □

From the counter triangle $\mathbf{C}_\alpha$ a basis for $\bar{\mathscr{A}}$ composed of α differences and f_α are easily obtained. To calculate f_α one need only count the nonzero rows of the counter triangle $\mathbf{C}_\alpha$. To obtain a basis for α contrasts simply choose one entry (i, j) such that $c_{ij} = 1$ from each nonzero row i and put $\alpha_i - \alpha_j$ in the basis.

The results we have obtained for $\bar{\mathscr{A}}$ and the method of constructing a basis for $\bar{\mathscr{A}}$ carry over, with only slight modifications, to $\bar{\mathscr{B}}$, the vector space of estimable linear parametric functions involving only the β parameters. Specifically, $\bar{\mathscr{B}}$ is also the vector space of estimable β contrasts, where a *β contrast* is any parametric function of the form $\sum_{j=1}^{b} v_j \beta_j$ satisfying $\sum_{j=1}^{b} v_j = 0$. Furthermore, a parametric function of the form $\sum_{j=1}^{b} v_j \beta_j$ is estimable if and only if $\sum_{j \in J_k} v_j = 0$ for $k = 1, 2, \ldots, s$. The dimension of $\bar{\mathscr{B}}$ is $f_\beta = b - s = r(\mathbf{X}) - a$.

A basis for $\bar{\mathscr{B}}$ can be constructed from estimable β differences. A β difference $\beta_i - \beta_j (1 \leqslant i, j \leqslant b)$ is estimable if and only if the (i, j) element of $\mathbf{M}'\mathbf{M}$ is nonzero. We define the counter triangle $\mathbf{C}_\beta$ to be the subdiagonal portion of $\mathbf{M}'\mathbf{M}$ with all nonzero elements replaced by ones and associate with $\mathbf{C}_\beta$ the set $I_\beta = \{i\colon\ c_{ij}^{(\beta)} = 1 \text{ for some } j,\ 1 \leqslant j < i\}$. Then the set of β differences $\{\beta_i - \beta_{j(i)}\colon\ i \in I_\beta\}$ constitutes a basis for the vector space $\bar{\mathscr{B}}$ of estimable β contrasts.

$\bar{\mathscr{A}}$ and $\bar{\mathscr{B}}$ are independent subspaces of $\bar{\Theta}$, the vector space of estimable linear parametric functions. Since

$$\dim \bar{\Theta} = \dim \Omega = r(\mathbf{X}) = f_\alpha + f_\beta + s,$$

a basis for $\bar{\Theta}$ could consist of the bases for $\bar{\mathscr{A}}$ and $\bar{\mathscr{B}}$ and s additional linearly independent parametric functions in $\bar{\Theta}$. Let $i(k) = \min\{i\colon\ i \in I_k\}$ and $j(k) = \min\{j; j \in I_k\}$ for $k = 1, 2, \ldots, s$. Then

$$\{\mu + \alpha_{i(k)} + \beta_{j(k)}\colon k = 1, 2, \ldots, s\} \tag{5.7}$$

is a set of s estimable linear parametric functions that are linearly independent among themselves and of the elements of $\bar{\mathscr{A}}$ and $\bar{\mathscr{B}}$. Let $\bar{\mathscr{E}}$ denote the vector space spanned by this set and let $f_c = \dim \bar{\mathscr{E}} = s$. Then the set in (5.7) is a basis for $\bar{\mathscr{E}}$. Another basis for $\bar{\mathscr{E}}$ is the set

$$\{\mu + \alpha_{i(1)} + \beta_{j(1)}, \alpha_{i(1)} - \alpha_{i(k)} + \beta_{j(1)} - \beta_{j(k)}\colon \qquad k = 2, 3, \ldots, s\}.$$

We have $\bar{\Theta}=\bar{\mathscr{A}}\oplus\bar{\mathscr{B}}\oplus\bar{\mathscr{E}}$ and a basis for $\bar{\Theta}$ can be obtained as the union of any three bases for $\bar{\mathscr{A}}$, $\bar{\mathscr{B}}$, and $\bar{\mathscr{E}}$, taken respectively.

Now we will illustrate how the methods just introduced can be used to construct a basis for $\bar{\Theta}$.

EXAMPLE 5.5 Consider an additive two-way model with $a=8$, $b=6$, and an incidence matrix $\mathbf{N}$ as follows:

$$\mathbf{N}=\begin{array}{c}\\1\\2\\3\\4\\5\\6\\7\\8\end{array}\begin{array}{c}\begin{array}{cccccc}\mathrm{I}&\mathrm{II}&\mathrm{III}&\mathrm{IV}&\mathrm{V}&\mathrm{VI}\end{array}\\\begin{bmatrix}0&0&2&0&1&0\\2&4&0&0&0&0\\0&0&3&0&0&0\\0&3&0&1&0&2\\0&0&0&0&2&0\\0&0&0&0&4&0\\0&0&0&5&0&1\\0&0&1&0&2&0\end{bmatrix}\end{array}.$$

The final matrix obtained from $\mathbf{N}$ by the R process is

$$\mathbf{M}=\begin{bmatrix}0&0&1&0&1&0\\1&1&0&1&0&1\\0&0&1&0&1&0\\1&1&0&1&0&1\\0&0&1&0&1&0\\0&0&1&0&1&0\\1&1&0&1&0&1\\0&0&1&0&1&0\end{bmatrix}.$$

From $\mathbf{M}$ we see that, for instance, $\mu+\alpha_7+\beta_2$ is estimable, whereas $\mu+\alpha_7+\beta_3$ is not estimable. We can also see

$$I_1=\{1,3,5,6,8\},\qquad J_1=\{3,5\},$$

$$I_2=\{2,4,7\},\qquad J_2=\{1,2,4,6\}.$$

The counter triangle for the α differences is

$$
\mathbf{C}_\alpha = \begin{array}{c|ccccccc}
 & 1 & 2 & 3 & 4 & 5 & 6 & 7 \\
\hline
2 & 0 & & & & & & \\
3 & 1 & 0 & & & & & \\
4 & 0 & 1 & 0 & & & & \\
5 & 1 & 0 & 1 & 0 & & & \\
6 & 1 & 0 & 1 & 0 & 1 & & \\
7 & 0 & 1 & 0 & 1 & 0 & 0 & \\
8 & 1 & 0 & 1 & 0 & 1 & 1 & 0
\end{array}
$$

and $I_\alpha = \{3, 4, 5, 6, 7, 8\}$.

From $\mathbf{C}_\alpha$ we see that, for instance, $\alpha_1 - \alpha_6$ is estimable, whereas $\alpha_2 - \alpha_6$ is not estimable. $f_\alpha = \#(I_\alpha) = 6$ and one choice of a basis for $\bar{\mathscr{A}}$ is

$$\{\alpha_1 - \alpha_3, \alpha_2 - \alpha_4, \alpha_1 - \alpha_5, \alpha_1 - \alpha_6, \alpha_2 - \alpha_7, \alpha_1 - \alpha_8\}.$$

We can easily compute $r(\mathbf{X}) = b + f_\alpha = 6 + 6 = 12$ and $f_\beta = r(\mathbf{X}) - a = 12 - 8 = 4$.

The counter triangle for the β differences is

$$
\mathbf{C}_\beta = \begin{array}{c|ccccc}
 & 1 & 2 & 3 & 4 & 5 \\
\hline
2 & 1 & & & & \\
3 & 0 & 0 & & & \\
4 & 1 & 1 & 0 & & \\
5 & 0 & 0 & 1 & 0 & \\
6 & 1 & 1 & 0 & 1 & 0
\end{array}
$$

Table 5.3 Source of Variation and Degrees of Freedom for Example 5.4

Source	Degrees of Freedom	
Regression	12	
Fitting overall mean		1
Blocks unadjusted (adjusted for mean)		5
Treatments		6
Fitting overall mean		1
Treatment unadjusted (adjusted for mean)		7
Blocks		4
Residuals	21	
Total	33	

and $I_\beta=\{2, 4, 5, 6\}$, confirming that $f_\beta=\#(I_\beta)=4$ and one choice of a basis for $\bar{\mathscr{B}}$ is

$$\{\beta_1-\beta_2, \beta_1-\beta_4, \beta_3-\beta_5, \beta_1-\beta_6\}.$$

Here $f_c=s=2$ and a basis for $\bar{\mathscr{E}}$ is

$$\{\mu+\alpha_1+\beta_3, \alpha_1-\alpha_2-\beta_1+\beta_3\}.$$

We summarize the degrees of freedom accompanying the above results in Table 5.3.

5.5 COMPUTATIONAL ASPECTS OF THE PROBLEM

In this section we present a method to illustrate a possible approach for obtaining some pertinent sum of squares associated with the model under consideration, and as a by-product obtaining a solution to the normal equations as well as the covariance associated with the solution obtained.

We employ the SS$(-)$ notation introduced in Chapter 2, that is, SS$(-)$ denotes the regression sum of squares for fitting the model indicated by the parameters enclosed in the parentheses. Since there are no restrictions on the parameters, observe that the quantities such as SS(μ, α) and SS(α) denote the same sum of squares. Additionally, we use the notation

$$\text{SS}(\alpha\,|\,\beta)=\text{SS}(\alpha, \beta)-\text{SS}(\beta) \quad \text{and} \quad \text{SS}(\beta\,|\,\alpha)=\text{SS}(\alpha, \beta)-\text{SS}(\alpha),$$

so SS$(\alpha\,|\,\beta)$ denotes the SS for the α effects adjusted for the β effects or the SS for the α effects after fitting the β effects, with a similar interpretation for SS$(\beta\,|\,\alpha)$. Observe that of the SS just introduced, the only ones which require work (i.e., actual matrix inversion) to obtain are SS(α, β), SS$(\alpha\,|\,\beta)$, and SS$(\beta\,|\,\alpha)$. Thus, if one can obtain any one of these last three indicated SS, then the remaining two are easily obtained. In the following we concentrate on obtaining SS$(\beta\,|\,\alpha)$. Alternatively, we could just as easily describe the procedure via SS$(\alpha\,|\,\beta)$, and in fact the actual choice of which SS to obtain should be dictated by which of the numbers f_α and f_β is the smaller.

Specifically, we describe a procedure to obtain the following quantities:

1 A random vector $\hat{\boldsymbol{\theta}}$ satisfying the normal equations, that is, $\mathbf{X}'\mathbf{X}\hat{\boldsymbol{\theta}}=\mathbf{X}'\mathbf{Y}$. Various restrictions that may be imposed upon a solution to the normal equations will also be indicated.

2 The matrix $\mathbf{V}$ satisfying

$$\text{Cov}(\hat{\boldsymbol{\theta}})=\sigma^2\mathbf{V} \tag{5.8}$$

where $\hat{\boldsymbol{\theta}}$ is the random vector described in (1).

3 The SS$(\beta\,|\,\alpha)$.

The random vector $\hat{\boldsymbol{\theta}}$ will be uniquely determined by the restrictions $\sum_i \hat{\alpha}_i = 0$ and $\sum_j \hat{\beta}_j = 0$ only when $r(\mathbf{X}) = a + b - 1$. Whether $r(\mathbf{X}) = a + b - 1$ or not, however, we obtain solutions to the normal equations as well as providing enough information so that the associated covariance matrices may be easily obtained by matrix multiplications.

Rather than use the original parametrization (5.3), we will use a more tractable alternative parametrization. We will use the parametrization introduced in the following theorem.

THEOREM 5.7 Suppose $f_\beta > 0$ and let $\boldsymbol{\Lambda}'\boldsymbol{\beta}$ denote an $f_\beta \times 1$ vector of estimable parametric functions which constitute a basis for all estimable β contrasts. Let $\mathbf{G} = (\mathbf{I} - \mathbf{P}_\mathrm{A})\mathbf{B}\boldsymbol{\Lambda}$ where $\mathbf{P}_\mathrm{A}$ denotes the orthogonal projection on $R(\mathbf{A})$. Then an unconstrained full column rank parametrization for Ω is

$$E(\mathbf{Y}) = \mathbf{A}\boldsymbol{\psi} + \mathbf{G}\boldsymbol{\xi} = (\mathbf{A}, \mathbf{G})\begin{pmatrix}\boldsymbol{\psi}\\ \boldsymbol{\xi}\end{pmatrix} = \mathbf{W}\boldsymbol{\delta}, \tag{5.9}$$

where $\boldsymbol{\psi} \in R^a$, $\boldsymbol{\xi} \in R^{f_\beta}$, and $\boldsymbol{\delta} \in R^m$.

Proof We begin by showing that

$$\bar{\mathscr{B}} = \{\mathbf{v}'\boldsymbol{\beta}\colon \mathbf{v} \in R[\mathbf{B}'(\mathbf{I} - \mathbf{P}_\mathrm{A})]\}.$$

The parametric function $\mathbf{v}'\boldsymbol{\beta}$ is estimable if and only if there exists some vector $\mathbf{x}$ in R^N such that $\mathbf{v} = \mathbf{B}'\mathbf{x}$ and $\mathbf{0} = \mathbf{1}'\mathbf{x} = \mathbf{A}'\mathbf{x}$. Suppose $\mathbf{v}'\boldsymbol{\beta}$ is estimable. Then $\mathbf{v} = \mathbf{B}'\mathbf{x}$ where $\mathbf{A}'\mathbf{x} = \mathbf{0}$. This means that $\mathbf{x} \in N(\mathbf{A}') = R(\mathbf{A})^\perp = R(\mathbf{I} - \mathbf{P}_\mathrm{A})$, so $\mathbf{x} = (\mathbf{I} - \mathbf{P}_\mathrm{A})\mathbf{y}$ for some vector $\mathbf{y}$ and $\mathbf{v} = \mathbf{B}'\mathbf{x} = \mathbf{B}'(\mathbf{I} - \mathbf{P}_\mathrm{A})\mathbf{y}$, that is $\mathbf{v} \in R[\mathbf{B}'(\mathbf{I} - \mathbf{P}_\mathrm{A})]$. Suppose now that $\mathbf{v} \in R[\mathbf{B}'(\mathbf{I} - \mathbf{P}_\mathrm{A})]$. Then $\mathbf{v} = \mathbf{B}'(\mathbf{I} - \mathbf{P}_\mathrm{A})\mathbf{y}$ for some vector $\mathbf{y}$. Let $\mathbf{x} = (\mathbf{I} - \mathbf{P}_\mathrm{A})\mathbf{y}$. Then $\mathbf{v} = \mathbf{B}'\mathbf{x}$ where $\mathbf{A}'\mathbf{x} = \mathbf{A}'(\mathbf{I} - \mathbf{P}_\mathrm{A})\mathbf{y} = \mathbf{0}$ and $\mathbf{v}'\boldsymbol{\beta}$ is estimable. Now

$$R(\mathbf{W}) = R[\mathbf{A}, (\mathbf{I} - \mathbf{P}_\mathrm{A})\mathbf{B}\boldsymbol{\Lambda}] \subset R[\mathbf{A}, (\mathbf{I} - \mathbf{P}_\mathrm{A})\mathbf{B}] = R(\mathbf{A}, \mathbf{B}) = R(\mathbf{X}).$$

But $\mathbf{A}'\mathbf{G} = \mathbf{A}'(\mathbf{I} - \mathbf{P}_\mathrm{A})\mathbf{B}\boldsymbol{\Lambda} = \mathbf{0}$, so

$$\begin{aligned} r(\mathbf{W}) &= r(\mathbf{A}) + r[(\mathbf{I} - \mathbf{P}_\mathrm{A})\mathbf{B}\boldsymbol{\Lambda}] \\ &= a + r(\boldsymbol{\Lambda}) - \dim\{R(\boldsymbol{\Lambda}) \cap N[(\mathbf{I} - \mathbf{P}_\mathrm{A})\mathbf{B}]\}. \end{aligned}$$

Suppose $\boldsymbol{\lambda} \in R(\boldsymbol{\Lambda}) \cap N[(\mathbf{I} - \mathbf{P}_\mathrm{A})\mathbf{B}]$. Then $\boldsymbol{\lambda}'\boldsymbol{\beta} \in \bar{\mathscr{B}}$ because $\bar{\mathscr{B}} = \{\boldsymbol{\lambda}'\boldsymbol{\beta}\colon \boldsymbol{\lambda} \in R(\boldsymbol{\Lambda})\}$. But we also have $\bar{\mathscr{B}} = \{\mathbf{v}'\boldsymbol{\beta}\colon \mathbf{v} \in R[\mathbf{B}'(\mathbf{I} - \mathbf{P}_\mathrm{A})]\}$, so $\boldsymbol{\lambda} = \mathbf{B}'(\mathbf{I} - \mathbf{P}_\mathrm{A})\mathbf{x}$ for some vector $\mathbf{x}$. But $(\mathbf{I} - \mathbf{P}_\mathrm{A})\mathbf{B}\boldsymbol{\lambda} = \mathbf{0}$, so

$$0 = \mathbf{x}'(\mathbf{I} - \mathbf{P}_\mathrm{A})\mathbf{B}\boldsymbol{\lambda} = \mathbf{x}'(\mathbf{I} - \mathbf{P}_\mathrm{A})'\mathbf{B}\mathbf{B}'(\mathbf{I} - \mathbf{P}_\mathrm{A})\mathbf{x} = \boldsymbol{\lambda}'\boldsymbol{\lambda}$$

and $\boldsymbol{\lambda} = \mathbf{0}$. Thus $\dim\{R(\boldsymbol{\Lambda}) \cap N[(\mathbf{I} - \mathbf{P}_\mathrm{A})\mathbf{B}]\} = 0$.

Because $r(\Lambda)=f_\beta$, where $f_\beta=r(\mathbf{X})-a$, we then have

$$r(\mathbf{W})=a+f_\beta=r(\mathbf{X})=m.$$

So $R(\mathbf{W})=R(\mathbf{X})$ and (5.9) is an unconstrained parametrization for Ω. Furthermore, $\mathbf{W}$ has full column rank. □

From the proof of Theorem 5.7 it also follows that $R(\Lambda)=R(\mathbf{B}'(\mathbf{I}-\mathbf{P}_\mathbf{A}))$.

For parametrization (5.9), normal equations are $\mathbf{W}'\mathbf{W}\hat{\boldsymbol{\delta}}=\mathbf{W}'\mathbf{Y}$, where $\mathbf{W}'\mathbf{W}$ is nonsingular. Furthermore, the matrices $\mathbf{A}'\mathbf{A}$ and $\mathbf{G}'\mathbf{G}$ are also nonsingular and $\mathbf{A}'\mathbf{G}=\mathbf{0}$, so the normal equations for (5.9) reduce to the two separate equations:

$$\mathbf{A}'\mathbf{A}\hat{\boldsymbol{\psi}}=\mathbf{A}'\mathbf{Y} \quad \text{and} \quad \mathbf{G}'\mathbf{G}\hat{\boldsymbol{\xi}}=\mathbf{G}'\mathbf{Y}.$$

With respect to the original parametrization (5.3), the quantity $SS(\beta|\alpha)$ is the deviation sum of squares for testing the linear hypothesis $H_0:\Lambda'\boldsymbol{\beta}=\mathbf{0}$. This is the hypothesis that all estimable linear parametric functions involving only the β parameters are zero. $SS(\beta|\alpha)$ is also called the sum of squares for the β effects adjusted for the α effects.

Using the SS notation for both the original and the alternative parametrizations, we have $SS(\alpha)=SS(\psi)$ and $SS(\alpha, \beta)=SS(\psi, \xi)=SS(\psi)+SS(\xi)$, so $SS(\beta|\alpha)=SS(\xi)$. $SS(\xi)$ is the deviation sum of squares for testing the linear hypothesis $H_0:\boldsymbol{\xi}=\mathbf{0}$. So $SS(\beta|\alpha)$ is also the deviation sum of squares for testing the linear hypothesis $H_0:E(\mathbf{Y})\in R(\mathbf{A})$, the hypothesis that $E(Y_{ijk})$ has the form $\mu+\alpha_i$. If $f_\beta=0$, then $SS(\beta|\alpha)=SS(\xi)=0$. This special case will be discussed later in the chapter. Until then, we assume that $f_\beta\geqslant 1$.

We now present methods for calculating $SS(\psi)$, $SS(\xi)$, $\hat{\boldsymbol{\theta}}$, and $\mathbf{V}$, and for constructing likelihood ratio tests of linear hypothesis on $\boldsymbol{\theta}$.

Let $\hat{\boldsymbol{\psi}}$ denote the solution to the normal equations for the ψ parameters, $\mathbf{A}'\mathbf{A}\hat{\boldsymbol{\psi}}=\mathbf{A}'\mathbf{Y}$. Now $\mathbf{A}'\mathbf{A}=\text{diag}(n_{1.}, n_{2.}, \ldots, n_{a.})$ and $\mathbf{A}'\mathbf{Y}=(Y_{1..}, Y_{2..}, \ldots, Y_{a..})'$, so:

(a) $\hat{\boldsymbol{\psi}}=(\hat{\psi}_1, \ldots, \hat{\psi}_a)'=(\mathbf{A}'\mathbf{A})^{-1}\mathbf{A}'\mathbf{Y}, \ \hat{\psi}_i=\dfrac{Y_{i..}}{n_{i.}}$ for $i=1, 2, \ldots, a$.

(b) $\text{Cov}(\hat{\boldsymbol{\psi}})=\sigma^2(\mathbf{A}'\mathbf{A})^{-1}=\sigma^2\,\text{diag}\left(\dfrac{1}{n_{1.}}, \ldots, \dfrac{1}{n_{a.}}\right).$

(c) $SS(\psi)=\hat{\boldsymbol{\psi}}'\mathbf{A}'\mathbf{A}\hat{\boldsymbol{\psi}}=\sum_{i=1}^{a} n_{i.}\hat{\psi}_i^2.$ (5.10)

(d)
$$\begin{aligned} E(\hat{\boldsymbol{\psi}})&=(\mathbf{A}'\mathbf{A})^{-1}\mathbf{A}'(\mathbf{1}\mu+\mathbf{A}\boldsymbol{\alpha}+\mathbf{B}\boldsymbol{\beta})\\ &=\mathbf{1}\mu+\boldsymbol{\alpha}+(\mathbf{A}'\mathbf{A})^{-1}\mathbf{N}\boldsymbol{\beta}\\ &\doteq\boldsymbol{\psi}\end{aligned}$$

Let $\hat{\boldsymbol{\xi}}$ denote the solution to the reduced normal equations for the ξ parameters, $\mathbf{G}'\mathbf{G}\hat{\boldsymbol{\xi}} = \mathbf{G}'\mathbf{Y}$. Since $\mathbf{P}_{\mathbf{A}}^2 = \mathbf{P}_{\mathbf{A}} = \mathbf{A}(\mathbf{A}'\mathbf{A})^{-1}\mathbf{A}'$ we have:

$$
\begin{aligned}
&\text{(a)}\quad \mathbf{G}'\mathbf{G} = \boldsymbol{\Lambda}'(\mathbf{B}'\mathbf{B} - \mathbf{B}'\mathbf{A}(\mathbf{A}'\mathbf{A})^{-1}\mathbf{A}'\mathbf{B})\boldsymbol{\Lambda} = \boldsymbol{\Lambda}'(\mathbf{B}'\mathbf{B} - \mathbf{N}'(\mathbf{A}'\mathbf{A})^{-1}\mathbf{N})\boldsymbol{\Lambda} \text{ and}\\
&\qquad \mathbf{G}'\mathbf{Y} = \boldsymbol{\Lambda}'(\mathbf{B}'\mathbf{Y} - \mathbf{B}'\mathbf{A}(\mathbf{A}'\mathbf{A})^{-1}\mathbf{A}'\mathbf{Y}) = \boldsymbol{\Lambda}'(\mathbf{B}'\mathbf{Y} - \mathbf{N}'\hat{\boldsymbol{\psi}}).\\
&\text{(b)}\quad \hat{\boldsymbol{\xi}} = (\mathbf{G}'\mathbf{G})^{-1}\mathbf{G}'\mathbf{Y} \quad \text{and} \quad \mathrm{Cov}(\hat{\boldsymbol{\xi}}) = \sigma^2(\mathbf{G}'\mathbf{G})^{-1}. \qquad (5.11)\\
&\text{(c)}\quad \mathrm{SS}(\xi) = \hat{\boldsymbol{\xi}}'\mathbf{G}'\mathbf{G}\hat{\boldsymbol{\xi}} = \hat{\boldsymbol{\xi}}'\mathbf{G}'\mathbf{Y}.\\
&\text{(d)}\quad E(\hat{\boldsymbol{\xi}}) = (\mathbf{G}'\mathbf{G})^{-1}\mathbf{G}'\mathbf{B}\boldsymbol{\beta} \doteq \boldsymbol{\xi}.
\end{aligned}
$$

With regard to the above quantities, several points should be noted. Observe that the only real computational effort required is the inversion of the $f_\beta \times f_\beta$ matrix $\mathbf{G}'\mathbf{G}$. It is easily seen at this point which of $\mathrm{SS}(\alpha|\beta)$ or $\mathrm{SS}(\beta|\alpha)$ should be obtained first. That is, one should first obtain the adjusted SS for the parameter set with the fewer degrees of freedom. Regarding the matrix **G**, note that only $\mathbf{G}'\mathbf{G}$ and $\mathbf{G}'\mathbf{Y}$ need be calculated. To calculate $\mathbf{G}'\mathbf{G}$, one needs only $\mathbf{N}$, $\boldsymbol{\Lambda}'$, $\mathbf{B}'\mathbf{B}$, and $(\mathbf{A}'\mathbf{A})^{-1}$, and note that $\mathbf{B}'\mathbf{B}$ and $\mathbf{A}'\mathbf{A}$ are diagonal matrices consisting of the $n_{.j}$'s and the $n_{i.}$'s, respectively. To calculate $\mathbf{G}'\mathbf{Y}$, one needs only $\boldsymbol{\Lambda}$, $\mathbf{B}'\mathbf{Y}$, $\mathbf{N}$, and $\hat{\boldsymbol{\psi}}$, and note that $\hat{\boldsymbol{\psi}}$ has been previously obtained and that the entries of the vector $\mathbf{B}'\mathbf{Y}$ are the $Y_{.j.}$'s. When $r(\mathbf{X}) = a+b-1$ a $\boldsymbol{\Lambda}$ matrix is easily obtained, and when $r(\mathbf{X}) < a+b-1$ the results of the estimability section can be used to construct an appropriate $\boldsymbol{\Lambda}$.

It should be noted that the unconstrained full column rank parametrization (5.9) is used only as an intermediate step in getting least squares solutions for parameters of the unconstrained model (5.3). So, having obtained the unique solutions for $\boldsymbol{\psi}$ and $\boldsymbol{\xi}$, one wants to get solutions for $\boldsymbol{\alpha}$ and $\boldsymbol{\beta}$. But before doing this it is necessary to find suitable correspondances between the two parametrizations. Note, that a suitable correspondence between $\boldsymbol{\beta}$ functions and $\boldsymbol{\xi}$ functions can be established by projecting the righthand sides of (5.3) and (5.9) by the operator $\mathbf{I} - \mathbf{P}_{\mathbf{A}}$. This yields the correspondance

$$(\mathbf{I} - \mathbf{P}_{\mathbf{A}})\mathbf{B}\boldsymbol{\beta} \doteq \mathbf{G}\boldsymbol{\xi}.$$

Hence, on account of Theorem 3.1,

$$(\mathbf{I} - \mathbf{P}_{\mathbf{A}})\mathbf{B}\tilde{\boldsymbol{\beta}} = \mathbf{G}\hat{\boldsymbol{\xi}} \qquad (= (\mathbf{I} - \mathbf{P}_{\mathbf{A}})\mathbf{B}\boldsymbol{\Lambda}\hat{\boldsymbol{\xi}}).$$

Solving the equality for $\tilde{\boldsymbol{\beta}}$ we obtain

$$
\begin{aligned}
\tilde{\boldsymbol{\beta}} &= [\mathbf{B}'(\mathbf{I} - \mathbf{P}_{\mathbf{A}})\mathbf{B}]^{-}\mathbf{B}'\mathbf{G}\hat{\boldsymbol{\xi}}\\
&= \boldsymbol{\Lambda}[\boldsymbol{\Lambda}'\mathbf{B}'(\mathbf{I} - \mathbf{P}_{\mathbf{A}})\mathbf{B}\boldsymbol{\Lambda}]^{-1}\boldsymbol{\Lambda}'\mathbf{B}'(\mathbf{I} - \mathbf{P}_{\mathbf{A}})\mathbf{B}\boldsymbol{\Lambda}\hat{\boldsymbol{\xi}} = \boldsymbol{\Lambda}\hat{\boldsymbol{\xi}},
\end{aligned}
$$

since $r[\boldsymbol{\Lambda}'\mathbf{B}'(\mathbf{I} - \mathbf{P}_{\mathbf{A}})\mathbf{B}\boldsymbol{\Lambda}] = r[\mathbf{B}'(\mathbf{I} - \mathbf{P}_{\mathbf{A}})\mathbf{B}]$ and Lemma 2.2.5 in Rao and Mitra (1971) is applicable.

We now obtain a solution $\tilde{\boldsymbol{\beta}}$ to the reduced normal equations which will be used in constructing a least squares solution for $\boldsymbol{\theta}$.

THEOREM 5.8 Let $\hat{\boldsymbol{\xi}}$ be least squares solution for $\boldsymbol{\xi}$ and let $\tilde{\boldsymbol{\beta}}=\boldsymbol{\Lambda}\hat{\boldsymbol{\xi}}$. Then $\tilde{\boldsymbol{\beta}}$ has the following properties:

(a) $\tilde{\boldsymbol{\beta}}$ is a solution to the reduced normal equations for the $\boldsymbol{\beta}$ parameter vector: $\mathbf{B}'(\mathbf{I}-\mathbf{P}_A)\mathbf{B}\tilde{\boldsymbol{\beta}}=\mathbf{B}'(\mathbf{I}-\mathbf{P}_A)\mathbf{Y}$.

(b) $\tilde{\boldsymbol{\beta}}$ is invariant under any choice of the basis matrix $\boldsymbol{\Lambda}$.

(c) $E(\tilde{\boldsymbol{\beta}})=\boldsymbol{\Lambda}(\mathbf{G}'\mathbf{G})^{-1}\mathbf{G}'\mathbf{B}\boldsymbol{\beta}\doteq\boldsymbol{\Lambda}\boldsymbol{\xi}$.

(d) $\text{Cov}(\hat{\boldsymbol{\beta}})=\sigma^2\mathbf{V}_{\boldsymbol{\beta}}$ where $\mathbf{V}_{\boldsymbol{\beta}}=\boldsymbol{\Lambda}(\mathbf{G}'\mathbf{G})^{-1}\boldsymbol{\Lambda}'$.

(e) $\text{Cov}(\hat{\boldsymbol{\psi}}, \tilde{\boldsymbol{\beta}})=\mathbf{0}$, that is, the random vectors $\hat{\boldsymbol{\psi}}$ and $\tilde{\boldsymbol{\beta}}$ are statistically independent.

Proof

(a) $$\mathbf{G}'\mathbf{G}\hat{\boldsymbol{\xi}}=\mathbf{G}'\mathbf{Y}\Rightarrow\boldsymbol{\Lambda}'\mathbf{B}'(\mathbf{I}-\mathbf{P}_A)\mathbf{B}'\boldsymbol{\Lambda}\hat{\boldsymbol{\xi}}=\boldsymbol{\Lambda}'\mathbf{B}'(\mathbf{I}-\mathbf{P}_A)\mathbf{Y}$$
$$\Rightarrow\mathbf{P}_{\Lambda}\mathbf{B}'(\mathbf{I}-\mathbf{P}_A)\mathbf{B}'\tilde{\boldsymbol{\beta}}=\mathbf{P}_{\Lambda}\mathbf{B}'(\mathbf{I}-\mathbf{P}_A)\mathbf{Y},$$
where $\mathbf{P}_{\Lambda}=\boldsymbol{\Lambda}(\boldsymbol{\Lambda}'\boldsymbol{\Lambda})^{-1}\boldsymbol{\Lambda}'$ is the orthogonal projection on $R(\boldsymbol{\Lambda})$. But we saw in the proof of Theorem 5.7 that $R(\boldsymbol{\Lambda})=R[\mathbf{B}'(\mathbf{I}-\mathbf{P}_A)]=$ $=R[\mathbf{B}'(\mathbf{I}-\mathbf{P}_A)\mathbf{B}]$. Hence, $\mathbf{P}_{\Lambda}\mathbf{B}'(\mathbf{I}-\mathbf{P}_A)=\mathbf{B}'(\mathbf{I}-\mathbf{P}_A)$ and we have
$$\mathbf{B}'(\mathbf{I}-\mathbf{P}_A)\mathbf{B}\tilde{\boldsymbol{\beta}}=\mathbf{B}'(\mathbf{I}-\mathbf{P}_A)\mathbf{Y}.$$

(b) From (5.11b) we have $\hat{\boldsymbol{\xi}}=(\mathbf{G}'\mathbf{G})^{-1}\mathbf{G}'\mathbf{Y}$. Suppose $\boldsymbol{\Lambda}_1$ and $\boldsymbol{\Lambda}_2$ are two basis matrices for $R(\boldsymbol{\Lambda})$, and let $\hat{\boldsymbol{\beta}}_1=\boldsymbol{\Lambda}_1\hat{\boldsymbol{\xi}}$ and $\hat{\boldsymbol{\beta}}_2=\boldsymbol{\Lambda}_2\hat{\boldsymbol{\xi}}$. Fix $\mathbf{Y}=\mathbf{y}$ and observe that
$$(\tilde{\boldsymbol{\beta}}_1-\tilde{\boldsymbol{\beta}}_2)=(\boldsymbol{\Lambda}_1-\boldsymbol{\Lambda}_2)(\mathbf{G}'\mathbf{G})^{-1}\mathbf{G}'\mathbf{y}\in R(\boldsymbol{\Lambda}).$$
But
$$\mathbf{B}'(\mathbf{I}-\mathbf{P}_A)\mathbf{B}(\tilde{\boldsymbol{\beta}}_1-\tilde{\boldsymbol{\beta}}_2)=\mathbf{0},$$
so $(\tilde{\boldsymbol{\beta}}_1-\tilde{\boldsymbol{\beta}}_2)\in N[(\mathbf{I}-\mathbf{P}_A)\mathbf{B}]=R[\mathbf{B}'(\mathbf{I}-\mathbf{P}_A)]^{\perp}=R(\boldsymbol{\Lambda})^{\perp}$, and hence $\tilde{\boldsymbol{\beta}}_1-\tilde{\boldsymbol{\beta}}_2=\mathbf{0}$. Since this is true for every outcome $\mathbf{y}$ on $\mathbf{Y}$, we obtain $\tilde{\boldsymbol{\beta}}_1=\tilde{\boldsymbol{\beta}}_2$.

(c) From (5.11d) we have $E(\hat{\boldsymbol{\xi}})=(\mathbf{G}'\mathbf{G})^{-1}\mathbf{G}'\mathbf{B}\boldsymbol{\beta}\doteq\boldsymbol{\xi}$ so $E(\tilde{\boldsymbol{\beta}})=\boldsymbol{\Lambda}E(\hat{\boldsymbol{\xi}})=$ $\boldsymbol{\Lambda}(\mathbf{G}'\mathbf{G})^{-1}\mathbf{G}'\mathbf{B}\boldsymbol{\beta}\doteq\boldsymbol{\Lambda}\boldsymbol{\xi}$.

(d) $\text{Cov}(\hat{\boldsymbol{\beta}})=\boldsymbol{\Lambda}\,\text{Cov}(\hat{\boldsymbol{\xi}})\boldsymbol{\Lambda}'=\sigma^2\boldsymbol{\Lambda}(\mathbf{G}'\mathbf{G})^{-1}\boldsymbol{\Lambda}'$ from (5.11b).

(e) $$\text{Cov}(\hat{\boldsymbol{\psi}}, \tilde{\boldsymbol{\beta}})=\text{Cov}[(\mathbf{A}'\mathbf{A})^{-1}\mathbf{A}'\mathbf{Y}, \boldsymbol{\Lambda}(\mathbf{G}'\mathbf{G})^{-1}\mathbf{G}'\mathbf{Y}]$$
$$=\sigma^2(\mathbf{A}'\mathbf{A})^{-1}\mathbf{A}'\mathbf{G}(\mathbf{G}'\mathbf{G})^{-1}\boldsymbol{\Lambda}'$$
$$=\mathbf{0}$$
because $\mathbf{A}'\mathbf{G}=\mathbf{0}$ □

Although there are many solutions to the reduced normal equations for the $\boldsymbol{\beta}$ vector, the solution given above has some rather interesting properties. We have seen that the space of estimable β contrasts is $\{\boldsymbol{\lambda}'\boldsymbol{\beta}: \sum_{j \in J_k} \lambda_j = 0, k=1, \ldots, s\}$. However, this space may also be represented as $\{\boldsymbol{\lambda}'\boldsymbol{\beta}: \boldsymbol{\lambda} \in R(\boldsymbol{\Lambda})\}$ and therefore

$$R(\boldsymbol{\Lambda}) = \left\{\boldsymbol{\lambda}: \sum_{j \in J_k} \lambda_j = 0, \qquad k=1, \ldots, s\right\}.$$

Since $\tilde{\boldsymbol{\beta}} = \Lambda\tilde{\boldsymbol{\xi}}$, $\tilde{\boldsymbol{\beta}} \ldots R(\Lambda)$ for every $\mathbf{Y}$ and

$$\sum_{j \in J_k} \tilde{\beta}_j = 0, \qquad k=1, \ldots, s.$$

Furthermore, $\tilde{\boldsymbol{\beta}}$ is the unique solution to the reduced normal equations for $\boldsymbol{\beta}$, subject to these constraints. To see this, represent these constraints by $\boldsymbol{\Delta}'\boldsymbol{\beta} = \mathbf{0}$. Then $R[\mathbf{B}'(\mathbf{I} - \mathbf{P}_{\mathbf{A}})] = R(\boldsymbol{\Lambda})$, $R(\boldsymbol{\Lambda}) \cap R(\boldsymbol{\Delta}) = \{\mathbf{0}\}$, and $r(\boldsymbol{\Lambda}, \boldsymbol{\Delta}) = f_\beta + s = b$.

Now we can get a solution for α. For this let us project the right-hand sides of (5.3) and (5.9) by the operator $\mathbf{P_A}$. We get

$$\mathbf{1}\mu + \mathbf{A}\boldsymbol{\alpha} + \mathbf{P_A B}\boldsymbol{\beta} \doteq \mathbf{A}\boldsymbol{\psi},$$

which gives the equality

$$\mathbf{A}\tilde{\boldsymbol{\alpha}} = \mathbf{A}\hat{\boldsymbol{\psi}} - \mathbf{P_A B}\tilde{\boldsymbol{\beta}} - \mathbf{1}\tilde{\mu}.$$

Solving it for $\tilde{\boldsymbol{\alpha}}$ we obtain

$$\tilde{\boldsymbol{\alpha}} = \hat{\boldsymbol{\psi}} - (\mathbf{A}'\mathbf{A})^{-1}\mathbf{N}\tilde{\boldsymbol{\beta}} - \mathbf{1}\tilde{\mu},$$

for any $\tilde{\mu}$, which in particular can be taken equal to zero.

Certainly, solutions of that type can be obtained separately for each connected portion of the experiment, treated as a subexperiment, as it will be described in Section 5.7.

We now obtain a particular least squares $\hat{\boldsymbol{\theta}}$ for $\boldsymbol{\theta}$:

Let $\hat{\boldsymbol{\psi}}$ and $\hat{\boldsymbol{\beta}}$ be defined as before and set $\tilde{\boldsymbol{\alpha}} = \hat{\boldsymbol{\psi}} - (\mathbf{A}'\mathbf{A})^{-1}\mathbf{N}\hat{\boldsymbol{\beta}}$. Then the following statements are true:

1 The random vector $\tilde{\boldsymbol{\theta}}' = (\mathbf{0}, \tilde{\boldsymbol{\alpha}}', \tilde{\boldsymbol{\beta}}')$ is a solution to the reduced normal equations, $\mathbf{X}'\mathbf{X}\tilde{\boldsymbol{\theta}} = \mathbf{X}'\mathbf{Y}$.

2 The vector $\tilde{\boldsymbol{\theta}}$ in (1) is the unique solution to the following minimization problem:

$$\min \|\mathbf{Y} - \mathbf{1}\mu - \mathbf{A}\boldsymbol{\alpha} - \mathbf{B}\boldsymbol{\beta}\|^2$$

subject to

(a) $\mu = 0$
(b) $\boldsymbol{\alpha} \in R^a$ and
(c) $\sum_{j \in J_k} \beta_j = 0$, $k=1, \ldots, s$;

where $\| - \|$ denotes the usual norm on R^N.

3 $\text{Cov}(\tilde{\boldsymbol{\alpha}})=\sigma^2\mathbf{V}_\alpha$ where $\mathbf{V}_\alpha=(\mathbf{A}'\mathbf{A})^{-1}+(\mathbf{A}'\mathbf{A})^{-1}\mathbf{N}\mathbf{V}_\beta\mathbf{N}'(\mathbf{A}'\mathbf{A})^{-1}$, $\mathbf{V}_\beta=\boldsymbol{\Lambda}(\mathbf{G}'\mathbf{G})^{-1}\boldsymbol{\Lambda}'$.

4 $\text{Cov}(\tilde{\boldsymbol{\alpha}}, \tilde{\boldsymbol{\beta}})=\sigma^2\mathbf{V}_{\alpha\beta}$ where $\mathbf{V}_{\alpha\beta}=-(\mathbf{A}'\mathbf{A})^{-1}\mathbf{N}\mathbf{V}_\beta$.

Observe that the only computations required to obtain the above solution to the normal equations and the associated covariance matrix are matrix multiplication and the inversion of the matrix $\mathbf{G}'\mathbf{G}$. Moreover, recall that $(\mathbf{G}'\mathbf{G})^{-1}$ is also needed to obtain $\text{SS}(\beta|\alpha)=\text{SS}(\xi)$.

We have obtained a solution $\tilde{\boldsymbol{\theta}}$ for $\boldsymbol{\theta}$ as well as its covariance matrix. From $\tilde{\boldsymbol{\theta}}$ any solution to normal equations is easily obtained. To see this note that $\hat{\boldsymbol{\theta}}$ will be a solution for the normal equations if and only if $\mathbf{X}\hat{\boldsymbol{\theta}}=\mathbf{X}\tilde{\boldsymbol{\theta}}$. This observation leads to the fact that any $\hat{\boldsymbol{\theta}}=(\mu, \alpha_1, \ldots, \alpha_a, \hat{\beta}_1, \ldots, \hat{\beta}_b)'$ will be a solution for the normal equations if and only if

$$\hat{\mu}+\hat{\alpha}_i+\hat{\beta}_j=\tilde{\alpha}_i+\tilde{\beta}_j \quad \text{for } i\in I_k, \quad j\in J_k, \quad k=1, \ldots, s.$$

From the form of the above conditions several points may be noted. Observe first that up to $s+1$ linearly independent constraints may be imposed upon a $\hat{\boldsymbol{\theta}}$ solution. Of course these $s+1$ linear constraints cannot be completely arbitrary. Any least squares solution $\hat{\boldsymbol{\theta}}$ for $\boldsymbol{\theta}$ satisfies $\boldsymbol{\Delta}'\hat{\boldsymbol{\theta}}=\mathbf{0}$ as well so long as the matrix $\boldsymbol{\Delta}$ is such that $R(\boldsymbol{\Delta})\cap R(\mathbf{X}')=\{\mathbf{0}\}$. Also, when $s+1$ linearly independent restrictions are imposed, the solution $\hat{\boldsymbol{\theta}}$ is unique and must necessarily be of the form $\mathbf{T}\tilde{\boldsymbol{\theta}}$ for some matrix $\mathbf{T}$; hence, the covariance matrix associated with the $\hat{\boldsymbol{\theta}}$ solution is easily described in terms of the covariance matrix associated with $\tilde{\boldsymbol{\theta}}$. As an example of the type of restrictions which might be imposed as well as the solution in terms of the $\tilde{\boldsymbol{\theta}}$ vector, consider the following restrictions:

$$\sum_{i=1}^{a} u_i\hat{\alpha}_i=0 \quad \text{and} \sum_{j\in J_k} w_j\hat{\beta}_j=0 \quad \text{for } k=1, \ldots, s;$$

where

$$\sum_{i=1}^{a} u_i=1 \quad \text{and} \quad \sum_{j\in J_k} w_j=1 \quad \text{for } k=1, \ldots, s.$$

For these restrictions the solution $\hat{\boldsymbol{\theta}}$ in terms of $\tilde{\boldsymbol{\theta}}$ is easily seen to be the following:

$$\hat{\mu}=\sum_i u_i\tilde{\alpha}_i+\sum_{k=1}^{s}\left(\sum_{i\in I_k} u_i\right)\left(\sum_{j\in J_k} w_j\tilde{\beta}_j\right),$$

$$\left.\begin{aligned} \hat{\alpha}_i&=\tilde{\alpha}_i+\sum_{j\in J_k} w_j\tilde{\beta}_j-\hat{\mu}, && i\in I_k \\ \hat{\beta}_h&=\tilde{\beta}_h-\sum_{j\in J_k} w_j\tilde{\beta}_j, && h\in J_k \end{aligned}\right\} k=1, \ldots, s. \tag{5.12}$$

Furthermore, it is clear that $\hat{\boldsymbol{\theta}}$ is of the form $\mathbf{T}\tilde{\boldsymbol{\theta}}$ so that $\text{Cov}(\hat{\boldsymbol{\theta}})=\mathbf{T}\,\text{Cov}(\tilde{\boldsymbol{\theta}})\mathbf{T}'$ is also easily described.

5.6 PURE ERROR AND LACK OF FIT

The residual sum of squares for parametrization (5.3) is $\|\mathbf{Y}\|^2-\text{SS}(\alpha,\beta)$ and will be denoted by $\hat{R}$. $\hat{R}$ is often partitioned into two components in the following manner:

$$\hat{R}=\sum_{ijk}(Y_{ijk}-\bar{Y}_{ij.})^2+\left[\hat{R}-\sum_{ijk}(Y_{ijk}-\bar{Y}_{ij.})^2\right],$$

where $\bar{Y}_{ij.}=Y_{ij.}/n_{ij}$ for $n_{ij}\neq 0$ and where of course the sum is over only those subscripts ijk for which $n_{ij}\neq 0$. The first component in the above partition is generally called the *pure error* and is associated with $N-q$ degrees of freedom where q denotes the number of nonzero n_{ij}'s. The second component in the above partition is associated with $q-m$ degrees of freedom and is generally called the sum of squares attributable to *lack of fit*, or in the more common terminology of the two-way classification model, is the sum of squares for interaction.

Some Special Situations

When $f_\beta=0$, there are no nonzero estimable functions involving only the β parameters and the sum of squares $\text{SS}(\beta|\alpha)$ is zero. Additionally, since $\mathbf{P}_\Lambda=\mathbf{0}$ the selection of $\tilde{\boldsymbol{\beta}}=\mathbf{0}$ and $\mathbf{V}_\beta=\mathbf{0}$ will make the previous discussion consistent.

When $r(\mathbf{X})=a+b-1$, then $s=1$ and we have a fully connected design. In this situation $f_\alpha=a-1$ and $f_\beta=b-1$. Additionally, imposing the conditions $\sum_i\hat{\alpha}_i=0$ and $\sum_j\hat{\beta}_j=0$ will lead to a unique solution $\hat{\boldsymbol{\theta}}$ for the normal equations. For this solution it is clear that $\hat{\boldsymbol{\beta}}=\tilde{\boldsymbol{\beta}}$; and $\hat{\mu}=\sum_{i=1}^a\tilde{\alpha}_i/a$ and $\hat{\alpha}_i=\tilde{\alpha}_i-\hat{\mu}$. Also, it might be noted that $\tilde{\boldsymbol{\alpha}}$ will possess properties similar to those discussed for $\tilde{\boldsymbol{\beta}}$. For example, $E(\tilde{\boldsymbol{\alpha}})=\mathbf{P}_\Gamma\boldsymbol{\alpha}$ where $\mathbf{P}_\Gamma$ denotes the orthogonal projection on $R(\boldsymbol{\Gamma})$ with $\boldsymbol{\Gamma}$ being such that the components of $\boldsymbol{\Gamma}'\boldsymbol{\alpha}$ constitute a basis for the estimable α contrasts.

5.7 AN ALTERNATIVE METHOD OF ESTIMATION

Let s, $I_1,\ldots,I_s$, $J_1,\ldots,J_s$, $a_1,\ldots,a_s$, and $b_1,\ldots,b_s$ be defined as before. Additionally, for each $k=1,\ldots,s$ let $\mathbf{N}_k$ denote the $a_k\times b_k$ incidence matrix composed of the n_{ij}'s such that $i\in I_k$ and $j\in J_k$ and let $\mathbf{Y}_{(k)}$ denote the random vector composed of the Y_{ijh}'s corresponding to incidence matrix $\mathbf{N}_k$. Also, for $k=1, \ldots, s$ let $\mathbf{X}_k=(\mathbf{1}, \mathbf{A}_k, \mathbf{B}_k)$ be the design matrix associated with the incidence matrix $\mathbf{N}_k$. It is easily seen that the range of the matrix

$$\begin{bmatrix} \mathbf{X}_1 & \mathbf{0} & & \mathbf{0} \\ \mathbf{0} & \mathbf{X}_2 & \cdots & \mathbf{0} \\ \vdots & \vdots & & \vdots \\ \mathbf{0} & \mathbf{0} & \cdots & \mathbf{X}_s \end{bmatrix} \tag{5.13}$$

is the same as the range of the design matrix $\mathbf{X}$ in (5.3) from which (5.13) is obtainable (after an appropriate permutation of rows and columns) associated with the incidence matrix $\mathbf{N}$. Thus, a reparametrization for our problem may be obtained via the matrix in (5.13). Hence our original problem may be viewed as s smaller subproblems, each of which is an additive two-way classification model and each of which is connected.

Utilizing the comments in the previous paragraph, for each $k=1, \ldots, s$ let $\boldsymbol{\theta}'_{(k)}=(\mu_k, \boldsymbol{\alpha}'_{(k)}, \boldsymbol{\beta}'_{(k)})$ where $\boldsymbol{\alpha}_{(k)}$ is an $a_k \times 1$ vector consisting of the α_i parameters for which $i \in I_k$ and $\beta_{(k)}$ is a $b_k \times 1$ vector consisting of the β_j parameters for which $j \in J_k$. Thus, the original parametrization of $E(\mathbf{Y})$ may be viewed as

$$E(\mathbf{Y}_{(k)})=\mathbf{X}_k\boldsymbol{\theta}_{(k)}, \qquad k=1, \ldots, s.$$

Furthermore, it is true for each of these s subproblems that $r(\mathbf{X}_k)=a_k+b_k-1$. Hence estimability considerations and considerations for degrees of freedom are for each subproblem as would typically be expected in an additive two-way classification model.

As just noted, the expectation of $\mathbf{Y}$ may be reparametrized and viewed as s separate subproblems. Using the reduction notation in conjunction with the parameters introduced for the subproblems, the following observations should be noted:

1 SS(α), SS(β), and SS(α, β) are additive over the subproblems:

$$\text{SS}(\alpha, \beta)=\sum_{k=1}^{s} \text{SS}(\alpha_{(k)}, \beta_{(k)}).$$

2 SS($\alpha|\beta$) and SS($\beta|\alpha$) are additive over the subproblems:

$$\text{SS}(\beta|\alpha)=\sum_{k=1}^{s} \text{SS}(\beta_{(k)}|\alpha_{(k)}).$$

Additionally, the residual sum of squares and the partition of the residual sum of squares are additive over the s subproblems.

Relationships between solutions for the normal equations for each of the s subproblems and a solution for the full set of normal equations as described for parametrization (5.3) are easily obtained. In fact, the following relationships must hold:

$$\hat{\mu}_k+\hat{\alpha}_i+\hat{\beta}_j=\tilde{\mu}+\tilde{\alpha}_i+\tilde{\beta}_j; \qquad i\in I_k; \qquad j\in J_k; \qquad k=1,\ldots,s. \quad (5.14)$$

This follows of course since the s subproblems when viewed as a single linear model is a reparametrization for the mean vector of $\mathbf{Y}$. From the relationships in (5.14), a $\tilde{\boldsymbol{\theta}}$ may be obtained from $\hat{\boldsymbol{\theta}}_{(1)}, \hat{\boldsymbol{\theta}}_{(2)}, \ldots, \hat{\boldsymbol{\theta}}_{(s)}$ and vice versa. To illustrate, let us suppose the $\hat{\boldsymbol{\theta}}_{(k)}$'s are such that

$$\mathbf{X}_k'\mathbf{X}_k\hat{\boldsymbol{\theta}}_{(k)}=\mathbf{X}_k'\mathbf{Y}_{(k)}$$

$$\sum_{i\in I_k}\hat{\alpha}_i=0 \qquad k=1,\ldots,s,$$

$$\sum_{j\in J_k}\hat{\beta}_j=0$$

and that $\tilde{\boldsymbol{\theta}}$ is such that

$$\mathbf{X}'\mathbf{X}\tilde{\boldsymbol{\theta}}=\mathbf{X}'\mathbf{Y}$$

$$\sum_{i=1}^{a}\tilde{\alpha}_i=0$$

$$\sum_{j\in J_k}\tilde{\beta}_j=0, \qquad k=1,\ldots,s.$$

Then $\tilde{\boldsymbol{\theta}}$ is unique (subject to the indicated constraints) and the $\tilde{\boldsymbol{\beta}}$ portion is exactly the same as $\tilde{\boldsymbol{\beta}}$ described in Section 5.5. The relationship between the above normal equations solutions is as follows:

1 $\hat{\beta}_j=\tilde{\beta}_j, \qquad j=1,2,\ldots,b.$

2 $\hat{\mu}_k=\tilde{\mu}+\dfrac{(\sum_{i\in I_k}\tilde{\alpha}_i)}{a_k}, \qquad k=1,\ldots,s.$

$\hat{\alpha}_i=\tilde{\mu}+\tilde{\alpha}_i-\hat{\mu}_k$ for $i\in I_k, \quad k=1,\ldots,s.$

3 $\tilde{\mu}=\dfrac{\sum_{k=1}^{s}a_k\hat{\mu}_k}{a}$

$\tilde{\alpha}_i=\hat{\mu}_k+\hat{\alpha}_i-\tilde{\mu}$ for $i\in I_k, k=1,\ldots,s.$

Thus, given either the $\hat{\boldsymbol{\theta}}_{(k)}$ solutions or the $\tilde{\boldsymbol{\theta}}$ solution, it is seen that the other solution may easily be obtained. It should also be noted that the covariance matrices for the $\hat{\boldsymbol{\theta}}_{(k)}$ solutions are easily obtained from a knowledge of the covariance matrix for the $\tilde{\boldsymbol{\theta}}$ solution, and vice versa. It might also be noted that the $\hat{\boldsymbol{\theta}}_{(k)}$ vectors are statistically independent.

By considering a reparametrization of $E(\mathbf{Y})$ when $r(\mathbf{X})<a+b-1$, as described in this section, several advantages accrue. The most obvious of

these advantages is of course that to obtain all the information described in the previous method one need only solve s subproblems, each of which is much simpler than the full problem. That is, the advantage basically reduces to the fact that the kth subproblem may be solved by inverting a square matrix whose size is equal to the minimum of the numbers $a_k - 1$ and $b_k - 1$; whereas the other procedure required inverting a square matrix whose size was the smaller of the two numbers f_α and f_β. Additionally, partitions of the various sums of squares are obtained as a by-product.

EXAMPLE 5.6 Let us use the methods of the previous sections to analyze a set of data from an experiment with an incidence matrix as in Example 5.5. Suppose the observations on the cells are as in Table 5.4.

Table 5.4 Hypothetical Data for the 8 × 6 Incidence Matrix of Example 5.5

	1	2	3	4	5	6
1	—	—	2.3 4.5	—	1.6	—
2	4.2 7.1	3.6 2.4 1.9 4.6	—	—	—	—
3	—	—	3.9 7.9 8.5	—	—	—
4	—	5.6 6.1 2.1	—	3.4	—	5.2 6.9
5	—	—	—	—	9.1 11.9	—
6	—	—	—	—	3.5 4.2 6.8 6.7	—
7	—	—	—	8.7 12.9 6.4 8.6 2.3	—	1.4
8	—	—	2.4	—	4.2 1.5	—

We first analyze the connected subexperiment with the incidence matrix determined by I_1 and J_1:

$$\mathbf{N}_1 = \begin{bmatrix} 2 & 1 \\ 3 & 0 \\ 0 & 2 \\ 0 & 4 \\ 1 & 2 \end{bmatrix}.$$

Since $2 < 5$, one should begin by solving for the SS for β effects adjusted for α effects and for the β effect estimates. The matrices and vectors involved in the computations are

$$\Lambda_1' = (1, -1),$$

$$\mathbf{A}_1'\mathbf{A}_1 = \text{diag}\{3, 3, 2, 4, 3\},$$

$$\mathbf{B}_1'\mathbf{B}_1 = \text{diag}\{6, 9\},$$

$$\mathbf{A}_1'\mathbf{Y}_1 = \begin{bmatrix} 8.4 \\ 20.3 \\ 21.0 \\ 21.2 \\ 8.1 \end{bmatrix},$$

$$\mathbf{B}_1'\mathbf{Y}_1 = \begin{bmatrix} 29.5 \\ 49.5 \end{bmatrix}.$$

We then obtain:

$$\hat{\psi}_1 = (\mathbf{A}_1'\mathbf{A}_1)^{-1}\mathbf{A}_1'\mathbf{Y}_1 = \begin{bmatrix} 2.8000 \\ 6.7667 \\ 10.5000 \\ 5.3000 \\ 2.7000 \end{bmatrix},$$

$$\text{SS}(\hat{\psi}_1) = \hat{\psi}_1'\mathbf{A}_1'\mathbf{Y}_1 = 515.6133,$$

$$\mathbf{G}_1'\mathbf{G}_1 = \frac{16}{3},$$

$$\mathbf{G}_1'\mathbf{Y}_1 = 1.8000,$$

$$\hat{\xi}_1 = \frac{3}{16}(1.8000) = 0.3375,$$

$$\text{SS}(\beta_{(1)}) = \mathbf{Y}_1'\mathbf{B}_1(\mathbf{B}_1'\mathbf{B}_1)^{-1}\mathbf{B}_1'\mathbf{Y}_1 = 417.2917,$$

$$SS(\beta_{(1)}|\alpha_{(1)}) = SS(\xi_1) = \hat{\boldsymbol{\xi}}_1'\mathbf{G}_1'\mathbf{Y}_1 = 0.6075,$$

$$SS(\alpha_{(1)}|\beta_{(1)}) = SS(\psi_1) + SS(\beta_{(1)}|\alpha_{(1)}) - \mathbf{Y}_1'\mathbf{B}_1(\mathbf{B}_1'\mathbf{B}_1)^{-1}\mathbf{B}_1'\mathbf{Y}$$
$$= 98.9291,$$

$$\tilde{\boldsymbol{\beta}}_{(1)} = \boldsymbol{\Lambda}_1\hat{\boldsymbol{\xi}}_1 = \begin{bmatrix} 0.3375 \\ -0.3375 \end{bmatrix},$$

$$\tilde{\boldsymbol{\alpha}}_{(1)} = \hat{\boldsymbol{\psi}}_1 - (\mathbf{A}_1'\mathbf{A}_1)^{-1}\mathbf{N}_1\tilde{\boldsymbol{\beta}}_{(1)} = \begin{bmatrix} 2.6875 \\ 6.4292 \\ 10.8375 \\ 5.6375 \\ 2.8125 \end{bmatrix}.$$

We next analyze the connected subexperiment with the incidence matrix determined by I_2 and J_2:

$$\mathbf{N}_2 = \begin{bmatrix} 2 & 4 & 0 & 0 \\ 0 & 3 & 1 & 2 \\ 0 & 0 & 5 & 1 \end{bmatrix}.$$

Since $3 < 4$, one should begin by solving for the SS for α effects adjusted for β effects and for the α effect estimates. The matrices and vectors involved in the computations are:

$$\boldsymbol{\Lambda}_2' = \begin{bmatrix} 1 & -1 & 0 \\ 1 & 0 & -1 \end{bmatrix},$$

$$\mathbf{A}_2'\mathbf{A}_2 = \text{diag}\{6, 6, 6\},$$

$$\mathbf{B}_2'\mathbf{B}_2 = \text{diag}\{2, 7, 6, 3\},$$

$$\mathbf{A}_2'\mathbf{Y}_2 = \begin{bmatrix} 23.8 \\ 29.3 \\ 40.3 \end{bmatrix},$$

$$\mathbf{B}_2'\mathbf{Y}_2 = \begin{bmatrix} 11.3 \\ 26.3 \\ 42.3 \\ 13.5 \end{bmatrix}.$$

Interchanging the roles of **A** with **B**, we then obtain:

$$\hat{\boldsymbol{\psi}}_2 = (\mathbf{B}_2'\mathbf{B}_2)^{-1}\mathbf{B}_2'\mathbf{Y}_2 = \begin{bmatrix} 5.6500 \\ 3.7571 \\ 2.0500 \\ 4.5000 \end{bmatrix},$$

$$\mathrm{SS}(\hat{\psi}_2)=\hat{\psi}_2'\mathbf{B}_2'\mathbf{Y}_2=521.6191,$$

$$\mathbf{G}_2'\mathbf{G}_2=\mathbf{\Lambda}_2'(\mathbf{A}_2'\mathbf{A}_2-\mathbf{N}_2(\mathbf{B}_2'\mathbf{B}_2)^{-1}\mathbf{N}_2')\mathbf{\Lambda}_2$$

$$=\begin{bmatrix}\frac{117}{14} & \frac{27}{14}\\[2pt] \frac{27}{14} & \frac{45}{14}\end{bmatrix},$$

$$\mathbf{G}_2'\mathbf{Y}_2=\mathbf{\Lambda}_2'(\mathbf{A}_2'\mathbf{Y}_2-\mathbf{N}_2\hat{\psi}_2)=\begin{bmatrix}-4.5071\\ -3.0786\end{bmatrix},$$

$$\hat{\boldsymbol{\xi}}_2=(\mathbf{G}_2'\mathbf{G}_2)^{-1}\mathbf{G}_2'\mathbf{Y}_2=\begin{bmatrix}-0.3694\\ -0.7361\end{bmatrix},$$

$$\mathrm{SS}(\alpha_{(2)})=\mathbf{Y}_2'\mathbf{A}_2(\mathbf{A}_2'\mathbf{A}_2)^{-1}\mathbf{A}_2'\mathbf{Y}_2=508.17,$$

$$\mathrm{SS}(\alpha_{(2)}\,|\,\beta_{(2)})=\mathrm{SS}(\xi_2)=\hat{\boldsymbol{\xi}}_2'\mathbf{G}_2'\mathbf{Y}_2=3.9313,$$

$$\mathrm{SS}(\beta_{(2)}\,|\,\alpha_{(2)})=\mathrm{SS}(\hat{\psi}_2)+\mathrm{SS}(\alpha_{(2)}\,|\,\beta_{(2)})-\mathbf{Y}_2'\mathbf{A}_2(\mathbf{A}_2'\mathbf{A}_2)^{-1}\mathbf{A}_2'\mathbf{Y}_2$$

$$=17.3804,$$

$$\tilde{\boldsymbol{\alpha}}_{(2)}=\mathbf{\Lambda}_2\hat{\boldsymbol{\xi}}_2=\begin{bmatrix}-1.1056\\ 0.3694\\ 0.7361\end{bmatrix},$$

$$\tilde{\boldsymbol{\beta}}_{(2)}=\hat{\psi}_2-(\mathbf{B}_2'\mathbf{B}_2)^{-1}\mathbf{N}_2'\tilde{\boldsymbol{\alpha}}_{(2)}=\begin{bmatrix}6.7556\\ 4.2306\\ 6.3750\\ 4.0083\end{bmatrix}.$$

Combining the results of the analyses of the two subexperiments:

$$\mathrm{SS}(\alpha\,|\,\beta)=\mathrm{SS}(\alpha_{(1)}\,|\,\beta_{(1)})+\mathrm{SS}(\alpha_{(2)}\,|\,\beta_{(2)})=102.8604,$$

$$\mathrm{SS}(\beta\,|\,\alpha)=\mathrm{SS}(\beta_{(1)}\,|\,\alpha_{(1)})+\mathrm{SS}(\beta_{(2)}\,|\,\alpha_{(2)})=17.9974.$$

For comparison with the estimates calculated by the computer program at the end of this chapter we will impose the restrictions

$$\sum_{j\in J_1}\hat{\beta}_j=0,\qquad \sum_{j\in J_2}\hat{\beta}_j=0,\qquad \sum_{i=1}^{8}\hat{\alpha}_i=0.$$

The first constraint is already met. To meet the second, calculate from $\tilde{\boldsymbol{\beta}}_{(2)}$ the deviation

$$\hat{\boldsymbol{\beta}}_{(2)}=\tilde{\boldsymbol{\beta}}_{(2)}-\tilde{\mu}_2\mathbf{1}=\begin{bmatrix}1.4132\\ -1.1118\\ 1.0326\\ -1.3340\end{bmatrix}$$

where $\tilde{\mu}_2 = \frac{1}{4}\mathbf{1}'\tilde{\boldsymbol{\beta}}_{(2)} = 5.3424$. Now set

$$\hat{\boldsymbol{\beta}} = \begin{bmatrix} 1.4132 \\ -1.1118 \\ 0.3375 \\ 1.0326 \\ -0.3375 \\ -1.3340 \end{bmatrix}.$$

Next in accordance with Eqs. (5.14), one should solve

$$\hat{\boldsymbol{\alpha}}_{(1)} = \tilde{\boldsymbol{\alpha}}_{(1)} - \hat{\mu}\mathbf{1},$$
$$\hat{\boldsymbol{\alpha}}_{(2)} = \tilde{\boldsymbol{\alpha}}_{(2)} + (\tilde{\mu}_2 - \hat{\mu})\mathbf{1}.$$

Then

$$\begin{aligned} 0 &= \mathbf{1}'\hat{\boldsymbol{\alpha}}_{(1)} + \mathbf{1}'\hat{\boldsymbol{\alpha}}_{(2)} \\ &= \mathbf{1}'\tilde{\boldsymbol{\alpha}}_{(1)} - 5\hat{\mu} + \mathbf{1}'\tilde{\boldsymbol{\alpha}}_{(2)} + 3(\tilde{\mu}_2 - \hat{\mu}) \\ &= 28.4042 + 16.0272 - 8\hat{\mu}, \end{aligned}$$

so $\hat{\mu} = 5.5539$, and

$$\hat{\boldsymbol{\alpha}} = \begin{bmatrix} -2.8664 \\ -1.3171 \\ 0.8753 \\ 0.1579 \\ 5.2836 \\ 0.0836 \\ 0.5246 \\ -2.7414 \end{bmatrix}.$$

Now let us present the essential estimates obtained for Example 5.6 in form of the table:

	1	2	3	4	5	6	Total	Mean
1	—	—	3.0250	—	2.3500	—	5.3750	2.6875
2	5.6500	3.1250	—	5.2694	—	2.9027	16.9471	4.2368
3	—	—	6.7667	—	6.0917	—	12.8584	6.4292
4	7.1250	4.6000	—	6.7444	—	4.3777	22.8471	5.7118
5	—	—	11.1750	—	10.5000	—	21.6750	10.8375
6	—	—	5.9750	—	5.3000	—	11.2750	5.6375
7	7.4917	4.9667	—	7.1111	—	4.7444	24.3139	6.0785
8	—	—	3.1500	—	2.4750	—	5.6250	2.8125
Total	20.2667	12.6917	30.0917	19.1249	26.7167	12.0248	120.9165	
Mean	6.7556	4.2306	6.0183	6.3750	5.3433	4.0083		

or, in form of the table

	3	5	1	2	4	6	Total	Mean
1	3.0250	2.3500	—	—	—	—	5.3750	2.6875
3	6.7667	6.0917	—	—	—	—	12.8584	6.4292
5	11.1750	10.5000	—	—	—	—	21.6750	10.8375
6	5.9750	5.3000	—	—	—	—	11.2750	5.6375
8	3.1500	2.4750	—	—	—	—	5.6250	2.8125
2	—	—	5.6500	3.1250	5.2694	2.9027	16.9471	4.2368
4	—	—	7.1250	4.6000	6.7444	4.3777	22.8471	5.7118
7	—	—	7.4917	4.9667	7.1111	4.7444	24.3139	6.0785
Total	30.0917	26.7167	20.2667	12.6917	19.1249	12.0248	120.9165	
Mean	6.0183	5.3433	6.7556	4.2306	6.3750	4.0083		

The entries of any of these tables give BLUE's of the cell expectations, that is, of

$$\mu+\alpha_i+\beta_j, \quad \text{where } i\in I_k,\ j\in J_k,\ k=1, 2.$$

From this, it is immediately seen what is estimated by the marginal means and how much they are comparable. For example, the mean 2.6875 estimates $\mu+\alpha_1+(\beta_3+\beta_5)/2$ and the difference $6.4292-2.6875$ estimates $\alpha_3-\alpha_1$, while 4.2368 estimates $\mu+\alpha_2+(\beta_1+\beta_2+\beta_4+\beta_6)/4$ and thus $4.2368-2.6875$ estimates $\alpha_2-\alpha_1+(\beta_1+\beta_2+\beta_4+\beta_6)/4-(\beta_3+\beta_5)/2$. This shows evidently that differences between α parameters are unbiasedly estimated within the connected portions of the experiment only, as stated in Theorem 5.5.

In fact, all estimates of expectation functions that are of interest in the analysis are obtainable from a table as above.

5.8 HYPOTHESIS TESTS

The discussion thus far has centered upon obtaining $SS(\beta|\alpha)$ as well as obtaining a solution for the normal equations and their associated covariance matrix. This emphasis is motivated by the possibility of making hypothesis tests without refitting another model. That is, suppose it is desired to test H_0: $\mathbf{H}'\boldsymbol{\theta}=\boldsymbol{\gamma}$ where the components of $\mathbf{H}'\boldsymbol{\theta}$ are estimable parametric functions. If $\hat{\boldsymbol{\theta}}$ is a least squares solution for $\boldsymbol{\theta}$ and if $\mathbf{V}$ is such that $\mathrm{Cov}(\hat{\boldsymbol{\theta}})=\sigma^2\mathbf{V}$, then the deviation sum of squares for testing H_0 is

$$\hat{R}_0-\hat{R}=(\mathbf{H}'\hat{\boldsymbol{\theta}}-\boldsymbol{\gamma})'\mathbf{V}_{\mathbf{H}}^{-}(\mathbf{H}'\hat{\boldsymbol{\theta}}-\boldsymbol{\gamma}),$$

where $\mathbf{V}_{\mathbf{H}}^{-}$ is any g inverse of $\mathbf{V}_{\mathbf{H}}=\mathbf{H}'\mathbf{V}\mathbf{H}$. It should also be observed that if the components of $\mathbf{H}'\boldsymbol{\theta}$ are linearly independent parametric functions (in the unconstrained parameter situation this is equivalent to the rows of $\mathbf{H}'$ being linearly independent), then $\mathbf{V_H}$ is invertible and $m-m_0=r(\mathbf{V_H})$ is equal to the number of rows in the matrix $\mathbf{H}'$.

We now give an example to illustrate the complete procedure required for the analysis of variance for a two-way classification with missing observations. This example is in fact a summary for the entire chapter.

EXAMPLE 5.7 Consider an additive two-way classification model with $a=8$, $b=6$, and an incidence matrix $\mathbf{N}$ as follows:

$$\mathbf{N}=\mathbf{A}'\mathbf{B}=(n_{ij})=\begin{array}{c} \\ 1\\2\\3\\4\\5\\6\\7\\8 \end{array}\begin{array}{c} \begin{array}{cccccc}1&2&3&4&5&6\end{array} \\ \begin{bmatrix} 0&0&2&0&1&0\\ 2&4&0&0&0&0\\ 0&0&3&0&0&0\\ 0&3&0&1&0&2\\ 0&0&0&0&2&0\\ 0&0&0&0&4&0\\ 0&0&0&5&0&1\\ 0&0&1&0&2&0 \end{bmatrix}\end{array} \quad \begin{array}{c} \text{Totals} \\ n_{i..} \\ \begin{bmatrix}3\\6\\3\\6\\2\\4\\6\\3\end{bmatrix}\end{array}.$$

$$\begin{array}{ll}\text{Totals} & \\ n_{.j.} & [\,2\ \ 7\ \ 6\ \ 6\ \ 9\ \ 3\,]\quad [n_{...}=33]\end{array}$$

Suppose the observations for the corresponding cells in the incidence matrix are as in Table 5.5.

Step 1 Find the final matrix $\mathbf{M}=(m_{ij})$ from the incidence matrix $\mathbf{N}=(n_{ij})$ by the *R process*. Observe that the final matrix $\mathbf{M}$ is a matrix of the same dimension as the incidence matrix $\mathbf{N}$. The *R* process is defined as follows:

1. For each pair i, j, if $n_{ij}\neq 0$ set $m_{ij}=1$.
2. For each pair i, j, if there exists k, l such that $m_{il}=m_{kl}=m_{kj}=1$, then set $m_{ij}=1$. (Pictorially, we add the fourth corner whenever three corners of a rectangle appear in the matrix.)
3. Continue step 2, using the new nonzero m_{ij}'s as corners of new rectangles, until no more entries can be changed.

Table 5.5 Hypothetical Data for a Two-Way Classification with $a=8$ and $b=6$

	β_1	β_2	β_3	β_4	β_5	β_6	Totals $\mathbf{A'Y}$
α_1	—	—	$Y_{131}=2.3$ $Y_{132}=4.5$	—	1.6	—	$Y_{1..}=8.4$
α_2	4.2 7.1	3.6 2.4 1.9 4.6	—	—	—	—	23.8
α_3	—	—	3.9 7.9 8.5	—	—	—	20.3
α_4	—	5.6 6.1 2.1	—	3.4	—	5.2 6.9	29.3
α_5	—	—	—	—	9.1 11.9	—	21.0
α_6	—	—	—	—	3.5 4.2 6.8 6.7	—	21.2
α_7	—	—	—	8.7 12.9 6.4 8.6 2.3	—	1.4	40.3
α_8	—	—	2.4	—	4.2 1.5	—	8.1
Totals $\mathbf{B'Y}$	$[Y_{.1.}=$ 11.3	26.3	29.5	42.3	49.5	13.5]	$[Y_{...}=172.4]$

So we obtain

$$\mathbf{M}=\begin{bmatrix} 0 & 0 & 1 & 0 & 1 & 0 \\ 1 & 1 & 0 & 1 & 0 & 1 \\ 0 & 0 & 1 & 0 & 1 & 0 \\ 1 & 1 & 0 & 1 & 0 & 1 \\ 0 & 0 & 1 & 0 & 1 & 0 \\ 1 & 1 & 0 & 1 & 0 & 1 \\ 0 & 0 & 1 & 0 & 1 & 0 \end{bmatrix}.$$

From the final matrix **M** we see which cells are actually estimable and which ones are not. For example, in the incidence matrix **N**, it appeared that $\mu+\alpha_4+\beta_1$ is not estimable. By applying the R process, we found that we could estimate this cell, as seen from the final matrix **M**.

Step 2 To obtain a basis for the estimable α contrasts, we form the *counter triangle* $\mathbf{C}_\alpha$. The counter triangle $\mathbf{C}_\alpha$ is the lower portion of $\mathbf{MM}'$ with nonzero entries replaced by ones. So we obtain

$$\mathbf{C}_\alpha = \begin{array}{c|ccccccc} & 1 & 2 & 3 & 4 & 5 & 6 & 7 \\ \hline 2 & 0 & & & & & & \\ 3 & 1 & 0 & & & & & \\ 4 & 0 & 1 & 0 & & & & \\ 5 & 1 & 0 & 1 & 0 & & & \\ 6 & 1 & 0 & 1 & 0 & 1 & & \\ 7 & 0 & 1 & 0 & 1 & 0 & 0 & \\ 8 & 1 & 0 & 1 & 0 & 1 & 1 & 0 \end{array} .$$

From $\mathbf{C}_\alpha$ we find all *estimable* α contrasts. For instance, $\alpha_1-\alpha_3$ *is estimable*, whereas $\alpha_1-\alpha_2$ *is not estimable*. That is, whenever there is a one in $\mathbf{C}_\alpha$, it corresponds to an estimable α contrast. To obtain a basis for α contrasts simply choose one entry (i, j) such that $c_{ij}=1$ from each nonzero row i and put $\alpha_i-\alpha_j$ in the basis. For example, some possible choices for a basis for $\bar{\mathscr{A}}$ are:

I $\{\alpha_1-\alpha_3, \alpha_2-\alpha_4, \alpha_1-\alpha_5, \alpha_1-\alpha_6, \alpha_2-\alpha_7, \alpha_1-\alpha_8\}$,
or

II $\{\alpha_1-\alpha_3, \alpha_2-\alpha_4, \alpha_3-\alpha_5, \alpha_3-\alpha_6, \alpha_5-\alpha_7, \alpha_3-\alpha_8\}$,
or

III $\{\alpha_1-\alpha_3, \alpha_2-\alpha_4, \alpha_1-\alpha_5, \alpha_5-\alpha_6, \alpha_2-\alpha_7, \alpha_7-\alpha_8\}$,

and so on. Let us choose the first set. To calculate $f_\alpha=\dim \bar{\mathscr{A}}$, or the degrees of freedom for the α effect adjusted for the β effect, we count the nonzero rows of the counter triangle $\mathbf{C}_\alpha$ which is equal to 6. Thus, $f_\alpha=6$.

Step 3 Compute the rank of the design matrix **X**. This is easily obtained by

$$r(\mathbf{X})=b+f_\alpha=6+6=12.$$

Step 4 Compute f_β, the degrees of freedom for the β effect adjusted for the α effect.

$$f_\beta = r(\mathbf{X}) - a = 12 - 8 = 4.$$

An alternative procedure to find f_β is by forming the counter triangle $\mathbf{C}_\beta$. In this way we can also obtain a basis for the estimable β contrasts. The counter triangle $\mathbf{C}_\beta$ is the lower portion of $\mathbf{M}'\mathbf{M}$ with nonzero entries replaced by ones. The counter triangle for the β differences is

$$\mathbf{C}_\beta = \begin{array}{c|ccccc} & 1 & 2 & 3 & 4 & 5 \\ \hline 2 & 1 & & & & \\ 3 & 0 & 0 & & & \\ 4 & 1 & 1 & 0 & & \\ 5 & 0 & 0 & 1 & 0 & \\ 6 & 1 & 1 & 0 & 1 & 0 \end{array}.$$

From $\mathbf{C}_\beta$ we find $f_\beta = 4$ and one choice of a basis for $\bar{\mathscr{B}}$ to be

$$\{\beta_1 - \beta_2, \beta_1 - \beta_4, \beta_3 - \beta_5, \beta_1 - \beta_6\}.$$

Step 5 Obtain sum of squares for α effects, and the β effect adjusted for α. To do this we use the parametrization (5.9). That is, we use the unconstrained full rank parametrization for Ω

$$E(\mathbf{Y}) = \mathbf{A}\psi + \mathbf{G}\xi$$

where $\mathbf{G} = (\mathbf{I} - \mathbf{P}_\mathbf{A})\mathbf{B}\Lambda$, $\mathbf{P}_\mathbf{A}$ is the orthogonal projection on $R(\mathbf{A})$, and $\Lambda'\boldsymbol{\beta}$ is an $f_\beta \times 1$ vector of estimable parametric functions which constitute a basis for all estimable β contrasts. In this way the solution to normal equations reduces to two separate equations

$$\mathbf{A}'\mathbf{A}\hat{\psi} = \mathbf{A}'\mathbf{Y} \quad \text{and} \quad \mathbf{G}'\mathbf{G}\hat{\xi} = \mathbf{G}'\mathbf{Y}.$$

The matrix $\mathbf{A}'\mathbf{A}$ is a diagonal matrix with its diagonal elements $n_{i..}$ for $i = 1, \ldots, a$. These diagonal elements are available from the row totals of the incidence matrix. From Table 5.5 we can also obtain the entries of the vector $\mathbf{A}'\mathbf{Y}$. These entries are $Y_{i..}$'s. Hence,

$$\mathbf{A}'\mathbf{A} = \text{diag}(n_{1.}, n_{2.}, \ldots, n_{8.}) = \text{diag}(3, 6, 3, 6, 2, 4, 6, 3),$$

and

$$\mathbf{A}'\mathbf{Y} = (8.4, 23.8, 20.3, 29.3, 21.0, 21.2, 40.3, 8.1)'.$$

So we obtain

$$\begin{bmatrix} 3 & 0 & 0 & 0 & 0 & 0 & 0 & 0 \\ 0 & 6 & 0 & 0 & 0 & 0 & 0 & 0 \\ 0 & 0 & 3 & 0 & 0 & 0 & 0 & 0 \\ 0 & 0 & 0 & 6 & 0 & 0 & 0 & 0 \\ 0 & 0 & 0 & 0 & 2 & 0 & 0 & 0 \\ 0 & 0 & 0 & 0 & 0 & 4 & 0 & 0 \\ 0 & 0 & 0 & 0 & 0 & 0 & 6 & 0 \\ 0 & 0 & 0 & 0 & 0 & 0 & 0 & 3 \end{bmatrix} \hat{\psi} = \begin{bmatrix} 8.4 \\ 23.8 \\ 20.3 \\ 29.3 \\ 21.0 \\ 21.2 \\ 40.3 \\ 8.1 \end{bmatrix},$$

$$\hat{\psi} = (\mathbf{A}'\mathbf{A})^{-1}\mathbf{A}'\mathbf{Y} = \begin{bmatrix} 2.8000 \\ 3.9667 \\ 6.7667 \\ 4.8833 \\ 10.5000 \\ 5.3000 \\ 6.7167 \\ 2.7000 \end{bmatrix},$$

and

$$SS(\mu, \alpha) = SS(\psi) = \hat{\psi}'\mathbf{A}'\mathbf{Y} = 1023.7833.$$

To obtain $SS(\alpha|\mu)$ we first calculate $SS(\mu) = N\bar{Y}_{..}^2$. From Table 5.5 we have $Y_{..} = 172.4$. Hence $\bar{Y}_{..} = 172.4/33 = 5.2242$, and $\bar{Y}_{..}^2 = (5.2242)^2 = 27.2927$. Therefore $SS(\mu) = 33 \times 27.2927 = 900.6593$, and $SS(\alpha|\mu) = SS(\mu, \alpha) - SS(\mu) = 1023.7833 - 900.6593 = 123.1240$.

Now since $\mathbf{P}_\mathbf{A}^2 = \mathbf{P}_\mathbf{A} = \mathbf{A}(\mathbf{A}'\mathbf{A})^{-1}\mathbf{A}'$ it follows that

$$\mathbf{G}'\mathbf{G} = \mathbf{\Lambda}'[\mathbf{B}'\mathbf{B} - \mathbf{B}'\mathbf{A}(\mathbf{A}'\mathbf{A})^{-1}\mathbf{A}'\mathbf{B}]\mathbf{\Lambda} = \mathbf{\Lambda}'[\mathbf{B}'\mathbf{B} - \mathbf{N}'(\mathbf{A}'\mathbf{A})^{-1}\mathbf{N}]\mathbf{\Lambda}$$

and

$$\mathbf{G}'\mathbf{Y} = \mathbf{\Lambda}'[\mathbf{B}'\mathbf{Y} - \mathbf{B}'\mathbf{A}(\mathbf{A}'\mathbf{A})^{-1}\mathbf{A}'\mathbf{Y}] = \mathbf{\Lambda}'(\mathbf{B}'\mathbf{Y} - \mathbf{N}'\hat{\psi}).$$

The matrix $\mathbf{B}'\mathbf{B}$ is a diagonal matrix and its diagonal elements are $n_{.j}$'s. The entries of the vector $\mathbf{B}'\mathbf{Y}$ are the $Y_{.j}$'s. The matrix $\mathbf{B}'\mathbf{B}$ and the vector

$\mathbf{B}'\mathbf{Y}$ are both available in Table 5.5. So we have $\mathbf{B}'\mathbf{B}=\text{diag}(2, 7, 6, 6, 9, 3)$ and $\mathbf{B}'\mathbf{Y}=(11.3,\ 26.3,\ 29.5,\ 42.3,\ 49.5,\ 13.5)$. The matrix $\mathbf{\Lambda}'$ is an $f_\beta \times b$ matrix with its entries being the coefficients of estimable β contrasts obtained from the counter triangle $\mathbf{C}_\beta$ in step 4. Hence

$$\mathbf{\Lambda}'=\begin{bmatrix} 1 & -1 & 0 & 0 & 0 & 0 \\ 1 & 0 & 0 & -1 & 0 & 0 \\ 0 & 0 & 1 & 0 & -1 & 0 \\ 1 & 0 & 0 & 0 & 0 & -1 \end{bmatrix}.$$

Thus

$$\mathbf{G}'\mathbf{G}=\begin{bmatrix} 1 & -1 & 0 & 0 & 0 & 0 \\ 1 & 0 & 0 & -1 & 0 & 0 \\ 0 & 0 & 1 & 0 & -1 & 0 \\ 1 & 0 & 0 & 0 & 0 & -1 \end{bmatrix} \left(\begin{bmatrix} 2 & 0 & 0 & 0 & 0 & 0 \\ 0 & 7 & 0 & 0 & 0 & 0 \\ 0 & 0 & 6 & 0 & 0 & 0 \\ 0 & 0 & 0 & 6 & 0 & 0 \\ 0 & 0 & 0 & 0 & 9 & 0 \\ 0 & 0 & 0 & 0 & 0 & 3 \end{bmatrix} \right.$$

$$- \begin{bmatrix} 0 & 2 & 0 & 0 & 0 & 0 & 0 & 0 \\ 0 & 4 & 0 & 3 & 0 & 0 & 0 & 0 \\ 2 & 0 & 3 & 0 & 0 & 0 & 0 & 1 \\ 0 & 0 & 0 & 1 & 0 & 0 & 5 & 0 \\ 1 & 0 & 0 & 0 & 2 & 4 & 0 & 2 \\ 0 & 0 & 0 & 2 & 0 & 0 & 1 & 0 \end{bmatrix} \cdot \begin{bmatrix} 3 & 0 & 0 & 0 & 0 & 0 & 0 & 0 \\ 0 & 6 & 0 & 0 & 0 & 0 & 0 & 0 \\ 0 & 0 & 3 & 0 & 0 & 0 & 0 & 0 \\ 0 & 0 & 0 & 6 & 0 & 0 & 0 & 0 \\ 0 & 0 & 0 & 0 & 2 & 0 & 0 & 0 \\ 0 & 0 & 0 & 0 & 0 & 4 & 0 & 0 \\ 0 & 0 & 0 & 0 & 0 & 0 & 6 & 0 \\ 0 & 0 & 0 & 0 & 0 & 0 & 0 & 3 \end{bmatrix}^{-1}$$

$$\left. \begin{bmatrix} 0 & 0 & 2 & 0 & 1 & 0 \\ 2 & 4 & 0 & 0 & 0 & 0 \\ 0 & 0 & 3 & 0 & 0 & 0 \\ 0 & 3 & 0 & 1 & 0 & 2 \\ 0 & 0 & 0 & 0 & 2 & 0 \\ 0 & 0 & 0 & 0 & 4 & 0 \\ 0 & 0 & 0 & 5 & 0 & 1 \\ 0 & 0 & 1 & 0 & 2 & 0 \end{bmatrix} \right) \cdot \begin{bmatrix} 1 & 1 & 0 & 1 \\ -1 & 0 & 0 & 0 \\ 0 & 0 & 1 & 0 \\ 0 & -1 & 0 & 0 \\ 0 & 0 & -1 & 0 \\ 0 & 0 & 0 & -1 \end{bmatrix},$$

and

$$\mathbf{G'Y}=\begin{bmatrix}1 & -1 & 0 & 0 & 0 & 0\\ 1 & 0 & 0 & -1 & 0 & 0\\ 0 & 0 & 1 & 0 & -1 & 0\\ 1 & 0 & 0 & 0 & 0 & -1\end{bmatrix}$$

$$\times\left(\begin{bmatrix}11.3\\ 26.3\\ 29.5\\ 42.3\\ 49.5\\ 13.5\end{bmatrix}-\begin{bmatrix}0 & 2 & 0 & 0 & 0 & 0 & 0 & 0\\ 0 & 4 & 0 & 3 & 0 & 0 & 0 & 0\\ 2 & 0 & 3 & 0 & 0 & 0 & 0 & 1\\ 0 & 0 & 0 & 1 & 0 & 0 & 5 & 0\\ 1 & 0 & 0 & 0 & 2 & 4 & 0 & 2\\ 0 & 0 & 0 & 2 & 0 & 0 & 1 & 0\end{bmatrix}\begin{bmatrix}2.8000\\ 3.9667\\ 6.7667\\ 4.8833\\ 10.5000\\ 5.3000\\ 6.7167\\ 2.7000\end{bmatrix}\right).$$

After required calculations, we obtain

$$\begin{bmatrix}6.8333 & 2.1667 & 0.0000 & 1.6667\\ 2.1667 & 3.0000 & 0.0000 & 0.1667\\ 0.0000 & 0.0000 & 5.3333 & 0.0000\\ 1.6667 & 0.1667 & 0.0000 & 3.5000\end{bmatrix}\hat{\boldsymbol{\xi}}=\begin{bmatrix}7.5833\\ -0.4666\\ 1.7999\\ 6.3499\end{bmatrix}$$

$$\hat{\boldsymbol{\xi}}=(\mathbf{G'G})^{-1}\mathbf{G'Y}=\begin{bmatrix}1.1118\\ -1.0326\\ 0.3375\\ 1.3340\end{bmatrix},$$

and

$$\mathrm{SS}(\beta\,|\,\alpha)=\mathrm{SS}(\xi)=\hat{\boldsymbol{\xi}}'\mathbf{G'Y}=17.9912.$$

This quantity is the deviation sum of squares for testing the linear hypothesis H_0: $\boldsymbol{\Lambda}'\boldsymbol{\beta}=\mathbf{0}$. This is the hypothesis that all estimable linear parametric functions involving only the β parameters are zero.

Observe that we have first calculated the sum of squares for β effect adjusted for α effect. One should first obtain the adjusted SS for the parameter set with the fewer degrees of freedom. This will allow one to obtain the basic pertinent SS by inverting a matrix of dimension equal to $\min(f_\alpha, f_\beta)$.

Step 6 Obtain a solution $\tilde{\boldsymbol{\alpha}}$ and $\tilde{\boldsymbol{\beta}}$ to the reduced normal equations. Set $\tilde{\boldsymbol{\beta}}=\boldsymbol{\Lambda}\hat{\boldsymbol{\xi}}$. Then $\tilde{\boldsymbol{\beta}}$ obtained in this way is a solution to the reduced normal equations for the β parameter vector. Hence

$$\tilde{\boldsymbol{\beta}} = \begin{bmatrix} 1 & 1 & 0 & 1 \\ -1 & 0 & 0 & 0 \\ 0 & 0 & 1 & 0 \\ 0 & -1 & 0 & 0 \\ 0 & 0 & -1 & 0 \\ 0 & 0 & 0 & -1 \end{bmatrix} \begin{bmatrix} 1.1118 \\ -1.0326 \\ 0.3375 \\ 1.3340 \end{bmatrix} = \begin{bmatrix} 1.4132 \\ -1.1118 \\ 0.3375 \\ 1.0326 \\ -0.3375 \\ -1.3340 \end{bmatrix}.$$

Set $\tilde{\boldsymbol{\alpha}} = \hat{\boldsymbol{\psi}} - (\mathbf{A}'\mathbf{A})^{-1}\mathbf{N}\tilde{\boldsymbol{\beta}}$. Then $\tilde{\boldsymbol{\alpha}}$ obtained in this way is a solution to the reduced normal equations for the $\boldsymbol{\alpha}$ parameter vector. So

$$\tilde{\boldsymbol{\alpha}} = \begin{bmatrix} 2.8000 \\ 3.9667 \\ 6.7667 \\ 4.8833 \\ 10.5000 \\ 5.3000 \\ 6.7167 \\ 2.7000 \end{bmatrix} - \begin{bmatrix} 3 & 0 & 0 & 0 & 0 & 0 & 0 & 0 \\ 0 & 6 & 0 & 0 & 0 & 0 & 0 & 0 \\ 0 & 0 & 3 & 0 & 0 & 0 & 0 & 0 \\ 0 & 0 & 0 & 6 & 0 & 0 & 0 & 0 \\ 0 & 0 & 0 & 0 & 2 & 0 & 0 & 0 \\ 0 & 0 & 0 & 0 & 0 & 4 & 0 & 0 \\ 0 & 0 & 0 & 0 & 0 & 0 & 6 & 0 \\ 0 & 0 & 0 & 0 & 0 & 0 & 0 & 3 \end{bmatrix}^{-1} \begin{bmatrix} 0 & 0 & 2 & 0 & 1 & 0 \\ 2 & 4 & 0 & 0 & 0 & 0 \\ 0 & 0 & 3 & 0 & 0 & 0 \\ 0 & 3 & 0 & 1 & 0 & 2 \\ 0 & 0 & 0 & 0 & 2 & 0 \\ 0 & 0 & 0 & 0 & 4 & 0 \\ 0 & 0 & 0 & 5 & 0 & 1 \\ 0 & 0 & 1 & 0 & 2 & 0 \end{bmatrix}$$

$$\times \begin{bmatrix} 1.4132 \\ -1.1118 \\ 0.3375 \\ 1.0326 \\ -0.3375 \\ -1.3340 \end{bmatrix} = \begin{bmatrix} 2.6875 \\ 4.2368 \\ 6.4292 \\ 5.7118 \\ 10.8375 \\ 5.6375 \\ 6.0785 \\ 2.8125 \end{bmatrix}.$$

The random vector $\tilde{\boldsymbol{\theta}}' = (\mathbf{0}, \tilde{\boldsymbol{\alpha}}', \tilde{\boldsymbol{\beta}}')$ is a solution to the reduced normal equations, $\mathbf{X}'\mathbf{X}\tilde{\boldsymbol{\theta}} = \mathbf{X}'\mathbf{Y}$, and is the unique solution to the following minimization problem:

$$\min \|\mathbf{Y} - \mathbf{1}\mu - \mathbf{A}\boldsymbol{\alpha} - \mathbf{B}\boldsymbol{\beta}\|^2$$

subject to

$$\mu = 0$$

$$\boldsymbol{\alpha} \in R^a$$

$$\sum_j \beta_j = 0.$$

For comparison with the estimates calculated by the computer program at the end of this chapter, we will impose the restriction

$$\sum_{i=1}^{8} \hat{\alpha}_i = 0.$$

From $\tilde{\boldsymbol{\alpha}}$ calculate $\hat{\boldsymbol{\alpha}} = \tilde{\boldsymbol{\alpha}} - \tilde{\mu}\mathbf{1}$, where $\tilde{\mu} = \frac{1}{8}\mathbf{1}'\tilde{\boldsymbol{\alpha}} = 44.431/8 = 5.5539$. Now set

$$\hat{\boldsymbol{\alpha}} = \begin{bmatrix} 2.6875 \\ 4.2368 \\ 6.4292 \\ 5.7118 \\ 10.8375 \\ 5.6375 \\ 6.0785 \\ 2.8125 \end{bmatrix} - \begin{bmatrix} 5.5539 \\ 5.5539 \\ 5.5539 \\ 5.5539 \\ 5.5539 \\ 5.5539 \\ 5.5539 \\ 5.5539 \end{bmatrix} = \begin{bmatrix} -2.8664 \\ -1.3171 \\ 0.8753 \\ 0.1579 \\ 5.2836 \\ 0.0836 \\ 0.5246 \\ -2.7414 \end{bmatrix}.$$

Note that $\tilde{\boldsymbol{\beta}}$ is invariant under any choice of the basis matrix $\boldsymbol{\Lambda}$.

Step 7 Find the regression sum of squares. The regression sum of squares $SS(\theta)$ can now be obtained by

$$SS(\theta) = SS(\psi) + SS(\xi) = 1023.7833 + 17.9912 = 1041.7745.$$

Note that

$$\begin{aligned} SS(\alpha) &= SS(\mu, \alpha) = SS(\psi), \\ SS(\mu, \alpha, \beta) &= SS(\alpha, \beta) = SS(\psi, \xi) \\ &= SS(\psi) + SS(\xi). \end{aligned}$$

Step 8 Calculate the sums of squares for total and the residuals. The total sum of squares SS_{tot} is simply obtained by

$$SS_{tot} = \sum_i \sum_j \sum_k Y_{ijk}^2 = (2.3)^2 + (4.5)^2 + \cdots + (1.5)^2 = 1184.06,$$

and the residual sum of squares $\hat{R}$ is obtained by subtracting the regression sum of squares from SS_{tot}:

$$\hat{R} = SS_{tot} - SS(\theta) = 1184.06 - 1041.7745 = 142.2855.$$

If the partition of $\hat{R}$ into sums of squares for pure error and lack of fit is desired, calculate

$$\begin{aligned}SS_{\text{pure error}} &= \sum_{ijk}(Y_{ijk}-\bar{Y}_{ij.})^2\\ &=(2.3-3.40)^2+(4.5-3.40)^2+\cdots+(1.5-2.85)^2\\ &=110.40.\end{aligned}$$

This sum of squares has $N-q=33-14=19$ degrees of freedom. By subtraction we then obtain the sum of squares attributable to lack of fit. So

$$\begin{aligned}SS_{\text{lack of fit}} &= \hat{R}-SS_{\text{pure error}}\\ &=142.29-110.40\\ &=31.89\end{aligned}$$

with $q-m=14-12=2$ degrees of freedom.

Step 9 Obtain the sum of squares for α effects adjusted for β effects $SS(\alpha|\beta)$.

This quantity is obtained by changing the role of **A** and **G** in the parametrized model. That is,

$$E(\mathbf{Y})=\mathbf{B}\psi+\mathbf{G}\xi$$

where $\mathbf{G}=(\mathbf{I}-\mathbf{P_B})\mathbf{A}\boldsymbol{\Gamma}$, $\mathbf{P_B}$ is the orthogonal projection on $R(\mathbf{B})$, and $\boldsymbol{\Gamma}'\boldsymbol{\alpha}$ denotes an $f_\alpha\times 1$ vector of estimable parametric functions which constitute a basis for all estimable functions involving only the α effects.

To find $SS(\alpha|\beta)$ we only need to calculate $SS(\mu,\beta)$, and then by subtracting this quantity from the regression sum of squares (already calculated) we obtain $SS(\alpha|\beta)$. The reduce normal equations are

$$\mathbf{B}'\mathbf{B}\hat{\psi}=\mathbf{B}'\mathbf{Y} \quad \text{and} \quad \mathbf{G}'\mathbf{G}\hat{\xi}=\mathbf{G}'\mathbf{Y}.$$

So we obtain

$$\begin{bmatrix}2&0&0&0&0&0\\0&7&0&0&0&0\\0&0&6&0&0&0\\0&0&0&6&0&0\\0&0&0&0&9&0\\0&0&0&0&0&3\end{bmatrix}\hat{\psi}=\begin{bmatrix}11.3\\26.3\\29.5\\42.3\\49.5\\13.5\end{bmatrix}.$$

$$\hat{\psi}=\begin{bmatrix}5.6500\\3.7571\\4.9167\\7.0500\\5.5000\\4.5000\end{bmatrix}.$$

So

$$SS(\mu, \beta) = SS(\beta) = SS(\psi) = \hat{\psi}'\mathbf{B}'\mathbf{Y} = 938.9146,$$

and

$$SS(\beta|\mu) = 938.9146 - 900.6593 = 38.2553.$$

So we find

$$SS(\alpha|\beta) = SS(\theta) - SS(\beta)$$
$$= 1041.7745 - 938.9146 = 102.8599.$$

This sum of squares is the sum of squares associated with testing the hypothesis that $\mathbf{\Gamma}'\boldsymbol{\alpha} = \mathbf{0}$, that is, the hypothesis that all estimable functions involving only the α effects are zero. In the usual terminology this is a *testable hypothesis* in the sense that each component of the vector $\mathbf{\Gamma}'\boldsymbol{\alpha}$ is an estimable parametric function.

Step 10 Form the analysis of variance table. Let us name the α effects and β effects treatments and blocks, respectively. Then we have:

Source of Variation	Degrees of Freedom		Sums of Squares	
Regression	12		1041.77	
Fitting overall mean		1		900.66
Blocks unadjusted		5		38.25
Treatments		6		102.86
Fitting overall mean		1		900.66
Treatments unadjusted		7		123.12
Blocks		4		17.99
Residual	21		142.29	
pure error		19		110.40
Lack of fit		2		31.89
Total	33		1184.06	

Step 11 Find the covariance matrix for the block effects.

$$\text{Cov}(\tilde{\boldsymbol{\beta}}) = \sigma^2 \mathbf{V}_{\boldsymbol{\beta}}, \qquad \text{where } \mathbf{V}_{\boldsymbol{\beta}} = \mathbf{\Lambda}(\mathbf{G}'\mathbf{G})^{-1}\mathbf{\Lambda}'.$$

$$\mathbf{\Lambda}(\mathbf{G}'\mathbf{G})^{-1}\mathbf{\Lambda}' = \begin{bmatrix} 1 & 1 & 0 & 1 \\ -1 & 0 & 0 & 0 \\ 0 & 0 & 1 & 0 \\ 0 & -1 & 0 & 0 \\ 0 & 0 & -1 & 0 \\ 0 & 0 & 0 & -1 \end{bmatrix}$$

$$\times\begin{bmatrix} 0.2182 & -0.1522 & 0.0000 & -0.0966 \\ -0.1522 & 0.4404 & 0.0000 & 0.0515 \\ 0.0000 & 0.0000 & 0.1875 & 0.0000 \\ -0.0966 & 0.0515 & 0.0000 & 0.3293 \end{bmatrix}\begin{bmatrix} 1 & -1 & 0 & 0 & 0 & 0 \\ 1 & 0 & 0 & -1 & 0 & 0 \\ 0 & 0 & 1 & 0 & -1 & 0 \\ 1 & 0 & 0 & 0 & 0 & -1 \end{bmatrix}$$

$$=\begin{bmatrix} 0.5932 & 0.0307 & 0.0000 & -0.3397 & 0.0000 & -0.2841 \\ 0.0307 & 0.2182 & 0.0000 & -0.1522 & 0.0000 & -0.0966 \\ 0.0000 & 0.0000 & 0.1875 & 0.0000 & -0.1875 & 0.0000 \\ -0.3397 & -0.1522 & 0.0000 & 0.4404 & 0.0000 & 0.0515 \\ 0.0000 & 0.0000 & -0.1875 & 0.0000 & 0.1875 & 0.0000 \\ -0.2841 & -0.0966 & 0.0000 & 0.0515 & 0.0000 & 0.3293 \end{bmatrix}.$$

Now suppose it is desired to test a hypothesis of the form H_0: $\mathbf{H}'\boldsymbol{\beta}=\mathbf{0}$ where the components of $\mathbf{H}'\boldsymbol{\theta}$ are estimable parametric functions for the β effects. For making such a hypothesis test, we know that if $\tilde{\boldsymbol{\beta}}$ is any solution to the reduced normal equations and if $\mathbf{V_H}$ is such that $\sigma^2\mathbf{V_H}=\text{Cov}(\mathbf{H}'\tilde{\boldsymbol{\beta}})$, then the appropriate numerator sum of squares for this test is

$$(\mathbf{H}'\boldsymbol{\beta})'\mathbf{V}_{\mathbf{H}}^{-1}(\mathbf{H}'\boldsymbol{\theta}).$$

(If we take the components of $\mathbf{H}'\tilde{\boldsymbol{\beta}}$ to be linearly independent parametric functions then $\mathbf{V_H}$ is inversible.) The degrees of freedom associated to this sum of squares is equal to $r(\mathbf{V_H})$ which is equal to the number of rows in the matrix $\mathbf{H}'$.

Step 12 Test H_0: $\mathbf{H}'\boldsymbol{\beta}=\mathbf{0}$, with, for example,

$$\mathbf{H}'=\begin{bmatrix} 5 & -1 & -1 & -1 & -1 & -1 \\ 1 & -1 & 0 & 0 & 0 & 0 \\ 0 & 3 & -1 & -1 & -1 & 0 \end{bmatrix}.$$

In this case the estimate of the hypothesis vector is

$$\mathbf{H}'\tilde{\boldsymbol{\beta}}=\begin{bmatrix} 5 & -1 & -1 & -1 & -1 & -1 \\ 1 & -1 & 0 & 0 & 0 & 0 \\ 0 & 3 & -1 & -1 & -1 & 0 \end{bmatrix}\begin{bmatrix} 1.4132 \\ -1.1118 \\ 0.3375 \\ 1.0326 \\ -0.3375 \\ -1.3340 \end{bmatrix}=\begin{bmatrix} 8.4792 \\ 2.5250 \\ -4.3681 \end{bmatrix}.$$

Compute the covariance matrix of estimator for the hypothesis vector $\mathbf{V_H}$ as

$$\mathbf{V_H}=\mathbf{H'V_\beta H}=\begin{bmatrix}5 & -1 & -1 & -1 & -1 & -1\\ 1 & -1 & 0 & 0 & 0 & 0\\ 0 & 3 & -1 & -1 & -1 & 0\end{bmatrix}[\mathbf{V}_\beta,\text{ as in step 11}]\begin{bmatrix}5 & 1 & 0\\ -1 & -1 & 3\\ -1 & 0 & -1\\ -1 & 0 & -1\\ -1 & 0 & -1\\ -1 & 0 & 0\end{bmatrix}$$

$$=\begin{bmatrix}21.3542 & 3.3750 & 2.5903\\ 3.3750 & 0.7500 & -0.3750\\ 2.5903 & -0.3750 & 3.3171\end{bmatrix}.$$

The sum of squares for this hypothesis test is

$$SS_{H_0}=(\mathbf{H}'\tilde{\boldsymbol{\beta}})'\mathbf{V}_{\mathbf{H}}^{-1}(\mathbf{H}'\tilde{\boldsymbol{\beta}})$$

$$=[8.4792 \quad 2.5250 \quad -4.3681]\begin{bmatrix}3.1296 & -16.2221 & -4.2777\\ -16.2221 & 85.4994 & 22.3331\\ -4.2777 & 22.3331 & 6.1666\end{bmatrix}$$

$$\times\begin{bmatrix}8.4792\\ 2.5250\\ -4.3681\end{bmatrix}$$

$$=17.3825.$$

Similar calculations can be performed for a hypothesis specification of the form $\mathbf{G}'\boldsymbol{\alpha}=\mathbf{0}$.

See computer outputs at the end of this chapter for different hypotheses specifications on treatments and blocks. Covariance matrices between different estimators along with some additional information are also given for the same example.

5.9 REMARKS

For an additive two-way classification model with an arbitrary incidence matrix **N**, two basic problems were considered in Section 5.4. The problem of determining which cell expectations are estimable was considered first, and secondly the problem of determining a basis for the estimable functions involving only one effect was considered. The solutions proposed for these two problems were, respectively, the final matrix **M** and the counter triangle **C**.

Our approach of first obtaining the final matrix before attempting to answer more pertinent questions deserves some comment. Most questions regarding estimability, or the analysis of the data, are more easily answered by means of the final matrix than in terms of the incidence matrix. For example, determining the connected portions of the design [in the sense of Bose (1949)] is trivially answered from the final matrix; whereas a search procedure would probably be the alternative approach when only the incidence matrix is utilized; for example, see the adaption by Searle (1971) of the procedure described by Weeks and Williams (1964). Also, the R process used for obtaining the final matrix is very easily programmed for an electronic computer. Moreover, most data sets will not have too many missing cells and thus the R process will converge very rapidly (usually one iteration) and hence use very little computer time.

Although the final matrix $\mathbf{M}$ and the counter triangle $\mathbf{C}$ seem, for the model we considered, to have satisfactorily answered the two basic questions posed at the outset of Section 5.4, it is also possible to find a spanning set for the estimable parametric functions involving only a single effect with some additional effort by employing the design matrix $\mathbf{X}$ as opposed to the incidence matrix $\mathbf{N}$.

Suppose that $\mathbf{Y}$ is a random vector such that $E(\mathbf{Y}) \in \Omega$ where Ω is a subspace and assume $E(\mathbf{Y}) = \mathbf{X}\boldsymbol{\theta}$, $\boldsymbol{\theta}$ unknown, is a reparametrization for $E(\mathbf{Y})$, where $\mathbf{X}$, the design matrix consisting of zeros and ones is an $N \times p$ matrix of rank r and $\boldsymbol{\theta}$ is a $p \times 1$ vector of unknown parameters. Assume further that $\mathbf{X}\boldsymbol{\theta}$ is partitioned as given in Theorem 5.7.

From Theorem 5.7, we know that the parametric function $\boldsymbol{\nu}'\boldsymbol{\beta}$ is estimable if and only if there exists some vector $\mathbf{x}$ in R^N such that $\boldsymbol{\nu} = \mathbf{B}'\mathbf{x}$ and $\mathbf{0} = \mathbf{1}'\mathbf{x} = \mathbf{A}'\mathbf{x}$. Then $\boldsymbol{\nu} = \mathbf{B}'\mathbf{x}$ where $\mathbf{A}'\mathbf{x} = \mathbf{0}$. This means that $\mathbf{x} \in N(\mathbf{A}') = R(\mathbf{A}) = R(\mathbf{I} - \mathbf{P}_\mathbf{A})$, so $\mathbf{x} = (\mathbf{I} - \mathbf{P}_\mathbf{A})\mathbf{y}$ for some vector $\mathbf{y}$ and $\boldsymbol{\nu} = \mathbf{B}'\mathbf{x} = \mathbf{B}'(\mathbf{I} - \mathbf{P}_\mathbf{A})\mathbf{y} \in R[\mathbf{B}'(\mathbf{I} - \mathbf{P}_\mathbf{A})]$. Suppose now that $\boldsymbol{\nu} \in R[\mathbf{B}'(\mathbf{I} - \mathbf{P}_\mathbf{A})]$. Then $\boldsymbol{\nu} = \mathbf{B}'(\mathbf{I} - \mathbf{P}_\mathbf{A})\mathbf{y}$ for some vector $\mathbf{y}$. Let $\mathbf{x} = (\mathbf{I} - \mathbf{P}_\mathbf{A})\mathbf{y}$. Then $\boldsymbol{\nu} = \mathbf{B}'\mathbf{x}$ where $\mathbf{A}'\mathbf{x} = \mathbf{A}'(\mathbf{I} - \mathbf{P}_\mathbf{A})\mathbf{y} = \mathbf{0}$ and $\boldsymbol{\nu}'\boldsymbol{\beta}$ is estimable.

Now let $\mathbf{C}$ be a matrix such that $R(\mathbf{C}) = N(\mathbf{A}')$. From Theorem 5.7, we had

$$\bar{\mathscr{B}} = \{\boldsymbol{\nu}'\boldsymbol{\beta} \colon \boldsymbol{\nu} \in R(\mathbf{B}'\mathbf{C})\}.$$

Therefore, the crucial problem in determining the vector space of estimable parametric functions involving only β contrasts is to find the $N(\mathbf{A}')$. The column vectors of the matrix $\mathbf{C}$ given in the following proposition form a basis for the $N(\mathbf{A}')$. Once this basis is found we can found a spanning set of estimable β contrasts by multiplying $\mathbf{B}'$ by $\mathbf{C}$. In a similar way we can obtain the space of estimable parametric functions involving only the α effects. In general we can choose the matrix $\mathbf{I} - \mathbf{P}_\mathbf{A}$ for the matrix $\mathbf{C}$, as we have seen from Theorem 5.7. Here we give an alternative method for obtaining the matrix $\mathbf{C}$.

PROPOSITION Let $\mathbf{T}' = (\mathbf{T}'_1, \mathbf{T}'_2)$ be an $s \times N$ matrix. Suppose $\mathbf{T}'_1$ is nonsingular. Then the columns of the matrix

$$\mathbf{C} = \begin{bmatrix} -(\mathbf{T}'_1)^{-1}\mathbf{T}'_2 \\ \mathbf{I}_{N-s} \end{bmatrix}$$

form a basis for $N(\mathbf{T}')$.

Proof Let $\mathbf{u} = (\mathbf{u}_1, \mathbf{u}_2)'$ be a vector in R^N such that $\mathbf{u}_1 \in R^s$ and $\mathbf{u}_2 \in R^{N-s}$. To find the null space of $\mathbf{T}'$ we need to calculate the vector $\mathbf{u}$ such that $\mathbf{T}'\mathbf{u} = \mathbf{0}$, or $(\mathbf{T}'_1, \mathbf{T}'_2)\binom{\mathbf{u}_1}{\mathbf{u}_2} = \mathbf{0}$, which implies $\mathbf{T}'_1\mathbf{u}_1 + \mathbf{T}'_2\mathbf{u}_2 = \mathbf{0}$, or $\mathbf{T}'_1\mathbf{u}_1 = -\mathbf{T}'_2\mathbf{u}_2$, and $\mathbf{u}_1 = -(\mathbf{T}'_1)^{-1}\mathbf{T}'_2\mathbf{u}_2$. Since the nullity of $N(\mathbf{T}')$ is $N-s$, we must choose $(N-s)$ independent vectors for $\mathbf{u}_2$. An obvious choice for $\mathbf{u}_2$ is to choose $(1, 0, \ldots, 0)'$, $(0, 1, 0, \ldots, 0)'$, and so on. Hence the result. □

There are many algorithms available for finding such a matrix. Below we give a well-known algorithm which employs the householder transformation.

Householder Transformation

Triangular reduction. Let $\mathbf{A}$ be an $m \times n$ matrix and $m \geqslant n$. Then there exists a unitary matrix $\mathbf{B}$ of order m such that

$$\mathbf{BA} = \begin{pmatrix} \mathbf{T} \\ \mathbf{0} \end{pmatrix}$$

where $\mathbf{T}$ is an upper triangular matrix of order n and $\mathbf{0}$ is a null matrix of order $(m-n) \times n$.

We now consider the $\mathbf{A}'$ matrix and reduce it by a series of householder transformation (HT).

As step 0, eliminate all the redundant columns of the matrix $\mathbf{A}'$. Now consider the first column of $\mathbf{A}'$. If it has a nonzero (here a 1) element, apply (HT) to have a nonzero value in the first position and zero elsewhere in the first column. If all the elements of the first column are zero, move to the nearest column which has a nonzero value, say the ith, and apply HT to have a nonzero value in the first position and zero elsewhere in the jth column. Now omit the first row and repeat the process stated on the reduced matrix, and so on, till all the columns of the $\mathbf{A}'$ are covered. Then by a rearrangement of columns (i.e., by renaming the parameters), if necessary, the reduced matrix is of the form

$$\begin{pmatrix} \mathbf{P} & \mathbf{Q} \\ \mathbf{0} & \mathbf{0} \end{pmatrix}$$

where **P** is an upper triangular and nonsingular matrix of rank s and **Q** is a $s \times (N-s)$ matrix.

The column vectors of the matrix **C** below form a basis for $N(\mathbf{A}')$:

$$\mathbf{C} = \begin{bmatrix} -\mathbf{P}^{-1}\mathbf{Q} \\ \mathbf{I}_{N-s} \end{bmatrix}.$$

Note that when we interchange the columns of the matrix $\mathbf{A}'$ we have to record the accomplish exchange.

EXAMPLE 5.7 Consider Example 5.4 in which we had an additive two-way model with $a=8$ and $b=6$, and an incidence matrix $\mathbf{N}=(n_{ij})$ as follows:

$$\mathbf{N} = \begin{bmatrix} 0 & 0 & 2 & 0 & 1 & 0 \\ 2 & 4 & 0 & 0 & 0 & 0 \\ 0 & 0 & 3 & 0 & 0 & 0 \\ 0 & 3 & 0 & 1 & 0 & 2 \\ 0 & 0 & 0 & 0 & 2 & 0 \\ 0 & 0 & 0 & 0 & 4 & 0 \\ 0 & 0 & 0 & 5 & 0 & 1 \\ 0 & 0 & 1 & 0 & 2 & 0 \end{bmatrix}.$$

For estimability considerations relative to this two-way model

$$E(Y_{ijk}) = \mu + \alpha_i + \beta_j,$$

$$i=1, \ldots, 8, \qquad j=1, \ldots, 6, \quad k=1, \ldots, n_{ij}$$

we need only suppose the nonzero elements of **N** are ones and this would be exactly the same matrix obtained by changing all nonzero entries in **N** to ones. Let **N*** denote such a matrix.

In this way we will not have repeated rows in the reduced design matrix **X***. So we have

$$\mathbf{N}^* = \begin{bmatrix} 0 & 0 & 1 & 0 & 1 & 0 \\ 1 & 1 & 0 & 0 & 0 & 0 \\ 0 & 0 & 1 & 0 & 0 & 0 \\ 0 & 1 & 0 & 1 & 0 & 1 \\ 0 & 0 & 0 & 0 & 1 & 0 \\ 0 & 0 & 0 & 0 & 1 & 0 \\ 0 & 0 & 0 & 1 & 0 & 1 \\ 0 & 0 & 1 & 0 & 1 & 0 \end{bmatrix}.$$

The reduce design matrix $\mathbf{X}^*$ for $\mathbf{N}^*$ (with no repeated rows) is

$$
\begin{array}{c}
\begin{array}{ccccccccc|cccccc}
\mu & \alpha_1 & \alpha_2 & \alpha_3 & \alpha_4 & \alpha_5 & \alpha_6 & \alpha_7 & \alpha_8 & \beta_1 & \beta_2 & \beta_3 & \beta_4 & \beta_5 & \beta_6
\end{array} \\
\mathbf{X}^* = \left[\begin{array}{ccccccccc|cccccc}
1 & 1 & 0 & 0 & 0 & 0 & 0 & 0 & 0 & 0 & 0 & 1 & 0 & 0 & 0 \\
1 & 1 & 0 & 0 & 0 & 0 & 0 & 0 & 0 & 0 & 0 & 0 & 0 & 1 & 0 \\
1 & 0 & 1 & 0 & 0 & 0 & 0 & 0 & 0 & 1 & 0 & 0 & 0 & 0 & 0 \\
1 & 0 & 1 & 0 & 0 & 0 & 0 & 0 & 0 & 0 & 1 & 0 & 0 & 0 & 0 \\
1 & 0 & 0 & 1 & 0 & 0 & 0 & 0 & 0 & 0 & 0 & 1 & 0 & 0 & 0 \\
1 & 0 & 0 & 0 & 1 & 0 & 0 & 0 & 0 & 0 & 1 & 0 & 0 & 0 & 0 \\
1 & 0 & 0 & 0 & 1 & 0 & 0 & 0 & 0 & 0 & 0 & 0 & 1 & 0 & 0 \\
1 & 0 & 0 & 0 & 1 & 0 & 0 & 0 & 0 & 0 & 0 & 0 & 0 & 0 & 1 \\
1 & 0 & 0 & 0 & 0 & 1 & 0 & 0 & 0 & 0 & 0 & 0 & 0 & 1 & 0 \\
1 & 0 & 0 & 0 & 0 & 0 & 1 & 0 & 0 & 0 & 0 & 0 & 0 & 1 & 0 \\
1 & 0 & 0 & 0 & 0 & 0 & 0 & 0 & 1 & 0 & 0 & 0 & 1 & 0 & 0 \\
1 & 0 & 0 & 0 & 0 & 0 & 0 & 1 & 0 & 0 & 0 & 0 & 0 & 0 & 1 \\
1 & 0 & 0 & 0 & 0 & 0 & 0 & 0 & 1 & 0 & 0 & 1 & 0 & 0 & 0 \\
1 & 0 & 0 & 0 & 0 & 0 & 0 & 0 & 1 & 0 & 0 & 0 & 0 & 1 & 0
\end{array}\right]. \\
\qquad\qquad\qquad \mathbf{A} \qquad\qquad\qquad\qquad\qquad \mathbf{B}
\end{array}
$$

Suppose we want to find the vector space of estimable parametric functions involving the β effects. So we are required to find the $N(\mathbf{A}')$. Now consider the matrix $\mathbf{A}'$ given below:

$$
\mathbf{A}' = \begin{bmatrix}
1 & 1 & 1 & 1 & 1 & 1 & 1 & 1 & 1 & 1 & 1 & 1 & 1 & 1 \\
1 & 1 & 0 & 0 & 0 & 0 & 0 & 0 & 0 & 0 & 0 & 0 & 0 & 0 \\
0 & 0 & 1 & 1 & 0 & 0 & 0 & 0 & 0 & 0 & 0 & 0 & 0 & 0 \\
0 & 0 & 0 & 0 & 1 & 0 & 0 & 0 & 0 & 0 & 0 & 0 & 0 & 0 \\
0 & 0 & 0 & 0 & 0 & 1 & 1 & 1 & 0 & 0 & 0 & 0 & 0 & 0 \\
0 & 0 & 0 & 0 & 0 & 0 & 0 & 0 & 1 & 0 & 0 & 0 & 0 & 0 \\
0 & 0 & 0 & 0 & 0 & 0 & 0 & 0 & 0 & 1 & 0 & 0 & 0 & 0 \\
0 & 0 & 0 & 0 & 0 & 0 & 0 & 0 & 0 & 0 & 1 & 1 & 0 & 0 \\
0 & 0 & 0 & 0 & 0 & 0 & 0 & 0 & 0 & 0 & 0 & 0 & 1 & 1
\end{bmatrix}
$$

In $\mathbf{A}'$ there is a nonzero element in position (1, 1). By subtracting row 2 from row 1, we obtain

$$\begin{bmatrix}
1 & 1 & 1 & 1 & 1 & 1 & 1 & 1 & 1 & 1 & 1 & 1 & 1 & 1 \\
\hline
0 & 0 & -1 & -1 & -1 & -1 & -1 & -1 & -1 & -1 & -1 & -1 & -1 & -1 \\
0 & 0 & 1 & 1 & 0 & 0 & 0 & 0 & 0 & 0 & 0 & 0 & 0 & 0 \\
0 & 0 & 0 & 0 & 1 & 0 & 0 & 0 & 0 & 0 & 0 & 0 & 0 & 0 \\
0 & 0 & 0 & 0 & 0 & 1 & 1 & 1 & 0 & 0 & 0 & 0 & 0 & 0 \\
0 & 0 & 0 & 0 & 0 & 0 & 0 & 0 & 1 & 0 & 0 & 0 & 0 & 0 \\
0 & 0 & 0 & 0 & 0 & 0 & 0 & 0 & 0 & 1 & 0 & 0 & 0 & 0 \\
0 & 0 & 0 & 0 & 0 & 0 & 0 & 0 & 0 & 0 & 1 & 1 & 0 & 0 \\
0 & 0 & 0 & 0 & 0 & 0 & 0 & 0 & 0 & 0 & 0 & 0 & 1 & 1
\end{bmatrix}.$$

If we continue this process, after some iterations we obtain

$$\left[\begin{array}{cccccccc|cccccc}
1 & 1 & 1 & 1 & 1 & 1 & 1 & 1 & 1 & 1 & 1 & 1 & 1 & 1 \\
0 & 1 & 0 & 0 & 0 & 0 & 0 & 0 & 1 & 0 & 0 & 0 & 0 & 0 \\
0 & 0 & 1 & 0 & 0 & 0 & 0 & 0 & 0 & 0 & 0 & 0 & 0 & 1 \\
0 & 0 & 0 & 1 & 0 & 0 & 0 & 0 & 0 & 0 & 0 & 0 & 0 & 0 \\
0 & 0 & 0 & 0 & 1 & 0 & 0 & 0 & 0 & 1 & 1 & 0 & 0 & 0 \\
0 & 0 & 0 & 0 & 0 & 1 & 0 & 0 & 0 & 0 & 0 & 0 & 0 & 0 \\
0 & 0 & 0 & 0 & 0 & 0 & 1 & 0 & 0 & 0 & 0 & 0 & 0 & 0 \\
0 & 0 & 0 & 0 & 0 & 0 & 0 & 1 & 0 & 0 & 0 & 1 & 0 & 0 \\
\hline
0 & 0 & 0 & 0 & 0 & 0 & 0 & 0 & 0 & 0 & 0 & 0 & 0 & 0
\end{array}\right] = \left[\begin{array}{c|c} \mathbf{P} & \mathbf{Q} \\ \hline \mathbf{0} & \mathbf{0} \end{array}\right].$$

$$\boxed{1 \quad 3 \quad 13 \quad 5 \quad 6 \quad 9 \quad 10 \quad 11 \qquad 4 \quad 7 \quad 8 \quad 12 \quad 2 \quad 14}$$

The numbers inside the closed block indicate the column exchange. So we obtain

$$\begin{bmatrix} -\mathbf{P}^{-1}\mathbf{Q} \\ \mathbf{I}_6 \end{bmatrix} = \left[\begin{array}{cccccc|c}
0 & 0 & 0 & 0 & -1 & 0 & 1 \\
-1 & 0 & 0 & 0 & 0 & 0 & 3 \\
0 & 0 & 0 & 0 & 0 & -1 & 13 \\
0 & 0 & 0 & 0 & 0 & 0 & 5 \\
0 & -1 & -1 & 0 & 0 & 0 & 6 \\
0 & 0 & 0 & 0 & 0 & 0 & 9 \\
0 & 0 & 0 & 0 & 0 & 0 & 10 \\
0 & 0 & 0 & -1 & 0 & 0 & 11 \\
1 & 0 & 0 & 0 & 0 & 0 & 4 \\
0 & 1 & 0 & 0 & 0 & 0 & 7 \\
0 & 0 & 1 & 0 & 0 & 0 & 8 \\
0 & 0 & 0 & 1 & 0 & 0 & 12 \\
0 & 0 & 0 & 0 & 1 & 0 & 2 \\
0 & 0 & 0 & 0 & 0 & 1 & 14
\end{array}\right].$$

The matric **C** [the columns of which form a basis for the $N(\mathbf{A}')$] is obtained by appropriate permutation of rows as

$$\mathbf{C}=\begin{bmatrix} 0 & 0 & 0 & 0 & -1 & 0 \\ 0 & 0 & 0 & 0 & 1 & 0 \\ -1 & 0 & 0 & 0 & 0 & 0 \\ 1 & 0 & 0 & 0 & 0 & 0 \\ 0 & 0 & 0 & 0 & 0 & 0 \\ 0 & -1 & -1 & 0 & 0 & 0 \\ 0 & 1 & 0 & 0 & 0 & 0 \\ 0 & 0 & 1 & 0 & 0 & 0 \\ 0 & 0 & 0 & 0 & 0 & 0 \\ 0 & 0 & 0 & 0 & 0 & 0 \\ 0 & 0 & 0 & -1 & 0 & 0 \\ 0 & 0 & 0 & 1 & 0 & 0 \\ 0 & 0 & 0 & 0 & 0 & -1 \\ 0 & 0 & 0 & 0 & 0 & 1 \end{bmatrix}.$$

A spanning set for the vector space of estimable β contrasts is obtained by

$$(\mathbf{B}'\mathbf{C})'=\begin{bmatrix} -1 & 1 & 0 & 0 & 0 & 0 \\ 0 & -1 & 0 & 1 & 0 & 0 \\ 0 & -1 & 0 & 0 & 0 & 1 \\ 0 & 0 & 0 & -1 & 0 & 1 \\ 0 & 0 & -1 & 0 & 1 & 0 \\ 0 & 0 & -1 & 0 & 1 & 0 \end{bmatrix}.$$

After row reduction we obtain a basis for β effects as

$$\mathbf{\Lambda}'=\begin{bmatrix} -1 & 1 & 0 & 0 & 0 & 0 \\ 0 & -1 & 0 & 1 & 0 & 0 \\ 0 & 0 & 0 & -1 & 0 & 1 \\ 0 & 0 & -1 & 0 & 1 & 0 \end{bmatrix}.$$

Note that in actual practice, where approximations have to be made in numerical computations, it may be difficult to decide whether a particular value, at any stage of reduction, is a real deviation from zero or is a rounding off error in place of zero. So this way of obtaining a basis (or even a spanning set) for a single effect involves rounding off errors, and consequently it may lead to the wrong computation for rank of the design matrix. Hence it is not recommended for two-way classification models.

Estimability of μ Under Some Restriction

Consider the same model introduced in Section (5.3), except suppose there are restrictions on the parameters of the form

$$\sum_i u_i\alpha_i=0 \quad \text{and} \quad \sum_j w_j\beta_j=0,$$

where the sum of the u_i's and the sum of the w_j's are nonzero, for convenience we will suppose $\sum_i u_i=1$ and $\sum_j w_j=1$. When there are no restrictions on the parameters none of the individual parameters are estimable; however, with restrictions as above each of the individual parameters is estimable provided that $r(\mathbf{X})=a+b-1$. In the event that $r(\mathbf{X})<a+b-1$, then individually some of the parameters may be estimable whereas others are not. We now give a necessary and sufficient condition, in the presence of the above restrictions, for the parameter μ to be estimable.

Before answering the question raised in the previous paragraph, we define precisely what is meant by a parametric function being estimable when there are restrictions on the parameters. Suppose for the moment there are known restrictions on $\boldsymbol{\theta}$ of the form $\boldsymbol{\Delta}'\boldsymbol{\theta}=\mathbf{0}$. In this case the domain of a parametric function is restricted to the set of all $\boldsymbol{\theta}$ satisfying $\boldsymbol{\Delta}'\boldsymbol{\theta}=\mathbf{0}$. A parametric function $\boldsymbol{\lambda}'\boldsymbol{\theta}$ is said to be estimable provided there is a vector $\mathbf{t}\in R^N$ such that

$$E(\mathbf{t}'\mathbf{Y})=\mathbf{t}'\mathbf{X}\boldsymbol{\theta}=\boldsymbol{\lambda}'\boldsymbol{\theta}$$

for all vectors $\boldsymbol{\theta}$ satisfying restrictions $\boldsymbol{\Delta}'\boldsymbol{\theta}=\mathbf{0}$. In other words, $\boldsymbol{\lambda}'\boldsymbol{\theta}$ is estimable if it is a linear combination of the estimable cell expectations $E(Y_{ijk})=\mu+\alpha_i+\beta_j$ [in fact one need only use $E(Y_{ijk})$ for which $n_{ij}\neq 0$], taking into account the relations $\boldsymbol{\Delta}'\boldsymbol{\theta}=\mathbf{0}$.

PROPOSITION The parameter μ is estimable if and only if

$$\sum_{i\in I_k} u_i=\sum_{j\in J_k} w_j \quad \text{for } k=1,\ldots,s,$$

where the I_k's, the J_k's, and s are defined as in Section 5.4.

Proof Suppose $\sum_{I_k} u_i=\sum_{J_k} w_j=$(say) d_k for $k=1,\ldots,s$. For

$$i\in I_k,\quad j\in J_k,\quad \text{set } c_{ij}=\begin{cases} \dfrac{u_iw_j}{d_k} & \text{if } d_k\neq 0. \\[2ex] \dfrac{u_i}{b_k}+\dfrac{w_j}{a_k} & \text{if } d_k=0. \end{cases}$$

For these pairs (i, j), $\mu+\alpha_i+\beta_j$ is estimable. If $d_k \neq 0$, then

$$\begin{aligned}\sum_{I_k}\sum_{J_k} c_{ij}(\mu+\alpha_i+\beta_j) &= \frac{1}{d_k}\sum_{I_k} u_i \sum_{J_k} w_j \mu \\ &\quad + \frac{1}{d_k}\sum_{J_k} w_j \sum_{I_k} u_i\alpha_i + \frac{1}{d_k}\sum_{I_k} u_i \sum_{J_k} w_j\beta_j \\ &= \left(\sum_{J_k} w_j\right)\mu + \sum_{I_k} u_i\alpha_i + \sum_{J_k} w_j\beta_j.\end{aligned}$$

If $d_k = 0$, then

$$\begin{aligned}\sum_{I_k}\sum_{J_k} c_{ij}(\mu+\alpha_i+\beta_j) &= \frac{1}{b_k}\sum_{J_k}\left(\sum_{I_k} u_i\right)\mu + \frac{1}{a_k}\sum_{I_k}\left(\sum_{J_k} w_j\right)\mu \\ &\quad + \frac{1}{b_k}\sum_{J_k}\left(\sum_{I_k} u_i\alpha_i\right) + \frac{1}{a_k}\sum_{I_k}\left(\sum_{J_k} w_j\right)\alpha_i \\ &\quad + \frac{1}{b_k}\sum_{J_k}\left(\sum_{I_k} u_i\right)\beta_j + \frac{1}{a_k}\sum_{I_k}\left(\sum_{J_k} w_j\beta_j\right) \\ &= 0+0+\sum_{I_k} u_i\alpha_i + 0 + 0 + \sum_{J_k} w_j\beta_j \\ &= \left(\sum_{J_k} w_j\right)\mu + \sum_{I_k} u_i\alpha_i + \sum_{J_k} w_j\beta_j.\end{aligned}$$

Now

$$\begin{aligned}\sum_k\sum_{I_k}\sum_{J_k} c_{ij}(\mu+\alpha_i+\beta_j) &= \sum_k\sum_{J_k} w_j\mu + \sum_k\sum_{I_k} u_i\alpha_i + \sum_k\sum_{J_k} w_j\beta_j \\ &= \sum_j w_j\mu - \sum_i u_i\alpha_i + \sum_j w_j\beta_j \\ &= \mu.\end{aligned}$$

Therefore μ is estimable.

Conversely, suppose μ is estimable. Then there exist coefficients c_{ij} such that $\mu = \sum_i \sum_j c_{ij}(\mu+\alpha_i+\beta_j)$, where $c_{ij} \neq 0$ only if $\mu+\alpha_i+\beta_j$ is estimable, that is only if $i \in I_k$, $j \in J_k$ for some $k = 1, \ldots, s$. In the original unrestricted model we have

$$\mu + c_\alpha\left(\sum_i u_i\alpha_i\right) + c_\beta\left(\sum_j w_j\beta_j\right) = \sum_i\sum_j c_{ij}(\mu+\alpha_i+\beta_j)$$

for some coefficients c_α and c_β. Comparing coefficients, we get

$$1 = \sum_i\sum_j c_{ij}, \qquad c_\alpha u_i = \sum_j c_{ij} \quad \text{for } i = 1, \ldots, a,$$

$$c_\beta w_j=\sum_i c_{ij} \quad \text{for } j=1, \ldots, b.$$

So $c_\alpha=c_\alpha \sum_i u_i=\sum_i c_\alpha u_i=\sum_i \sum_j c_{ij}=1$ and, similarly, $c_\beta=1$. Hence $u_i=\sum_j c_{ij}=\sum_{J_k} c_{ij}$ for $i \in I_k$ and $w_j=\sum_i c_{ij}=\sum_{I_k} c_{ij}$ for $j \in J_k$, $k=1,\ldots,s$. Now $\sum_{I_k} u_i=\sum_{I_k}\sum_{J_k} c_{ij}=\sum_{J_k}\sum_{I_k} c_{ij}=\sum_{J_k} w_j$ for $k=1, \ldots, s$.

From the above proposition it is easily established whether or not the parameter μ is estimable. Two typical situations which might be noted are the following:

First, suppose

$$u_i=\frac{1}{a}, \qquad i=1, \ldots, a \quad \text{and} \quad w_j=\frac{1}{b}, \qquad j=1, \ldots, b.$$

For these restrictions examples are easily constructed for which μ is not estimable. Secondly, suppose

$$u_i=\frac{n_{i.}}{N}, \qquad i=1, \ldots, a \quad \text{and} \quad w_j=\frac{n_{.j}}{N}, \qquad j=1, \ldots, b.$$

For these restrictions it follows from the proposition that μ is always estimable.

PROGRAM TWOWAY

General Description

This program performs the calculations required for analyzing a set of data following an additive two-way classification model. All notation used in the main body of the chapter will be utilized without further comment. For convenience the row effects $\alpha_1, \alpha_2, \ldots, \alpha_a$ will be designated as treatments and the column effects $\beta_1, \beta_2, \ldots, \beta_b$ will be designated as blocks. The only restriction on the number of observations per cell is that each row, as well as each column, must have at least one observation. Currently the program is limited by $a, b \leqslant 50$ although this could easily be changed.

1 The input for this program includes:

(a) Dimensions of the incidence matrix and the number of hypotheses to be tested.
(b) Number of observations per cell, that is, the incidence matrix.
(c) Vector of observations.
(d) Specification of the various hypotheses to be tested.

2 The output of this program includes:

(a) Initial (incidence) and final matrix.
(b) Table of sums of squares and degrees of freedom.
(c) Estimable treatment and block differences.
(d) A basis for the estimable treatment contrasts and a basis for the estimable block contrasts.
(e) Normal equation solution and associated covariance matrix.
(f) Relevant information for each hypothesis specification.

Input Order and Format Specifications

The input is in a file which name is choosen by the user, the file extension has to be ".DAT", and the file name is asked by the program. The format is free.

1 The dimensions of the incidence matrix, that is, a and b, and the number of hypotheses to be tested.

2 The rows of the incidence matrix. The format is free.

3 The vector of observations, one per line, and ordered in a row-wise fashion [i.e., the observations for cell (1, 1) are in row 1 those for cell (1, 2) are in row 2, etc.]

4 Using the program it is possible to obtain the sum of squares associated with a hypothesis of the form $\mathbf{\Delta}'\boldsymbol{\alpha}=\mathbf{0}$, or of the form $\mathbf{\Gamma}'\boldsymbol{\beta}=\mathbf{0}$, provided the hypothesis is testable, that is, provided the components of $\mathbf{\Delta}'\boldsymbol{\alpha}$, or of $\mathbf{\Gamma}'\boldsymbol{\beta}$, are estimable. (This is the user's responsibility.) The program supposes that the rows of the hypothesis matrix, that is, the matrix $\mathbf{\Gamma}'$, are linearly independent. The input for each such desired hypothesis has the following form:

(a) The hypothesis type (1 for treatments, 2 for blocks) and the number of rows in the hypothesis matrix.
(b) The hypothesis matrix, one entry per line, and in a row-wise fashion.

For each hypothesis desired the above input is required.

Output Description

The output is in a file which name is the same as the input files, but the extension is ".IMP".

Following is a description of the output from the program:

1 The initial or incidence matrix and the final matrix obtained from the R process.

2 Table of analysis which has the following structure:

Source of variation	Degrees of Freedom		Sums of Squares	
Regression	m		$SS(\alpha, \beta)$	
Fitting overall mean		1		$SS(\mu)$
Blocks unadjusted		$b-1$		$SS(\beta\mid\mu)$
Treatments		f_α		$SS(\alpha\mid\beta)$
Fitting overall mean		1		$SS(\mu)$
Treatments unadjusted		$a-1$		$SS(\alpha\mid\mu)$
Blocks		f_β		$SS(\beta\mid\alpha)$
Residual	$N-m$		$\hat{R}$	
Pure error		$N-q$		$\sum_{ijk}(Y_{ijk}-\bar{Y}_{ij})^2$
Interaction		$q-m$		subtraction
Total	N		$\mathbf{Y'Y}$	

3 If the final matrix $\mathbf{M}$ does not consist of all ones, the counter triangle for both treatment and block effects is given. The symbol "*" in the counter triangle is used in place of a one as described in Section 5.4. For example, if $a=3$ the following counter triangle for treatment effects might occur:

	1	2
2	0	
3	*	0.

Such a counter triangle means that $\alpha_1-\alpha_3$ is estimable whereas $\alpha_1-\alpha_2$ and $\alpha_2-\alpha_3$ are not estimable. In this situation the program also gives a basis in terms of estimable differences for both the estimable treatment contrasts and the estimable block contrasts.

4 A solution $\hat{\boldsymbol{\theta}}$ to the normal equations as well as the covariance matrix associated with the solution. If $f_\alpha \leqslant f_\beta$, the solution to the normal equations has the following properties:

$$\sum_{i\in I_k}\hat{\alpha}_i=0 \quad \text{for } k=1,\ldots,s \quad \text{and} \quad \sum_{j=1}^{b}\hat{\beta}_j=0.$$

If $f_\alpha > f_\beta$, the solution has the above conditions reversed, that is,

$$\sum_{i=1}^{a} \hat{\alpha}_i = 0 \quad \text{and} \quad \sum_{j \in J_k} \hat{\beta}_j = 0 \qquad \text{for } k = 1, \ldots, s.$$

In the output the α parameters are represented as T's.

5 For a hypothesis specification of the form $\mathbf{\Delta}'\boldsymbol{\alpha} = \mathbf{0}$, the following information is given:

(a) The estimate of $\mathbf{\Delta}'\boldsymbol{\alpha}$, that is, $\mathbf{\Delta}'\hat{\boldsymbol{\alpha}}$.
(b) $\mathbf{V}_\Delta$ is such that $\text{Cov}(\mathbf{\Delta}'\hat{\boldsymbol{\alpha}}) = \sigma^2\mathbf{V}$.
(c) $\hat{\boldsymbol{\alpha}}'\mathbf{\Delta}\mathbf{V}_\Delta^{-1}\mathbf{\Delta}'\hat{\boldsymbol{\alpha}}$, that is, the sum of squares associated with the hypothesis.

Similar output is given for a hypothesis specification of the form $\mathbf{\Gamma}'\boldsymbol{\beta} = \mathbf{0}$.

BIBLIOGRAPHICAL NOTES

5.1 In analysing a data set following a *designed experiment*, Allen and Wishart (1930) seem to have been the first to have considered the problem of missing observations. They provided two formulae for estimating the value of a single missing observation for a randomized block and a Latin square design, respectively. The estimated value of the missing observation is placed on the missing plot, and the usual analysis of variance carried out on the *now complete data*, provided the number of degrees of freedom for the error is reduced by one.

The principal argument given by Allen and Wishart for a single plot missing in a randomized block design, in brief, is as follows: the yield Y depends on the block factor b and the treatment factor t. Let $Y = b_i + t_j + k$ be the function, where k is constant throughout, b changes from block to block, and t from treatment to treatment, and let $b = 0$ and $t = 0$ in the block and for the treatment which contains the missing plot, respectively, so that k is actually the estimated yield of the missing plot, and b and t represent deviations from it due to changes in block and treatment in the other plots.

If there are s treatments and n blocks, then we have $ns - 1$ actual yields y, and we want to determine the constants so that $\sum (y - Y)^2$ is minimum over all existing plots. That is, we want to minimize

$$\sum_{ij}^{ns-1} (y - b_i - t_j - k)^2,$$

over b_i, t_j, and k. Differentiating with respect to three constants and summing over $ns - 1$ plots for k, $(n - 1)$ for blocks, and $(s - 1)$ for treatments, the estimated value of k is found to be

$$k = \frac{(n+s-1)S - s.S_t - nS_b}{(n-1)(s-1)}$$

where S_t is the sum of treatment totals not including the treatment of which one plot is missing. S_b is the sum of block totals not including the block with missing plot, and S is the sum of $ns-1$ existing plots. In a similar way they gave the estimated value of k for Latin square design.

In (1933) Yates in his article "The analysis of replicated experiments when the field results are incomplete" stated that

> The whole problem can be dealt with most successfully by *estimating the yields of the missing plots.* Such estimates, if properly chosen, when included in the treatment means make these latter efficient estimates of the treatment differences, free from any bias due to other effects such as fertility differences which the experiment was designed to eliminate. After estimating the yields of the missing plots, the ordinary procedure of the analysis of variance suitable to orthogonal experiments may be followed, and though not strictly correct it will be shown that it gives quite satisfactory results in ordinary cases provided the number of degrees of freedom for error is reduced by the number of plots missing. The significance of the results is always slightly exaggerated, though quite negligibly so when only a few values are missing.

He then showed that if each missing observation is replaced by an additional parameter, and if this parameter is estimated by the principle of *least squares,* then the actual parameters of the design can then be calculated in a straightforward fashion by treating the estimated missing value as an observation. Yates' reasoning can be summarized as follows:

If the yields are assumed to be made up of additive functions of the blocks and treatments, and an error term, so that

$$Y_{ij} = k + t_i + b_j + e_{ij},$$

then the analysis of variance may be regarded as the process of finding the most likely values of the constants k, $t_1, \ldots, t_p$, $b_1, \ldots, b_q$, and the errors associated with them, such that

$$\sum_{ij} (Y_{ij} - k - t_i - b_j)^2$$

is minimized. Now suppose some observations are missing. Assuming for the moment that the most likely values $k, t_1, \ldots, t_p, b_1, \ldots, b_q$ of the constants have been found by the above method, put $Y_{uv} = \hat{k} + \hat{t}_u + \hat{b}_v$ for each missing plot. Then Y_{uv} may be taken as an estimate of the yield of the *uv*th missing plot. If we complete the table of plot yields with these estimates and then perform an ordinary analysis of variance, we shall in fact minimize

$$\sum_{i,j}(Y_{ij}-k-t_i-b_j)^2+\sum_{u,v}(Y_{uv}-k-t_u-b_v)^2$$

where the first summation is taken over all the existing plots, and the second over all those of which the yields are missing; this function is the part of the sum of squares alloted to error in the analysis of variance. Equating partial derivatives to zero would thus provide a set of equations for the missing values. In doing so, Yates found an estimate for a single value missing a randomized block design with p treatments and of blocks to be

$$x=\frac{pP+qQ-T}{(p-1)(q-1)}$$

where P and Q are sums of the yields of all plots receiving the same treatment and in the same block as missing plot, and T is the total yield.

Briefly, Yates showed that the correct values to insert are those for which, when formally completed experiment is analyzed, *the residuals in the missing plots are zero.* Yates' criterion of "estimating the missing observation(s)" by *minimization of the residual sum of squares* becomes "*the criterion*" and is recommended in nearly all major texts on experimental design, such as Federer (1955), Cochran and Cox (1957), and John (1971). See also Taylor (1948), Nelder (1954), Norton (1955), Thompson (1956), and Marshall (1968).

To see Yates' method in linear model terminology, consider the linear model

$$\mathbf{Y}=\mathbf{X}\boldsymbol{\theta}+\boldsymbol{\varepsilon} \tag{2.1}$$

where $\mathbf{Y}$ is an $n\times 1$ vector of observations, $\mathbf{X}$ is an $n\times p$ design matrix of rank r, $\boldsymbol{\theta}$ is a $p\times 1$ vector of unknown parameters, and the elements of $\boldsymbol{\varepsilon}$ are assumed to be independently distributed with means 0 and variance σ^2. If $r<p$ "usual constraints" may be placed on the vector $\boldsymbol{\theta}$ to ensure that least squares estimate is unique.

Now suppose m out of n observations are missing. Without lost in generality, we may assume that the last observations in the data vector are missing. Let $\mathbf{Y}_1$ and $\mathbf{Y}_2$ denote the actual and missing vector of observations. Thus we may write (2.1) in partitioned form as

$$\begin{bmatrix}\mathbf{Y}_1\\ \mathbf{Y}_2\end{bmatrix}=\begin{bmatrix}\mathbf{X}_1\\ \mathbf{X}_2\end{bmatrix}\boldsymbol{\theta}+\begin{bmatrix}\boldsymbol{\varepsilon}_1\\ \boldsymbol{\varepsilon}_2\end{bmatrix}.$$

The sum of squares residuals

$$\begin{aligned} S(\boldsymbol{\theta}, \mathbf{Y}_2) &= (\mathbf{Y} - \mathbf{X}\boldsymbol{\theta})'(\mathbf{Y} - \mathbf{X}\boldsymbol{\theta}) \\ &= \|\mathbf{Y}_1 - \mathbf{X}_1\boldsymbol{\theta}\|^2 + \|\mathbf{Y}_2 - \mathbf{X}_2\boldsymbol{\theta}\|^2 \\ &= S_1(\boldsymbol{\theta}) + S_2(\mathbf{Y}_2, \boldsymbol{\theta}) \end{aligned} \tag{2.2}$$

must now be minimized with respect to both $\boldsymbol{\theta}$, the vector parameter, and $\mathbf{Y}_2$, the vector of unknown replacing the missing values. Yates' method is based on minimization of $S(\boldsymbol{\theta}, \mathbf{Y}_2)$ first with respect to $\boldsymbol{\theta}$ (in this way we obtain the least squares solutions as if there had been no missing observations, but the estimator and the residual sum of squares of the solution will now be both function of $\mathbf{Y}_2$) and then with respect to $\mathbf{Y}_2$.

However, as recommended by Yates if there is more than one missing observation, the formula for one missing value is used *iteratively*, starting with a guessed values for all but one missing cell. The iterations are continued until the residuals in the missing cells are negligible.

Yates (1936) adopted the same principle for Latin squares with cases in which one row, or treatment, or a row and column, or either and a treatment, are missing. In a subsequent paper, Yates and Hale (1939) considered the analysis of Latin squares when two or more rows, columns, or treatments are missing using the same principle.

Cornish (1940a, b), (1941), and (1944) extended Yates' method to cover a wide range of experimental designs, and generalized the method for the recovery of interblock information. When several observations are missing in the case of balance incomplete block design and youden square, Cornish (1940) points out: when more than one value is missing there will be two or more unknowns and the process of minimizing the residual sum of squares will yield as many equations as there are unknowns, analogous to the simultaneous equations of partial regression. The form assumed by these equations, however, depends on the *structural relationships between the treatments and blocks from which the values are missing* (an important concept that can be used for characterization of incomplete block designs with missing observations). For a discussion of missing plots in split plot designs, see Anderson (1946), and an interesting note on "missing-plot" techniques by Finney (1946).

Healy and Westmacott (1956) proposed an *iterative* method for estimating the missing observations. With this method, initial guesses are inserted for the missing values, the analysis is performed, and then from each missing cell the calculated residual is subtracted from the guessed values. Once these missing value estimates have been obtained for the full model, the complete analysis is performed with these inserted values, the degrees of freedom for the residual sum of squares being reduced by the number of missing values. The iterative method of Healy and Westmacott (1956) has been expressed in mathematical terms by Jaech (1966), Preece (1971), and Sclove (1972). See also Pearce and Jeffers (1971) and Shearer (1973).

Hartley (1956) has given an alternative technique when there is just a single missing observation. If there is one missing observation, one can find the least squares estimate for that cell by analysing the data three times, using three equally spaced values, for example a_0, a_1, and a_2 so that $a_i = a_{i-1} + 1$, for the missing cell. The least squares estimate of the missing cell is then given by "the universal formula" involving only the three values a_0, a_1, a_2 and the three residual sum of squares obtained. The degrees of freedom for residual sum of squares is reduced by 1. Hartley (1956) recommended an *iterative* procedure when more than one cell is missing.

Bartlett (1937) in analysing a winter feeding trial data, introduced an alternative noniterative procedure for handling missing observations. The data are augmented with 0 values corresponding to the m missing observations, and the influence of these values on the analysis of the augmented data is removed by performing an analysis of *covariance* on m pseudo-concomitant variables, the ith pseudovariables taking the value 1 in the position corresponding to the ith missing observations and 0 otherwise. The regression coefficient derived in the analysis is the negative estimate of the missing observations. Bartlett (1937) noted that his method must be *theoretically identical with Yates' method of obtaining missing values and the corresponding exact tests, but for those who have become familiar with the use of covariance, it may sometimes be helpful to realize this.* Bartlett's method has been mentioned in a number of papers and textbooks. See, for example, Tocher (1952), Quenouille (1953), Coons (1957), Wilkinson (1957, 1960), Seber (1966), Rubin (1972), Haseman and Gaylor (1973), John and Prescott (1975), and Seber (1977).

De Lury (1946) provided a *direct* and simple procedure for determining the value of the missing observations for Latin squares designs. De Lary uses the fact that if an unknown is formally substituted for the missing value, then the unknown can be equated to its estimated value, the estimates of the parameters being treated as functions of the unknown value. The equations derived from the formally complete data by minimizing the residual sum of squares can then be solved for the unknown. Observe that in De Lury's method the minimization of (2.2) is first with respect to $\mathbf{Y}_2$ and then $\boldsymbol{\theta}$ (Kendall and Stuart (1975)], and the estimated values are determined according to the linear model which is to be fitted to data.

Following De Lury's (1946) direct method, Wilkinson (1958a) provided explicit formulae for many standard designs. In (2.2) let $\hat{\boldsymbol{\theta}}$ be a least squares solution of $\boldsymbol{\theta}$; that is, $S_1(\boldsymbol{\theta})$ is minimized at $\boldsymbol{\theta} = \hat{\boldsymbol{\theta}}$, where $\hat{\boldsymbol{\theta}}$ is a solution of the normal equations $\mathbf{X}_1'\mathbf{X}_1\boldsymbol{\theta}=\mathbf{X}_1'\mathbf{Y}_1$. Then $S(\boldsymbol{\theta}, \mathbf{Y}_2)$ will be minimized with respect to both $\mathbf{Y}_2$ and $\boldsymbol{\theta}$ if we set $\boldsymbol{\theta}=\hat{\boldsymbol{\theta}}$ and $\mathbf{Y}_2=\hat{\mathbf{Y}}_2=\mathbf{X}_2\hat{\boldsymbol{\theta}}$. Since $S_2(\mathbf{Y}_2, \boldsymbol{\theta})=\mathbf{0}$ we have

$$S_1(\hat{\boldsymbol{\theta}})=\min_{\boldsymbol{\theta},\mathbf{Y}_2}\|\mathbf{Y}-\mathbf{X}\boldsymbol{\theta}\|^2=\min_{\boldsymbol{\theta}}\left(\min_{\mathbf{Y}_2}\|\mathbf{Y}-\mathbf{X}\boldsymbol{\theta}\|^2\right).$$

So we obtain $\hat{\mathbf{Y}}_2$ as a least square estimate of $\mathbf{Y}_2$. See Kruskal (1960) for coordinate-free approach to the problem.

The model for the existing data is $\mathbf{Y}_1=\mathbf{X}_1\boldsymbol{\theta}+\boldsymbol{\varepsilon}$. Therefore the exact least squares solution based on the observed data are

$$\hat{\boldsymbol{\theta}}=(\mathbf{X}_1'\mathbf{X}_1)^-\mathbf{X}_1'\mathbf{Y}_1,$$

where $(\mathbf{X}_1'\mathbf{X}_1)^-$ is a generalized inverse of $(\mathbf{X}_1'\mathbf{X}_1)$ satisfying $(\mathbf{X}_1'\mathbf{X}_1)(\mathbf{X}_1'\mathbf{X}_1)^- \times(\mathbf{X}_1'\mathbf{X}_1)=(\mathbf{X}_1'\mathbf{X}_1)$. Then provided $\mathbf{X}_2\boldsymbol{\theta}$ are *estimable*, the best linear unbiased estimator of $\mathbf{X}_2\boldsymbol{\theta}$ is $\mathbf{X}_2\hat{\boldsymbol{\theta}}$. Here $\hat{\mathbf{Y}}_2=\mathbf{X}_2\hat{\boldsymbol{\theta}}$ is called a least squares solution of $\mathbf{Y}_2$. If the rank of $\mathbf{X}_1$ and $\mathbf{X}$ are the same, the constraints initially imposed on $\boldsymbol{\theta}$ are sufficient to ensure the uniqueness of $\hat{\boldsymbol{\theta}}$. This is only the case when the design matrix is of maximal rank, which is equivalent to saying that all cell expectations *are estimable*. If however, the design matrix is not of maximal rank, then by minimizing the residual sum of squares with respect to the unknown parameters $\boldsymbol{\theta}$ and the unknown values $\mathbf{Y}_2$, after some algebra, we obtain

$$[\mathbf{I}-\mathbf{X}_2(\mathbf{X}'\mathbf{X})^-\mathbf{X}_2']\hat{\mathbf{Y}}_2=\mathbf{X}_2(\mathbf{X}'\mathbf{X})^-\mathbf{X}_1'\mathbf{Y}_1 \tag{2.3}$$

or the expected values of the missing observations are seen to be

$$\hat{\mathbf{Y}}_2=\mathbf{R}^-\mathbf{X}_2(\mathbf{X}'\mathbf{X})^-\mathbf{X}_1'\mathbf{Y}_1$$

where $\mathbf{R}=[\mathbf{I}-\mathbf{X}_2(\mathbf{X}'\mathbf{X})^-\mathbf{X}_2']$, or $\mathbf{C}$ matrix as called by Tocher. Tocher (1952) suggests that

> first a standard analysis is performed with the missing observations given zero values. From this analysis the expected values of the missing observations are estimated and then "corrected" by the matrix factor **C**. Finally the experiment is analysed with those corrected values for the missing observations.

In brief, one should set

$$\mathbf{X}_1'\mathbf{Y}_1=(\mathbf{X}_1', \mathbf{X}_2')\begin{bmatrix}\mathbf{Y}_1\\ \mathbf{0}\end{bmatrix}=\mathbf{X}'\mathbf{Y},$$

and finds

$$\hat{\mathbf{Y}}_2=[\mathbf{I}_m-\mathbf{X}_2(\mathbf{X}'\mathbf{X})^-\mathbf{X}_2']^-\mathbf{X}_2'\hat{\boldsymbol{\theta}}(\mathbf{0}) \tag{2.4}$$

where $\hat{\boldsymbol{\theta}}(\mathbf{0})$ is the estimate of $\boldsymbol{\theta}$ obtained from the data assuming the missing observations have zero values.

However, Tocher (1952) recommended that, in practice, if several observations are missing it may be advantageous to use the procedure for a single missing observation for each observation in turn, repeating the cycle until the changes are negligible. See also Draper (1961) for comparison of Tocher's method with that of Yates (1933), and Kendall and Stuart (1976).

Now consider the analysis of covariance model

$$\mathbf{Y}_0 = \mathbf{X}\boldsymbol{\theta} + \mathbf{Z}\boldsymbol{\gamma} + \boldsymbol{\varepsilon} \tag{2.5}$$

or

$$\begin{bmatrix} \mathbf{Y}_1 \\ \mathbf{0} \end{bmatrix} = \begin{bmatrix} \mathbf{X}_1 \\ \mathbf{X}_2 \end{bmatrix} \boldsymbol{\theta} + \begin{bmatrix} \mathbf{0} \\ \mathbf{I}_m \end{bmatrix} \boldsymbol{\gamma} + \boldsymbol{\varepsilon}$$

where we put m zeros for the missing values in the $\mathbf{Y}$ vector, and $\boldsymbol{\gamma}$ is an $m \times 1$ vector of unknown values. Each γ_i is the coefficient of a dummy variable introduced by a covariate which is zero for all observations except the ith missing observation where it takes the value of 1. Using the method of two-step least squares, we obtain a least squares solution of $\boldsymbol{\gamma}$ to be a solution of

$$\mathbf{Z}'[\mathbf{I}_n - \mathbf{X}(\mathbf{X}'\mathbf{X})^- \mathbf{X}']\mathbf{Z}\hat{\boldsymbol{\gamma}} = \mathbf{Z}'[\mathbf{I}_n - \mathbf{X}(\mathbf{X}'\mathbf{X})^- \mathbf{X}']\mathbf{Y}_0.$$

Since $\mathbf{Z}'\mathbf{Y}_0 = \mathbf{0}$, it follows that

$$\mathbf{Z}'[\mathbf{I}_n - \mathbf{X}(\mathbf{X}'\mathbf{X})^- \mathbf{X}']\mathbf{Z}\hat{\boldsymbol{\gamma}} = -\mathbf{Z}'\mathbf{X}(\mathbf{X}'\mathbf{X})^- \mathbf{Y}_0,$$

or

$$\mathbf{I}_m - \mathbf{X}_2(\mathbf{X}'\mathbf{X})^- \mathbf{X}_2'\hat{\boldsymbol{\gamma}} = -\mathbf{X}_2(\mathbf{X}'\mathbf{X})^- \mathbf{X}_1'\mathbf{Y}_1. \tag{2.6}$$

From (2.3) it follows that $\hat{\mathbf{Y}}_2 = -\hat{\boldsymbol{\gamma}} = \mathbf{R}^- \mathbf{X}_2(\mathbf{X}'\mathbf{X})^- \mathbf{X}_1'\mathbf{Y}_1$, which is to be expected because of the equivalence between the Yates' method and the method of two-step least squares. See Barlett (1937).

Wilkinson (1958a) observes that the ith column of the matrix on the left-hand side of (2.3), the matrix $\mathbf{R}$, is the vector of residuals obtained by analysing a dummy data vector with 1 in the ith missing observation and zero elsewhere, and the right-hand side in (2.3) consists of minus the residuals $\boldsymbol{\rho}$ in the missing cells when the observed data $\mathbf{Y}_1$ are augmented by fictitious values, say zeros [Rubin (1972)], in the missing cells. Thus, if m observations are missing in a designed experiment, one may perform a series of $m+1$ standard analyses with the data vectors as described by recording the residuals in the missing cells and solve the equations

$$\mathbf{R}\hat{\mathbf{Y}}_2 = -\boldsymbol{\rho}.$$

For comparison between covariance method and the Wilkinson's method see Wilkinson (1960) and John and Prescott (1975). The essential difference as noted by Jarrett (1978) is on the number of analyses required by both methods.

Rubin (1972) presented a simple algorithm based on the analysis of covariance which finds the $m \times m$ matrix $\mathbf{R}$, and the $m \times 1$ vector $\boldsymbol{\rho}$. Based on Rubin's (1972) algorithm, one only needs a subroutine to find residuals and an additional subroutine to invert an $m \times m$ symmetric matrix $\mathbf{R}$. However, as noted by Rubin (1972), if the matrix $\mathbf{R}$ is singular his method produces no solution. Rubin adds that some thought reveals that a singular $\mathbf{R}$ corresponds to a solution in which one is trying *to estimate a parameter for which there is no data*; in other words, an attempt is being made *to estimate a nonestimable parameter*. See also Rubin (1976), Smith (1981), and Williams, Ratcliff, and Speed (1981).

Haseman and Gaylor (1973), following Yates method, give a simple non-iterative procedure for obtaining missing value estimates by solving a set of simultaneous linear equations which yields to direct formulae. However, they noted that

> in order to obtain a solution to a set of simultaneous linear equations the design matrix must be "*connected*." With disconnected data the matrix of coefficients, $\mathbf{R}$, is singular. In this case the missing values can be estimated by applying the algorithm within group of connected data with a reduced number of rows and/or columns.

A similar remark is made by Haseman and Gaylor (1973) for estimation of missing values in a p-way crossed classification.

Jarrett (1978) gives a detail comparison of Wilkinson (1958a, b) exact method, Rubin (1972) covariance procedure, and the three iterative methods proposed by Yates (1933), Healy and Westmacott (1956), and Preece (1971). Jarrett (1978) shows that the three iterative methods effectively give a power series expansion for the inverse of the $\mathbf{R}$ matrix. Jarrett (1978) outlines the results by Wilkinson (1958b) and adds

> the rank of $\mathbf{R}$ is $(m-l)$ if and only if the rank of $\mathbf{X}_1$ is l less than the rank of $\mathbf{X}$, and the correct number of residual degrees of freedom is the number of observations $(N-m)$ minus the rank of $\mathbf{X}_1$, namely $(r-l)$. Jarrett (1978) adds, if $\mathbf{R}$ is *singular*, any *generalized inverse* $\mathbf{R}^-$ will give the correct residual sum of squares, the parameter estimates, and the covariance matrix will not be in general unique, although the estimate $\mathbf{c}'\boldsymbol{\theta}$ and its variance calculated from
>
> $$\mathrm{Var}(\hat{\boldsymbol{\theta}}) = \sigma^2[(\mathbf{X}'\mathbf{X})^- + (\mathbf{X}'\mathbf{X})^- \mathbf{X}_2'\mathbf{R}^- \mathbf{X}_2(\mathbf{X}'\mathbf{X})^{-\prime})$$
>
> and
>
> $$\mathrm{Var}(\mathbf{c}'\boldsymbol{\beta}) = \sigma^2(\mathbf{c}(\mathbf{X}'\mathbf{X})^- \mathbf{c} + \mathbf{c}'(\mathbf{X}'\mathbf{X})^- \mathbf{X}_2'\mathbf{R}^- \mathbf{X}_2(\mathbf{X}'\mathbf{X})^{-\prime}\mathbf{c})$$

will be unique and correct provided $\mathbf{c}'\hat{\boldsymbol{\beta}}$ is *estimable*, i.e., provided $\mathbf{c}'(\mathbf{X}'\mathbf{X})^{-}\mathbf{X}_2'\mathbf{A}=\mathbf{0}$ where $\mathbf{A}$ is an $m\times l$ matrix of rank l such that $\mathbf{RA}=\mathbf{0}$.

Recently, Li (1982) presented a self-study approach for handling the analysis of two-way classification experiments with unbalanced data. Li's (1982) approach is based on solving simultaneous linear equations.

The methods just described for estimating the missing values are only a partial solution to the problem since all require the inversion of a singular matrix when the original design matrix $\mathbf{X}$ is not of maximal rank. Moreover, in order to obtain the degrees of freedom for each effect separately, it appears that one needs to find the rank of the design matrix $\mathbf{X}_1$ of the observed data, and thus two subroutines are required, one to construct the design matrix $\mathbf{X}_1$, and another that finds the rank $(\mathbf{X}_1)$. When enough observations are missing, the dimensions of the $\mathbf{R}$ matrix become larger (if more than 50% of observations are missing arbitrary) than that of $\mathbf{X}_1$ and hence we gain nothing by working (finding) on the $\mathbf{R}$ matrix with any of the above methods. Additionally, it may be very well possible to estimate a nonestimable parametric function in attempting to find a generalized inverse of $\mathbf{R}$, namely, $\mathbf{R}^{-}$. Note that an additional subroutine is required to find such a generalized inverse, without any rounding off errors.

To make these points clear, consider the following two-way additive classification with the incidence matrix as

$\mathbf{N}=$	β_1	β_2	β_3
α_1	1	0	0
α_2	0	1	1
α_3	0	1	0

with the corresponding observations as

	β_1	β_2	β_3
α_1	5.6	0	0
α_2	0	6.3	4.9
α_3	0	7.5	0

Using Rubin's (1972) noniterative method we obtain for the missing values equations $\mathbf{R}\hat{\mathbf{Y}}_2 = -\boldsymbol{\rho}$,

$$\begin{bmatrix} 0.444 & -0.222 & 0.111 & 0.111 & 0.111 \\ -0.222 & 0.444 & 0.111 & 0.111 & -0.222 \\ 0.111 & 0.111 & 0.444 & -0.222 & 0.111 \\ 0.111 & 0.111 & -0.222 & 0.444 & -0.222 \\ 0.111 & -0.222 & 0.111 & -0.222 & 0.444 \end{bmatrix} \hat{\mathbf{Y}}_2 = \begin{bmatrix} 3.77 \\ 0.80 \\ 2.90 \\ 1.67 \\ 1.43 \end{bmatrix}.$$

The 5×5 matrix **R** is singular and therefore to obtain "intelligent" least squares estimates for the missing observations we are required a generalized inverse of **R**. Using Wilkinson (1958a) direct method for determining equations for the missing values, we obtain

$$\begin{bmatrix} 4 & -2 & 1 & 1 & 1 \\ -2 & 4 & 1 & 1 & -2 \\ 1 & 1 & 4 & -2 & 1 \\ 1 & 1 & -2 & 4 & -2 \\ 1 & -2 & 1 & -2 & 4 \end{bmatrix} \hat{\mathbf{Y}}_2 = \begin{bmatrix} 33.9 \\ 7.2 \\ 26.1 \\ 15.0 \\ 12.9 \end{bmatrix}.$$

Again we must obtain a g inverse for the matrix on the left-hand side of the above equation.

From the above example we can conclude the main difference between those methods that find a least squares estimate for the missing cells is in how to obtain these estimated values. Although they sound mathematically correct, apart from some technical difficulties that may be involved for computer programs, they may mislead the applied users in the sense that if enough observations are missing, not all the parametric functions are estimable, and hence they estimate a cell which is not estimable. In fact, if some cells are not estimable, what is really obtained by these methods is not even solutions to the normal equations, and therefore no attempt should be made (nor it does not make any sense) to obtain such so-called estimates to the missing cells.

Another major problem with these methods is that it is not so clear how one may obtain the correct degrees of freedom for each effect given the fact that the design matrix is not of maximal rank. If one has to use the design matrix at each stage of the computation, why not then start with the design matrix and carry the problem all the way to the end. As everyone knows this procedure is exact and does not require any more computer programs than the above mentioned methods. It appears that the methods given by Haseman and Gaylor (1973) and that of Rubin (1972) are correct in the sense that they will not continue with the process if their **R** matrix is singular. It is worth mentioning that in the Haseman and Gaylor (1973) method, the concept of connectedness is only true for two-way classification method and *not* for higher dimension. The author knows of no method up to date that provides a necessary and sufficient condition for the connectedness of N-way classification designs which can be programmed for electronic computers.

5.2 See John (1980). See also Arthanari and Dodge (1981) for references on BIB designs with repeated blocks and the mathematical programming approach to the problem.

5.3 Utilizing the incidence matrix **N** to obtain the estimability information seems to have been first considered by Bose (1949) for an additive two-way model. For an additive two-way classification model (block by treatment, with arbitrary incidence) Bose introduced the notion of *connectedness*, and via this concept answered the question of whether every treatment contrast is estimable. In Bose's terminology a treatment α_i is said to be associated with a block β_j if the treatment is contained in the block β_j, that is, there is at least one observation in the (i, j) subclass. Two treatments, two blocks, or a treatment and a block, are said to be connected if it is possible to pass from one to the other by means of a chain consisting alternately of blocks and treatments such that any two adjacent members of the chain are associated. And a design is said to be a connected design if every block and treatment of the design are connected to each other. Bose (1949) then proved that the additive two-way model is connected if and only if every treatment contrast is estimable.

Chakrabarti (1963) defines a design to be connected if its **C** matrix has rank $a-1$, where

$$\mathbf{C}=(\mathbf{A}'\mathbf{A})-\mathbf{N}(\mathbf{B}'\mathbf{B})^{-1}\mathbf{N}',$$

and has proved that his definition of connected designs is equivalent to that of Bose (1949). Chakrabarti's (1963) paper contains many important results on the **C** matrix and is considered a major contribution to the theory of connected designs.

Weeks and Williams (1964) treated the additive n-way classification model. They defined the design points of such a model to be connected if all simple contrasts (i.e., differences of two levels of the same factor) are estimable, and defined two design points to be nearly identical if the n-tuples corresponding to them are equal in all except one component. Using the idea of nearly identical design points, Weeks and Williams described a procedure for determining connectedness. However, as Weeks and Williams (1964) pointed out in their errata, their condition for data to be connected is sufficient but not necessary. This is easily seen by considering an additive three-way model

$$E(Y_{ijk})=\mu+\alpha_i+\beta_j+\gamma_k,$$

where $i=1, 2, j=1, 2$, and $k=1, 2$ and with data occurring in cells (1, 1, 2), (2, 1, 1), (2, 2, 2), and (1, 2, 1). This is a 1/2 replication of a 2^3 factorial. No pairs of these observations are nearly identical, but $\alpha_1-\alpha_2$, $\beta_1-\beta_2$, $\gamma_1-\gamma_2$ are estimable. For instance, we can write $\gamma_1-\gamma_2$ as

$$(-\tfrac{1}{2})[(\mu+\alpha_1+\beta_1+\gamma_2)-(\mu+\alpha_1+\beta_2+\gamma_1)+(\mu+\alpha_2+\beta_2+\gamma_2) -(\mu+\alpha_2+\beta_1+\gamma_1)].$$

Therefore the data in the above model are connected, but have no property of being nearly identical, so that Weeks and Williams' procedure fails to provide any information.

Srivastava and Anderson (1970) also discussed the concept of connectedness in additive n-way classification models. Their definition of connectedness is equivalent to that of Weeks and Williams (1964) but stated in a slightly different form as: "the design is said to be completely connected if and only if all the linear contrasts within each factor are estimable." They defined a chain connecting two levels of a factor to be a sequence of occupied cells such that the alternating sum of the corresponding cell expectations is a nonzero multiple of the difference of the two levels. Then they established a theorem that a simple contrast is estimable if and only if there is a chain connecting the two levels involved in the contrast. They gave no algorithm for finding such chains and it seems that in any such algorithm there would be no upper bound on the number of sequences of occupied cells that must be looked at in order to find a chain.

A graphical presentation of classification data of arbitrary incidence is contained in an unpublished paper by Mexas (1972). The possibility of extending Bose's theorem to more than two factors has been considered. By a counterexample Mexas showed that pairwise connectedness is not sufficient for maximality of rank in an additive three-way classification model.

Searle (1971) mentions that "the general problem of finding necessary conditions for main effect differences to be estimable [for an additive model having more than two factors] remains as yet unsolved." Calinski (1971) gave an iterative formula for analysing two-way classification models possessing any arbitrary pattern. But it appears that one is uncertain about the number of steps needed for the required accuracy.

Recently, Birkes, Dodge, Hartmann, and Seely (1972a) presented general and complete results for estimability considerations in an additive two-way classification model which are easily programmed for electronic computers. They introduced an algorithm, the *R process*, which determines what cell expectations are estimable. Furthermore they gave a method for determining a *basis* for the estimable functions involving only one effect; for determining ranks of matrices pertinent to considerations for degrees of freedom; and for determining which portions of the design are connected.

Eccleston and Hedayat (1974) classified the family of connected designs into three subclasses: locally connected, globally connected, and pseudo-globally connected designs. A locally connected design is one in which not all the observations participate in the estimation. A globally connected design is one in which all observations participate in the estimation, and finally, a pseudoglobally connected design is a compromise between locally and globally connected designs. Butz (1982) provides a combinatorial approach to the problem of connectivity in multifactor designs.

5.4–5.8 The material in these sections is from Birkes, Dodge, Hartman, and Seely (1972). See also Birkes, Dodge, and Seely (1976).

5.9 See Exercise 6, Chapter 6 of Seely's (1979) lecture notes on general linear hypotheses. Similar results are obtained by Lejeune (1980).

REFERENCES

Afifi, A. A. and Elashoff, R. M. (1966). Missing observations in multivariate statistics, I. Review of the literature. *J. Am. Statist. Assn.* **61**, 595–604.

Afifi, A. A. and Elashoff, R. M. (1967). Missing observations in multivariate statistics, II. Point estimates in simple linear regression. *J. Am. Statist. Assn.*, **62**, 10–29.

Afifi, A. A. and Elashoff, R. M. (1969a). Missing observations in multivariate statistics, III. Large sample analysis of simple linear regression. *J. Am. Statist. Assn.* **64**, 337–358.

Afifi, A. A. and Elashoff, R. M. (1969b). Missing observations in multivariate statistics, IV. A note on simple linear regression. *J. Am. Statist. Assn.* **64**, 359–365.

Allan, F. E. and Wishart, J. (1930). A method of estimating the yield of a missing plot in field experimental work. *J. Agric. Sci.* **20**, part 3, 399–406.

Anderson, R. L. (1946). Missing-plot Techniques. *Biometrics* **2**, 41–47.

Anderson, R. L. (1960). Some remarks on the design and analysis of factorial experiments. *In Contributions to probability and statistics: Essays in honor of Harold Hotelling*, I. Olkin et al., Eds. Stanford: Stanford University Press, pp. 35–36.

Arthanari, T. S. and Dodge, Y. (1981). *Mathematical Programming in Statistics*. New York: Wiley.

Atiqullah, M. (1961). On a property of balanced designs. *Biometrika* **18**, 215.

Baird, H. R. and Kramer, C. Y. (1960). Analysis of variance of a balanced incomplete block design with missing observations. *Appl. Statist.* **9**, 189–198.

Bargmann, R. E. (1977). Multiple factor factorial experiments, unbalanced and with missing cells. Comparison of hierarchical analysis with other approaches. Amer. Statist. Assn. Proc. of Statistical Computing Section.

Bartlett, M. S. (1937). Some examples of statistical methods of research in agriculture and applied biology. *J. Roy. Statist. Soc. Suppl.* **4**, 137–183.

Baten, W. D. (1952). Variances of differences between means when there are two missing values in randomized block designs. *Biometrics* **8**, 42–50.

Berger, M. P. F. (1979). A FORTRAN IV program for the estimation of missing data. *Behavior Research Methods and Instrumentation* **II**, 395–396.

Biggers, J. D. (1959). The estimation of missing and mixed-up observations in several experimental designs. *Biometrika* **46**, 91–105.

Birkes, D., Dodge, Y., Hartmann, N., and Seely, J. (1972). Estimability and analysis for additive two-way classification models with missing observations. Oregon State University. Dept. of Statistics. Technical Report No. 30.

Birkes, D., Dodge, Y., and Seely, J. (1972). Estimability and analysis for additive three-way classification models with missing observations. Oregon State University. Dept. of Statistics. Technical Report No. 33.

Birkes, D., Dodge, Y., and Seely, J. (1976). Spanning sets for estimable contrasts in classification models. *Ann. Statist.* **4**, 86–107.

Birkes, D. and Seely, J. (1976). Three way classification designs of resolutions III, IV, and V. Tech. Report 50, Oregon State University, Corvallis, Oregon.

Boddy, R. and Goldsmith, P. L. (1973). Critical analysis of factorial experiments and orthogonal fractions. *Appl. Statist.* **22**, 141–160.

Bose, R. C. (1949). Least square aspects of analysis of variance. *Inst. Statist.*, Mimeo Series 9, Chapel Hill, North Carolina.

Bose, R. C. and Nair, K. R. (1939). Partially balanced incomplete block designs. *Sankhya* **4**, 337.

Bradley, E. H. (1968). Multiple classification analysis for arbitrary experimental arrangements. *Technometrics* **10**, 13–27.

Bruce, G. R., Scott, D. T., and Carter, M. W. (1980). Estimation and hypothesis testing in linear models—a reparametrization approach to the cell means model. *Commun. Statist. Theor. Math.* **A9**, 131–150.

Buck, S. F. (1960). A method of estimation of missing values in multivariate data suitable for use with an electronic computer. *J. Roy Statist. Soc.* **B22**, 302–307.

Burdick, D. S., Herr, D. G., O'Fallon, W. M., and O'Neill, B. V. (1974). Exact methods in the unbalanced, two-way analysis of variance—A geometric view. *Commun. Statist.* **3**(6), 581–595.

Butz, L. (1982). *Connectivity in Multifactor Designs: A Combinatorial Approach.* Berlin: Heldermann Verlag.

Calinski, T. (1971). On some desirable patterns in block designs. *Biometrics* **27**, 275–292.

Calinski, T. (1977). On the notion of balance in block designs. *In Recent Developments in Statistics*, J. R. Barra et al., Eds. Amsterdam: North-Holland.

Chakrabarti, M. C. (1963). On the *C*-matrix in design of experiments. *J. Indian Statist. Assn.* **1**, 8–23.

Cochran, W. G. (1947). Some consequences when the assumptions for the analysis of variance are not satisfied. *Biometrics* **3**, 22–38.

Cochran, W. G. and Cox, G. (1957). *Experimental Designs.* New York: Wiley.

Coons, I. (1957). The analysis of covariance as a missing plot technique. *Biometrics* **13**, 387–405.

Cornish, E. A. (1940a). The estimation of missing values in incomplete randomized block experiments. *Ann. Eug.* **10**, 112–118.

Cornish, E. A. (1940b). The estimation of missing values in quasi-factorial designs. *Ann. Eug.* **10**, 137–143.

Cornish, E. A. (1940c). The analysis of quasi-factorial designs with incomplete data: Incomplete randomized Bl. *J. Aust. Inst. Agric. Sci.* **6**, 31–39.

Cornish, E. A. (1941). The analysis of quasi-factorial designs with incomplete data: lattice squares. *J. Aust. Inst. Agric. Sci.* **7**, 19–26.

Cornish, E. A. (1944). The recovery of inter-block information in quasi-factorial designs with incomplete data. 2. Lattice squares. *Aust. Coun. Sci. Ind. Res. Bull.*, 175.

Corsten, L. C. A. (1958). Vectors, a tool in statistical regression theory. *Meded. Landbouwhogesch. Wageningen* **58**, 1–92.

Cramer, E. M. (1972). Missing values in experimental design models. *Am. Statist.* **26**(4), 58.

Das, M. N. (1954). Missing plots and randomized block design with balanced incompleteness. *J. Indian Soc. Agric. Statist.* **6**, 58–76.

Das, M. N. (1955a). Latin squares with several missing plots. *J. Indian Soc. Agric. Statist.* **7**, 46–56.

Das, M. N. (1955b). Missing plots in partially balanced and other incomplete block designs. *J. Indian Soc. Agric. Statist.* **7**, 111–126.

Das, M. N. (1956). Analysis of covariance in incomplete block designs with or without missing plots. *J. Indian Soc. Agric. Statist.* **8**, 76–83.

Dash, S. P. (1980). Validity of F-test in a randomized block design with missing data. *Biometrical J.* **22**, 407–411.

De Lury, D. B. (1946). The analysis of lating squares when some observations are missing. *J. Am. Statist. Assn.* **41**, 370.

Dodge, Y. (1974). Estimability considerations of N-way classification experimental arrangements with missing observations. Ph.D. thesis, Oregon State University, Corvallis.

Dodge, Y. (1976). Estimability considerations for 2^n factorial experiments with missing observations. Technical Report 7606, Indian Statist. Inst., New Delhi.

Dodge, Y. and Shah, K. R. (1977). Estimation of parameters in latin squares and greco-latin squares with missing observations. *Commun. Statist.* **6**, A, 1465–1472.

Dodge, Y. and Shah, K. R. (1977). On a measure of usefulness for fractionally replicated designs. *Proc. Inst. Statist. Inst.*, New Dehli, 156–159.

Dodge, Y. and Majumdar, D. (1979). An algorithm for finding least square generalized inverses for classification models with arbitrary patterns. *J. Statist. Comp. Simul.* **9**, 1–17.

Draper, N. R. (1961). Missing values in response surface designs. *Technometrics* **3**, 389–398.

Draper, N. R. and Stoneman, D. M. (1963). Estimating missing values in unreplicated two-level factorial and fractional factorial designs. University of Wisconsin, Technical Report No. 20.

Eccleston, J. A. (1972). On the theory of connected design. Ph.D. thesis, Cornell University.

Eccleston, J. A. and Hedayat, A. S. (1974). On the theory of connected designs: characterisation and optimality. *Ann. Statist.* **2**, 1238–1255.

Eccleston, J. A. and Russel, K. (1975). Connectedness and orthogonality in multi factor designs. *Biometrika* **62**, 341–345.

Elston, R. C. and Bush, N. (1964). The hypotheses that can be tested when there are interactions in an analysis of variance model. *Biometrics* **20**, 681–698.

Fairfied-Smith, H. (1957). Note: missing plot estimates. *Biometrika* **13**, 115.

Federer, W. T. (1955). *Experimental Design, Theory, and Application.* New York MacMillan.

Federer, W. T. (1957). Variance and covariance analyses for unbalanced classifications. *Biometrics* **13**, 333–362.

Federer, W. T. and Zelen, M. (1966). Analysis of multifactor classifications with unequal numbers of observations. *Biometrics* **22**, 525–552.

Finkbeiner, C. (1979). Estimation for the multiple factor model when data are missing. *Psychometrika* **44**, 409–420.

Finney, D. J. (1946). Standard errors of yield adjusted for regression on an independent measurement. *Biometrics* **2**, 53–55.

Finney, D. J. (1946). A note on "missing-plot techniques." *Biometrics* **2**, 94.

Frane, J. W. (1976). Some simple procedures for handling missing data in multivariate analysis. *Psychometrika* **41**, 409–415.

Frane, J. W. (1978). Missing data and BMDP: some pragmatic approaches. *Proc. Statist. Comp. Am. Statist. Assn.* 27–33.

Freund, R. J. (1980). The case of the missing cell. *Am. Statist.* **34**, 94–98.

Ghosh, S. (1979). On robustness of designs against incomplete data. *Sankhya* **B40**, 204–208.

Glenn, W. A. and Kramer, C. Y. (1958). Analysis of variance of a randomized block design with missing observations. *Appl. Statist.* **7**, 173–185.

Goldman, A. J. and Zelen, M. (1964). Weak generalized inverses and minimum variance linear unbiased estimation. *J. Res. Natl. Bur. Stand.* **62B**, 151–172.

Goodnight, J. H. (1978). Hypothesis testing in multi-way. ANOVA models. *Proc. of Comp. Sci. and Statistics: 10th Symposium on the Interface* **10**, 48–53.

Goulden, C. H. (1939). Methods of statistical analysis. New York: Wiley.

Haitovsky, Y. (1966). Unbiased multiple regression coefficients estimated from one-way classifications are unknown. *J. Am. Statist. Assn.* **61**, 720–728.

Haitovsky, Y. (1968). Missing data in regression analysis. *J. Roy. Statist. Soc.* **B30**, 67–82.

Haitovsky, Y. (1969). Estimation of regression equations when a block of observations is missing. *J. Am. Statist. Assn.* **64**, 195–196 (abstract).

Hartley, H. O. (1951). The fitting of polynomials to equidistant data with missing values. *Biometrika* **38**, 410–413.

Hartley, H. O. (1956). A plan for programming analysis of variance for general purpose computers. *Biometrics* **12**, 110–122.

Hartley, H. O. (1958). Maximum likelihood estimation from incomplete data. *Biometrics* **14**, 174–194.

Haseman, J. K. and Gaylor, D. W. (1973). An algorithm for non-iterative estimation of multiple missing values for crossed classifications. *Technometrics* **15**, 631–636.

Healy, M. J. R. (1952). The analysis of lattice designs when a variety is missing. *Emp. J. Exp. Agric.* **20**, 220–226.

Healy, M. J. R. and Westmacott, M. (1956). Missing values in experiments analized on automatic computers. *Appl. Statist.* **5**, 203–206.

Hedayat, A. and Federer, W. T. (1974). Pairwise and variance balanced incomplete block designs. *Ann. Inst. Statist. Math.* **26**, 331.

Hemmerle, W. J. (1974). Nonorthogonal analysis of variance using iterative improvement and balanced residuals. *J. Am. Statist. Assn.* **69**, 772–778.

Hemmerle, W. J. and Downs, B. W. (1978). Nonhomogeneous variances in the mixed AOV model; maximum likelihood estimation. *Contributions to Survey Sampling and Appl. Statist.*, 153–172.

Henderson, C. R. and McAllister, A. J. (1978). The missing subclass problem in twoway fixed models. *Journal of Animal Science* **46**, 1125–1137.

Herr, D. G. (1976). A geometric characterization of connectedness in a two-way design. *Biometrika* **63**, 1, 93–100.

Hinkelmann, K. (1968). Missing values in partial diallel cross experiments. *Biometrics* **24**, 903–913.

Hocking, R. R. (1978). Discussion. *Proc. Statist. Comp. (J. Am. Statist. Assn.)*, 39.

Hocking, R. R. and Speed, F. M. (1975). A full rank analysis of some linear model problems. *J. Am. Statist. Assn.* **70**, 706–712.

Hocking, R. R., Hackney, O. P., and Speed, F. M. (1978). The analysis of linear models with unbalanced data. In *Papers in Honor of H.O. Hartley*, H. A. David, Ed. New York: Academic Press.

Hocking, R. R., Speed, F. M., and Coleman, A. T. (1980). Hypotheses to be tested with unbalanced data. *Commun. Statist. Theor. Math.* **A9**, 117–129.

Hoyle, M. H. (1971). Spoilt data—an introduction and bibliography. *J. Roy. Statist. Soc A* **134**, 429–439.

Irwin, J. O. (1931). Mathematical theorems involved in the analysis of variance. *J. Roy. Statist. Soc.* **94**, 284–300.

Ishikawa, E. (1951). On the condition for the randomized block method and studies on the missing plots. *Ann. Rep. Gakugei Fac., Iwata Univ.* **3**, 4–8.

Ishikawa, E. (1956). A note on formulas for finding estimates for any number of missing plots. *Rep. Statist. Appl. Res. (JUSE)* **4**(1), 33–55.

Jaech, J. L. (1966). An alternative approach to missing value estimation. *Am. Statist.* **20**(5), 27–29.

Jarrett, R. G. (1978). The analysis of designed experiments with missing observations. *Appl. Statist.* **27**, 38–46.

Jennings, E. (1977). Fixed effects analysis of variance with unequal cell sizes. *J. Exp. Educ.* **46**, 42–51.

Jennings, E. and Ward, J. H. (1982). Hypothesis identification in case of the missing cell. *Am. Statist.* **36**, 25–27.

John, J. A. and Prescott, P. (1975). Estimating missing values in experiments. *Appl. Statist.* **24**, 190–192.

John, P. W. M. (1971). *Statistical Design and Analysis of Experiments.* New York: Macmillan.

John, P. W. M. (1979). Missing points in 2^n and $2^{(n-k)}$ factorial designs. *Technometrics* **21**, 225–228.

John, P. W. M. (1980). *Incomplete Block Designs.* New York: Dekker.

Jones, R. M. (1959). On a property of incomplete blocks. *J. Roy. Statist. Soc. B***21**, 172–179.

Kempthorne, O. (1952). The design and analysis of experiments. New York: Wiley.

Kendall, M. G. and Stuart, A. (1976). *The Advanced Theory of Statistics*, 3rd ed. London: Charles Griffin E., Comp.

Khargonkar, S. A. (1948). The estimation of missing plot values in split plot and strip trials. *J. Indian Soc. Agric. Statist.* **1**, 147–161.

Khatri, C. G. and Shah, K. R. (1976). Some designs for two-way elimination of heterogeneity. *Sankhya Ser. B* **37**, 418–428.

Kramer, C. Y. and Glass, S. (1963). Analysis of variance of a latin square design with missing observations. *Appl. Statist.* **9**, 43–50.

Kruskal, W. (1960). The coordinate free approach to Gauss–Markov estimation and its application to missing and estra observations. *Proc. Fourth Berkeley Symp. Math. Stat. Probab.* **1**, 435–451.

Kshirsager, A. M. (1971). Bias due to missing plots. *Am. Statist.* **25**(1), 47–50.

Lejeune, M. (1980). Resolution des plans d'expérience non-connexes. *Revue belge de statistique*, *d'informatique et de recherche opérationnelle* **21**, 3–23.

Levy, K. J., Naruly, S., and Abrami, P. (1973). An empirical comparison of the methods of least squares and unweighted means for the analysis of disproportionate cell data. *Int. Statist. Rev.* **3**, 335–338.

Li, C. C. (1982). *Analysis of Unbalanced Data. A Pre-Program Introduction.* New York: Cambridge University Press.

Li, C. C. and Mazumdar, S. (1981). A type of orthogonal contrasts for unbalanced data. *Biometrical Journal* **23**(7), 645–651.

Lindstrom, R. (1970). Estimability of effects and interactions in an n-way cross classification experiment with missing data. Unpublished Ph.D. thesis. Oklahoma State University.

Little, R. J. A. (1976). Comments on inference and missing data. *Biometrika* **63**, 590–592.

Little, R. J. A. and Rubin, D. (1983). On jointly estimating parameters and missing data by maximizing the complete-data likelihood. *Am. Statist.* **37**, 218–220.

Lord, F. M. (1955). Estimation of parameters from incomplete data. *J. Am. Statist. Assn.* **50**, 870–876.

Margolin, B. H. (1969). Resolution IV fractional factorial designs. *J. Roy. Statist. Soc. B* **31**, 514–523.

Marshall, J. (1968). On Yates's approximation for the missing value problem in model 1 analysis of variance. Unpublished Ph.D. thesis submitted to the University of Chicago.

Matthai, A. (1951). Estimation of parameters from incomplete data with application to design of sample surveys. *Sankhya* **11**, 145–152.

Metha, J. S. and Gurland, J. (1969). Some properties and an application of a statistic arising in testing correlation. *Ann. Math. Statist.* **40**, 1736–1745.

Mexas, A. G. (1972). Analysis of classification data of arbitrarily incomplete data. Universidad Autonoma de Guerrero. Chilpancingo, Mexico.

Nair, K. R. (1940). The application of the technique of analysis of covariance to field experiments with several missing or mixed-up plots. *Sankhya* **4**, 581–588.

Nair, K. R. (1944). The recovery of inter-block information in incomplete block designs. *Sankhya* **6**, 383.

Nair, K. R. and Rao, C. R. (1941). Confounded designs for asymmetrical factorial experiments. *Science and Culture* **7**, 313.

Nair, K. R. and Rao, C. R. (1942). Confounded designs for $K \times p^m \times q^n$ type of factorial experiment. *Science and Culture* **7**, 361.

Nair, K. R. and Rao, C. R. (1942). A note on partially balanced incomplete block designs. *Science and Culture* **7**, 516.

Nair, K. A. and Rao, C. R. (1942). Incomplete block designs for experiments involving several groups of varieties. *Science and Culture* **7**, 625.

Nair, K. R. and Rao, C. R. (1948). Confounding in asymmetrical factorial experiments. *J. Roy. Statist. Soc. B* **10**, 109

Nelder, J. A. (1954). A note on missing plot values. *Biometrics* **10**, 400–401.

Nelder, J. A. (1965). The analysis of randomized experiments with orthogonal block structure (Parts I and II). *Proc. Roy. Soc.* **A283**, 147–178.

Nordheim, E. V. (1978). Obtaining information from nonrandomly missing data. *Proc. Statist. Comp. (Am. Statist. Assn.)*, 34–38.

Norton, H. W. (1955). A further note on missing data. *Biometrics* **11**, 110.

Orchard, T. and Woodbury, M. A. (1972). A missing information principle: theory and applications. *Proc. 6th Berkeley Symp.* **1**, 697–715.

Paik, U. B. and Federer, W. T. (1974). Analysis of nonorthogonal n-way classifications. *Ann. Statist.* **2**, 1000–1021.

Pearce, S. C. (1965). *Biological Statistics in an Introduction*, Section 7.3. New York: McGraw-Hill.

Pearce, S. C. (1974). The estimation of treatment means by designed experiments. *Appl. Statist.* **23**(1), 22–25.

Pearce, S. C. (1983). *The Agricultural Field Experiment*. New York: Wiley.

Pearce, S. C. and Jeffers, J. R. N. (1971). Block designs and missing data. *J. Roy. Statist. Soc. B*, **33**, 131–136.

Pearce, S. C., Calinski, T., and Marshall, T. F. (1974). The basic contrast of an experimental design with special reference to the analysis of data. *Biometrika* **61**(3), 449–460.

Pearson, E. S. (1931). The analysis of variance in cases of non-normal variation. *Biometrika* **23**, 114.

Preece, D. A. (1971). Iterative procedures for missing values in experiments. *Technometrics* **13**, 743–753.

Preece, D. A. and Gower, J. C. (1974). An iterative computer procedure for mixed-up values in experiments. *Appl. Statist.* **23**, 73–74.

Quenouille, M. H. (1948). The analysis of covariance and non-orthogonal comparisons. *Biometrics* **4**, 240–246.

Quenouille, M. H. (1953). *The Design and Analysis of Experiment*. London: Griffin.

Raghavarao, D. (1971). *Construction and Combinatorial Problems in Design of Experiments*. New York: Wiley.

Raghavarao, D. (1981). Missing observations in general linear model. (abstract No. 177–39). IMS Bulletin, 10, No. 5 (58).

Raghavarao, D. and Federer, W. T. (1975). On connectedness in two-way elimination of heterogeneity designs. *Ann. Statist.* **3**, 730–735.

Rao, C. R. (1947). General methods of analysis for incomplete block designs. *J. Am. Statist. Assn.* **42**, 541–561.

Rao, C. R. (1947). On the linear combination of observations and general theory of least squares. *Sankhya* **7**, 237.

Rao, C. R. (1962). A note on a generalized inverse of a matrix with applications to problems in mathematical statistics. *J. Roy. Statist. Soc. B* **24**, 152–158.

Rao, C. R. (1973). *Linear Statistical Inference and Its Applications*, (2nd ed. New York: Wiley.

Rees, D. H. (1966). The analysis of variance of designs with many non-orthogonal classifications. *J. Roy. Statist. Soc. B* **28**, 110–117.

Rubin, D. B. (1972). A non-iterative algorithm for least squares estimation of missing values in any analysis of variance design. *Appl. Statist.* **21**, 136–141.

Rubin, D. B. (1974). Characterizing the estimation of parameters in incomplete data problems. *J. Am. Statist. Assn.* **69**, 467–474.

Rubin, D. B. (1976). Inference and missing data. *Biometrika* **63**, 581–590.

Rubin, D. B. (1972). A non-iterative algorithm for least squares estimation of missing values in any analysis of variance design. *Appl. Statist* **21**, 136–141.

Rubin, D. B. (1976). Noniterative least squares estimates, standard errors and F-tests for analysis of variance with missing data. *J. Roy. Statist. Soc. B* **38**, 270–274.

Scheffé, H. (1959). *The Analysis of Variance*. New York: Wiley.

Sclove, S. L. (1972). On missing value estimation in experimental design models. *Am. Statist.* **26**(2), 25–26.

Searle, S. R. (1971). *Linear Models.* New York: Wiley.

Seber, G. A. F. (1964). Orthogonality in analysis of variance. *Ann. Math. Statist.* **35**, 705–710.

Seber, G. A. F. (1964a). Linear hypotheses and induced tests. *Biometrika* **51**, 41–47.

Seber, G. A. F. (1964b). The linear hypothesis and idempotent matrices. *J. Roy Statist. Soc. B* **26**, 261–266.

Seber, G. A. F. (1966). Linear hypothesis: A general theory. Griffin's statistical Monographs, No. 19, Griffin, London.

Seber, G. A. F. (1977). *Linear Regression Analysis.* New York: Wiley, Section 10.2.

Seely, J. (1970). Linear spaces and unbiased estimation. *Ann. Math. Statist.* **41**, 1725–1734.

Seely, J. (1979). Parametrizations and correspondences in linear models. Technical Report 72. Department of Statistics, Oregon State University.

Shah, K. R. and Dodge, Y. (1977). On connectedness of designs. *Sankhya* **39**, 284.

Shah, K. R. and Khatri, C. G. (1973). Connectedness in row-column designs. *Commun. Statist.* **2**, 571–573.

Shearer, P. R. (1973). Missing data in quantitative designs. *Appl. Statist.* **22**, 135–140.

Sirotnik, K. (1971). On the meaning of the mean in Aoua (or the case of the missing degree of freedom). *Am. Statist.* **25**(4), 36–37.

Smith, H. F. (1950). Error variance of treatment contrasts in an experiment with missing observations (with special references to incomplete latin squares). *Indian J. Agric. Statist.* **2**, 111–124.

Smith, H. F. (1957). Missing plot estimates. *Biometrics* **13**, 115–118.

Smith, P. L. (1981). The use of analysis of covariance to analyse data from designed experiments with missing or mixed-up values. *Appl. Statist.* **30**, 1–8.

Snedecor, G. W. and Cochran, W. G. (1980). *Statistical Methods. Ames, Iowa:* IOWA State University Press.

Speed, F. M., Hocking, R. R. and Coleman, A. H. (1977). Three factor factorial analysis with unequal number of observations and missing cells. *J. Am. Statist. Assn., Proceeding of Statistical Computation.*

Speed, F. M., Hocking, R. R., and Hackney, O. P. (1978). Methods of analysis of linear models with unbalanced data. *J. Am. Statist. Assn.* **72**, 105–112.

Srivastava, J. N. and Anderson, D. A. (1970). Some basic properties of multidimensional partially balanced designs. *Ann. Math. Statist.* **41**, 1438–1445.

Stevens, W. L. (1948). Statistical analysis of a non-orthogonal tri-factorial experiment. *Biometrika* **35**, 346–367.

Taylor, J. (1948). Errors of treatment comparisons when observations are missing. *Nature* **162**, 262–263.

Thompson, H. R. (1956). Extensions to missing plot techniques. *Biometrics* **12**, 241–244.

Tjur, T. (1984). Analysis of variance models in orthogonal designs. *Int. Statist. Rev.* **52**(1), 33–81.

Tocher, K. D. (1952). The design and analysis of block experiments. *J. Roy. Statist. Soc. B* **14**, 45–100.

Urquhart, N. S. and Weeks, D. L. (1978). Linear models in messy data: some problems and alternatives. *Biometrics* **34**, 696–705.

Urquhart, N. S., Weeks, D. L., and Henderson, C. R. (1973). Estimation associated with linear models: A revisitation. *Commun. Statist.* **1**, 303–330.

Webb, S. R. (1968). Non-orthogonal designs of even resolution. *Technometrics* **10**, 291–299.

Webb, S. R. (1971). Small incomplete factorial experiment designs for two- and three level factors. *Technometrics* **13**, 243–256.

Weeks, D. L. and Williams, D. R. (1964). A note on the determination of connectedness in an N-way cross classification. *Technometrics* **6**, 319–324. Errata, *Technometrics* **7**, 281.

Wilkinson, G. N. (1957). An analysis of paired comparisons designs with incomplete repetitions. *Biometrika* **44**, 97–113.

Wilkinson, G. N. (1957). The analysis of covariance with incomplete data. *Biometrics* **13**, 363–372.

Wilkinson, G. N. (1958a). Estimation of missing values for the analysis of incomplete data. *Biometrics* **14**, 257–286.

Wilkinson, G. N. (1958b). The analysis of variance and derivation of standard errors for incomplete data. *Biometrics* **14**, 360–384.

Wilkinson, G. N. (1960). Comparison of missing value procedures. *Aust. J. Statist.* **2**, 53–65.

Wilkinson, G. N. (1970). A general recursive procedure for analysis of variance. *Biometrika* **57**, 19–46.

Williams, E. R. (1977). Iterative analysis of generalized lattice designs. *Aust. J. Statist.* **19**(1), 39–42.

Williams, E. R. and Ratcliff, D. (1980). The analysis of covariance in multi-stratum experiments. In *Compstat 1980: Proceedings in Computational Statistics*. Berlin: Physica-Verlag.

Williams, E. R., Ratcliff, D., and Speed, T. P. (1981). Estimating missing values in multi-stratum experiments. *Appl. Statist.* **30**, 71–72.

Woodbury, M. A. and Siler, W. (1966). Factor analysis with missing data. *Ann. N.Y. Acad. Sci.* **128**, 746–754.

Wright, G. M. (1956). Missing values in factorial experiments. *Nature London* **178**, 1481.

Wynn, H. P. (1977). Combinatorial characteristics of certain. 2 × J × K three-way layouts. *Communications in Statist.* **10**, 945–953.

Yates, F. (1933). The principles of orthogonality and confounding in replicated experiments. *J. Agric. Sci.* **23**, part 1, 108–145.

Yates, F. (1933). The analysis of replicated experiments when the field results are incomplete. *Emp. J. Exp. Agric.* **1**, 129–142.

Yates, F. (1934). The analysis of multiple classifications with unequal numbers in the different classes. *J. Am. Statist. Assn.* **29**, 52–66.

Yates, F. (1935). Complex experiments. *Supplement to the J. Roy. Statist. Soc.* **2**, 181–247.

Yates, F. (1936). Balanced incomplete randomized blocks. *Ann. Eug.* **7**, part 2, 121–140.

Yates, F. (1936). Incomplete latin squares. *J. Agric. Sci.* **26**, part 2, 301–315.

Yates, F. (1936). The formation of latin squares for use in field experiments. *Emp. J. Exp. Agric.* **1**, 235–244.

Yates, F. (1937). The design and analysis of factorial experiments. Technical Communication No. 35, Imperial Bureau of Soil Science, London.

Yates, F. and Hale, R. W. (1939). The analysis of latin squares when two or more rows, columns or treatments are missing. *Suppl. J. Roy. Statist. Soc.* **6**, 67–79.

Zelen, M. (1957). The analysis of covariance for incomplete block designs. *Biometrics* **13**, 309–332.

Zyskind, G. (1967). On canonical forms, nonnegative covariance matrices and best and simple least squares linear estimators in linear models. *Ann. Math. Statist.* **38**, 1092–1109.

PROGRAM TWOWAY

```
C
C
C         THIS INTERACTIVE PROGRAM PERFORMS THE CALCULATIONS
C         REQUIRED FOR ANALYZING A SET OF DATA FOLLOWING AN ADDITIVE
C         TWOWAY CLASSIFICATION MODEL WITH MISSING OBSERVATIONS,
C         USING THE INCIDENCE MATRIX IN.
C         RESTRICTIONS:
C         GIVEN A TWO-WAY INCIDENCE MATRIX IN, EACH ROW AS WELL AS
C         EACH COLUMN MUST HAVE AT LEAST ONE OBSERVATION.
C
C         INPUT:
C         THE FIRST INPUT LINE CONSIST OF DIMENSION OF INCIDENCE
C         MATRIX IN (NxM) AND NUMBER OF HYPOTHESES (NH) TO BE
C         TESTED. NEXT INPUT LINES ARE THE ENTRIES OF THE INCIDENCE
C         MATRIX ROW BY ROW. THIS FOLLOWS BY THE ACTUAL
C         OBSERVATIONS. FOR EACH NON EMPTY CELL, THE ACTUAL
C         OBSERVATIONS ARE READ IN IN ONE LINE WITH A BLANK SPACE
C         BETWEEN TWO OBSERAVTIONS. THE NON EMPTY CELLS ARE READ
C         IN FROM LEFT TO RIGHT, RWO WISE.
C         THE HYPOTHESIS ASSOCIATED WITH TREATMENT IS SPECIFIED
C         BY NTYPE 1. IN THE SAME LINE, NUMBER OF ROWS ASSOCIATED
C         WITH THIS HYPOTHESIS WILL BE READ IN. THE ROWS OF THE
C         HYPOTHESIS MATRIX FOR TREATMENT WILL BE READ IN FROM
C         LEFT TO RIGHT AND EACH ELEMENT OCCUPIES A ROW SUCCES-
C         SIVELY. FOR BLOCK HYPOTHESIS USE NTYPE 2.
C         SEE INPUT FOR FIRST EXAMPLE.
C
C
C                   STRUCTURE
C
C
C         VARIABLES
C         COVM      OUTPUT: VARIANCE OF MEAN ESTIMATOR
C         IDFB      OUTPUT: BLOCK DEGREES OF FREEDOM
C         IDFT      OUTPUT: TREATMENT DEGREES OF FREEDOM
C         IDFG      OUTPUT: REGRESSION DEGREES OF FREEDOM
C         IPURE     OUTPUT: PURE ERROR DEGREES OF FREEDOM
C         IDSSBU    OUTPUT: BLOCK UNADJUSTED DEGREES OF FREEDOM
C         IDSSTU    OUTPUT: TREATMENT UNADJUSTED  DEGREES OF FREEDOM
C         IRES      OUTPUT: RESIDUAL DEGREES OF FREEDOM
C         ITB       OUTPUT: INTERACTION DEGREES OF FREEDOM
C         ITOT      OUTPUT: TOTAL DEGREES OF FREEDOM
C         N         INPUT: DIMENSION OF INCIDENCE MATRIX
C         NH        INPUT: NUMBER OF HYPOTHESIS
C         NTYPE     INPUT: TYPE OF HYPOTHESIS:
C                          1    TREATMENT
C                          2    BLOCK
C         NR        INPUT: NUMBER OF ROWS OF HYPOTHESIS MATRIX
C         M         INPUT: DIMENSION OF INCIDENCE MATRIX
C         SSB       OUTPUT: BLOCK SUMS OF SQUARES
C         SSBU      OUTPUT: BLOCK UNADJUSTED SUMS OF SQUARES
C         SSINT     OUTPUT: INTERACTION SUMS OF SQUARES
C         SSH       OUTPUT: SUMS OF SQUARES FOR HYPOTHESIS
C         SSM       OUTPUT: FITTING OVERALL MEAN
C         SSPURE    OUTPUT: PURE ERROR SUMS OF SQUARES
C         SSREG     OUTPUT: REGRESSION SUMS OF SQUARES
C         SSRES     OUTPUT: RESIDUAL SUMS OF SQUARES
```

```
C         SST         OUTPUT: TREATMENT SUMS OF SQUARES
C         SSTU        OUTPUT: TREATMENT UNADJUSTED SUMS OF SQUARES
C         YMEAN       OUTPUT: ESTIMATE OF MEAN
C         YSS         OUTPUT: TOTAL SUMS OF SQUARES
C
C         ARRAYS
C         BETA        OUTPUT: ESTIMATES OF BLOCK EFFECTS
C         COVB        OUTPUT: COVARIANCE MATRIX FOR BLOCK ESTIMATORS
C         COVMB       OUTPUT: COVARIANCE BETWEEN MEAN AND BLOCK ESTIMATORS
C         COVMT       OUTPUT: COVARIANCE BETWEEN MEAN AND TREATMENT ESTIMATORS
C         COVT        OUTPUT: COVARIANCE MATRIX FOR TREATMENT ESTIMATORS
C         COVTB       OUTPUT: COVARIANCE BETWEEN TREATMENT AND BLOCK ESTIMATORS
C         DELPT       OUTPUT: ESTIMATES OF HYPOTHESIS VECTOR
C         DELTA       INPUT: HYPOTHESIS MATRIX
C         IBASB       OUTPUT: BASIS FOR ESTIMABLE LINEAR COMBINATION OF
C                             BLOCK EFFECTS
C         IBASB2      OUTPUT: BASIS FOR ESTIMABLE LINEAR COMBINATION OF
C                             BLOCK EFFECTS
C         IBAST       OUTPUT: BASIS FOR ESTIMABLE LINEAR COMBINATION OF
C                             TREATMENT EFFECTS
C         IBAST2      OUTPUT: BASIS FOR ESTIMABLE LINEAR COMBINATION OF
C                             ESTIMABLE TREATMENT EFFECTS
C         ICB         OUTPUT: ESTIMABLE BLOCK DIFFERENCES
C         ICT         OUTPUT: ESTIMABLE TREATMENT DIFFERENCES
C         INT         OUTPUT: FINAL MATRIX
C         IN          INPUT: INCIDENCE MATRIX
C         TAU         OUTPUT: ESTIMATES OF TREATMENT EFFECTS
C         Y           INPUT: DATA
C
C         SUBROUTINES
C         DEGREE            COMPUTES ALL DEGREES OF FREEDOM
C         ESTIM             COMPUTES ALL SUMS OF SQUARES, ESTIMATORS, ETC
C         HYPO              TREATS THE HYPOTHESES
C
C
C
          PROGRAM TWOWAY
          PARAMETER IA=50
          DIMENSION IN(IA,IA),INT(IA,IA),ICB(IA,IA),ICT(IA,IA),
          1 QLAMT(IA,IA),QLAMB(IA,IA),YSUM(IA,IA),Y(IA),SS(IA,IA),
          2 BETA(IA),COVT(IA,IA),COVB(IA,IA),IBAST2(IA),
          3 IBAST(IA),IBASB(IA),COVTB(IA,IA),COVMT(IA),COVMB(IA),
          4 DELTA(IA,IA),DELPT(IA),QNTR(IA,IA),YSUMTR(IA,IA),
          5  QN(IA,IA),TAU(IA),IBASB2(IA),CVX(IA,IA)
          CHARACTER WEB*7
C
C         GIVE A NAME FOR INPUT FILE WITHOUT FILE EXTENSION
C
          TYPE 51
51        FORMAT(//,1X,'NAME OF INPUT FILE ? ',$)
          ACCEPT 52,WEB
52        FORMAT(A)
          OPEN(UNIT=10,FILE=WEB//'.DAT',STATUS='OLD')
          OPEN(UNIT=20,FILE=WEB//'.IMP',STATUS='NEW')
```

```
C
C       READ IN THE DIMENSION OF INCIDENCE MATRIX AND
C       NUMBER OF HYPOTHESES
C
        READ(10,*) N,M,NH
        IF(N.GT.IA.OR.M.GT.IA) THEN
        TYPE 56
56      FORMAT(//,1X,'THE INCIDENCE MATRIX IS TOO BIG')
        STOP
        END IF
        WRITE(20.1)
1      FORMAT(//,25X,'ANALYSIS OF VARIANCE FOR ADDITIVE TWO-WAY',
        1 /,25X,'CLASSIFICATION MODEL WITH MISSING OBSERVATIONS',///,
        2 <M+4>X,'INITIAL MATRIX',/)
C
C       READ IN THE INCIDENCE MATRIX
C
C
C       THE SUBROUTINE OUTONE GIVES AN OUTPUT OF THE INCIDENCE
C       MATRIX IN WITH A MAXIMUM OF 100 CHARACTERS FOR ONE
C       LINE. THEREFORE ONE INCIDENCE MATRIX ROW CAN TAKE
C       MORE THAN ONE OUTPUT LINE.
C
        DO I=1,N
                READ(10,*)(IN(I,L),L=1,M)
                CALL OUTONE(IN,I,M,IA)
                DO J=1,M
                        QN(I,J)=IN(I,J)
                END DO
        END DO
        YTOT=0
        SSPURE=0
        YSS=0
        DO I=1,N
        DO 4 J=1,M
                YSUM(I,J)=0.
                SS(I,J)=0.
                IF(IN(I,J).EQ.0) GOTO 4
C
C       READ IN OBSERVATIONS
C
                READ(10,*) (Y(K),K=1,IN(I,J))
                DO K=1,IN(I,J)
                        YSUM(I,J)=YSUM(I,J)+Y(K)
                        SS(I,J)=SS(I,J)+Y(K)**2
                        YTOT=YTOT+Y(K)
                END DO
                IF(IN(I,J).EQ.1) GOTO 5
                SSPURE=SSPURE+SS(I,J)-YSUM(I,J)**2/FLOAT(IN(I,J))
5               YSS=YSS+SS(I,J)
4       CONTINUE
        END DO
        WRITE(20,3)
3       FORMAT(///,<M+4>X,'FINAL MATRIX',/)
```

```
C
C         THE SUBROUTINE DEGREE CALCULATES DEGREES OF FREEDOM FOR
C         REGRESSION, EACH EFFECT, RESIDUAL, TOTAL, PURE ERROR
C         AND INTERACTION. IT ALSO FINDS THE BASIS FOR TREATMENTS
C         AND BLOCKS, AND COUNTER TRIANGLE FOR TREATMENT AND BLOCK
C         RESPECTIVELY.
C
          CALL DEGREE(IN,INT,ICB,ICT,M,N,IDFB,IDFT,ITOT,IRES,IPURE,
          1 ITB,ICON,QLAMT,QLAMB,IBAST,IBAST2,IBASB,IBASB2)
          SSM=YTOT**2/FLOAT(ITOT)
          CYSS=YSS-SSM
          DO I=1,N
                  CALL OUTONE(INT,I,M,IA)
          END DO
          IF(IDFT.LE.IDFB) GOTO 6
          DO I=1,N
          DO J=1,M
                  YSUMTR(J,I)=YSUM(I,J)
                  QNTR(J,I)=FLOAT(IN(I,J))
          END DO
          END DO
C
C         THE SUBROUTINE ESTIM CALCULATES SUMS OF SQUARES FOR TOTAL,
C         TREATMENT ADJUSTED, TREATMENT UNADJUSTED, BLOCK ADJUSTED,
C         BLOCK UNADJUSTED, RESIDUAL. IT ALSO COMPUTES ESTIMATE OF
C         MEAN ,OF BLOCK EFFECTS, OF TREATMENT EFFECTS, COVARIANCE
C         BETWEEN MEAN AND TREATMENTS ESTIMATORS, COVARIANCE
C         BETWEEN MEAN AND BLOCK ESTIMATORS, COVARIANCE MATRIX
C         FOR TREATMENTS ESTIMATORS, COVARIANCE MATRIX FOR BLOCK
C         ESTIMATORS AND COVARIANCE BETWEEN TREATMENT AND BLOCK
C         ESTIMATORS.
C
          CALL ESTIM(QNTR,QLAMB,QLAMT,M,N,IDFB,IDFT,IRES,YSUMTR,YSS,BETA,
          1 TAU,COVB,COVT,SSTU,SSB,SSBU,SST,SSRES,SIGSQ,YMEAN,CVX,COVMB,
          2 COVMT,COVM,IFAUX)
          IF(IFAUX.EQ.1) GOTO 49
          DO I=1,N
          DO J=1,M
                  COVTB(I,J)=CVX(J,I)
          END DO
          END DO
          GOTO 7
6         CALL ESTIM(QN,QLAMT,QLAMB,N,M,IDFT,IDFB,IRES,YSUM,YSS,TAU,BETA,
          1 COVT,COVB,SSBU,SST,SSTU,SSB,SSRES,SISGQ,YMEAN,COVTB,COVMT,
          2 COVMB,COVM,IFAUX)
          IF(IFAUX.EQ.1) GOTO 49
7         SSINT=SSRES-SSPURE
          SSREG=SST+SSBU
          SSBU=SSBU-SSM
          SSTU=SSTU-SSM
          IDFG=IDFT+M
          IDSSBU=M-1
          IDSSTU=N-1
```

```
      WRITE(20,8) IDFG,SSREG,SSM,IDSSBU,SSBU
8     FORMAT('1',///,1X,64('*'),/,28X,'*',12X,'*',/,3X,'SOURCE OF ',
     1 'VARIATION',6X,'*',4X,'D.F.',4X,'*',3X,'SUMS OF SQUARES',/,
     2 28X,'*',12X,'*',/,1X,64('*'),/,28X,'*',12X,'*',/,2X,'REGRESSION'
     3 ,16X,'*',2X,I3,7X,'*',2X,E15.7,/,28X,'*',12X,'*',/,28X,'*',12X
     4 ,'*',/,6X,'FITTING OVERALL MEAN',2X,'*',8X,'1',3X,'*',6X,E15.7,
     5 /,28X,'*',12X,'*',/,6X,'BLOCKS UNADJUSTED',5X,'*',6X,I3,3X,'*'
     6 ,6X,E15.7,/,28X,'*',12X,'*')
      WRITE(20,9) IDFT,SST,SSM,IDSSTU,SSTU,IDFB,SSB,IRES,SSRES
9     FORMAT(6X,'TREATMENTS',12X,'*',6X,I3,3X,'*',6X,E15.7,/,28X,'*',
     1 12X,'*',/,28X,'*',12X,'*',/,6X,'FITTING OVERALL MEAN',2X,'*',
     2 8X,'1',3X,'*',6X,E15.7,/,28X,'*',12X,'*',/,6X,'TREATMENTS ',
     3 'UNADJUSTED *',6X,I3,3X,'*',6X,E15.7,/,28X,'*',12X,'*',/,6X,
     4 'BLOCKS',16X,'*',6X,I3,3X,'*',6X,E15.7,/,28X,'*',12X,'*',/,
     5 28X,'*',12X,'*',/,2X,'RESIDUAL'18X,'*',2X,I3,7X,'*',2X,E15.7,
     6 /,28X,'*',12X,'*',/,28X,'*',12X,'*')
      WRITE(20,10) IPURE,SSPURE,ITB,SSINT,ITOT,YSS
10    FORMAT(6X,'PURE ERROR',12X,'*',6X,I3,3X,'*',6X,E15.7,/,28X,'*',
     1 12X,'*',/,6X,'INTERACTION',11X,'*',6X,I3,3X,'*',6X,E15.7,/,
     2 28X,'*',12X,'*',/,1X,64('*'),/,28X,'*',12X,'*',/,2X,'TOTAL',
     3 21X,'*',2X,I3,7X,'*',2X,E15.7)
      WRITE(20,11) YMEAN,COVM
11    FORMAT('1',///,3X,'ESTIMATE OF MEAN = ',F13.6,///,3X,'VARIANCE ',
     1 'OF MEAN ESTIMATOR',/,1X,F11.5,///,3X,'COVARIANCE BETWEEN ',
     2 'MEAN AND TREATMENT ESTIMATORS',/)
C
C     THE SUBROUTINE OUTTWO GIVES AN OUTPUT OF COVARIANCE BETWEEN
C     MEAN AND TREATMENT ESTIMATORS WITH A MAXIMUM OF 100 CHARACTERS
C     FOR ONE LINE.
C
      CALL OUTTWO(COVMT,N,IA)
      WRITE(20,53)
53    FORMAT(///,3X,'COVARIANCE BETWEEN MEAN AND BLOCK ESTIMATORS',/)
      CALL OUTTWO(COVMB,M,IA)
      N1=N-1
      IF(N1.GE.33) GOTO 54
      IF(IDFT.EQ.N1) GOTO 12
      WRITE(20,13) (I,I=1,N1)
13    FORMAT('1',///,1X,'ESTIMABLE TREATMENT DIFFERENCES',/,/,3X,<N1>I3)
      DO I=2,N
      L=I-1
      WRITE(20,14) I,(ICT(I,J),J=1,L)
14    FORMAT(1X,I2,<L>A3)
      END DO
54    WRITE(20,15)
15    FORMAT(///,2X,'BASIS FOR ESTIMABLE LINEAR COMBINATION',/,2X,
     1 'OF TREATMENT EFFECTS',/)
      DO K=1,IDFT
            I1=IBAST2(K)
            I2=IBAST(I1)
            WRITE(20,16) I2,I1
16          FORMAT(1X,'T(',I2,') - T(',I2,')')
      END DO
      GOTO 17
12    WRITE(20,18)
18    FORMAT('1',///,1X,'ALL TREATMENT DIFFERENCES ARE ESTIMABLE')
```

```
17        WRITE(20,19)
19        FORMAT(///,1X,'ESTIMATES OF TREATMENT EFFECTS',/)
          DO K=1,N
                  WRITE(20,20) K,TAU(K)
20                FORMAT(1X,'T(',I2,') = ',F13.6)
          END DO
          WRITE(20,21)
21        FORMAT('1',///,1X,'COVARIANCE MATRIX FOR TREATMENT ESTIMATORS'
          1 ,/)
C         THE SUBROUTINE OUTHREE GIVES AN OUTPUT OF COVARIANCE MATRIX
C         FOR TREATMENT ESTIMATORS WITH A MAXIMUM OF 100 CHARACTERS
C         FOR ONE LINE. THEREFORE, ONE MATRIX ROW CAN TAKE MORE THEN
C         ONE OUTPUT LINE.
C
          DO I=1,N
                  CALL OUTTHREE(COVT,I,N,IA)
          END DO
          M1=M-1
          IF(M1.GE.33) GOTO 55
          IF(IDFB.EQ.M1) GOTO 23
          WRITE(20,24) (J,J=1,M1)
24        FORMAT('1',///,1X,'ESTIMABLE BLOCK DIFFERENCES',//,3X,<M1>I3)
          DO J1=2,M
                  L=J1-1
                  WRITE(20,14) J1,(ICB(J1,J2),J2=1,L)
          END DO
55        WRITE(20,25)
25        FORMAT(///,1X,'BASIS FOR ESTIMABLE LINEAR COMBINATION',/,1X,
          1 ' OF BLOCK EFFECTS',/)
          DO K=1,IDFB
                  J1=IBASB2(K)
                  J2=IBASB(J1)
                  WRITE(20,26) J2,J1
26                FORMAT(1X,'B(',I2,') - B(',I2,')')
          END DO
          GOTO 27
23        WRITE(20,47)
47        FORMAT('1',///,1X,'ALL BLOCK DIFFERENCES ARE ESTIMABLE')
27        WRITE(20,28)
28        FORMAT(///,1X,'ESTIMATES OF BLOCK EFFECTS',/)
          DO K=1,M
                  WRITE(20,29) K,BETA(K)
29                FORMAT(1X,'B(',I2,') = ',F13.6)
          END DO
          WRITE(20,30)
30        FORMAT('1',///,1X,'COVARIANCE MATRIX FOR BLOCK ESTIMATORS',/)
          DO I=1,M
                  CALL OUTTHREE(COVB,I,M,IA)
          END DO
          WRITE(20,46)
46        FORMAT('1',///,1X,'COVARIANCE BETWEEN TREATMENT AND BLOCK ',
          1 'ESTIMATORS',/)
          DO I=1,N
                  CALL OUTTHREE(COVTB,I,M,IA)
          END DO
```

```
      IF(NH.EQ.0) GOTO 31
C
C     TREATMENT OF THE HYPOTHESES
C
      DO 32 K=1,NH
C
C     READ IN TYPE OF HYPOTHESIS AND NUMBER OF ROWS NEEDED
C
            READ(10,*) NTYPE,NR
            IF(NTYPE.EQ.1) GOTO 35
            IF(NTYPE.EQ.2) GOTO 36
            WRITE(20,37)
37          FORMAT(//,1X,'  HYPOTHESIS SPECIFICATION ERROR',/)
            GOTO 32
36          WRITE(20,38)
38          FORMAT('1',///,1X,'BLOCK HYPOTHESIS',///)
            DO I=1,NR
                  DELPT(I)=0.
                  DO J=1,M
C
C     READ IN HYPOTHESIS ELEMENTS
C
                        READ(10,*) DELTA(I,J)
                        DELPT(I)=DELPT(I)+BETA(J)*DELTA(I,J)
                  END DO
                  CALL OUTTHREE(DELTA,I,M,IA)
            END DO
C
C     THE SUBROUTINE OUTTHREE GIVE AN OUTPUT OF HYPOTHESIS
C
            GOTO 48
35          WRITE(20,41)
41          FORMAT('1',///,1X,'TREATMENT HYPOTHESIS',///)
            DO I=1,NR
                  DELPT(I)=0.
                  DO J=1,N
C
C     READ IN HYPOTHESIS ELEMENTS
C
                        READ(10,*) DELTA(I,J)
                        DELPT(I)=DELPT(I)+TAU(J)*DELTA(I,J)
                  END DO
C
C     THE SUBROUTINE OUTTHREE GIVE AN OUTPUT OF HYPOTHESIS
C
                  CALL OUTTHREE(DELTA,I,N,IA)
            END DO
48          WRITE(20,42)
42          FORMAT(///,1X,'ESTIMATE OF HYPOTHESIS VECTOR',//)
            CALL OUTTWO(DELPT,NR,IA)
            WRITE(20,43)
43          FORMAT(//,1X,'COVARIANCE MATRIX OF ESTIMATOR FOR ',
     1 'HYPOTHESIS VECTOR',/)
```

```
C
C        SUBROUTINE HYPO COMPUTES COVARIANCE MATRIX FOR BLOCK
C        OR TREATMENT ESTIMATORS AND SUM OF SQUARES
C
                CALL HYPO(DELTA,DELPT,COVB,COVT,SSH,N,M,NR,NTYPE,IFAUX)
                IF(IFAUX.EQ.1) GOTO 49
                DO I=1,NR
                        CALL OUTTHREE(DELTA,I,NR,IA)
                END DO
                WRITE(20,44) SSH
44              FORMAT(//,1X,'SUM OF SQUARES = ',F17.7)
32       CONTINUE
49       IF(IFAUX.NE.1) GOTO 31
C
C        STOP TEST WHEN ERROR APPEARS
C
         WRITE(20,50)
50       FORMAT(//,1X,'UNABLE TO COMPUTE INVERSE')
31       CLOSE(UNIT=10)
         CLOSE(UNIT=20)
         STOP
         END
C
C
C
C
         SUBROUTINE DEGREE(IN,INT,ICB,ICT,M,N,IDFB,IDFT,ITOT,IRES,IPURE,
         1 ITB,ICON,QLAMT,QLAMB,IBAST,IBAST2,IBASB,IBASB2)
         PARAMETER IA=50
         DIMENSION IN(IA,IA),INT(IA,IA),ICB(IA,IA),ICT(IA,IA),INS(IA,IA),
         1 QLAMT(IA,IA),QLAMB(IA,IA),IBAST(IA),IBASB(IA),IBAST2(IA),IBASB2(IA)
         COMMON /DATA/F(IA),ISTAR,IZERO
         DATA ISTAR,IZERO/'  *','  0'/
         ITOT=0
         ICELL=0
C
C        INITIALIZE INS MATRIX FOR R-PROCESS
C
         DO I=1,N
         DO J=1,M
                 ITOT=ITOT+IN(I,J)
                 INS(I,J)=IN(I,J)
                 IF(INS(I,J).GT.0) INS(I,J)=1
                 ICELL=ICELL+INS(I,J)
         END DO
         END DO
         IPURE=ITOT-ICELL
C
C        R-PROCESS
C
1        DO I=1,N
         DO L=1,M
                 IL=0
                 DO K=1,N
                 DO J=1,M
                         IL=IL+INS(I,J)*INS(K,J)*INS(K,L)
                 END DO
```

```
          END DO
          IF(IL.GT.0) INT(I,L)=1
      END DO
      END DO
      INSZ=0
      INTZ=0
      DO I=1,N
      DO L=1,M
          INSZ=INSZ+INS(I,L)
          INTZ=INTZ+INT(I,L)
      END DO
      END DO
C
C     STOP TEST FOR R-PROCESS
C
      IF(INSZ.EQ.INTZ) GOTO 3
C
C     CALCULATION OF M*M' IN ORDER TO FIND DEGREES OF FREEDOM
C
      DO I=1,N
      DO J=1,M
          INS(I,J)=INT(I,J)
      END DO
      END DO
      GOTO 1
3     DO J1=2,M
          L=J1-1
          DO J2=1,L
              JJ=0
              DO K=1,N
                  JJ=JJ+INT(K,J1)*INT(K,J2)
              END DO
              IF(JJ.GT.0) ICB(J1,J2)=1
          END DO
      END DO
      IDFB=0
      DO J1=2,M
          L=J1-1
          ICC=0
          DO J2=1,L
              ICC=ICC+ICB(J1,J2)
          END DO
          IF(ICC.GT.0) ICC=1
          IDFB=IDFB+ICC
      END DO
      DO I1=2,N
          L=I1-1
          DO I2=1,L
              II=0
              DO K=1,M
                  II=II+INT(I1,K)*INT(I2,K)
              END DO
              IF(II.GT.0) ICT(I1,I2)=1
          END DO
      END DO
```

```
C
C         OTHER DEGREES OF FREEDOM ARE FOUND BY SUBTRACTION
C
          IDFT=IDFB+N-M
          IRES=ITOT-IDFB-N
          ITB=IRES-IPURE
          ICON=M-1-IDFB
C
C         COUNTER TRIANGLE FOR BLOCK
C
          DO I1=2,N
                  L=I1-1
                  DO 13 I2=1,L
                          IF(ICT(I1,I2).EQ.0) GOTO 13
                          IBAST(I1)=I2
                          GOTO 6
13                CONTINUE
6                 DO 5 I2=1,L
                          IF(ICT(I1,I2).EQ.0) GOTO 4
                          ICT(I1,I2)=ISTAR
                          GOTO 5
4                         ICT(I1,I2)=IZERO
5                 CONTINUE
          END DO
C
C         COUNTER TRIANGLE FOR TREATMENT
C
          DO J1=2,M
                  L=J1-1
                  DO 12 J2=1,L
                          IF(ICB(J1,J2).EQ.0) GOTO 12
                          IBASB(J1)=J2
                          GOTO 11
12                CONTINUE
11                DO 8 J2=1,L
                          IF(ICB(J1,J2).EQ.0) GOTO 7
                          ICB(J1,J2)=ISTAR
                          GOTO 8
7                         ICB(J1,J2)=IZERO
8                 CONTINUE
          END DO
C
C         CALCULATION OF TREATMENT BASIS
C
          K=0
          DO 9 I1=2,N
                  IF(IBAST(I1).EQ.0) GOTO 9
                  K=K+1
                  IBAST2(K)=I1
                  I2=IBAST(I1)
                  QLAMT(K,I1)=-1.0
                  QLAMT(K,I2)=1.0
9         CONTINUE
C
C         CALCULATION OF BLOCK BASIS
C
          K=0
```

```
      DO 10 J1=2,M
              IF(IBASB(J1).EQ.0) GOTO 10
              K=K+1
              IBASB2(K)=J1
              J2=IBASB(J1)
              QLAMB(K,J1)=-1.0
              QLAMB(K,J2)=1.0
10    CONTINUE
      RETURN
      END
C
C
C
C
      SUBROUTINE ESTIM(QN,QLAMT,QLAMB,N,M,IDFT,IDFB,IRES,YSUM,YSS,
     1 TAU,BETA,COVT,COVB,SSBU,SST,SSTU,SSB,SSRES,SIGSQ,YMEAN,COVTB,
     2 COVMT,COVMB,COVM,IFAUX)
      PARAMETER IA=50,IA2=IA**2
      DIMENSION QN(IA,IA),QLAMT(IA,IA),TAU(IA),COVT(IA,IA),APY(IA),
     1 APA(IA),BPY(IA),BPB(IA),BPBI(IA),R(IA2),RR(IA2),S(IA2),
     2 APAI(IA),COVB(IA,IA),QLAMB(IA,IA),BETA(IA),Q(IA2),P(IA2),
     3 ET(IA),COVTB(IA,IA),COVMT(IA),COVMB(IA),VV(IA2),V(IA2),
     4 W(IA2),YSUM(IA,IA),WW(IA2)
      COMMON /DATA/F(IA)
      DATA F/IA*1.0/
C
C     CALCULATION OF A'A, A'Y, B'B B'Y AND INVERSE MATRIX
C
      DO I=1,N
              APA(I)=0.
              APY(I)=0.
              DO J=1,M
                      APY(I)=APY(I)+YSUM(I,J)
                      APA(I)=APA(I)+QN(I,J)
              END DO
              APAI(I)=1./APA(I)
      END DO
      DO J=1,M
              BPY(J)=0.
              BPB(J)=0.
              DO I=1,N
                      BPY(J)=BPY(J)+YSUM(I,J)
                      BPB(J)=BPB(J)+QN(I,J)
              END DO
              BPBI(J)=1./BPB(J)
      END DO
C
C     DESCRIPTION OF EACH SEPARATE CALCULATION IS NOT POSSIBLE
C     BECAUSE THE DIFFERENT COMPUTATIONS ARE MIXTED
C
      CALL ARRAY(QN,S,N,M,1,IA)
      CALL MPRO(S,BPBI,R,N,M,M,2,IA2,IA,IA2)
      CALL MTRA(S,P,N,M,IA2,IA2)
      CALL MPRO(R,P,S,N,M,N,1,IA2,IA2,IA2)
      CALL MSUB(APA,S,WW,N,N,2,IA,IA2,IA2)
      CALL MPRO(R,BPY,S,N,M,1,1,IA2,IA,IA2)
      CALL MSUB(APY,S,W,N,1,1,IA,IA2,IA2)
```

```
CALL ARRAY(QLAMT,R,IDFT,N,1,IA)
CALL MTRA(R,RR,IDFT,N,IA2,IA2)
CALL MPRO(WW,RR,S,N,N,IDFT,1,IA2,IA2,IA2)
CALL MPRO(R,S,Q,IDFT,N,IDFT,1,IA2,IA2,IA2)
CALL ARRAY(COVT,Q,IDFT,IDFT,2,IA)
CALL MINV(COVT,IDFT)
CALL ARRAY(COVT,Q,IDFT,IDFT,1,IA)
CALL MPRO(R,W,V,IDFT,N,1,1,IA2,IA2,IA2)
CALL MPRO(Q,V,W,IDFT,IDFT,1,1,IA2,IA2,IA2)
CALL MPRO(RR,W,TAU,N,IDFT,1,1,IA2,IA2,IA)
CALL MPRO(RR,Q,S,N,IDFT,IDFT,1,IA2,IA2,IA2)
CALL MPRO(S,R,Q,N,IDFT,N,1,IA2,IA2,IA2)
CALL ARRAY(COVT,Q,N,N,2,IA)
CALL MTRA(W,S,IDFT,1,IA2,IA2)
CALL MPRO(S,V,W,1,IDFT,1,1,IA2,IA2,IA2)
CALL MPRO(BPBI,P,WW,M,M,N,3,IA,IA2,IA2)
CALL MPRO(WW,TAU,P,M,N,1,1,IA2,IA,IA2)
CALL MPRO(BPBI,BPY,S,M,M,1,3,IA,IA,IA2)
CALL MSUB(S,P,ET,M,1,1,IA2,IA2,IA)
CALL MTRA(BPY,VV,M,1,IA,IA2)
CALL MPRO(VV,S,V,1,M,1,1,IA2,IA2,IA2)
SST=W(1)
SSBU=V(1)
SSRES=YSS-SSBU-SST
SIGSQ=SSRES/FLOAT(IRES)
CALL ARRAY(QN,S,N,M,1,IA)
CALL MPRO(APAI,APY,S,N,N,1,3,IA,IA,IA2)
CALL MTRA(APY,VV,N,1,IA,IA2)
CALL MPRO(VV,S,V,1,N,1,1,IA2,IA2,IA2)
SSTU=V(1)
SSB=SSBU+SST-SSTU
CALL MPRO(F,ET,V,1,M,1,1,IA,IA,IA2)
YMEAN=V(1)/FLOAT(M)
CALL MPRO(F,F,R,M,1,M,4,IA,IA,IA2)
SC=1./FLOAT(M)
CALL SMPY(R,SC,S,M,M,IA2,IA2)
CALL MSUB(F,S,R,M,M,2,IA,IA2,IA2)
CALL MPRO(R,ET,BETA,M,M,1,1,IA2,IA,IA)
CALL SMPY(F,SC,V,M,1,IA,IA2)
CALL SMPY(V,-1.0,VV,M,1,IA2,IA2)
CALL MPRO(WW,Q,P,M,N,N,1,IA2,IA2,IA2)
CALL MPRO(VV,P,COVMT,1,M,N,1,IA2,IA2,IA).
CALL MTRA(WW,S,M,N,IA2,IA2)
CALL MPRO(P,S,RR,M,N,M,1,IA2,IA2,IA2)
CALL MPRO(Q,S,P,N,N,M,1,IA2,IA2,IA2)
CALL MPRO(P,R,S,N,M,M,1,IA2,IA2,IA2)
CALL SMPY(S,-1.0,P,N,M,IA2,IA2)
CALL ARRAY(COVTB,P,N,M,2,IA)
CALL MADD(BPBI,RR,S,M,IA)
CALL MPRO(S,V,P,M,M,1,1,IA2,IA2,IA2)
CALL MPRO(V,P,W,1,M,1,1,IA2,IA2,IA2)
COVM=W(1)
CALL MPRO(S,R,P,M,M,M,1,IA2,IA2,IA2)
CALL MPRO(R,P,S,M,M,M,1,IA2,IA2,IA2)
```

```
      CALL ARRAY(COVB,S,M,M,2,IA)
      CALL MPRO(V,P,COVMB,1,M,M,1,IA2,IA2,IA)
      RETURN
      END
C
C
C
C
C
      SUBROUTINE HYPO(DELTA,DELPT,COVB,COVT,SSH,N,M,NR,NTYPE,IFAUX)
      PARAMETER IA=50,IA2=IA**2
      DIMENSION BETA(IA2),DELTA(IA,IA),R(IA2),S(IA2),DELPT(IA),
      1 COVB(IA,IA),COVT(IA,IA),P(IA,IA),V(IA),VV(IA),Q(IA2)
      IF(NTYPE.EQ.2) GOTO 1
      K=N
      CALL ARRAY(COVT,S,N,N,1,IA)
      GOTO 2
1     K=M
      CALL ARRAY(COVB,S,M,M,1,IA)
2     CALL ARRAY(DELTA,R,NR,K,1,IA)
      CALL MPRO(R,S,BETA,NR,K,K,1,IA2,IA2,IA2)
      CALL MTRA(R,Q,NR,K,IA2,IA2)
      CALL MPRO(BETA,Q,S,NR,K,NR,1,IA2,IA2,IA2)
      CALL ARRAY(DELTA,S,NR,NR,2,IA)
      CALL ARRAY(P,S,NR,NR,2,IA)
      CALL MINV(P,NR)
      CALL ARRAY(P,S,NR,NR,1,IA)
      CALL MPRO(DELPT,S,V,1,NR,NR,1,IA,IA2,IA)
      CALL MPRO(V,DELPT,VV,NR,NR,1,1,IA,IA,IA)
      SSH=VV(1)
      RETURN
      END
C
C
C
      SUBROUTINE ARRAY(A,B,N,M,ICODE,IA)
C
C     SUBROUTINE ARRAY TRANSFORMS MATRIX INTO VECTOR IN USING
C     ICODE 1 AND TRANSFORMS VECTOR INTO MATRIX IN USING ICODE 2
C
      DIMENSION A(IA,IA),B(IA**2)
      IF(ICODE.EQ.2) GOTO 1
      K=1
      DO I=1,N
      DO L=1,M
              B(K)=A(I,L)
              K=K+1
      END DO
      END DO
      GOTO 2
1     I=1
      L=1
      DO K=1,N*M
              A(I,L)=B(K)
              IF(L.LT.M) GOTO 3
              L=0
              I=I+1
```

```
3             L=L+1
        END DO
2       RETURN
        END
C
C
C
        SUBROUTINE MADD(A,B,C,N,IA)
C
C       SUBROUTINE MADD ADDS A DIAGONAL MATRIX A WITH A NORMAL MATRIX B
C
        DIMENSION A(IA),B(IA**2),C(IA**2)
        K=0
        DO 2 I=1,N**2
              IF(I.EQ.1+K*(N+1)) GOTO 1
              C(I)=B(I)
              GOTO 2
1             C(I)=A(K+1)+B(I)
              K=K+1
2       CONTINUE
        RETURN
        END
C
C
C
        SUBROUTINE MINV(A,M)
C
C       SUBROUTINE MINV INVERSES MATRIX A
C       FOR HIGHIER PRECISION USE FOR EXAMPLE IMSL SUBROUTINE OR
C       OTHER SYSTEM SUBROUTINES
C       ERROR TESTS ARE PROVIDED
C
        PARAMETER IA=50
        DIMENSION A(IA,IA),B(IA,IA),C(IA,IA)
        IF(M.EQ.0.OR.(M.EQ.1.AND.A(1,1).EQ.0.)) THEN
              TYPE *,'ERROR IN COMPUTING INVERSE'
              STOP
        END IF
        IF(M.EQ.1) THEN
              A(1,1)=1./A(1,1)
              RETURN
        END IF
        DO I=1,M
        DO K=1,M
              C(I,K)=A(I,K)
        END DO
        END DO
        CALL DET(C,M,DETER)
        IF(DETER.EQ.0.) THEN
        TYPE *,'ERROR IN COMPUTING INVERSE'
        STOP
        END IF
        IF(M.EQ.2) GOTO 6
        DO I=1,M
        DO 5 K=1,M
              J1=1
              J2=0
```

```
               DO 1 J=1,M
               DO 2 L=1,M
                     IF(J.EQ.I) GOTO 1
                     IF(L.EQ.K) GOTO 2
                     J2=J2+1
                     IF(J2.LE.M-1) GOTO 3
                     J2=1
                     J1=J1+1
3                    C(J1,J2)=A(J,L)
2              CONTINUE
1              CONTINUE
               IF(M.NE.2) GOTO 7
               DE=C(1,1)
               GOTO 8
7              CALL DET(C,M-1,DE)
8              B(K,I)=DE/DETER*(-1)**(I+K)
5        CONTINUE
         END DO
         DO I=1,M
         DO J=1,M
               A(I,J)=B(I,J)
         END DO
         END DO
         GOTO 9
6        B1=A(2,2)/DETER
         B2=-A(1,2)/DETER
         B3=-A(2,1)/DETER
         B4=A(1,1)/DETER
         A(1,1)=B1
         A(1,2)=B2
         A(2,1)=B3
         A(2,2)=B4
9        RETURN
         END
C
C
C
         SUBROUTINE DET(C,N,TOTAL)
C
C        SUBROUTINE DET COMPUTES DETERMINANT OF MATRIX C
C        IT WORKS IN DOUBLE PRECISION
C        ERROR TESTS ARE PROVIDED
C
         PARAMETER IB=50
         DIMENSION C(IB,IB)
         DOUBLE PRECISION A(IB,IB),RAP
         DO I=1,N
               DO K=1,N
                     A(I,K)=C(I,K)
               END DO
         END DO
         RAP=0.
         SIGNE=1.
         TOTAL=1.
         DO I=N,2,-1
         DO 1 L=I-1,1,-1
               IF(A(L,I).EQ.0.) GOTO 1
```

```
               IF(A(I,I).NE.0.) GOTO 4
               DO I1=L,1,-1
                      IND=I1
                      IF(A(I1,I).NE.0.) GOTO 3
               END DO
               TYPE *,'ERROR IN COMPUTING INVERSE'
               STOP
3              DO I1=1,N
                      B=A(IND,I1)
                      A(IND,I1)=A(I,I1)
                      A(I,I1)=B
               END DO
               SIGNE=(-1.)**(IND+I)*SIGNE
4              RAP=A(L,I)/A(I,I)
               DO K=I-1,1,-1
                      A(L,K)=A(L,K)-RAP*A(I,K)
               END DO
1       CONTINUE
        END DO
        DO I=1,N
               TOTAL=TOTAL*A(I,I)
        END DO
        TOTAL=TOTAL*SIGNE
        END
C
C
C
        SUBROUTINE MPRO(A,B,C,N,II,M,ICODE,ICA,ICB,ICC)
C
C       SUBROUTINE MPRO MULTIPLICATES A MATRIX A BY A MATRIX B
C       FOUR CASES APPEARS:
C       ICODE 1: MULTIPLICATION OF TWO NORMAL MATRIX
C       ICODE 2: MULTIPLICATION OF A NORMAL MATRIX BY A DIAGONAL
C                MATRIX
C       ICODE 3: MULTIPLICATION OF A DIAGONAL MATRIX BY A NORMAL
C                MATRIX
C       ICODE 4: MULTIPLICATION OF TWO VECTORS IN ORDER TO FORM
C                A MATRIX
C
        DIMENSION A(ICA),B(ICB),C(ICC)
        IF(ICODE.NE.1) GOTO 1
        IPAS=1
        DO K=1,N
        DO J=1,M
               L=0
               C(IPAS)=0.
               DO I=1+II*(K-1),II*K
                      C(IPAS)=A(I)*B(L*M+J)+C(IPAS)
                      L=L+1
               END DO
               IPAS=IPAS+1
        END DO
        END DO
        GOTO 2
1       IF(ICODE.NE.2) GOTO 3
        K=1
```

```
      DO I=1,N*II
              C(I)=A(I)*B(K)
              K=K+1
              IF(K.GT.II) K=1
      END DO
      GOTO 2
3     IF(ICODE.NE.3) GOTO 5
      K=1
      J=0
      DO 4 I=1,N*M
              C(I)=A(K)*B(I)
              J=J+1
              IF(J.LT.M) GOTO 4
              K=K+1
              J=0
4     CONTINUE
      GOTO 2
5     IPAS=1
      DO I=1,M
      DO K=1,N
              C(IPAS)=A(K)*B(I)
              IPAS=IPAS+1
      END DO
      END DO
2     RETURN
      END
C
C
C
      SUBROUTINE MSUB(A,B,C,N,M,ICODE,ICA,IA,IB)
C
C     SUBROUTINE MSUB SUBTRACT A MATRIX B FROM A MATRIX A
C     ICODE 1 DEUX NORMAL MATRIX ARE PRESENT
C     ICODE 2 A DIAGONAL AND A NORMAL MATRIX ARE PRESENT
C
      DIMENSION A(ICA),B(IA),C(IB)
      K=0
      IF(ICODE.NE.1) GOTO 3
      DO I=1,N*M
              C(I)=A(I)-B(I)
      END DO
      GOTO 4
3     DO 2 I=1,N*M
              IF(I.EQ.1+K*(M+1)) GOTO 1
              C(I)=-B(I)
              GOTO 2
1             C(I)=A(K+1)-B(I)
              K=K+1
2     CONTINUE
4     RETURN
      END
C
C
C
```

```
      SUBROUTINE MTRA(A,B,N,M,IC,ID)
C
C     SUBROUTINE MTRA TRANSPOSES A MATRIX A INTO B
C
      DIMENSION A(IC),B(ID)
      K=0
      IPAS=1
      DO 1 I=1,M*N
              B(I)=A(IPAS+K*M)
              K=K+1
              IF(K.LT.N) GOTO 1
              IPAS=IPAS+1
              K=0
1     CONTINUE
      RETURN
      END
C
C
C
      SUBROUTINE SMPY(A,COEF,B,N,M,IA,IB)
C
C     SUBROUTINE SMPY MULTIPLICATES A VECTOR BY A NUMBER (COEF)
C
      DIMENSION A(IA),B(IB)
      DO I=1,M*N
              B(I)=A(I)*COEF
      END DO
      RETURN
      END
C
C
C
      SUBROUTINE OUTONE(N,I,M,IA)
      DIMENSION N(IA,IA)
      K=0
      L=20
      KTEST=0
      IF(M.LE.20) GOTO 1
      ZM=FLOAT(M)
      KTEST=INT(ZM/20.)
      IF(KTEST.EQ.M/20) GOTO 2
      KTEST=INT(ZM/20.)-1
2     K=1
      DO ID=1,KTEST
              WRITE(20,3) (N(I,IE),IE=K,L)
3             FORMAT(1X,20(2X,I3))
              K=K+20
              L=L+20
      END DO
      K=K-1
1     IDIF=M-KTEST*20
      IF(IDIF.LE.0) GOTO 4
      WRITE(20,5)(N(I,J),J=K+1,M)
5     FORMAT(1X,<IDIF>(2X,I3))
4     RETURN
      END
C
```

```
C
C
        SUBROUTINE OUTTWO(Z,M,IA)
        DIMENSION Z(IA)
        K=0
        L=7
        KTEST=0
        IF(M.LE.7) GOTO 1
        ZM=FLOAT(M)
        KTEST=INT(ZM/7.)
        IF(KTEST.EQ.M/7) GOTO 2
        KTEST=INT(ZM/7.)-1
2       K=1
        DO ID=1,KTEST
                WRITE(20,3) (Z(IE),IE=K,L)
3               FORMAT(1X,7(3X,F11.5))
                K=K+7
                L=L+7
        END DO
        K=K-1
1       IDIF=M-KTEST*7
        IF(IDIF.LE.0) GOTO 4
        WRITE(20,5)(Z(J),J=K+1,M)
5       FORMAT(1X,<IDIF>(3X,F11.5))
4       RETURN
        END
C
        SUBROUTINE OUTTHREE(Z,I,M,IA)
        DIMENSION Z(IA,IA)
        K=0
        L=7
        KTEST=0
        IF(M.LE.7) GOTO 1
        ZM=FLOAT(M)
        KTEST=INT(ZM/7.)
        IF(KTEST.EQ.M/7) GOTO 2
        KTEST=INT(ZM/7.)-1
2       K=1
        DO ID=1,KTEST
                WRITE(20,3) (Z(I,IE),IE=K,L)
3               FORMAT(/,1X,7(3X,F11.5))
                K=K+7
                L=L+7
        END DO
        K=K-1
1       IDIF=M-KTEST*7
        IF(IDIF.LE.0) GOTO 4
        WRITE(20,5)(Z(I,J),J=K+1,M)
5       FORMAT(/,1X,<IDIF>(3X,F11.5))
4       RETURN
```

INPUT DATA

```
8 6 3
0 0 2 0 1 0
2 4 0 0 0 0
0 0 3 0 0 0
0 3 0 1 0 2
0 0 0 0 2 0
0 0 0 0 4 0
0 0 0 5 0 1
0 0 1 0 2 0
2.3 4.5
1.6
4.2 7.1
3.6 2.4 1.9 4.6
3.9 7.9 8.5
5.6 6.1 2.1
3.4
5.2 6.9
9.1 11.9
3.5 4.2 6.8 6.7
8.7 12.9 6.4 8.6 2.3
1.4
2.4
4.2 1.5
1 3
4.
-1.
-1.
-1.
-1.
0.
0.
0.
0.
1.
-1.
0.
0.
0.
0.
0.
2.
-1.
-1.
0.
0.
0.
0.
0.
1 2
0.
6.
-1.
-1.
-1.
-1.
-1.
-1.
```

```
0.
0.
0.
3.
-1.
-1.
-1.
0.
2 3
5.
-1.
-1.
-1.
-1.
-1.
1.
-1.
0.
0.
0.
0.
0.
3.
-1.
-1.
-1.
0.
```

ANALYSIS OF VARIANCE FOR ADDITIVE TWO-WAY
CLASSIFICATION MODEL WITH MISSING OBSERVATIONS

INITIAL MATRIX

0	0	2	0	1	0
2	4	0	0	0	0
0	0	3	0	0	0
0	3	0	1	0	2
0	0	0	0	2	0
0	0	0	0	4	0
0	0	0	5	0	1
0	0	1	0	2	0

FINAL MATRIX

0	0	1	0	1	0
1	1	0	1	0	1
0	0	1	0	1	0
1	1	0	1	0	1
0	0	1	0	1	0
0	0	1	0	1	0
1	1	0	1	0	1
0	0	1	0	1	0

SOURCE OF VARIATION	D.F.	SUMS OF SQUARES
REGRESSION	12	0.1041775E+04
FITTING OVERALL MEAN	1	0.9006592E+03
BLOCKS UNADJUSTED	5	0.3825555E+02
TREATMENTS	6	0.1028604E+03
FITTING OVERALL MEAN	1	0.9006592E+03
TREATMENTS UNADJUSTED	7	0.1231243E+03
BLOCKS	4	0.1799167E+02
RESIDUAL	21	0.1422849E+03
PURE ERROR	19	0.1103971E+03
INTERACTION	2	0.3188780E+02
TOTAL	33	0.1184060E+04

ESTIMATE OF MEAN = 5.553906

VARIANCE OF MEAN ESTIMATOR
0.04077

COVARIANCE BETWEEN MEAN AND TREATMENT ESTIMATORS

-0.00692 -0.02493 -0.02254 -0.01191 0.04517 0.01392 -0.00149
0.00871

COVARIANCE BETWEEN MEAN AND BLOCK ESTIMATORS

0.03103 -0.00803 0.02344 -0.02192 -0.02344 -0.00109

ESTIMABLE TREATMENT DIFFERENCES

	1	2	3	4	5	6	7
2	0						
3	*	0					
4	0	*	0				
5	*	0	*	0			
6	*	0	*	0	*		
7	0	*	0	*	0	0	
8	*	0	*	0	*	*	0

BASIS FOR ESTIMABLE LINEAR COMBINATION
OF TREATMENT EFFECTS

T(1) - T(3)
T(2) - T(4)
T(1) - T(5)
T(1) - T(6)
T(2) - T(7)
T(1) - T(8)

ESTIMATES OF TREATMENT EFFECTS

T(1) = -2.866407
T(2) = -1.317101
T(3) = 0.875261
T(4) = 0.157899
T(5) = 5.283594
T(6) = 0.083593
T(7) = 0.524567
T(8) = -2.741406

COVARIANCE MATRIX FOR TREATMENT ESTIMATORS

0.32723	-0.00892	0.05119	-0.02195	-0.14152	-0.11027	-0.03236	-0.06340
-0.00892	0.35226	0.00670	-0.01493	-0.06101	-0.02976	-0.21979	-0.02455
0.05119	0.00670	0.52515	-0.00632	-0.25090	-0.21965	-0.01674	-0.08944
-0.02195	-0.01493	-0.00632	0.20122	-0.07403	-0.04278	-0.00364	-0.03757
-0.14152	-0.06101	-0.25090	-0.07403	0.55640	0.08765	-0.08445	-0.03215
-0.11027	-0.02976	-0.21965	-0.04278	0.08765	0.36890	-0.05320	-0.00090
-0.03236	-0.21979	-0.01674	-0.00364	-0.08445	-0.05320	0.45816	-0.04799
-0.06340	-0.02455	-0.08944	-0.03757	-0.03215	-0.00090	-0.04799	0.29598

ESTIMABLE BLOCK DIFFERENCES

	1	2	3	4	5
2	*				
3	0	0			
4	*	*	0		
5	0	0	*	0	
6	*	*	0	*	0

BASIS FOR ESTIMABLE LINEAR COMBINATION
OF BLOCK EFFECTS

B(1) - B(2)
B(1) - B(4)
B(3) - B(5)
B(1) - B(6)

ESTIMATES OF BLOCK EFFECTS

B(1) = 1.413195
B(2) = -1.111805
B(3) = 0.337500
B(4) = 1.032639
B(5) = -0.337500
B(6) = -1.334028

COVARIANCE MATRIX FOR BLOCK ESTIMATORS

0.59317	0.03067	0.00000	-0.33970	0.00000	-0.28414
0.03067	0.21817	0.00000	-0.15220	0.00000	-0.09664
0.00000	0.00000	0.18750	0.00000	-0.18750	0.00000
-0.33970	-0.15220	0.00000	0.44039	0.00000	0.05150
0.00000	0.00000	-0.18750	0.00000	0.18750	0.00000
-0.28414	-0.09664	0.00000	0.05150	0.00000	0.32928

COVARIANCE BETWEEN TREATMENT AND BLOCK ESTIMATORS

-0.03103	0.00803	-0.08594	0.02192	0.08594	0.00109
-0.24920	-0.14764	-0.02344	0.23662	0.02344	0.16023
-0.03103	0.00803	-0.21094	0.02192	0.21094	0.00109
0.10496	-0.04348	-0.02344	0.00745	0.02344	-0.06894
-0.03103	0.00803	0.16406	0.02192	-0.16406	0.00109
-0.03103	0.00803	0.16406	0.02192	-0.16406	0.00109
0.29941	0.15097	-0.02344	-0.35366	0.02344	-0.09672
-0.03103	0.00803	0.03906	0.02192	-0.03906	0.00109

TREATMENT HYPOTHESIS

```
    4.00000      -1.00000      -1.00000      -1.00000      -1.00000      0.00000      0.00000
    0.00000
    0.00000       1.00000      -1.00000       0.00000       0.00000      0.00000      0.00000
    0.00000
    2.00000      -1.00000      -1.00000       0.00000       0.00000      0.00000      0.00000
    0.00000
```

ESTIMATE OF HYPOTHESIS VECTOR

```
  -16.46528      -2.19236      -5.29097
```

COVARIANCE MATRIX OF ESTIMATOR FOR HYPOTHESIS VECTOR

```
    7.03935      -0.24884       3.24884
   -0.24884       0.86400       0.05266
    3.24884       0.05266       2.03067
```

SUM OF SQUARES =   63.8277054

TREATMENT HYPOTHESIS

```
    0.00000       6.00000      -1.00000      -1.00000      -1.00000      -1.00000      -1.00000
   -1.00000
    0.00000       0.00000       0.00000       3.00000      -1.00000      -1.00000      -1.00000
    0.00000
```

ESTIMATE OF HYPOTHESIS VECTOR

```
  -12.08611      -5.41806
```

COVARIANCE MATRIX OF ESTIMATOR FOR HYPOTHESIS VECTOR

```
   17.46297       2.07870
    2.07870       3.81713
```

SUM OF SQUARES =   12.8008976

BLOCK HYPOTHESIS

5.00000	-1.00000	-1.00000	-1.00000	-1.00000	-1.00000
1.00000	-1.00000	0.00000	0.00000	0.00000	0.00000
0.00000	3.00000	-1.00000	-1.00000	-1.00000	0.00000

ESTIMATE OF HYPOTHESIS VECTOR

8.47917	2.52500	-4.36805

COVARIANCE MATRIX OF ESTIMATOR FOR HYPOTHESIS VECTOR

21.35417	3.37500	2.59028
3.37500	0.75000	-0.37500
2.59028	-0.37500	3.31713

SUM OF SQUARES = 17.3840466

INPUT DATA

```
4 5 0
1 1 0 0 0
0 0 0 1 2
1 0 1 0 0
0 1 0 0 0
2
3
1
4 5
3
2
4
```

ANALYSIS OF VARIANCE FOR ADDITIVE TWO-WAY CLASSIFICATION MODEL WITH MISSING OBSERVATIONS

INITIAL MATRIX

1	1	0	0	0
0	0	0	1	2
1	0	1	0	0
0	1	0	0	0

FINAL MATRIX

1	1	1	0	0
0	0	0	1	1
1	1	1	0	0
1	1	1	0	0

SOURCE OF VARIATION	D.F.	SUMS OF SQUARES
REGRESSION	7	0.8350000E+02
FITTING OVERALL MEAN	1	0.7200000E+02
BLOCKS UNADJUSTED	4	0.1050000E+02
TREATMENTS	2	0.1000000E+01
FITTING OVERALL MEAN	1	0.7200000E+02
TREATMENTS UNADJUSTED	3	0.2333336E+01
BLOCKS	3	0.9166664E+01
RESIDUAL	1	0.4999999E+00
PURE ERROR	1	0.5000000E+00
INTERACTION	0	-0.1192093E-06
TOTAL	8	0.8400000E+02

ESTIMATE OF MEAN = 2.700000

VARIANCE OF MEAN ESTIMATOR
0.18000

COVARIANCE BETWEEN MEAN AND TREATMENT ESTIMATORS

0.00000 0.00000 -0.20000 0.20000

COVARIANCE BETWEEN MEAN AND BLOCK ESTIMATORS

0.02000 -0.18000 0.22000 0.02000 -0.08000

```
ESTIMABLE TREATMENT DIFFERENCES

     1  2  3
  2  0
  3  *  0
  4  *  0  *


BASIS FOR ESTIMABLE LINEAR COMBINATION
OF TREATMENT EFFECTS

T( 1) - T( 3)
T( 1) - T( 4)


ESTIMATES OF TREATMENT EFFECTS

T( 1) =     -0.666667
T( 2) =      0.000000
T( 3) =      0.333333
T( 4) =      0.333333
```

COVARIANCE MATRIX FOR TREATMENT ESTIMATORS

0.44444	0.00000	-0.22222	-0.22222
0.00000	0.00000	0.00000	0.00000
-0.22222	0.00000	1.11111	-0.88889
-0.22222	0.00000	-0.88889	1.11111

```
ESTIMABLE BLOCK DIFFERENCES

     1  2  3  4
  2  *
  3  *  *
  4  0  0  0
  5  0  0  0  *


BASIS FOR ESTIMABLE LINEAR COMBINATION
OF BLOCK EFFECTS

B( 1) - B( 2)
B( 1) - B( 3)
B( 4) - B( 5)


ESTIMATES OF BLOCK EFFECTS

B( 1) =     -0.033333
B( 2) =      0.966667
B( 3) =     -1.033333
B( 4) =     -1.700000
B( 5) =      1.800000
```

COVARIANCE MATRIX FOR BLOCK ESTIMATORS

0.55778	-0.24222	0.02444	-0.22000	-0.12000
-0.24222	0.95778	-0.77556	-0.02000	0.08000
0.02444	-0.77556	1.49111	-0.42000	-0.32000
-0.22000	-0.02000	-0.42000	0.78000	-0.12000
-0.12000	0.08000	-0.32000	-0.12000	0.48000

COVARIANCE BETWEEN TREATMENT AND BLOCK ESTIMATORS

-0.11111	-0.11111	0.22222	0.00000	0.00000
0.00000	0.00000	0.00000	0.00000	0.00000
-0.24444	0.75556	-0.91111	0.20000	0.20000
0.35556	-0.64444	0.68889	-0.20000	-0.20000

CHAPTER 6

Additive Three-Way Classification with Missing Observations: Estimability and Analysis

Strange, is it not? That of the myriads who
Before us pass'd the door of Darkness through,
Not one returns to tell us of the Road,
which to discover we must travel too.

KHAYYAM NAISHAPURI-RUBAIYAT

6.1 INTRODUCTION

Chapter 5 presented a process, called the R process, which when applied to the incidence matrix $\mathbf{N}$ of a factorial experiment with an additive two-way classification model produces a matrix $\mathbf{M}$ identifying all estimable cell expectations. Furthermore, $\mathbf{M}$ can be used to construct a basis for $\bar{\Theta}$, the vector space of estimable linear parametric functions. In particular, we can obtain bases for $\bar{\mathscr{A}}$, $\bar{\mathscr{B}}$, and $\bar{\mathscr{E}}$ where $\bar{\Theta}=\bar{\mathscr{A}}\oplus\bar{\mathscr{B}}\oplus\bar{\mathscr{E}}$, $\bar{\mathscr{A}}$ is the subspace of estimable α contrasts, $\bar{\mathscr{B}}$ is the subspace of estimable β contrasts, and $\bar{\mathscr{E}}$ is the subspace of estimable linear parametric functions that are not functions of only the α or the β contrasts.

Chapter 5 also presented a method of constructing least squares estimators of the parameters of the model. These least squares estimators can be used to construct BLUEs of estimable parametric functions and to construct likelihood ratio tests of linear hypotheses.

In this chapter we consider estimability and analysis for the additive three-way classification model with missing observations. We will find that the R process can be applied to obtain a sufficient condition for a cell expectation to be estimable. The R process will sometimes indicate that cell expectations are estimable. If the R process does not indicate this, we can still proceed to obtain a spanning set of the estimable linear parametric functions. As part of this, we will obtain spanning sets of the vector spaces consisting of estimable contrasts involving single parameter types.

We shall then obtain least squares estimators of the model parameters. As in Chapter 5, these can be used to obtain BLUEs of estimable parametric functions and likelihood ratio tests of linear hypotheses.

We begin our discussion first by briefly considering the Latin square design as an example of an incomplete three-way classification design in which the position of the occupied cells are prespecified. Later, we provide results on estimability and analysis for an additive three-way classification model with arbitrary incidence, that is, with an arbitrary data pattern among the cells.

6.2 LATIN SQUARES

We have seen that in the design that we called a complete three-way classification design there is at least one observation in each cell. Often this is not practical or possible to accommodate all possible treatment combinations, and we must adopt a design in which there are no observations in some of the cells. The balanced incomplete block design is an example of an incomplete two-way classification design in which observations are selected according to some pattern. The design matrix associated with the two-dimensional $a \times b$ incidence matrix $\mathbf{N}$ is of maximal rank, that is, $r(\mathbf{X}) = a + b - 1$.

The Latin square design is an incomplete three-way classification design in which all three factors are at the same number v levels, and observations are taken on only v^2 of the v^3 possible treatment combinations according to the pattern described below:

A *Latin square* is an arrangement of v elements, each repeated v times, in a square matrix of order v, in such a way that each element appears exactly once in each row and in each column. Let L be a Latin square. Let L_{ij} represent the i, jth element of L, $i, j = 1, 2, \ldots, v$. Consider $v = 4$.

$$L = \begin{matrix} 1 & 2 & 3 & 4 \\ 4 & 1 & 2 & 3 \\ 3 & 4 & 1 & 2 \\ 2 & 3 & 4 & 1 \end{matrix}.$$

L as given is a Latin square. From this example, we see that Latin squares of all sizes exist. If the columns of a Latin square are permuted the result is again a Latin square. This is true for rows as well as numbers.

Originally Latin square designs were introduced in agricultural experimentation. In these designs the treatments are grouped into replicates in two different ways (double grouping). Each treatment occurs once and once only in each row and each column. By double grouping we eliminate from the errors all differences among rows and equally among columns.

Suppose four varieties of corn were to be compared and that we have a piece of land that can be subdivided into a 4×4 grid of 16 plots. Suppose also that the ground moisture varies across the field in one direction, and soil fertility occurs in the other direction. Each variety would be planted in four plots, in such a way that it appears once in each row and once in each column.

The Latin square arrangement shown in Figure 6.1 has the property that each row and each column receives each treatment (represented by numbers 1, 2, 3, 4) exactly once.

Columns

	1	2	3	4
	2	3	4	1
Rows	3	4	1	2
	4	1	2	3

Figure 6.1 4×4 Latin square design for variety experiment.

In this way we can eliminate (here soil heterogeneity) to a large extent the contribution to the experimental environment by using this design when the size of the square is not too large.

The Latin square design can be used in any three-factor experiment where each factor is at v levels.

Let the rows of the square matrix denote the levels of the first factor, the columns the levels of the second factor, and the numbers the levels of the third factor. A complete factorial experiment with three factors, each at four levels for example, calls for $4^3 = 64$ observations. If we take a Latin square of size 4, then we have a set of $4^2 = 16$ out of 64 observations with the property that each level of any factor appears exactly once with each level of each of the other factors in the 16-treatment combinations.

Such a reduction in the number of observations can be very attractive. The advantage of this design over the complete three-way classification design is that only $1/v$ times as many observations (v^3) are required.

The single three-dimensional pattern of a Latin square can be alternatively presented in a three two-dimensional matrices. For example, let us consider the following Latin square of size 3 with rows as levels of factor A, columns as levels of factor B, and numbers as the levels of factor C:

Columns

Rows	1	2	3
	2	3	1
	3	1	2

The occupied cells of the $3 \times 3 \times 3$ incidence matrix $\mathbf{N} = (n_{ijk})$ are presented by ones in the following configuration:

		Factor C: 1			Factor C: 2			Factor C: 3		
		Factor B			Factor B			Factor B		
		1	2	3	1	2	3	1	2	3
Factor A	1	1	0	0	0	1	0	0	0	1
	2	0	0	1	1	0	0	0	1	0
	3	0	1	0	0	0	1	1	0	0

We assume a completely additive model with no interaction terms in developing the analysis of variance of the Latin square of the form

$$Y_{ijk} = \mu + \alpha_i + \beta_j + \gamma_k + e_{ijk}$$

where Y_{ijk} is the observation made at the ith level of the first factor, the jth level of the second factor, and kth level of the third factor, and α_i, β_j, and γ_k corresponding to the main effects of A, B, and C, respectively. The analysis of this model is a special case of the additive three-way classification models with arbitrary patterns which we shall consider in the following sections.

6.3 ADDITIVE THREE-WAY CLASSIFICATIONS WITH ARBITRARY PATTERNS

In this chapter we assume that $\{Y_{ijkh}\}$ is a collection of N independent and normally distributed random variables each having a common unknown variance σ^2 and an expectation of the form

$$E(Y_{ijkh}) = \mu + \alpha_i + \beta_j + \gamma_k, \tag{6.1}$$

where i, j, k range from 1 to a, b, c, respectively, and for given i, j, k the index h ranges from 1 to n_{ijk}. It is to be understood that when $n_{ijk}=0$ no random variables with the first three subscripts i, j, k occur in the collection. It is further assumed, and without loss in generality, that

$$\begin{aligned} n_{i..} &= \sum_{jk} n_{ijk} \neq 0 \quad \text{for } i=1,\ldots,a, \\ n_{.j.} &= \sum_{ik} n_{ijk} \neq 0 \quad \text{for } j=1,\ldots,b, \\ n_{..k} &= \sum_{ij} n_{ijk} \neq 0 \quad \text{for } k=1,\ldots,c. \end{aligned} \tag{6.2}$$

Thus, we suppose that the Y_{ijkh}'s follow an additive fixed-effect three-way classification model with no restrictions on the unknown parameters.

Associated with the assumed model we define the incidence matrix $\mathbf{N}$ to be a three-dimensional $(a \times b \times c)$ array consisting of the n_{ijk}'s. Also $\mathbf{Y}$ denotes the $N \times 1$ vector consisting of the Y_{ijkh}'s ordered lexicographically and $\mathbf{1}$, $\mathbf{A}$, $\mathbf{B}$, $\mathbf{C}$ and $\mathbf{X}=(\mathbf{1}, \mathbf{A}, \mathbf{B}, \mathbf{C})$ denote matrices defined so that

$$E(\mathbf{Y}) = \mathbf{1}\mu + \mathbf{A}\boldsymbol{\alpha} + \mathbf{B}\boldsymbol{\beta} + \mathbf{C}\boldsymbol{\gamma} = \mathbf{X}\boldsymbol{\theta}, \tag{6.3}$$

where $\boldsymbol{\alpha}=(\alpha_1,\ldots,\alpha_a)'$, $\boldsymbol{\beta}=(\beta_1,\ldots,\beta_b)'$, $\boldsymbol{\gamma}=(\gamma_1,\ldots,\gamma_c)'$, and $\boldsymbol{\theta}=(\mu, \boldsymbol{\alpha}', \boldsymbol{\beta}', \boldsymbol{\gamma}')'$.

6.4 ESTIMABILITY

A linear combination of the parameters $\mu, \alpha_1, \ldots, \alpha_a, \beta_1, \ldots, \beta_b, \gamma_1, \ldots, \gamma_c$ in the additive three-way classification model (6.1) is a linear parametric function. Such a function is estimable if and only if it is a linear combination of the expectations $\mu+\alpha_i+\beta_j+\gamma_k$ of those cells for which observations are available. So one spanning set for $\bar{\Theta}$ is

$$\{\mu+\alpha_i+\beta_j+\gamma_k : n_{ijk} > 0\}. \tag{6.4}$$

However, this spanning set is not useful for determining what linear hypotheses on $\boldsymbol{\theta}$ correspond to the various reduction sums of squares $SS(\gamma|\alpha, \beta)$, $SS(\beta|\alpha)$, $SS(\alpha)$, and so on. Furthermore, we may wish to know which linear parametric functions involving only one parameter type are estimable.

Let $\bar{\mathscr{A}}$ denote the set of estimable linear parametric functions involving only the α parameters. It is easy to see that Theorem 5.2 also holds for the additive three-way classification model, and that $\bar{\mathscr{A}}$ is also the vector space of estimable α-contrasts. Let $\bar{\mathscr{B}}$ and $\bar{\mathscr{G}}$ denote the sets of estimable linear parametric functions involving only the β and the γ parameters, respectively. Then $\bar{\mathscr{B}}$ and $\bar{\mathscr{G}}$ are the vector spaces of estimable β contrasts and estimable γ contrasts, respectively.

$\bar{\mathscr{A}}$, $\bar{\mathscr{B}}$, and $\bar{\mathscr{G}}$ are linearly independent subspaces of $\bar{\Theta}$, the space of estimable linear parametric functions. Therefore, $\bar{\Theta}$ can be expressed as a direct sum

$$\bar{\Theta} = \bar{\mathscr{A}} \oplus \bar{\mathscr{B}} \oplus \bar{\mathscr{G}} \oplus \bar{\mathscr{E}}, \tag{6.5}$$

where the subspace $\bar{\mathscr{E}}$ is not uniquely determined. The goal of this section is to present procedures that provide bases for these subspaces.

If we could determine that all cell expectations were estimable, then we would have

$$\begin{aligned}
\bar{\mathscr{A}} &= \text{span}\{\alpha_1-\alpha_2, \alpha_1-\alpha_3, \ldots, \alpha_1-\alpha_a\},\\
\bar{\mathscr{B}} &= \text{span}\{\beta_1-\beta_2, \beta_1-\beta_3, \ldots, \beta_1-\beta_b\},\\
\bar{\mathscr{G}} &= \text{span}\{\gamma_1-\gamma_2, \gamma_1-\gamma_3, \ldots, \gamma_1-\gamma_c\},\\
\bar{\mathscr{E}} &= \text{span}\{\mu+\alpha_1+\beta_1+\gamma_1\}.
\end{aligned} \tag{6.6}$$

Furthermore, these spanning sets would then be bases and the dimensions of the subspaces would be $f_\alpha = a-1$, $f_\beta = b-1$, $f_\gamma = c-1$, and $f_c = 1$. Note that the subspace $\bar{\mathscr{E}}$ is not unique.

In the case of missing observations, we may be able to determine that all cell expectations are estimable by applying an extension of the R process presented in Chapter 5 to the three-way classification model. The R process is a procedure applied to a two-dimensional matrix $\mathbf{W}$ with non-negative integers as entries to obtain a matrix $\mathbf{M}$ with the same dimensions as $\mathbf{W}$ and having zeros and ones as entries.

The rows and columns of $\mathbf{W}$ may be indexed by any two finite index sets, say R and C, and the entries of $\mathbf{W}$ are denoted by w_{rc} for $r \in R$ and $c \in C$. The rows, columns, and entries of $\mathbf{M}$ are indexed exactly like $\mathbf{W}$. (In Chapter 5, we had $R = \{1, 2, \ldots, a\}$ and $C = \{1, 2, \ldots, b\}$, the levels of factors α and β, respectively). The R process applied to the matrix $\mathbf{W}$ is defined as follows:

1 For all $r \in R$ and $c \in C$, set m_{rc} equal to zero or 1 according to whether w_{rc} is zero or nonzero.

2 Change any zero m_{rc} to 1 if there exists $s \in R$ and $d \in C$ such that $m_{rd} = m_{sd} = m_{sc} = 1$. (Pictorially, we add the fourth corner whenever three corners of a rectangle appear in the matrix $\mathbf{M}$.)

3 Continue step 2, using both the original and the new nonzero m_{rc}'s as corners of new rectangles, until no more entries can be changed.

The matrix **M** which results from applying the R process to a two-dimensional matrix **W** is called the *final matrix* of the R process obtained from **W**. Once an entry of **M** is set to 1, it will never change. Since **M** has a finite number of entries, the R process must terminate.

The R process can be extended to a three-dimensional matrix **W** with nonnegative integers as entries and finite index sets I, J, and K:

1 Let **M** be the matrix with the same dimensions as **W** and having zeros as entries where **W** has zeros and ones where **W** was positive integers.

2 Form the two-dimensional matrix $\mathbf{M}^{(1)}$ from **M** with row index set $R=J\times K$ and column index set $C=I$. Apply the R process to $\mathbf{M}^{(1)}$ obtaining the final matrix, also labeled $\mathbf{M}^{(1)}$. Construct a (possibly) new three-dimensional matrix **M** with index sets I, J, and K from $\mathbf{M}^{(1)}$.

3 Form the two-dimensional matrix $\mathbf{M}^{(2)}$ from **M** with $R=I\times K$ and $C=J$. Apply the R process to $\mathbf{M}^{(2)}$, obtaining the final matrix $\mathbf{M}^{(2)}$, and then construct a new matrix **M** with index sets I, J, and K from $\mathbf{M}^{(2)}$.

4 Form $\mathbf{M}^{(3)}$ from **M** with $R=I\times J$ and $C=K$. Apply the R process to $\mathbf{M}^{(3)}$, obtaining the final matrix $\mathbf{M}^{(3)}$, and construct **M** from $\mathbf{M}^{(3)}$.

5 Repeat steps 2–4 until **M** remains unchanged over these three steps.

This process will be called the $R3$ *process*, and the resulting matrix **M** will be called the *final matrix* of the $R3$ process obtained from **W**. The $R3$ process must terminate for the same reasons that the R process must terminate. The $R3$ process is easily generalized to higher dimensions.

We apply the $R3$ process to the $a\times b\times c$ incidence matrix **N** of the three-way classification with missing observations. We take the index sets $I=\{1, 2, \ldots, a\}$, $J=\{1, 2, \ldots, b\}$, and $K=\{1, 2, \ldots, c\}$ to correspond to the sets of levels of the three factors (the subscripts of the α, β, and γ parameters, respectively). Let **M** denote the $a\times b\times c$ final matrix of the $R3$ process obtained from the incidence matrix **N**. Then we have the following sufficient condition for a cell expectation to be estimable:

THEOREM 6.1 If $m_{ijk}=1$, then $\mu+\alpha_i+\beta_j+\gamma_k$ is estimable.

Proof If $m_{ijk}=1$ in step 1 of the $R3$ process, then $\mu+\alpha_i+\beta_j+\gamma_k$ is estimable. Let $m_{ijk}^{(1)}$ denote the $[(j, k), i]$ entry of $\mathbf{M}^{(1)}$. If $m_{ijk}^{(1)}=1$ at the beginning of the first execution of step 2, then $m_{ijk}=1$ and $\mu+\alpha_i+\beta_j+\gamma_k$ is estimable. If an entry $m_{cjk}^{(1)}$ is set equal to 1 at some point in step 2, then there exist $(v, w)\in J\times K$ and $u\in I$ such that $m_{ivw}^{(1)}=m_{ujk}^{(1)}=m_{uvw}^{(1)}=1$ at that point. Now

$$\mu+\alpha_i+\beta_j+\gamma_k=(\mu+\alpha_i+\beta_v+\gamma_w)-(\mu+\alpha_u+\beta_v+\gamma_w)+(\mu+\alpha_u+\beta_j+\gamma_k),$$

and by applying an induction argument on the number of steps of the R process required to get to this point, we see that $\mu+\alpha_i+\beta_j+\gamma_k$ is estimable. Then if $m_{ijk}=1$ at the end of step 2 of the $R3$ process, $\mu+\alpha_i+\beta_j+\gamma_k$ is estimable.

Applying a similar argument, we see that the same must hold true at the end of the first execution of steps 3 and 4 of the $R3$ process. Using another induction argument, we see that the desired result must hold true at the end of any execution of steps 2, 3 or 4, and consequently at the end of the $R3$ process. □

Note that Theorem 6.1 provides only a sufficient condition for $\mu+\alpha_i+\beta_j+\gamma_k$ to be estimable, whereas the corresponding result in the two-way classification model was both necessary and sufficient. For many incidence matrices, however, the $R3$ process will provide a final matrix composed of all ones, indicating that all cell expectations are estimable. This is especially likely when there are few missing observations.

EXAMPLE 6.1 For an $a\times b\times c$ incidence matrix $\mathbf{N}$ let $\mathbf{N}_1, \mathbf{N}_2, \ldots, \mathbf{N}_c$ denote the γ levels of $\mathbf{N}$, that is, $\mathbf{N}_k$ is an $a\times b$ matrix with entries n_{ijk}. Consider the following $4\times 3\times 3$ incidence matrix $\mathbf{N}$ exhibited in terms of the three γ levels:

$$\mathbf{N}_1=\begin{bmatrix}2&0&0\\0&0&0\\0&1&0\\1&0&0\end{bmatrix},\ \mathbf{N}_2=\begin{bmatrix}0&1&1\\1&0&0\\0&0&2\\0&1&0\end{bmatrix},\ \mathbf{N}_3=\begin{bmatrix}0&0&0\\0&2&0\\0&0&0\\0&0&0\end{bmatrix}.$$

We shall apply the $R3$ process to $\mathbf{N}$. Step 1 merely changes all nonzero entries to ones. In step 2 the initial and final versions of $\mathbf{M}^{(1)}$ are

$$\begin{array}{r} \\ (j,k)=(1,1)\\(2,1)\\(3,1)\\(1,2)\\(2,2)\\(3,2)\\(1,3)\\(2,3)\\(3,3)\end{array}
\begin{array}{c} i=1\ \ 2\ \ 3\ \ 4\\ \begin{bmatrix}1&0&0&1\\0&0&1&0\\0&0&0&0\\0&1&0&0\\1&0&0&1\\1&0&1&0\\0&0&0&0\\0&1&0&0\\0&0&0&0\end{bmatrix}\end{array}
\text{ and }
\begin{bmatrix}1&0&1&1\\1&0&1&1\\0&0&0&0\\0&1&0&0\\1&0&1&1\\1&0&1&1\\0&0&0&0\\0&1&0&0\\0&0&0&0\end{bmatrix}.$$

At the end of step 2, we have

$$\mathbf{M}_1=\begin{bmatrix}1&1&0\\0&0&0\\1&1&0\\1&1&0\end{bmatrix},\ \mathbf{M}_2=\begin{bmatrix}0&1&1\\1&0&0\\0&1&1\\0&1&1\end{bmatrix},\ \mathbf{M}_3=\begin{bmatrix}0&0&0\\0&1&0\\0&0&0\\0&0&0\end{bmatrix}.$$

After step 3, we obtain

$$\mathbf{M}_1=\begin{bmatrix}1&1&1\\0&0&0\\1&1&1\\1&1&1\end{bmatrix},\ \mathbf{M}_2=\begin{bmatrix}1&1&1\\1&1&1\\1&1&1\\1&1&1\end{bmatrix},\ \mathbf{M}_3=\begin{bmatrix}0&0&0\\1&1&1\\0&0&0\\0&0&0\end{bmatrix}.$$

Step 4 fills $\mathbf{M}$ with ones. We conclude that all cell expectations are estimable, the model matrix $\mathbf{X}$ has full column rank $r(\mathbf{X})=a+b+c-2$, and that (6.6) provides bases for $\bar{\mathscr{A}}$, $\bar{\mathscr{B}}$, $\bar{\mathscr{G}}$, and $\bar{\Theta}$.

EXAMPLE 6.2 Consider the following $6\times6\times5$ incidence matrix $\mathbf{N}$:

$$\mathbf{N}_1=\begin{bmatrix}0&0&0&3&0&1\\2&0&0&0&0&0\\0&0&0&0&0&0\\0&0&0&0&2&0\\0&0&1&0&0&0\\0&0&0&0&0&0\end{bmatrix},\ \mathbf{N}_4=\begin{bmatrix}0&0&0&0&0&0\\0&0&0&0&0&0\\0&0&2&0&0&0\\0&0&0&0&0&0\\0&0&0&0&0&0\\0&0&0&0&0&1\end{bmatrix},$$

$$\mathbf{N}_2=\begin{bmatrix}0&0&0&0&1&0\\0&0&1&0&0&0\\0&2&0&0&0&0\\0&0&0&0&0&0\\0&0&0&0&0&0\\0&0&0&2&0&0\end{bmatrix},\ \mathbf{N}_5=\begin{bmatrix}0&0&0&0&0&0\\0&0&0&0&0&0\\0&0&0&0&0&0\\0&0&0&2&0&1\\1&0&0&0&0&0\\0&0&0&0&0&0\end{bmatrix}.$$

$$\mathbf{N}_3=\begin{bmatrix}0&1&0&0&0&0\\0&0&0&0&0&0\\0&0&0&0&0&0\\0&0&0&0&0&0\\0&0&0&0&2&0\\3&0&0&0&0&0\end{bmatrix},$$

Let us apply the $R3$ process to $\mathbf{N}$.

Again, step 1 merely changes all nonzero entries to ones. The first iteration of step 2 makes no changes. After the first iteration of step 3, cells (6, 6, 2) and (6, 4, 4) are filled with ones. After the first iteration of step 4, cells (3, 3, 2), (1, 5, 4), (2, 3, 4), and (3, 2, 4) are filled. After the second iteration of step 2, cells (3, 1, 1), (2, 2, 2), and (2, 2, 4) are filled. After the second iteration of step 3, cells (5, 2, 1) and (1, 3, 3) are filled. The next two steps make no changes, and so we know that we have the final matrix **M**.

$$\mathbf{M}_1 = \begin{bmatrix} 0 & 0 & 0 & 1 & 0 & 1 \\ 1 & 0 & 0 & 0 & 0 & 0 \\ 1 & 0 & 0 & 0 & 0 & 0 \\ 0 & 0 & 0 & 0 & 1 & 0 \\ 0 & 1 & 1 & 0 & 0 & 0 \\ 0 & 0 & 0 & 0 & 0 & 0 \end{bmatrix}, \mathbf{M}_4 = \begin{bmatrix} 0 & 0 & 0 & 0 & 1 & 0 \\ 0 & 1 & 1 & 0 & 0 & 0 \\ 0 & 1 & 1 & 0 & 0 & 0 \\ 0 & 0 & 0 & 0 & 0 & 0 \\ 0 & 0 & 0 & 0 & 0 & 0 \\ 0 & 0 & 0 & 1 & 0 & 1 \end{bmatrix},$$

$$\mathbf{M}_2 = \begin{bmatrix} 0 & 0 & 0 & 0 & 1 & 0 \\ 0 & 1 & 1 & 0 & 0 & 0 \\ 0 & 1 & 1 & 0 & 0 & 0 \\ 0 & 0 & 0 & 0 & 0 & 0 \\ 0 & 0 & 0 & 0 & 0 & 0 \\ 0 & 0 & 0 & 1 & 0 & 1 \end{bmatrix}, \mathbf{M}_5 = \begin{bmatrix} 0 & 0 & 0 & 0 & 0 & 0 \\ 0 & 0 & 0 & 0 & 0 & 0 \\ 0 & 0 & 0 & 0 & 0 & 0 \\ 0 & 0 & 0 & 1 & 0 & 1 \\ 1 & 0 & 0 & 0 & 0 & 0 \\ 0 & 0 & 0 & 0 & 0 & 0 \end{bmatrix}.$$

$$\mathbf{M}_3 = \begin{bmatrix} 0 & 1 & 1 & 0 & 0 & 0 \\ 0 & 0 & 0 & 0 & 0 & 0 \\ 0 & 0 & 0 & 0 & 0 & 0 \\ 0 & 0 & 0 & 0 & 0 & 0 \\ 0 & 0 & 0 & 0 & 1 & 0 \\ 1 & 0 & 0 & 0 & 0 & 0 \end{bmatrix},$$

The $R3$ process has not filled **M** with ones. We shall return to this example later.

As we saw in Example 6.2, the $R3$ process may not produce a final matrix **M** which is entirely filled with ones. If not, further work is needed to find bases for $\bar{\mathscr{A}}$, $\bar{\mathscr{B}}$, $\bar{\mathscr{G}}$, and $\bar{\Theta}$.

We now describe a method for obtaining a spanning set for $\bar{\mathscr{G}}$. We then discuss the problem of extracting a basis from this spanning set. Spanning sets for $\bar{\mathscr{A}}$ and $\bar{\mathscr{B}}$ can be found similarly. Finally, we obtain a spanning set for a space $\bar{\mathscr{E}}$ satisfying (6.4).

Consider the γ levels $\mathbf{M}_1, \mathbf{M}_2, \ldots, \mathbf{M}_c$ of **M**, where $\mathbf{M}_k$ is an $a \times b$ matrix consisting of elements m_{ijk}. Because the $R3$ process has been applied, either $\mathbf{M}_k = \mathbf{M}_h$ or else $m_{ijk} m_{ijh} = 0$ for all i, j.

Define the relation $\sim$ on $\{1, 2, \ldots, c\}$ by $k \sim h$ if $\mathbf{M}_k = \mathbf{M}_h$. It is easy to see that $\sim$ is an equivalence relation. Let $E_k = \{h: k \sim h\}$ denote the set of indices of γ that are equivalent to k. The set E_k is an equivalence set or equivalence class and the set $\{1, 2, \ldots, c\}$ can be partitioned into, say, s equivalence classes. Let S be a complete set of representatives for these equivalence classes. Of course, $s = \#(S)$.

The assumptions (6.2) guarantee that no $\mathbf{M}_k$ consists only of zeros. Then if $k \sim h$, there are some i and j such that $m_{ijk} = m_{ijh} = 1$. Then by Theorem 6.1 both $\mu + \alpha_i + \beta_j + \gamma_k$ and $\mu + \alpha_i + \beta_j + \gamma_h$ are estimable, and so is

$$\gamma_k - \gamma_h = (\mu + \alpha_i + \beta_j + \gamma_k) - (\mu + \alpha_i + \beta_j + \gamma_h).$$

We have proved the following:

THEOREM 6.2 If $k \sim h$, then $\gamma_k - \gamma_h$ is estimable.

Let $D_\gamma = \{\gamma_k - \gamma_h: k \in S, h \in E_k\}$. Then D_γ is a set of $c - s$ linearly independent functions in $\mathscr{G}$. The elements of D_γ are said to be *directly estimable* from the final matrix $\mathbf{M}$. The vector space spanned by the basis D_γ will be denoted $\bar{\mathscr{D}}$, and will be called the space of directly estimable γ contrasts.

Suppose $k \sim h$. Then a linear parametric function $\boldsymbol{\lambda}'\boldsymbol{\theta}$ is estimable if and only if the parametric function obtained by substituting γ_k for γ_h in $\boldsymbol{\lambda}'\boldsymbol{\theta}$ is estimable. Therefore, for further considerations of estimability we may restrict our attention to $\{\gamma_k: k \in S\}$.

In the same fashion, we may define equivalence relations on the sets $\{1, 2, \ldots, a\}$ and $\{1, 2, \ldots, b\}$ where, for instance, $i \sim l$ in $\{1, 2, \ldots, a\}$ if there are j, k such that $m_{ijk} = m_{ljk} = 1$. Statements about estimability similar to those in the previous paragraphs can be made.

We now work with the $a \times b \times s$ submatrix of $\mathbf{M}$ consisting of the s levels $\{\mathbf{M}_k: k \in S\}$. For each i, j, there is at most one $k \in S$ such that $m_{ijk} \neq 0$.

Suppose that $\boldsymbol{\lambda}'\boldsymbol{\gamma} = \sum_{k \in S} \lambda_k \gamma_k$ is a nonzero estimable γ contrast. By the definition of estimability we can write

$$\boldsymbol{\lambda}'\boldsymbol{\gamma} = \sum_{i=1}^{a} \sum_{j=1}^{b} \sum_{k=1}^{c} c_{ijk}(\mu + \alpha_i + \beta_j + \gamma_k)$$

where $c_{ijk} = 0$ if $m_{ijk} = 0$. For each $h \notin S$, find $t \in S$ such that $h \sim t$ and substitute γ_t for γ_h in the preceding equation. Since the coefficient of γ_h in $\boldsymbol{\lambda}'\boldsymbol{\gamma}$ is 0, the equality is preserved. For each $k \in S$, set $d_{ijk} = \sum_{h \sim k} c_{ijh}$. If $m_{ijk} = 0$, then $m_{ijh} = 0$ for all $h \sim k$ and so $d_{ijk} = 0$. Thus

$$\boldsymbol{\lambda}'\boldsymbol{\gamma} = \sum_{i=1}^{a} \sum_{j=1}^{b} \sum_{k \in S} d_{ijk}(\mu + \alpha_i + \beta_j + \gamma_k)$$

where $d_{ijk} = 0$ if $m_{ijk} = 0$.

Set $D_{ij}=\sum_{k\in S} d_{ijk}$. Either $D_{ij}=0$ or $D_{ij}=d_{ijt}$ where t is the unique index such that $m_{ijt}\neq 0$. Because the coefficients of the α_i and β_j in $\boldsymbol{\lambda}'\boldsymbol{\gamma}$ are all zero, we see that $\sum_j D_{ij}=0$ for all i and $\sum_i D_{ij}=0$ for all j. There must exist some nonzero coefficient $d_{i_1 j_1 k_1}$. Then there exists some pair (i_1, j_1) such that $D_{i_1 j_1}=d_{i_1 j_1 k_1}\neq 0$. Since $\sum_j D_{i_1 j}=0$, there is some j_2, $j_2\neq j_1$, with $D_{i_1 j_2}\neq 0$. Since $\sum_i D_{ij_2}=0$, there is some i_2, $i_2\neq i_1$, with $D_{i_2 j_2}\neq 0$. In this way we get a sequence $(i_1, j_1)(i_1, j_2)(i_2, j_2)(i_2, j_3)(i_3, j_3)\ldots$, where $i_p\neq i_{p+1}$ and $j_p\neq j_{p+1}$ for all p and $D_{ij}\neq 0$ for each (i, j) in the sequence.

Let p, q be the first two indices such that $j_p=j_q$ with $p<q$. In this case the sequence $(i_p, j_p)(i_p, j_{p+1})(i_{p+1}, j_{p+1})\cdots(i_{q-1}, j_{q-1})(i_{q-1}, j_q)$ yields a loop.

We define a *loop* to be a sequence of an even number $u(u>0)$ of pairs

$$(i_1, j_1)(i_2, j_2)\cdots(i_u, j_u)$$

such that:

1 The pairs are distinct.

2 $i_t=i_{t+1}$ for t add $(t=1, 3, \ldots, u-1)$.

3 $j_t=j_{t+1}$ for t even $(t=2, 4, \ldots, u-2)$.

4 $j_1=j_u$.

For example, (3, 3)(3, 1)(5, 1)(5, 2)(1, 2)(1, 4)(2, 4)(2, 3) is a loop. If we connect the corresponding entries of a two-dimensional matrix we obtain a picture of a rectilinear loop:

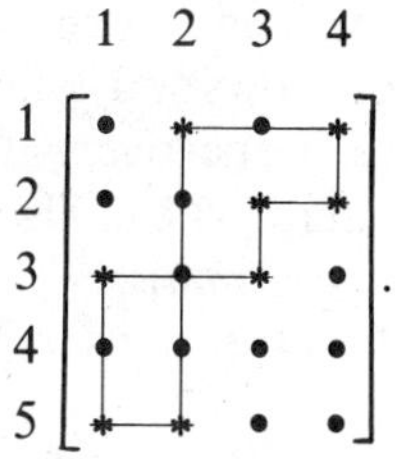

Above we said that the sequence $(i_p, j_p)(i_p, j_{p+1})\cdots(i_{q-2}, j_{q-1})(i_{q-1}, j_{q-1})$ (i_{q-1}, j_q) yields a loop. It is itself a loop if $i_{q-1}\neq i_p$. If $i_{q-1}=i_p$, then (i_{q-1}, j_{q-1}) $(i_p, j_{p+1})\cdots(i_{q-2}, j_{q-1})$ is a loop. For convenience let us assume (i_p, j_p) $(i_p, j_{p+1})\cdots(i_{q-1}, j_q)$ is a loop. For each (i_u, j_v) in this sequence, let k_{uv} be the unique index such that $d_{i_u j_v k_{uv}}\neq 0$. Then $m_{i_u j_v k_{uv}}\neq 0$, so $\mu+\alpha_{i_u}+\beta_{j_v}+\gamma_{k_{uv}}$ is estimable. Now $(\mu+\alpha_{i_p}+\beta_{j_p}+\gamma_{k_{pp}})-(\mu+\alpha_{i_p}+\beta_{j_{p+1}}+\gamma_{k_{p,p+1}})+\cdots-(\mu+\alpha_{i_{q-1}}+\beta_{j_q}+\gamma_{k_{q-1,q}})=\gamma_{k_{pp}}-\gamma_{k_{p,p+1}}+\cdots-\gamma_{k_{q-1,q}}$ is an estimable γ contrast; call it $\boldsymbol{\rho}'\boldsymbol{\gamma}$. The estimable γ contrast $\boldsymbol{\lambda}'\boldsymbol{\gamma}-d_{i_p j_p k_{pp}}\boldsymbol{\rho}'\boldsymbol{\gamma}$ can be expressed as a sum of estimable cell expectations with strictly fewer nonzero coefficients than the expression for $\boldsymbol{\lambda}'\boldsymbol{\gamma}$. Using induction we can argue that every estimable

γ contrast $\boldsymbol{\lambda}'\boldsymbol{\gamma}=\lambda_1\gamma_1+\cdots+\lambda_s\gamma_s$ can be expressed as a linear combination of estimable γ contrasts which can be derived from loops in the above manner.

Let $\bar{\mathbf{M}}=\sum_{k\in S} k\mathbf{M}_k$. Then for $k\in S$, $m_{ijk}=1$ if and only if $\bar{m}_{ij}=k$, where $\bar{m}_{ij}$ denotes the (i,j) entry of $\bar{\mathbf{M}}$. A loop $(i_1,j_1)\cdots(i_u,j_u)$ is called a *loop in* $\bar{\mathbf{M}}$ if $\bar{m}_{i_t j_t}\neq 0$ for all $t=1,2,\ldots,u$. We have just proved the following:

THEOREM 6.3 The vector space of estimable linear parametric functions involving γ_k, $k\in S$, is spanned by contrasts $\gamma_{k_1}-\gamma_{k_2}+\cdots-\gamma_{k_u}$ where $k_t=\bar{m}_{i_t j_t}$ and $(i_1,j_1)(i_2,j_2)\cdots(i_u,j_u)$ is a loop in $\bar{\mathbf{M}}$.

Let $\bar{\mathscr{S}}_\gamma$ denote the vector space of estimable linear parametric functions involving only γ_k, $k\in S$. Suppose $\boldsymbol{\lambda}'\boldsymbol{\gamma}$ is an estimable γ contrast. By using the directly estimable γ differences in D_γ, we can express $\boldsymbol{\lambda}'\boldsymbol{\gamma}$ as an estimable linear parametric function involving only γ_k, $k\in S$. So

$$\bar{\mathscr{G}}=\bar{\mathscr{D}}_\gamma+\bar{\mathscr{S}}_\gamma$$

and we have proved the following:

COROLLARY 6.1 The vector space $\bar{\mathscr{G}}$ of estimable γ contrasts is spanned by the directly estimable γ differences found from the final matrix $\mathbf{M}$ and the γ contrasts derived from the loops in $\bar{\mathbf{M}}$.

The matrix $\bar{\mathbf{M}}$ as defined is of dimensions $a\times b$. For obtaining the relevant loops in $\bar{\mathbf{M}}$ it is generally possible to work with a matrix of smaller dimensions. Such a matrix may be obtained by eliminating all redundant rows and columns from $\bar{\mathbf{M}}$. The redundant rows and columns in $\bar{\mathbf{M}}$ provide the directly estimable α and β differences, respectively. Furthermore, any γ contrast derived from a loop in the original $\bar{\mathbf{M}}$ can also be expressed as a γ contrast derived from a loop in the reduced matrix.

Two problems remain: to find the loops in $\bar{\mathbf{M}}$ and to extract a basis from a spanning set.

As an algorithm for deriving γ contrasts from the loops in $\bar{\mathbf{M}}$ we will use the *Q process*. For convenience we isolate a part of the Q process and call it the X process. The *X process* applied to a two-dimensional matrix $\mathbf{W}$ consists of the following procedure:

1 Find every entry of $\mathbf{W}$ which is the only nonzero entry in its row or column and change it to zero.
2 Continue this until each row and each column either is all zeros or has at least two nonzero entries.

We now describe the Q process. At the first stage put the submatrix of

$\bar{\mathbf{M}}$ consisting of the first two columns into a temporary working area $\mathbf{W}$. Apply the X process to $\mathbf{W}$. If the entire second column of $\mathbf{W}$ is zero (and hence $\mathbf{W}=\mathbf{0}$), then proceed to the second stage. Suppose there is a nonzero number k_1 in entry $(i_1, 2)$. Since the X process has been applied, there must be a nonzero number k_2 in entry $(i_1, 1)$. Similarly, there must be a nonzero number k_3 in an entry $(i_2, 1)$, $i_2 \neq i_1$, and also a nonzero number k_4 in entry $(i_2, 2)$. These entries form a loop in $\bar{\mathbf{M}}$, so $\gamma_{k_1}-\gamma_{k_2}+\gamma_{k_3}-\gamma_{k_4}$ is estimable. We include it in our spanning set.

Now change $\bar{m}_{i_2,2}=k_4$ to 0 both in $\mathbf{W}$ and permanently in $\bar{\mathbf{M}}$. If there was another loop in $\bar{\mathbf{M}}$ which involved $(i_2, 2)$, the corresponding γ contrast will not be lost. Where $(i_2, 2)$ occurred in the loop we can now put $(i_2, 1)(i_1, 1)(i_1, 2)$ or $(i_1, 2)(i_1, 1)(i_2, 1)$ to obtain another loop. (Actually it may not be a proper loop—this matter is dealt with more precisely in Lemma 6.1.) A suitable linear combination of the contrast derived from this loop and $\gamma_{k_1}-\gamma_{k_2}+\gamma_{k_3}-\gamma_{k_4}$ gives us the contrast that would have been derived from the "lost" loop.

Again apply the X process to $\mathbf{W}$. If there are any nonzero entries left, find a loop and derive a γ contrast as was done above. Then change the last entry in the loop to zero in both $\mathbf{W}$ and $\bar{\mathbf{M}}$. Apply the X process again and continue as above. Eventually a point is reached where $\mathbf{W}=\mathbf{0}$. This finishes the first stage.

To begin the second stage put the submatrix of $\bar{\mathbf{M}}$ consisting of the first three columns into a working area $\mathbf{W}$.

Let us suppose that we have proceeded through the first $p-1$ stages of the Q process and have thus obtained a spanning set for all γ contrasts which can be derived from loops in the first p columns of $\bar{\mathbf{M}}$. If $p=b$, the Q process has been completed. If $p<b$, begin the pth stage by putting the first $p+1$ columns of $\bar{\mathbf{M}}$ into a temporary working area $\mathbf{W}$. Apply the X process to $\mathbf{W}$. If the entire $(p+1)$th column of $\mathbf{W}$ is $\mathbf{0}$, then $\mathbf{W}=\mathbf{0}$, because every loop in $\mathbf{W}$ must involve column $p+1$ since there are now no loops in the first p columns of $\bar{\mathbf{M}}$. In this case proceed to the next stage. Suppose there is a nonzero number k_{10} in entry $(i_1, p+1)$. Since the X process has been applied, there must be another nonzero number k_{11} in row i_1, say in entry (i_1, j_1). Similarly, there must be another nonzero number k_{21} in column j_1, say in entry (i_2, j_1). Alternately look along rows and columns for the next nonzero number until one is found in column $p+1$, say in entry (i_u, j_u), $j_u=p+1$. One is assured that column $p+1$ is eventually reached by the fact that there are no loops in the first p columns. Then $(i_1, p+1)(i_1, j_1)(i_2, j_1)\cdots(i_u, j_{u-1})$ $(i_u, p+1)$ is a loop and $\gamma_{k_{10}}-\gamma_{k_{11}}+\gamma_{k_{21}}-\cdots+\gamma_{k_{u,u-1}}-\gamma_{k_{uu}}$ is estimable. We include it in our spanning set.

Now change $\bar{m}_{i_u,p+1}=k_{uu}$ to zero both in $\mathbf{W}$ and permanently in $\bar{\mathbf{M}}$. As indicated above, although certain loops may be "lost" by this change, it will not hinder us in obtaining our desired spanning set. This will be proved in the lemma below.

Apply the X process to the current matrix in $\mathbf{W}$. If any nonzero entries remain, find a loop and derive a γ contrast to include in our spanning set. Change the last entry in the loop to zero in both $\mathbf{W}$ and $\bar{\mathbf{M}}$. Apply the X process to $\mathbf{W}$ again. Continue this until $\mathbf{W}=\mathbf{0}$. This completes the $(p+1)$th stage. After $b-1$ stages the Q process has been completed.

LEMMA 6.1 Let $(i_1, j_1)(i_1, j_2)(i_2, j_2)\cdots(i_p, j_p)(i_p, j_1)$ and $(i'_1, j'_1)(i'_1, j'_2)(i'_2, j'_2)\cdots(i'_q, j'_q)(i'_q, j'_1)$ be two loops in $\bar{\mathbf{M}}$ and let $\boldsymbol{\lambda}'\boldsymbol{\gamma}=\gamma_{k_{11}}-\gamma_{k_{12}}+\gamma_{k_{22}}-\cdots+\gamma_{k_{pp}}-\gamma_{k_{p1}}$ and $\boldsymbol{\rho}'\boldsymbol{\gamma}=\gamma_{k'_{11}}-\gamma_{k'_{12}}+\gamma_{k'_{22}}-\cdots+\gamma_{k'_{qq}}-\gamma_{k'_{q1}}$ be the γ contrasts derived from them. Suppose $(i_p, j_1)=(i'_1, j'_1)$. Let $\bar{\mathbf{M}}_1$ be the matrix which results from changing the (i_p, j_1) entry of $\bar{\mathbf{M}}$ to 0. Then $\boldsymbol{\rho}'\boldsymbol{\gamma}$ is a linear combination of $\boldsymbol{\lambda}'\boldsymbol{\gamma}$ and of γ contrasts derived from loops in $\bar{\mathbf{M}}_1$.

Proof It suffices to show that $\boldsymbol{\lambda}'\boldsymbol{\gamma}+\boldsymbol{\rho}'\boldsymbol{\gamma}$ is a linear combination of γ contrasts derived from loops in $\bar{\mathbf{M}}_1$. Consider the sequence $(i_1, j_1)(i_1, j_2)\cdots(i_p, j_p)(i'_1, j'_2)(i'_2, j'_2)\cdots(i'_q, j'_q)(i'_q, j'_1)$. Since $i_p=i'_1$ and $j_1=j'_1$, this satisfies all the conditions for being a loop in $\bar{\mathbf{M}}_1$ except possibly condition 1. If it is a loop we are done because $\gamma_{k_{11}}-\gamma_{k_{12}}+\cdots+\gamma_{k_{pp}}-\gamma_{k'_{12}}+\gamma_{k'_{22}}-\cdots+\gamma_{k'_{qq}}-\gamma_{k'_{q1}}=\boldsymbol{\lambda}'\boldsymbol{\gamma}+\boldsymbol{\rho}'\boldsymbol{\gamma}$.

If not, let t be the largest integer and v the smallest integer such that $j_t=j'_v$. If $t=p$ and $v=2$, then $(i_p, j_p)=(i'_1, j'_2)$ and so $\gamma_{k_{pp}}-\gamma_{k'_{12}}=0$. Thus we can delete the pairs $(i_p, j_p)(i'_1, j'_2)$ from our sequence; now we can apply the induction argument stated below. If $t\neq p$ or $v\neq 2$ then $(i_t, j_t)(i_t, j_{t+1})\cdots(i_p, j_p)(i'_1, j'_2)\cdots(i'_{v-1}, j'_{v-1})(i'_{v-1}, j'_v)$ yields a loop (see the discussion preceding Theorem 6.3) from which can be derived $\gamma_{k_{tt}}-\gamma_{k_{t,t+1}}+\cdots+\gamma_{k_{pp}}-\gamma_{k'_{12}}+\cdots+\gamma_{k'_{v-1,v-1}}-\gamma_{k'_{v-1,v}}$. Delete $(i_t, j_t)\cdots(i'_{v-1}, j'_v)$ from the sequence to obtain the new sequence $(i_1, j_1)(i_1, j_2)\cdots(i_{t-1}, j_{t-1})(i_{t-1}, j_t)(i'_v, j'_v)(i'_v, j'_{v+1})\cdots(i'_q, j'_1)$. This also satisfies all the conditions for being a loop in $\bar{\mathbf{M}}_1$ except possibly condition 1. By an induction argument on the number of pairs in such a sequence, we can conclude that $\gamma_{k_{11}}-\gamma_{k_{12}}+\cdots-\gamma_{k_{t-1,t}}+\gamma_{k'_{vv}}-\cdots-\gamma_{k'_{q1}}$ is a linear combination of contrasts derived from loops in $\bar{\mathbf{M}}_1$. Adding these two γ contrasts together, we get $\boldsymbol{\lambda}'\boldsymbol{\gamma}+\boldsymbol{\rho}'\boldsymbol{\gamma}$. □

We have demonstrated the following:

THEOREM 6.4 The Q process obtains a spanning set for $\bar{\mathscr{S}}_\gamma$, the space of estimable γ contrasts involving only γ_k, $k\in S$.

Note that the Q process, as well as the $R3$ process, can be programmed for a computer using only logical and fixed point operations.

A computer program for selecting a basis for the estimable γ contrasts can easily be written using floating point arithmetic, but then the result is subject to round-off error. For small values of a, b, and s, it is possible to use fixed point arithmetic; large values introduce the danger of overflow.

First it should be observed that the γ differences found directly after the $R3$ process immediately yield a linearly independent set D_γ of $c-s$ differences and that the vector space $\bar{\mathscr{D}}_\gamma$ spanned by these differences has $\{0\}$ as intersection with the space $\bar{\mathscr{S}}_\gamma$ of estimable contrasts involving only γ_k, $k \in S$. Then $\bar{\mathscr{G}} = \bar{\mathscr{D}}_\gamma \oplus \bar{\mathscr{S}}_\gamma$ and our problem is reduced to finding a basis for this latter space.

Let Q_γ denote the set of estimable γ contrasts produced by the Q process and let $S=\{k_1, k_2, \ldots, k_s\}$. When $s=2$, a basis for $\bar{\mathscr{S}}_\gamma$ is $\gamma_{k_1}-\gamma_{k_2}$ or else $\gamma=\{0\}$ according to whether Q_γ contains a nonzero element or not. When $s=3$, search Q_γ for a nonzero contrast. If none is found, $\bar{\mathscr{S}}_\gamma=\{0\}$. If one is found, say $\boldsymbol{\lambda}'\boldsymbol{\gamma}$, then compare every other nonzero contrast in Q_γ with $\boldsymbol{\lambda}'\boldsymbol{\gamma}$. If they are all multiples of $\boldsymbol{\lambda}'\boldsymbol{\gamma}$, then $\boldsymbol{\lambda}'\boldsymbol{\gamma}$ is a basis for $\bar{\mathscr{S}}_\gamma$. If not $\gamma_{k_1}-\gamma_{k_2}$ and $\gamma_{k_1}-\gamma_{k_3}$ constitute a basis for $\bar{\mathscr{S}}_\gamma$. When $s \geqslant 4$, row reduction can be done in fixed point artihmetic, but there will be the danger of overflow unless a, b, and s are small. We will denote the dimension of $\bar{\mathscr{G}}$, the space of estimable γ contrasts, by f_γ.

After the $R3$ process, the whole procedure of finding direct differences, applying the Q process, and determining a basis could be done for estimable α contrasts and estimable β contrasts. This would tell one precisely which contrasts are contributing to the adjusted sums of squares for α effects and β effects. The dimensions of $\bar{\mathscr{A}}$ and $\bar{\mathscr{B}}$, the spaces of estimable α and β contrasts, will be denoted by f_α and f_β, respectively.

If at any point it is discovered that all contrasts in one effect are estimable, then bases for the estimable contrasts in the other two effects are easily found. Suppose, for instance, that all γ contrasts are estimable. Then the estimable α contrasts and β contrasts can be obtained by dealing with the model $E(Y_{ijh}) = \mu + \alpha_i + \beta_j$ where $i=1, \ldots, a$; $j=1, \ldots, b$; and $h=1, \ldots, n_{ij.}$, $n_{ij.} = \sum_k n_{ijk}$. The incidence matrix for this model is found by merging the γ levels of the original incidence matrix. Procedures for dealing with such a two-way model are discussed in Chapter 5.

Although in some cases it is more efficient to do so, it is not necessary to use the $R3$ process. All of the preceding is valid if we use the following matrix $\mathbf{M}$. First define a relation on $\{1, \ldots, c\}$ by $k \sim h$ if there exist u, v such that $n_{uvk} \neq 0$ and $n_{uvh} \neq 0$. Now set $m_{ijk}=1$ if $n_{ijh} \neq 0$ for some $h \sim k$ and set $m_{ijk}=0$ if $n_{ijk}=0$ for all $h \sim k$. If the relation $\sim$ is not an equivalence relation, then define a new relation, using $\mathbf{M}$ in place of $\mathbf{N}$, and form a new $\mathbf{M}$. Proceed in this fashion until the relation is in fact an equivalence relation. The last $\mathbf{M}$ matrix may be used in place of the final matrix.

EXAMPLE 6.2 (Continued) We see that $\mathbf{M}_2 = \mathbf{M}_4$, so $2 \sim 4$ and $\gamma_2 - \gamma_4$ is estimable. This is the only directly estimable γ difference, so $D_\gamma = \{\gamma_2 - \gamma_4\}$. We choose $S = \{1, 2, 3, 5\}$ and form $\bar{\mathbf{M}} = \mathbf{M}_1 + 2\mathbf{M}_2 + 3\mathbf{M}_3 + 5\mathbf{M}_5$:

$$\bar{\mathbf{M}} = \begin{bmatrix} 0 & 3 & 3 & 1 & 2 & 1 \\ 1 & 2 & 2 & 0 & 0 & 0 \\ 1 & 2 & 2 & 0 & 0 & 0 \\ 0 & 0 & 0 & 5 & 1 & 5 \\ 5 & 1 & 1 & 0 & 3 & 0 \\ 3 & 0 & 0 & 2 & 0 & 2 \end{bmatrix}.$$

From the redundant rows and columns of $\bar{\mathbf{M}}$ we see that $\alpha_2 - \alpha_3$, $\beta_2 - \beta_3$, and $\beta_4 - \beta_6$ are estimable, and $D_\alpha = \{\alpha_2 - \alpha_3\}$ and $D_\beta = \{\beta_2 - \beta_3, \beta_4 - \beta_6\}$. For obtaining a basis for $\mathscr{S}_\gamma$ we can deal with the reduced matrix

$$\bar{\mathbf{M}} = \begin{bmatrix} 0 & 3 & 1 & 2 \\ 1 & 2 & 0 & 0 \\ 0 & 0 & 5 & 1 \\ 5 & 1 & 0 & 3 \\ 3 & 0 & 2 & 0 \end{bmatrix}.$$

Let us apply the Q process to $\bar{\mathbf{M}}$. At the first stage we consider

$$\mathbf{W} = \begin{bmatrix} 0 & 3 \\ 1 & 2 \\ 0 & 0 \\ 5 & 1 \\ 3 & 0 \end{bmatrix}.$$

After applying the X process, we can find the loop (2, 2)(2, 1)(4, 1)(4, 2), which yields the estimable γ contrasts $\gamma_2 - \gamma_1 + \gamma_5 - \gamma_1 = -2\gamma_1 + \gamma_2 + \gamma_5$. We change entry (4, 2) in $\mathbf{W}$ and $\bar{\mathbf{M}}$ to zero and find that the X process now reduces the matrix $\mathbf{W}$ to $\mathbf{0}$. At the second stage we consider

$$\mathbf{W} = \begin{bmatrix} 0 & 3 & 1 \\ 1 & 2 & 0 \\ 0 & 0 & 5 \\ 5 & 0 & 0 \\ 3 & 0 & 2 \end{bmatrix}.$$

After applying the X process, we can find the loop (1, 3)(1, 2)(2, 2)(2, 1)(5, 1)(5, 3), which yields the contrast $\gamma_1 - \gamma_3 + \gamma_2 - \gamma_1 + \gamma_3 - \gamma_2 = 0$. We change entry (5, 3) of $\mathbf{W}$ and $\bar{\mathbf{M}}$ to zero and find that the X process now reduces the matrix to zero. At the third stage we consider

$$\mathbf{W}=\begin{bmatrix}0&3&1&2\\1&2&0&0\\0&0&5&1\\5&0&0&3\\3&0&0&0\end{bmatrix}.$$

The X process leaves the matrix

$$\mathbf{W}=\begin{bmatrix}0&3&1&2\\1&2&0&0\\0&0&5&1\\5&0&0&3\\0&0&0&0\end{bmatrix}.$$

We then find the loop (1, 4)(1, 3)(3, 3)(3, 4), which yields the contrast $\gamma_2-\gamma_1+\gamma_5-\gamma_1=-2\gamma_1+\gamma_2+\gamma_5$. We change entry (3, 4) of $\mathbf{W}$ and $\bar{\mathbf{M}}$ to zero and apply the X process to obtain

$$\mathbf{W}=\begin{bmatrix}0&3&0&2\\1&2&0&0\\0&0&0&0\\5&0&0&3\\0&0&0&0\end{bmatrix}.$$

This reveals the loop (1, 4)(1, 2)(2, 2)(2, 1)(4, 1)(4, 4), which yields the contrast $\gamma_2-\gamma_3+\gamma_2-\gamma_1+\gamma_5-\gamma_3=-\gamma_1+2\gamma_2-2\gamma_3+\gamma_5$. We change entry (4, 4) of $\mathbf{W}$ and $\bar{\mathbf{M}}$ to zero and find that the X process now reduces the matrix to zero. This completes the Q process.

By the corollary to Theorem 6.3 a spanning set for the vector space $\bar{\mathscr{G}}$ of estimable γ contrasts is $\{\gamma_2-\gamma_4, -2\gamma_1+\gamma_2+\gamma_5, 0, -2\gamma_1+\gamma_2+\gamma_5, -\gamma_1+2\gamma_2-2\gamma_3+\gamma_5\}$. In this case it is not difficult to see that a basis for $\bar{\mathscr{G}}$ is $\{\gamma_2-\gamma_4, -2\gamma_1+\gamma_2+\gamma_5, -\gamma_1+2\gamma_2-2\gamma_3+\gamma_5\}$ and $f_\gamma=3$.

We saw before that $D_\alpha=\{\alpha_2-\alpha_3\}$. For obtaining a basis for $\bar{\mathscr{A}}$, we choose $S=\{1, 2, 4, 5, 6\}$ and construct $\bar{\mathbf{M}}$ from the α levels of $\mathbf{M}$:

$$\bar{\mathbf{M}}=\begin{bmatrix}2&0&6&0&5\\5&2&1&2&0\\5&2&1&2&0\\1&6&0&6&4\\4&1&5&1&0\\1&6&0&6&4\end{bmatrix}.$$

The rows of $\bar{\mathbf{M}}$ correspond to β and the columns to γ. Eliminating the redundant rows and columns of $\bar{\mathbf{M}}$, we obtain

$$\bar{\mathbf{M}}=\begin{bmatrix}2&0&6&5\\5&2&1&0\\1&6&0&4\\4&1&5&0\end{bmatrix}.$$

Applying the Q process to $\bar{\mathbf{M}}$, we obtain the spanning set $\{\alpha_1+\alpha_2-\alpha_5-\alpha_6, -\alpha_1+\alpha_2+\alpha_4-\alpha_5, -\alpha_1-\alpha_2+\alpha_5+\alpha_6, -\alpha_2+\alpha_4-\alpha_5+\alpha_6, \alpha_1-\alpha_2-\alpha_4+\alpha_5\}$ for $\bar{\mathscr{S}}_\alpha$. Eliminating redundant elements in this set, we obtain the basis $\{\alpha_1+\alpha_2-\alpha_5-\alpha_6, \alpha_1-\alpha_2-\alpha_4+\alpha_5, \alpha_2-\alpha_4+\alpha_5-\alpha_6\}$ for $\bar{\mathscr{S}}_\alpha$. A basis for $\bar{\mathscr{A}}$, the space of estimable α contrasts is then $\{\alpha_2-\alpha_3, \alpha_1+\alpha_2-\alpha_5-\alpha_6, \alpha_1-\alpha_2-\alpha_4+\alpha_5, \alpha_2-\alpha_4+\alpha_5-\alpha_6\}$, and $f_\alpha=4$.

Now we obtain a basis for $\bar{\mathscr{B}}$. We already have $D_\beta=\{\beta_2-\beta_3, \beta_4-\beta_6\}$. We choose $S=\{1, 2, 4, 5\}$ and construct $\bar{\mathbf{M}}$ from the β levels of $\mathbf{M}$:

$$\bar{\mathbf{M}}=\begin{bmatrix}4&5&2&5&0\\1&2&0&2&0\\1&2&0&2&0\\5&0&0&0&4\\2&0&5&0&1\\0&4&1&4&0\end{bmatrix}.$$

Here the rows and columns of $\bar{\mathbf{M}}$ correspond to α and γ, respectively. After eliminating the redundant rows and columns of $\bar{\mathbf{M}}$, we have

$$\bar{\mathbf{M}}=\begin{bmatrix}4&5&2&0\\1&2&0&0\\5&0&0&4\\2&0&5&1\\0&4&1&0\end{bmatrix}.$$

We apply the Q process to $\bar{\mathbf{M}}$ and obtain the spanning set $\{\beta_1-\beta_2-\beta_4+\beta_5, -\beta_1+\beta_2+\beta_4-\beta_5, 2\beta_2-\beta_4-\beta_5, -\beta_1+\beta_2+\beta_4-\beta_5\}$ for $\bar{\mathscr{S}}_\beta$. A basis for $\bar{\mathscr{S}}_\beta$ is then $\{\beta_1-\beta_2-\beta_4+\beta_5, 2\beta_2-\beta_4-\beta_5\}$ and a basis for $\bar{\mathscr{B}}$, the space of estimable β contrasts, is

$$\{\beta_2-\beta_3, \beta_4-\beta_6, \beta_1-\beta_2-\beta_4+\beta_5, 2\beta_2-\beta_4-\beta_5\}, \qquad \text{with } f_\beta=4.$$

We have seen how the $R3$ and Q processes can be used to obtain spanning sets for $\bar{\mathscr{A}}, \bar{\mathscr{B}}$, and $\bar{\mathscr{G}}$. We can extract bases for these spaces from their spanning sets using row reduction. We now wish to determine a basis for some subspace $\bar{\mathscr{E}}$ satisfying (6.5).

We know that $\bar{\Theta}$ is spanned by the set

$$\{\mu+\alpha_i+\beta_j+\gamma_k: n_{ijk}>0\}.$$

Let the matrix **E** consist of the distinct rows of the model matrix **X** and **B** consist of rows corresponding to the basis elements for $\bar{\mathscr{A}}$, $\bar{\mathscr{B}}$, and $\bar{\mathscr{G}}$. We apply row reduction to the augmented matrix

$$\begin{bmatrix} \mathbf{B} \\ \hdashline \mathbf{E} \end{bmatrix} \text{ obtaining } \begin{bmatrix} \mathbf{B} \\ \hdashline \mathbf{E}^R \end{bmatrix}.$$

The nonzero rows in $\mathbf{E}^R$ then represent a basis for some space $\bar{\mathscr{E}}$ satisfying (6.5). One basis element will contain the parameter μ. The others, if any, will be *inseparable contrasts*, that is, estimable contrasts involving the α, β, and γ parameters, but not involving one type alone. If all cell expectations are estimable then (6.6) holds and there are no inseparable contrasts. Let f_c denote the dimension of $\bar{\mathscr{E}}$.

An inseparable contrast, if one exists, is a linear combination of nonestimable α, β, and γ differences. These nonestimable differences are said to be *aliased* with each other and the corresponding factors in the experiment are said to be *confounded*.

EXAMPLE 6.2 (Continued) Applying row reduction as just described, one may obtain the following basis elements for $\bar{\mathscr{E}}$:

$$\mu+\alpha_1+\beta_4+\gamma_1,$$

$$\alpha_1-\alpha_6-\beta_4+\beta_5=(\alpha_1-\alpha_6)-(\beta_4-\beta_5),$$

$$\beta_4-\beta_5-\gamma_1+\gamma_5=(\beta_4-\beta_5)-(\gamma_1-\gamma_5).$$

We see, for example, that the nonestimable differences $\alpha_1-\alpha_6$ and $\beta_4-\beta_5$ are aliased. Furthermore, we see that each of the three factors are confounded with the others. For this example, we have found that $f_\alpha=f_\beta=4$, $f_\gamma=f_c=3$, and

$$r(\mathbf{X})=\dim\bar{\Theta}=f_\alpha+f_\beta+f_\gamma+f_c=14.$$

The above calculations are summarized in Table 6.1a.

Table 6.1a can alternatively be presented as in Table 6.1b.

Or we may present Table 6.1a as in Table 6.1c.

We will discuss this example in detail at the end of this chapter.

Table 6.1a Source of Variations and Degrees of Freedom for Example 6.2

Source	Degrees of Freedom	
Regression	14	
Factor B unadjusted		6
Factor *B* adjusted for *A*		5
Factor *C* adjusted for *A* and *B*		3
Factor *B* unadjusted		6
Factor *C* adjusted for *B*		4
Factor *A* adjusted for *B* and *C*		4
Factor *C* unadjusted		5
Factor *A* adjusted for *C*		5
Factor *B* adjusted for *A* and *C*		4
Residual	14	
Total	28	

Table 6.1b Source of Variations and Degrees of Freedom for Example 6.2

Source	Degrees of Freedom	
Regression	14	
Factor *A* adjusted for *B* and *C*		4
Factor *B* adjusted for *A* and *C*		4
Factor *C* adjusted for *A* and *B*		3
Confounded		3
Residual	14	
Total	28	

Further Examples

Example 6.1 illustrated how the $R3$ process may show that all cell expectations are estimable and hence that all α, β, and γ contrasts are estimable. However, even when all cell expectations are estimable, the $R3$ process may fail to show that any differences are estimable. The following example illustrate this.

Table 6.1c An Alternative Presentation of Table 6.1a

Source	Degrees of Freedom	
Regression	14	
Fitting overall mean		1
Factor A adjusted for mean		5
Factor B adjusted for mean and A		5
Factor C adjusted for mean, A and B		3
Fitting overall mean		1
Factor B adjusted for mean		5
Factor C adjusted for mean and B		4
Factor A adjusted for mean, B and C		4
Fitting overall mean		1
Factor C adjusted for mean		4
Factor A adjusted for mean and C		5
Factor B adjusted for mean, A and C		4
Residual	14	
Total	28	

EXAMPLE 6.3 Consider the $2\times2\times2$ incidence matrix with γ-levels

$$\mathbf{N}_1=\begin{bmatrix}1 & 0\\ 0 & 1\end{bmatrix} \quad\text{and}\quad \mathbf{N}_2=\begin{bmatrix}0 & 1\\ 1 & 0\end{bmatrix}.$$

The $R3$ process fails to add any zeros to the incidence matrix, and the final matrix is identical to $\mathbf{N}$. Hence the $R3$ process fails to provide an estimable γ difference. We could apply the $R3$ process to the α levels of $\mathbf{N}$ and then to the β levels of $\mathbf{N}$, but neither would provide estimable α or β differences. Equivalently, we construct $\bar{\mathbf{M}}$ from the final matrix for the $R3$ process applied to the γ levels of $\mathbf{N}$ and see no redundant rows or columns:

$$\bar{\mathbf{M}}=\begin{bmatrix}1 & 2\\ 2 & 1\end{bmatrix}.$$

However, when we apply the Q process to $\bar{\mathbf{M}}$ we obtain the spanning set $\{2\gamma_1-2\gamma_2\}$ for $\bar{\mathscr{G}}$. Hence all γ contrasts are estimable. We then merge the γ levels of $\mathbf{N}$ to determine the estimable α and β contrasts, and obtain a 2×2 incidence matrix consisting of all ones. Therefore, all α and β contrasts are estimable. Hence $\mathbf{X}$ has maximal rank and all cell expectations are estimable.

The following example illustrates how questions about estimability can be answered without applying the $R3$ process. The procedure for obtaining a $\bar{\mathbf{M}}$ matrix without using the $R3$ process was described just prior to the continuation of Example 6.2.

EXAMPLE 6.4 Consider the $4\times3\times3$ design with incidence matrix $\mathbf{N}$ which is exhibited below in terms of the three γ levels:

$$\mathbf{N}_1=\begin{bmatrix}2&0&0\\0&0&0\\0&1&0\\1&0&1\end{bmatrix},\qquad \mathbf{N}_2=\begin{bmatrix}0&1&0\\1&0&0\\0&0&2\\0&1&0\end{bmatrix},\qquad \mathbf{N}_3=\begin{bmatrix}0&0&0\\0&2&0\\0&0&0\\0&0&0\end{bmatrix}.$$

There are no γ differences which are directly estimable from the matrix $\mathbf{N}$. Hence $\mathbf{M}$ is obtained from $\mathbf{N}$ simply by changing each nonzero element to a one. Now

$$\bar{\mathbf{M}}=\begin{bmatrix}1&2&0\\2&3&0\\0&1&2\\1&2&1\end{bmatrix}.$$

This is not the same $\bar{\mathbf{M}}$ that would be formed after the $R3$ process; nevertheless, the loops in the above $\bar{\mathbf{M}}$ will still provide a spanning set for the estimable γ contrasts. From $\bar{\mathbf{M}}$ we see that $\gamma_1-2\gamma_2+\gamma_3$ and $2(\gamma_1-\gamma_2)$ form a spanning set for the γ contrasts and since these two parametric functions are linearly independent they actually provide a basis. Thus, $f_\gamma=2$.

Now we wish to find $r(\mathbf{A},\mathbf{B})$. This is easily accomplished by supposing the γ effects are not present in the model and just considering a two-way model. The incidence matrix corresponding to the two-way model with just the α and β effects is

$$\mathbf{A}'\mathbf{B}=(n_{ij.})=\begin{bmatrix}2&1&0\\1&2&0\\0&1&2\\1&0&1\end{bmatrix}.$$

Note for estimability considerations relative to this two-way model that we need only suppose the nonzero elements of $\mathbf{A}'\mathbf{B}$ are ones, and this would be exactly the same matrix obtained by changing all nonzero entries in $\bar{\mathbf{M}}$ to ones.

The R process applied to $\mathbf{A}'\mathbf{B}$ results in a final matrix consisting of all ones. Hence the two-way model matrix $(\mathbf{1},\mathbf{A},\mathbf{B})$ has maximal rank and $r(\mathbf{A},\mathbf{B})=r(\mathbf{1},\mathbf{A},\mathbf{B})=a+b-1=4+3-1=6$. Then $r(\mathbf{X})=r(\mathbf{A},\mathbf{B})+f_\gamma=6+2=8=$

$a+b+c-2$ and $\mathbf{X}$ has maximal rank. Therefore, all cell expectations in the three-way model are estimable.

6.5 NORMAL EQUATIONS AND SUMS OF SQUARES

We now consider two reparametrizations of our three-way model for purposes of analysis. Both reparametrizations discussed are of full column rank. Using our previous results, these two reparametrizations can generally be formed without difficulty.

The two reparametrizations we consider lead naturally to the following partition of the regression sum of squares:

$$\text{SS}(\theta)=\text{SS}(\mu, \alpha)+\text{SS}(\beta\,|\,\mu, \alpha)+\text{SS}(\gamma\,|\,\mu, \alpha, \beta). \tag{6.7}$$

Note that there are $r=r(\mathbf{X})$ degrees of freedom associated with the regression sum of squares $\text{SS}(\theta)$ and a degrees of freedom associated with $\text{SS}(\mu, \alpha)$. Furthermore, note that f_γ, the dimension of the vector space of estimable γ contrasts, is the degrees of freedom associated with $\text{SS}(\gamma\,|\,\mu, \alpha, \beta)$ and let $\bar{f}_\beta$ denote the degrees of freedom associated with $\text{SS}(\beta\,|\,\mu, \alpha)$. In the event that either $f_\gamma=0$ or $\bar{f}_\beta=0$ the partition in (6.7) may be obtained via methods for two-way models. $f_\gamma=0$ implies $\text{SS}(\gamma\,|\,\mu, \alpha, \beta)=0$ and similarly $\bar{f}_\beta=0$ implies $\text{SS}(\beta\,|\,\mu, \alpha)=0$ and $\text{SS}(\gamma\,|\,\mu, \alpha, \beta)=\text{SS}(\gamma\,|\,\mu, \alpha)$. Thus, we shall suppose that f_γ and $\bar{f}_\beta$ are both nonzero. Let $\mathbf{\Lambda}'$ and $\mathbf{\Gamma}'$ be $f_\gamma \times c$ and $\bar{f}_\beta \times b$ matrices, respectively, such that $\mathbf{\Lambda}'\boldsymbol{\gamma}$ constitutes a basis for the estimable γ contrasts and $\mathbf{\Gamma}'\boldsymbol{\beta}$ would constitute a basis for the estimable β contrasts if there were no γ effects present in the model. The methods developed earlier in this chapter may be used to obtain f_γ and $\mathbf{\Lambda}'$, and $\mathbf{\Gamma}'$ and $\bar{f}_\beta$ are easily obtained by the methods given in Chapter 5.

The first reparametrization uses the matrices introduced in the following lemma:

LEMMA 6.2 Let $\mathbf{F}_1=\mathbf{A}$, $\mathbf{F}_2=\mathbf{B\Gamma}$, $\mathbf{F}_3=\mathbf{C\Lambda}$, and $\mathbf{F}=(\mathbf{F}_1, \mathbf{F}_2, \mathbf{F}_3)$, where $\mathbf{\Gamma}$ and $\mathbf{\Lambda}$ were defined before. Then

(a) $R(\mathbf{F}_1, \mathbf{F}_2)=R(\mathbf{A}, \mathbf{B})$, and

(b) $R(\mathbf{F})=R(\mathbf{X})$.

Proof (a) Clearly $R(\mathbf{A}, \mathbf{B\Gamma})\subset R(\mathbf{A}, \mathbf{B})$. Since $\mathbf{\Gamma}'\boldsymbol{\beta}$ is estimable,

$$R\begin{pmatrix}\mathbf{0}\\ \mathbf{\Gamma}\end{pmatrix}\subset R\begin{pmatrix}\mathbf{A}'\\ \mathbf{B}'\end{pmatrix}.$$

Thus there exists some matrix $\mathbf{H}$ such that $\mathbf{0}=\mathbf{A}'\mathbf{H}$ and $\mathbf{\Gamma}=\mathbf{B}'\mathbf{H}$. Suppose $\mathbf{w}\in R(\mathbf{A})\cap R(\mathbf{B\Gamma})$. Then there exist some $\mathbf{u}$ and $\mathbf{v}$ such that $\mathbf{w}=\mathbf{Au}=\mathbf{B\Gamma v}$. Then $\mathbf{0}=\mathbf{H}'\mathbf{Au}=\mathbf{H}'\mathbf{B\Gamma v}=\mathbf{\Gamma}'\mathbf{\Gamma v}$. Because $\boldsymbol{\beta}$ is unconstrained and the rows of $\mathbf{\Gamma}'\boldsymbol{\beta}$ form a basis, $\mathbf{\Gamma}'\mathbf{\Gamma}$ is invertible. This means that $\mathbf{v}=\mathbf{0}$ and therefore $R(\mathbf{A})\cap R(\mathbf{B\Gamma})=\{\mathbf{0}\}$, so $r(\mathbf{A},\mathbf{B\Gamma})=r(\mathbf{A})+r(\mathbf{B\Gamma})$. Furthermore,

$$\begin{aligned} r(\mathbf{B\Gamma}) &= r(\mathbf{\Gamma})-\dim\{R(\mathbf{\Gamma})\cap N(\mathbf{B})\} \\ &= r(\mathbf{\Gamma})-\dim\{R(\mathbf{\Gamma})\cap R(\mathbf{B}')^{\perp}\} \\ &= r(\mathbf{\Gamma})=\bar{f}_\beta \end{aligned}$$

because $R(\mathbf{\Gamma})\subset R(\mathbf{B}')$. Also, $r(\mathbf{A})=a$, so

$$r(\mathbf{A},\mathbf{B\Gamma})=a+\bar{f}_\beta=r(\mathbf{X})=r(\mathbf{A},\mathbf{B})$$

by Theorem 5.3 (with $\boldsymbol{\alpha}$ replaced by $\boldsymbol{\beta}$ and remembering that $\bar{f}_\beta$ is f_β for the two-way model). So $R(\mathbf{A},\mathbf{B\Gamma})=R(\mathbf{A},\mathbf{B})$.

(b) Part (b) follows in the same way as part (a) after relabeling:

$$(\mathbf{A},\mathbf{B\Gamma}) \text{ as } \mathbf{A} \quad\text{and}\quad \begin{pmatrix}\boldsymbol{\alpha}\\ \mathbf{\Gamma}'\boldsymbol{\beta}\end{pmatrix} \text{ as } \boldsymbol{\alpha}. \ \square$$

From Lemma 6.2b, we see that a reparametrization for this model is

$$E(\mathbf{Y})=\mathbf{F}_1\boldsymbol{\eta}_1+\mathbf{F}_2\boldsymbol{\eta}_2+\mathbf{F}_3\boldsymbol{\eta}_3=\mathbf{F}\boldsymbol{\eta}, \tag{6.8}$$

where $\boldsymbol{\eta}_1$, $\boldsymbol{\eta}_2$, and $\boldsymbol{\eta}_3$ are unknown parameter vectors of appropriate dimensions. This reparametrization is of full column rank, that is, $\mathbf{F}'\mathbf{F}$ is nonsingular. Additionally, if this model is fit in stages the partition of the regression sum of squares given in (6.7) is easily obtained upon noting that $\mathrm{SS}(\eta_1)=\mathrm{SS}(\mu,\alpha)$; $\mathrm{SS}(\eta_1,\eta_2)=\mathrm{SS}(\mu,\alpha,\beta)$; and $\mathrm{SS}(\theta)=\mathrm{SS}(\eta)$.

The normal equations for this parametrization are easily formed from a knowledge of $\mathbf{X}'\mathbf{X}$, $\mathbf{X}'\mathbf{Y}$, $\mathbf{\Gamma}$, and $\mathbf{\Lambda}$:

$$\mathbf{F}'\mathbf{F}\hat{\boldsymbol{\eta}}=\mathbf{F}'\mathbf{Y},$$

where

$$\mathbf{F}'\mathbf{F}=\begin{bmatrix}\mathbf{A}'\mathbf{A} & \mathbf{A}'\mathbf{B\Gamma} & \mathbf{A}'\mathbf{C\Lambda}\\ \mathbf{\Gamma}'\mathbf{B}'\mathbf{A} & \mathbf{\Gamma}'\mathbf{B}'\mathbf{B\Gamma} & \mathbf{\Gamma}'\mathbf{B}'\mathbf{C\Lambda}\\ \mathbf{\Lambda}'\mathbf{C}'\mathbf{A} & \mathbf{\Lambda}'\mathbf{C}'\mathbf{B\Gamma} & \mathbf{\Lambda}'\mathbf{C}'\mathbf{C\Lambda}\end{bmatrix}$$

and

$$\mathbf{F}'\mathbf{Y}=\begin{bmatrix}\mathbf{A}'\mathbf{Y}\\ \mathbf{\Gamma}'\mathbf{B}'\mathbf{Y}\\ \mathbf{\Lambda}'\mathbf{C}'\mathbf{Y}\end{bmatrix}.$$

Solutions for the normal equations corresponding to the original model (6.3) may be obtained from the solution $\hat{\boldsymbol{\eta}}$ for the normal equations for the reparametrized model (6.8), for example, any random vector $\hat{\boldsymbol{\theta}}$ satisfying $\mathbf{X}\hat{\boldsymbol{\theta}}=\mathbf{F}\hat{\boldsymbol{\eta}}$ also satisfies $\mathbf{X}'\mathbf{X}\hat{\boldsymbol{\theta}}=\mathbf{X}'\mathbf{Y}$. For example,

$$\begin{aligned} &\textbf{(a)}\quad \hat{\mu}=(\mathbf{1}'\hat{\boldsymbol{\eta}}_1)/a,\\ &\textbf{(b)}\quad \hat{\boldsymbol{\alpha}}=\hat{\boldsymbol{\eta}}_1-\mathbf{1}\hat{\mu},\\ &\textbf{(c)}\quad \hat{\boldsymbol{\beta}}=\boldsymbol{\Gamma}\hat{\boldsymbol{\eta}}_2, \text{ and}\\ &\textbf{(d)}\quad \hat{\boldsymbol{\gamma}}=\boldsymbol{\Lambda}\hat{\boldsymbol{\eta}}_3 \end{aligned} \tag{6.9}$$

is such a solution.

The solution described in (6.9) has the following properties:

$$\begin{aligned} &\textbf{(a)}\quad \textstyle\sum_i \alpha_i=0.\\ &\textbf{(b)}\quad \boldsymbol{\lambda}'\boldsymbol{\beta}=0 \text{ for all } \boldsymbol{\lambda}\in N(\boldsymbol{\Gamma}').\\ &\textbf{(c)}\quad \boldsymbol{\lambda}'\boldsymbol{\gamma}=0 \text{ for all } \boldsymbol{\lambda}\in N(\boldsymbol{\Lambda}'). \end{aligned} \tag{6.10}$$

When $\mathbf{X}$ is of maximal rank, conditions (6.10.b) and (6.10.c) reduce to $\sum_j \hat{\beta}_j=0$ and $\sum_k \hat{\gamma}_k=0$, respectively. Also note that $\hat{\boldsymbol{\theta}}$ in (6.9) is a very simple linear transformation of $\hat{\boldsymbol{\eta}}$ and hence $\text{Cov}(\hat{\boldsymbol{\theta}})$ is easily described in terms of the covariance matrix of $\hat{\boldsymbol{\eta}}$ which is $\sigma^2(\mathbf{F}'\mathbf{F})^{-1}$.

The second reparametrization uses the matrices introduced in the next lemma:

LEMMA 6.3 Let $\mathbf{F}_1$, $\mathbf{F}_2$, and $\mathbf{F}_3$ be as in Lemma 6.2, and set $\mathbf{G}_1=\mathbf{F}_1$, $\mathbf{G}_2=(\mathbf{I}-\mathbf{P}_1)\mathbf{F}_2$, $\mathbf{G}_3=(\mathbf{I}-\mathbf{P}_{12})\mathbf{F}_3$, and $\mathbf{G}=(\mathbf{G}_1, \mathbf{G}_2, \mathbf{G}_3)$, where $\mathbf{P}_1$ and $\mathbf{P}_{12}$ are the orthogonal projection matrices on $R(\mathbf{G}_1)$ and $R(\mathbf{G}_1, \mathbf{G}_2)$, respectively. Then

$$\begin{aligned} &\textbf{(a)}\quad R(\mathbf{G}_1, \mathbf{G}_2)=R(\mathbf{A}, \mathbf{B}),\\ &\textbf{(b)}\quad R(\mathbf{G})=R(\mathbf{X}), \text{ and}\\ &\textbf{(c)}\quad \mathbf{G}_i'\mathbf{G}_j=\mathbf{0} \text{ for } i\neq j=1, 2, 3. \end{aligned} \tag{6.11}$$

Proof (a) and (b) Using Lemma 6.2, we have

$$R(\mathbf{G}_1, \mathbf{G}_2)=R[\mathbf{F}_1, (\mathbf{I}-\mathbf{P}_1)\mathbf{F}_2]=R(\mathbf{F}_1, \mathbf{F}_2)=R(\mathbf{A}, \mathbf{B})$$

$$R(\mathbf{G})=R[\mathbf{F}_1, (\mathbf{I}-\mathbf{P}_1)\mathbf{F}_2, (\mathbf{I}-\mathbf{P}_{12})\mathbf{F}_3]=R(\mathbf{F})=R(\mathbf{X}).$$

(c) $\mathbf{G}_1'\mathbf{G}_2=\mathbf{F}_1'(\mathbf{I}-\mathbf{P}_1)\mathbf{F}_2=[(\mathbf{I}-\mathbf{P}_1)\mathbf{F}_1]'\mathbf{F}_2=\mathbf{0}$,
$\mathbf{G}_2'\mathbf{G}_1=(\mathbf{G}_1'\mathbf{G}_2)'=\mathbf{0}$,

$$\begin{aligned}
\mathbf{G}_1'\mathbf{G}_3 &= \mathbf{F}_1'(\mathbf{I}-\mathbf{P}_{12})\mathbf{F}_3 = [(\mathbf{I}-\mathbf{P}_{12})\mathbf{F}_1]'\mathbf{F}_3 = \mathbf{0},\\
\mathbf{G}_3'\mathbf{G}_1 &= (\mathbf{G}_1'\mathbf{G}_3)' = \mathbf{0},\\
\mathbf{G}_2'\mathbf{G}_3 &= \mathbf{G}_2'(\mathbf{I}-\mathbf{P}_{12})\mathbf{F}_3 = [(\mathbf{I}-\mathbf{P}_{12})\mathbf{G}_2]'\mathbf{F}_3 = \mathbf{0},\\
\mathbf{G}_3'\mathbf{G}_2 &= (\mathbf{G}_2'\mathbf{G}_3)' = \mathbf{0}. \quad \square
\end{aligned}$$

From (6.11.b) we see that a full rank reparametrization for the model is

$$E(\mathbf{Y}) = \mathbf{G}_1\boldsymbol{\xi}_1 + \mathbf{G}_2\boldsymbol{\xi}_2 + \mathbf{G}_3\boldsymbol{\xi}_3 = \mathbf{G}\boldsymbol{\xi}, \tag{6.12}$$

where $\boldsymbol{\xi}_1$, $\boldsymbol{\xi}_2$, and $\boldsymbol{\xi}_3$ are unknown parameter vectors of appropriate dimensions. The properties in (6.11) along with noting that $\mathbf{G}_1 = \mathbf{A}$ also imply that $SS(\xi_1) = SS(\mu, \alpha)$, $SS(\xi_2) = SS(\beta \mid \mu, \alpha)$, and $SS(\xi_3) = SS(\gamma \mid \mu, \alpha, \beta)$.

Property (6.11.c) implies that the solution $\hat{\boldsymbol{\xi}} = (\hat{\boldsymbol{\xi}}_1', \hat{\boldsymbol{\xi}}_2', \hat{\boldsymbol{\xi}}_3')'$ to $\mathbf{G}'\mathbf{G}\hat{\boldsymbol{\xi}} = \mathbf{G}'\mathbf{Y}$ is given by

$$\hat{\boldsymbol{\xi}}_i = (\mathbf{G}_i'\mathbf{G}_i)^{-1}\mathbf{G}_i'\mathbf{Y}, \qquad i = 1, 2, 3,$$

and that the random vectors $\hat{\boldsymbol{\xi}}_1$, $\hat{\boldsymbol{\xi}}_2$, and $\hat{\boldsymbol{\xi}}_3$ are statistically independent.

Note also that from the $\hat{\boldsymbol{\xi}}_i$'s the normal equation solution $\hat{\boldsymbol{\eta}} = (\mathbf{F}'\mathbf{F})^{-1}\mathbf{F}'\mathbf{Y}$ may be obtained in the following manner:

THEOREM 6.5 Consider parametrizations (6.8) and (6.12) and let $\hat{\boldsymbol{\eta}}$ and $\hat{\boldsymbol{\xi}}$ denote normal equation solutions for $\boldsymbol{\eta}$ and $\boldsymbol{\xi}$. Then

$$\begin{aligned}
&\textbf{(a)}\quad \hat{\boldsymbol{\eta}}_3 = \hat{\boldsymbol{\xi}}_3.\\
&\textbf{(b)}\quad \hat{\boldsymbol{\eta}}_2 = \hat{\boldsymbol{\xi}}_2 - (\mathbf{G}_2'\mathbf{G}_2)^{-1}\mathbf{G}_2'\mathbf{F}_3\hat{\boldsymbol{\eta}}_3.\\
&\textbf{(c)}\quad \hat{\boldsymbol{\eta}}_1 = \hat{\boldsymbol{\xi}}_1 - (\mathbf{F}_1'\mathbf{F}_1)^{-1}\mathbf{F}_1'(\mathbf{F}_2\hat{\boldsymbol{\eta}}_2 + \mathbf{F}_3\hat{\boldsymbol{\eta}}_3).
\end{aligned} \tag{6.13}$$

Proof We have $\mathbf{F}\hat{\boldsymbol{\eta}} \doteq \mathbf{G}\hat{\boldsymbol{\xi}}$ and therefore

$$\mathbf{F}_1\hat{\boldsymbol{\eta}}_1 + \mathbf{F}_2\hat{\boldsymbol{\eta}}_2 + \mathbf{F}_3\hat{\boldsymbol{\eta}}_3 = \mathbf{G}_1\hat{\boldsymbol{\xi}}_1 + \mathbf{G}_2\hat{\boldsymbol{\xi}}_2 + \mathbf{G}_3\hat{\boldsymbol{\xi}}_3. \tag{6.14}$$

Premultiplying by $\mathbf{G}_3'(\mathbf{I}-\mathbf{P}_{12}) \doteq \mathbf{G}_3'$ where $\mathbf{P}_{12}$ was defined in Lemma 6.3, we obtain

$$\mathbf{G}_3'\mathbf{G}_3\hat{\boldsymbol{\eta}}_3 = \mathbf{G}_3'\mathbf{G}_3\hat{\boldsymbol{\xi}}_3$$

and because $\mathbf{G}_3'\mathbf{G}_3$ is nonsingular, we obtain (a).

Premultiplying (6.14) by $\mathbf{G}_2'(\mathbf{I}-\mathbf{P}_1) = \mathbf{G}_2'$, we obtain

$$\mathbf{G}_2'\mathbf{G}_2\hat{\boldsymbol{\eta}}_2 + \mathbf{G}_2'\mathbf{F}_3\hat{\boldsymbol{\eta}}_3 = \mathbf{G}_2'\mathbf{G}_2\hat{\boldsymbol{\xi}}_2 + \mathbf{G}_2'(\mathbf{I}-\mathbf{P}_{12})\mathbf{F}_3\hat{\boldsymbol{\xi}}_3 = \mathbf{G}_2'\mathbf{G}_2\hat{\boldsymbol{\xi}}_2$$

and because $\mathbf{G}_2'\mathbf{G}_2$ is nonsingular, part (b) follows.

Premultiplying (6.14) by $\mathbf{F}_1'$, we have

$$\mathbf{F}_1'\mathbf{F}_1\hat{\boldsymbol{\eta}}_1 + \mathbf{F}_1'\mathbf{F}_2\hat{\boldsymbol{\eta}}_2 + \mathbf{F}_1'\mathbf{F}_3\hat{\boldsymbol{\eta}}_3 = \mathbf{F}_1'\mathbf{F}_1\hat{\boldsymbol{\xi}}_1$$

and because $\mathbf{F}_1'\mathbf{F}_1$ is nonsingular, we have part (c). $\square$

We can easily obtain the covariance matrix of $\hat{\eta}$ using (6.13) and the covariance matrices of $\hat{\xi}_1$, $\hat{\xi}_2$, and $\hat{\xi}_3$ since these three random vectors are independent.

The formation of the normal equations for ξ_1 and ξ_2 in parametrization (6.12) is fully discussed in Chapter 5. Once $\hat{\xi}_1$ and $\hat{\xi}_2$ have been obtained, we form the normal equation for ξ_3 in the following way: first note that $\mathbf{P}_{12}=\mathbf{P}_1+\mathbf{P}_2$ where $\mathbf{P}_2=\mathbf{G}_2(\mathbf{G}_2'\mathbf{G}_2)^{-1}\mathbf{G}_2'$ is the orthogonal projection matrix on $R(\mathbf{G}_2)$. Then

$$\begin{aligned}
\mathbf{G}_3'\mathbf{G}_3 &= \mathbf{\Lambda}'[\mathbf{C}'\mathbf{C}-\mathbf{C}'\mathbf{A}(\mathbf{A}'\mathbf{A})^{-1}\mathbf{A}'\mathbf{C}]\mathbf{\Lambda}-\mathbf{\Lambda}'\mathbf{C}'\mathbf{G}_2(\mathbf{G}_2'\mathbf{G}_2)^{-1}\mathbf{G}_2'\mathbf{C}\mathbf{\Lambda},\\
\mathbf{G}_3'\mathbf{Y} &= \mathbf{\Lambda}'(\mathbf{C}'\mathbf{Y}-\mathbf{C}'\mathbf{A}\hat{\xi}_1-\mathbf{C}'\mathbf{G}_2\hat{\xi}_2), \qquad (6.15)\\
\mathbf{C}'\mathbf{G}_2 &= \mathbf{C}'\mathbf{B}\mathbf{\Gamma}-\mathbf{C}'\mathbf{A}(\mathbf{A}'\mathbf{A})^{-1}\mathbf{A}'\mathbf{B}\mathbf{\Gamma}.
\end{aligned}$$

Thus, the normal equation for ξ_3 may be formed by knowing $\mathbf{\Gamma}$, $\mathbf{\Lambda}$, $\hat{\xi}_1$, $\hat{\xi}_2$, $(\mathbf{A}'\mathbf{A})^{-1}$ which is a diagonal matrix, $(\mathbf{G}_2'\mathbf{G}_2)^{-1}$, $\mathbf{C}'\mathbf{C}$, $\mathbf{C}'\mathbf{Y}$, and the two-way incidence matrices $\mathbf{A}'\mathbf{B}$, $\mathbf{A}'\mathbf{C}$, and $\mathbf{B}'\mathbf{C}$.

The solution $\hat{\boldsymbol{\theta}}$ given in (6.9) can also be expressed in terms of $\hat{\xi}$. Using (6.13), we have

$$\begin{aligned}
\hat{\boldsymbol{\gamma}} &= \mathbf{\Lambda}\hat{\xi}_3,\\
\hat{\boldsymbol{\beta}} &= \mathbf{\Gamma}[\hat{\xi}_2-(\mathbf{G}_2'\mathbf{G}_2)^{-1}\mathbf{G}_2'\mathbf{C}\hat{\boldsymbol{\gamma}}],\\
\hat{\mu} &= \mathbf{1}'[\hat{\xi}_1-(\mathbf{A}'\mathbf{A})^{-1}(\mathbf{A}'\mathbf{B}\hat{\boldsymbol{\beta}}+\mathbf{A}'\mathbf{C}\hat{\boldsymbol{\gamma}})]/a, \qquad (6.16)\\
\hat{\boldsymbol{\alpha}} &= \hat{\xi}_1-(\mathbf{A}'\mathbf{A})^{-1}(\mathbf{A}'\mathbf{B}\hat{\boldsymbol{\beta}}+\mathbf{A}'\mathbf{C}\hat{\boldsymbol{\gamma}})-\mathbf{1}\hat{\mu}.
\end{aligned}$$

We have discussed normal equation solutions and the partition of SS(θ) given in (6.7). Clearly, the particular partition in (6.7) is not crucial. That is, if the partition

$$\text{SS}(\theta)=\text{SS}(\mu, \gamma)+\text{SS}(\beta|\mu, \gamma)+\text{SS}(\alpha|\mu, \beta, \gamma) \qquad (6.17)$$

is desired our discussion is applicable with only a relabeling of the parameters or an obvious change in notation. If one desires a partition as in (6.17) it is not always the case, however, that the natural way to proceed is the most efficient. For example, suppose the partition in (6.17) is wanted and that $a=10$, $b=3$, and $c=2$. For notational convenience let us suppose $\mathbf{X}$ is of maximal rank. (The point will be clear whether $\mathbf{X}$ is of maximal rank or not.) If one proceeds in the natural fashion, a 2×2 matrix and a 9×9 matrix must be inverted to obtain the partition. Note that a sum of squares like SS(μ, α), SS(μ, β), or SS(μ, γ) are easily obtained with no need for actual matrix inversion. Suppose, however, you first obtain the partition in (6.7) which requires inverting a 2×2 and a 1×1 matrix. Next obtain the partition

$$\text{SS}(\mu, \beta, \gamma)=\text{SS}(\mu, \beta)+\text{SS}(\gamma|\mu, \beta) \qquad (6.18)$$

which requires inverting 1×1 matrix. Now the partition in (6.17) is easily obtained upon noting the identities

$$\mathrm{SS}(\mu,\beta,\gamma)=\mathrm{SS}(\mu,\gamma)+\mathrm{SS}(\beta|\mu,\gamma)\quad\text{and}\quad \mathrm{SS}(\theta)=\mathrm{SS}(\mu,\beta,\gamma)+\mathrm{SS}(\alpha|\mu,\beta,\gamma).$$

Thus, by obtaining the partitions in (6.7) and (6.18) the partition in (6.17) is obtained with less work than directly obtaining the partition in (6.17). As a general rule for obtaining a partition of SS(θ) as in (6.7) or (6.17) first partition SS(θ) according to a decreasing order of magnitude of the numbers a, b, c, for example, if $a>b\geqslant c$ the partition in (6.7) is appropriate, and then proceed with additive two-way analyses. This general rule will not always be optimal; when **X** is not of maximal rank there may be alternative computational procedures which are better, but for most situations the rule should prove satisfactory.

We illustrate reparametrizations (6.8) and (6.12) with two examples.

EXAMPLE 6.5 Suppose $a=b=c=4$, $n=8$, and $\bar{\mathbf{M}}$ for the γ parameters is

$$\bar{\mathbf{M}}=\begin{bmatrix}1&2&0&0\\2&4&0&0\\0&0&1&4\\0&0&4&3\end{bmatrix}.$$

There are two loops in $\bar{\mathbf{M}}$ yielding the spanning set $\gamma_1-2\gamma_2+\gamma_4$ and $\gamma_1+\gamma_3-2\gamma_4$ for $\mathscr{G}$. Since these two contrasts are linearly independent they form a basis for $\mathscr{G}$. Then $f_\gamma=2$ and one choice for Λ' is

$$\Lambda'=\begin{bmatrix}1&-2&0&1\\1&0&1&-2\end{bmatrix}.$$

To obtain $\bar{f}_\beta$ and Γ', we consider the model

$$E(Y_{ijh})=\mu+\alpha_i+\beta_j,$$

where i, $j=1$, 2, 3, 4 and $h=1,\ldots,n_{ij}$. For estimability considerations relative to this model we may suppose that the incidence matrix is $\bar{\mathbf{M}}$ with all nonzero entries changed to ones. Doing this and applying the methods of Chapter 5 we obtain the basis $\{\beta_1-\beta_2,\beta_3-\beta_4\}$ for the β contrasts relative to the two-way model. Thus $\bar{f}_\beta=2$ and we may choose

$$\Gamma'=\begin{bmatrix}1&-1&0&0\\0&0&1&-1\end{bmatrix}.$$

With these choices of Λ' and Γ', parametrization (6.8) is

$$E(\mathbf{Y})=\begin{bmatrix}1&0&0&0\\1&0&0&0\\0&1&0&0\\0&1&0&0\\0&0&1&0\\0&0&1&0\\0&0&0&1\\0&0&0&1\end{bmatrix}\boldsymbol{\eta}_1+\begin{bmatrix}1&0\\-1&0\\1&0\\-1&0\\0&1\\0&-1\\0&1\\0&-1\end{bmatrix}\boldsymbol{\eta}_2+\begin{bmatrix}1&1\\-2&0\\-2&0\\1&-2\\1&1\\1&-2\\1&-2\\0&1\end{bmatrix}\boldsymbol{\eta}_3.$$

EXAMPLE 6.6 Suppose we have a $3\times2\times2$ classification with the following observations on the indicated cells:

(1, 1, 1): 2, 3	(2, 1, 2): 5	(3, 1, 1): 6	(3, 2, 2):8
(1, 2, 2): 4, 7	(2, 2, 1): 4, 6	(3, 2, 1): 7	

Thus, we have the following incidence matrix **N** which is exhibited in terms of the γ levels:

$$\mathbf{N}_1=\begin{bmatrix}2&0\\0&2\\1&1\end{bmatrix}\quad\text{and}\quad\mathbf{N}_2=\begin{bmatrix}0&2\\1&0\\0&1\end{bmatrix}.$$

From this incidence matrix the corresponding model matrix is seen to be of maximal rank. That is, $\gamma_1-\gamma_2$ is directly estimable (n_{321} and n_{322} are both nonzero) and $\mathbf{A}'\mathbf{B}$ has no nonzero entries so that $r(\mathbf{X})=r(\mathbf{A}, \mathbf{B})+f_\gamma=4+1=5$.

The second reparametrization (6.12) requires solving three sets of normal equations to obtain the partition of SS(θ) given in (6.7). Since $a>b, c$, the partition is most easily obtained in the natural fashion. To obtain SS(μ, α) we consider the first set of normal equations and obtain

$$\begin{bmatrix}4&0&0\\0&3&0\\0&0&3\end{bmatrix}\hat{\boldsymbol{\xi}}_1=\begin{bmatrix}16\\15\\21\end{bmatrix}\quad\text{and}\quad\hat{\boldsymbol{\xi}}_1=(4, 5, 7)'.$$

Then $\text{SS}(\mu, \alpha)=\hat{\boldsymbol{\xi}}_1'\mathbf{G}_1'\mathbf{Y}=\hat{\boldsymbol{\xi}}_1'\mathbf{A}'\mathbf{Y}=286$.

To obtain SS($\beta|\mu$, α) we consider the second set of normal equations. Since **X** is of maximal rank we may choose $\boldsymbol{\Gamma}'=(1, -1)$. From (5.11.a) (with $\boldsymbol{\Gamma}'$ replacing $\boldsymbol{\Lambda}'$), we have

$$\begin{aligned}\mathbf{G}_2'\mathbf{G}_2&=\boldsymbol{\Gamma}'[\mathbf{B}'\mathbf{B}-\mathbf{B}'\mathbf{A}(\mathbf{A}'\mathbf{A})^{-1}\mathbf{A}'\mathbf{B}]\boldsymbol{\Gamma}=(28/3),\\ \mathbf{G}_2'\mathbf{Y}&=\boldsymbol{\Gamma}'(\mathbf{B}'\mathbf{Y}-\mathbf{B}'\mathbf{A}\hat{\boldsymbol{\xi}}_1)=(-8),\\ (28/3)\hat{\boldsymbol{\xi}}_2&=(-8)\quad\text{so}\quad\hat{\boldsymbol{\xi}}_2=(-6/7),\end{aligned}$$

and

$$\mathrm{SS}(\beta \mid \mu, \alpha)=\hat{\boldsymbol{\xi}}_2'\mathbf{G}_2'\mathbf{Y}=\frac{48}{7}.$$

Finally, to obtain $\mathrm{SS}(\gamma \mid \mu, \alpha, \beta)$ we construct the third set of normal equations using (6.15). Since $\mathbf{X}$ is of maximal rank, we may choose $\boldsymbol{\Lambda}'=(1, -1)$. Then

$$\mathbf{C}'\mathbf{G}_2=\left(\frac{4}{3}, -\frac{4}{3}\right)',$$

$$\mathbf{G}_3'\mathbf{G}_3=\left(\frac{60}{7}\right),$$

$$\mathbf{G}_3'\mathbf{Y}=\left(-\frac{40}{7}\right),$$

$$\left(\frac{60}{7}\right)\hat{\boldsymbol{\xi}}_3=\left(\frac{40}{7}\right) \quad \text{so} \quad \hat{\boldsymbol{\xi}}_3=\left(-\frac{2}{3}\right),$$

and

$$\mathrm{SS}(\gamma \mid \mu, \alpha, \beta)=\hat{\boldsymbol{\xi}}_3'\mathbf{G}_3'\mathbf{Y}=\frac{80}{21}.$$

We now have the partition of $\mathrm{SS}(\theta)$ given in (6.7). Any other similar partition may now be obtained via two-way model methods since we can get $\mathrm{SS}(\theta)$ by simply adding together the three sums of squares we have just computed.

Applying (6.16), we obtain a solution $\hat{\boldsymbol{\theta}}$ for our original parametrization:

$$\hat{\boldsymbol{\gamma}}=\left(-\frac{2}{3}, \frac{2}{3}\right)',$$

$$\hat{\boldsymbol{\beta}}=\left(-\frac{2}{3}, \frac{2}{3}\right)',$$

$$\hat{\mu}=\frac{16}{3},$$

and

$$\hat{\boldsymbol{\alpha}}=\left(-\frac{4}{3}, -\frac{1}{3}, \frac{5}{3}\right)'.$$

EXAMPLE 6.7 To illustrate the complete analysis of variance procedure, consider Example 6.2 with the following observations on the indicated cells:

(1, 2, 3): 3 (1, 4, 1): 3, 1, 2 (1, 5, 2): 6 (1, 6, 1): 2
(2, 1, 1): 3, 2 (2, 3, 2): 7
(3, 2, 2): 5, 7 (3, 3, 4): 8, 9
(4, 4, 5): 4, 5 (4, 5, 1): 3, 4 (4, 6, 5): 9
(5, 1, 5): 11 (5, 3, 1): 1 (5, 5, 3): 5, 5
(6, 1, 3): 5, 6, 7 (6, 4, 2): 4, 3 (6, 6, 4): 9

Thus, we have the following incidence matrix $\mathbf{N}$ which is exhibited in terms of the γ levels:

$$\mathbf{N}_1 = \begin{bmatrix} 0 & 0 & 0 & 3 & 0 & 1 \\ 2 & 0 & 0 & 0 & 0 & 0 \\ 0 & 0 & 0 & 0 & 0 & 0 \\ 0 & 0 & 0 & 0 & 2 & 0 \\ 0 & 0 & 1 & 0 & 0 & 0 \\ 0 & 0 & 0 & 0 & 0 & 0 \end{bmatrix}, \quad \mathbf{N}_4 = \begin{bmatrix} 0 & 0 & 0 & 0 & 0 & 0 \\ 0 & 0 & 0 & 0 & 0 & 0 \\ 0 & 0 & 2 & 0 & 0 & 0 \\ 0 & 0 & 0 & 0 & 0 & 0 \\ 0 & 0 & 0 & 0 & 0 & 0 \\ 0 & 0 & 0 & 0 & 0 & 1 \end{bmatrix},$$

$$\mathbf{N}_2 = \begin{bmatrix} 0 & 0 & 0 & 0 & 1 & 0 \\ 0 & 0 & 1 & 0 & 0 & 0 \\ 0 & 2 & 0 & 0 & 0 & 0 \\ 0 & 0 & 0 & 0 & 0 & 0 \\ 0 & 0 & 0 & 0 & 0 & 0 \\ 0 & 0 & 0 & 2 & 0 & 0 \end{bmatrix}, \quad \mathbf{N}_5 = \begin{bmatrix} 0 & 0 & 0 & 0 & 0 & 0 \\ 0 & 0 & 0 & 0 & 0 & 0 \\ 0 & 0 & 0 & 0 & 0 & 0 \\ 0 & 0 & 0 & 2 & 0 & 1 \\ 1 & 0 & 0 & 0 & 0 & 0 \\ 0 & 0 & 0 & 0 & 0 & 0 \end{bmatrix}.$$

$$\mathbf{N}_3 = \begin{bmatrix} 0 & 1 & 0 & 0 & 0 & 0 \\ 0 & 0 & 0 & 0 & 0 & 0 \\ 0 & 0 & 0 & 0 & 0 & 0 \\ 0 & 0 & 0 & 0 & 0 & 0 \\ 0 & 0 & 0 & 0 & 2 & 0 \\ 3 & 0 & 0 & 0 & 0 & 0 \end{bmatrix},$$

Suppose that a suitable linear model for analyzing the above data is, as described in Section 6.3,

$$E(Y_{ijkl}) = \mu + \alpha_i + \beta_j + \gamma_k$$

$$i = 1, \ldots, 6; \qquad j = 1, \ldots, 6; \qquad k = 1, \ldots, 5; \qquad l = 1, \ldots, n_{ijk}$$

where μ is a mean, α_i is the effect due to the ith level of the α factor, β_j is the effect due to the jth level of the β factor, γ_k is the effect due to the kth level of γ factor, and Y_{ijkl} is the lth observation in the (i, j, k) cell.

In matrix form we can write the above model as

$$E(\mathbf{Y}) = \mathbf{X}\boldsymbol{\theta}, \qquad \boldsymbol{\theta} = (\mu, \alpha_1, \ldots, \alpha_6, \beta_1, \ldots, \beta_6, \gamma_1, \ldots, \gamma_5)'$$

where the design matrix $\mathbf{X}$ is a 28×18 matrix:

$\mathbf{X} =$

$$\begin{array}{cccccccccccccccccc}
\mu & \alpha_1 & \alpha_2 & \alpha_3 & \alpha_4 & \alpha_5 & \alpha_6 & \beta_1 & \beta_2 & \beta_3 & \beta_4 & \beta_5 & \beta_6 & \gamma_1 & \gamma_2 & \gamma_3 & \gamma_4 & \gamma_5
\end{array}$$

$$\left[\begin{array}{cccccccccccccccccc}
1 & 1 & 0 & 0 & 0 & 0 & 0 & 0 & 1 & 0 & 0 & 0 & 0 & 0 & 0 & 1 & 0 & 0 \\
1 & 1 & 0 & 0 & 0 & 0 & 0 & 0 & 0 & 0 & 1 & 0 & 0 & 1 & 0 & 0 & 0 & 0 \\
1 & 1 & 0 & 0 & 0 & 0 & 0 & 0 & 0 & 0 & 1 & 0 & 0 & 1 & 0 & 0 & 0 & 0 \\
1 & 1 & 0 & 0 & 0 & 0 & 0 & 0 & 0 & 0 & 1 & 0 & 0 & 1 & 0 & 0 & 0 & 0 \\
1 & 1 & 0 & 0 & 0 & 0 & 0 & 0 & 0 & 0 & 0 & 1 & 0 & 0 & 1 & 0 & 0 & 0 \\
1 & 1 & 0 & 0 & 0 & 0 & 0 & 0 & 0 & 0 & 0 & 0 & 1 & 1 & 0 & 0 & 0 & 0 \\
1 & 0 & 1 & 0 & 0 & 0 & 0 & 1 & 0 & 0 & 0 & 0 & 0 & 1 & 0 & 0 & 0 & 0 \\
1 & 0 & 1 & 0 & 0 & 0 & 0 & 1 & 0 & 0 & 0 & 0 & 0 & 1 & 0 & 0 & 0 & 0 \\
1 & 0 & 1 & 0 & 0 & 0 & 0 & 0 & 0 & 1 & 0 & 0 & 0 & 0 & 1 & 0 & 0 & 0 \\
1 & 0 & 0 & 1 & 0 & 0 & 0 & 0 & 1 & 0 & 0 & 0 & 0 & 0 & 1 & 0 & 0 & 0 \\
1 & 0 & 0 & 1 & 0 & 0 & 0 & 0 & 1 & 0 & 0 & 0 & 0 & 0 & 1 & 0 & 0 & 0 \\
1 & 0 & 0 & 1 & 0 & 0 & 0 & 0 & 0 & 1 & 0 & 0 & 0 & 0 & 0 & 0 & 1 & 0 \\
1 & 0 & 0 & 1 & 0 & 0 & 0 & 0 & 0 & 1 & 0 & 0 & 0 & 0 & 0 & 0 & 1 & 0 \\
1 & 0 & 0 & 0 & 1 & 0 & 0 & 0 & 0 & 0 & 1 & 0 & 0 & 0 & 0 & 0 & 0 & 1 \\
1 & 0 & 0 & 0 & 1 & 0 & 0 & 0 & 0 & 0 & 1 & 0 & 0 & 0 & 0 & 0 & 0 & 1 \\
1 & 0 & 0 & 0 & 1 & 0 & 0 & 0 & 0 & 0 & 0 & 1 & 0 & 1 & 0 & 0 & 0 & 0 \\
1 & 0 & 0 & 0 & 1 & 0 & 0 & 0 & 0 & 0 & 0 & 1 & 0 & 1 & 0 & 0 & 0 & 0 \\
1 & 0 & 0 & 0 & 1 & 0 & 0 & 0 & 0 & 0 & 0 & 0 & 1 & 0 & 0 & 0 & 0 & 1 \\
1 & 0 & 0 & 0 & 0 & 1 & 0 & 1 & 0 & 0 & 0 & 0 & 0 & 0 & 0 & 0 & 0 & 1 \\
1 & 0 & 0 & 0 & 0 & 1 & 0 & 0 & 0 & 1 & 0 & 0 & 0 & 1 & 0 & 0 & 0 & 0 \\
1 & 0 & 0 & 0 & 0 & 1 & 0 & 0 & 0 & 0 & 0 & 1 & 0 & 0 & 0 & 1 & 0 & 0 \\
1 & 0 & 0 & 0 & 0 & 1 & 0 & 0 & 0 & 0 & 0 & 1 & 0 & 0 & 0 & 1 & 0 & 0 \\
1 & 0 & 0 & 0 & 0 & 0 & 1 & 1 & 0 & 0 & 0 & 0 & 0 & 0 & 0 & 1 & 0 & 0 \\
1 & 0 & 0 & 0 & 0 & 0 & 1 & 1 & 0 & 0 & 0 & 0 & 0 & 0 & 0 & 1 & 0 & 0 \\
1 & 0 & 0 & 0 & 0 & 0 & 1 & 1 & 0 & 0 & 0 & 0 & 0 & 0 & 0 & 1 & 0 & 0 \\
1 & 0 & 0 & 0 & 0 & 0 & 1 & 0 & 0 & 0 & 1 & 0 & 0 & 0 & 1 & 0 & 0 & 0 \\
1 & 0 & 0 & 0 & 0 & 0 & 1 & 0 & 0 & 0 & 1 & 0 & 0 & 0 & 1 & 0 & 0 & 0 \\
1 & 0 & 0 & 0 & 0 & 0 & 1 & 0 & 0 & 0 & 0 & 0 & 1 & 0 & 0 & 0 & 1 & 0
\end{array}\right].$$

$$[\mathbf{1}_N][\qquad \mathbf{A} \qquad][\qquad \mathbf{B} \qquad][\qquad \mathbf{C} \qquad]$$

For solutions to the normal equations and the analysis of variance computations we require **A′B**, **B′C**, **C′A**, their transposes, and **A′A**, **B′B** and **C′C** matrices. From the design matrix **X**, we see that

$$\mathbf{A'B}=\begin{bmatrix}0&1&0&3&1&1\\2&0&1&0&0&0\\0&2&2&0&0&0\\0&0&0&2&2&1\\1&0&1&0&2&0\\3&0&0&2&0&1\end{bmatrix},\quad \mathbf{A'A}=\begin{bmatrix}6&0&0&0&0&0\\0&3&0&0&0&0\\0&0&4&0&0&0\\0&0&0&5&0&0\\0&0&0&0&4&0\\0&0&0&0&0&6\end{bmatrix},$$

$$\mathbf{B'C}=\begin{bmatrix}2&0&3&0&1\\0&2&1&0&0\\1&1&0&2&0\\3&2&0&0&2\\2&1&2&0&0\\1&0&0&1&1\end{bmatrix},\quad \mathbf{B'B}=\begin{bmatrix}6&0&0&0&0&0\\0&3&0&0&0&0\\0&0&4&0&0&0\\0&0&0&7&0&0\\0&0&0&0&5&0\\0&0&0&0&0&3\end{bmatrix},$$

$$\mathbf{C'A}=\begin{bmatrix}4&2&0&2&1&0\\1&1&2&0&0&2\\1&0&0&0&2&3\\0&0&2&0&0&1\\0&0&0&3&1&0\end{bmatrix},\quad \mathbf{C'C}=\begin{bmatrix}9&0&0&0&0\\0&6&0&0&0\\0&0&6&0&0\\0&0&0&3&0\\0&0&0&0&4\end{bmatrix}.$$

However, we can obtain these matrices directly from the incidence matrix **N** with no effort.

Our aim is to find each element of the following regression sum of squares and their associated degrees of freedom.

$$\begin{aligned}\mathrm{SS}(\theta)&=\mathrm{SS}(\mu,\alpha)+\mathrm{SS}(\beta\,|\,\mu,\alpha)+\mathrm{SS}(\gamma\,|\,\mu,\alpha,\beta)\\&=\mathrm{SS}(\mu,\beta)+\mathrm{SS}(\gamma\,|\,\mu,\beta)+\mathrm{SS}(\alpha\,|\,\mu,\beta,\gamma)\\&=\mathrm{SS}(\mu,\gamma)+\mathrm{SS}(\alpha\,|\,\mu,\gamma)+\mathrm{SS}(\beta\,|\,\mu,\alpha,\gamma).\end{aligned}\tag{6.19}$$

This way of partitioning sums of squares is same as that of Table 6.1a. Since it is customary to give 1 degree of freedom to overall mean, we may present the table of analysis of variance similar to that of Table 6.1c, as in Table 6.2. Observe that,

$$\mathrm{SS}(\mu,\alpha)=\mathrm{SS}(\alpha),$$

and that

$$\mathrm{SS}(\alpha\,|\,\mu)=\mathrm{SS}(\mu,\alpha)-\mathrm{SS}(\mu).$$

Table 6.2 General Analysis of Variance Table for an Additive Three-Way Classification Model

Source			
Regression	r		$SS(\theta)$
Fitting overall mean		1	$SS(\mu)$
Factor A adjusted for mean		$a-1$	$SS(\alpha\|\mu)$
Factor B adjusted for mean and A		$\bar{f}_\beta$	$SS(\beta\|\mu, \alpha)$
Factor C adjusted for mean, A, and B		f_γ	$SS(\gamma\|\mu, \alpha, \beta)$
Fitting overall mean		1	$SS(\mu)$
Factor B adjusted for mean		$b-1$	$SS(\beta\|\mu)$
Factor C adjusted for mean and B		$\bar{f}_\gamma$	$SS(\gamma\|\mu, \beta)$
Factor A adjusted for mean, B, and C		f_α	$SS(\alpha\|\mu, \beta, \gamma)$
Fitting overall mean		1	$SS(\mu)$
Factor C adjusted for mean		$c-1$	$SS(\gamma\|\mu)$
Factor A adjusted for mean and C		$\bar{f}_\alpha$	$SS(\alpha\|\mu, \gamma)$
Factor B adjusted for mean, A, and C		f_β	$SS(\beta\|\mu, \alpha, \gamma)$
Residual	$N-r$		$\hat{R}$
Total	N		$\mathbf{Y'Y}$

Now consider the full rank reparametrization (6.12) as given in Lemma 6.3.

$$E(\mathbf{Y})=\mathbf{G}_1\boldsymbol{\xi}_1+\mathbf{G}_2\boldsymbol{\xi}_2+\mathbf{G}_3\boldsymbol{\xi}_3=\mathbf{G}\boldsymbol{\xi},$$

and let $\mathbf{G}_1=\mathbf{A}$, $\mathbf{G}_2=(\mathbf{I}-\mathbf{P}_1)\mathbf{B}\boldsymbol{\Gamma}$, and $\mathbf{G}_3=(\mathbf{I}-\mathbf{P}_{12})\mathbf{C}\boldsymbol{\Lambda}$, where $\mathbf{P}_1=\mathbf{A}(\mathbf{A}'\mathbf{A})^{-1}\mathbf{A}'$, $\mathbf{P}_{12}=\mathbf{P}_1+\mathbf{P}_2$, and $\mathbf{P}_2=\mathbf{G}_2(\mathbf{G}_2'\mathbf{G}_2)^{-1}\mathbf{G}_2'$. This reparametrization leads naturally to the first partition form of $SS(\theta)$ as given in (6.19).

To begin with let us present the n_{ijkl}'s, Y_{ijkl}'s, and the γ levels by the following 6×6 matrix, shown in Table 6.3.

From the above table we can obtain:

1. $SS(\mu, \alpha)$ and its associated degrees of freedom which is a.
2. $SS(\beta\,|\,\mu, \alpha)$ and its associated degrees $\bar{f}_\beta$.
3. The degrees of freedom for the γ effect, f_γ.

To obtain $SS(\mu, \alpha)$ we consider the first set of normal equations, that is

$$(\mathbf{G}_1'\mathbf{G}_1)\hat{\boldsymbol{\xi}}_1=\mathbf{G}_1'\mathbf{Y}.$$

Here $\mathbf{G}_1=\mathbf{A}$ and therefore we have

$$(\mathbf{A}'\mathbf{A})\hat{\boldsymbol{\xi}}_1=\mathbf{A}'\mathbf{Y}.$$

The ith element of the diagonal matrix $\mathbf{A}'\mathbf{A}$ is $n_{i...}$ and these elements are

Table 6.3 The **A′B** *Matrix Superimposed over the* $\bar{\mathbf{M}}_\gamma$ *(with* γ *Levels Being Present),* n_{ijkl}*'s, and the Observations,* Y_{ijkl}*'s.*

	β_1	β_2	β_3	β_4	β_5	β_6	Totals
α_1		γ_3 [a]$l=1$ [b]$Y_{1231}=3$		γ_1 3 3, 1, 2	γ_2 1 6	γ_1 1 2	$n_{1\ldots}=6$ $Y_{1\ldots}=17$
α_2	γ_1 2 3, 2		γ_2 1 7				3 12
α_3		γ_2 2 5, 7	γ_4 2 8, 9				4 29
α_4				γ_5 2 4, 5	γ_1 2 3, 4	γ_5 1 9	5 25
α_5	γ_5 1 11		γ_1 1 1		γ_3 2 5, 5		4 22
α_6	γ_3 3 5, 6, 7			γ_2 2 4, 3		γ_4 1 9	6 34
Totals	$n_{.1..}=6$ $Y_{.1..}=34$	3 15	4 25	7 22	5 23	3 20	$n_{....}=28$ $Y_{....}=139$

[a]Indicates the number of observations in each cell.
[b]Indicates the observation(s).

already available in row totals. The vector $\mathbf{A'Y}=(Y_{1\ldots}, Y_{2\ldots}, \ldots, Y_{6\ldots})'$ can also be found in the same column (row totals). Thus, we have

$$\begin{bmatrix} 6 & 0 & 0 & 0 & 0 & 0 \\ 0 & 3 & 0 & 0 & 0 & 0 \\ 0 & 0 & 4 & 0 & 0 & 0 \\ 0 & 0 & 0 & 5 & 0 & 0 \\ 0 & 0 & 0 & 0 & 4 & 0 \\ 0 & 0 & 0 & 0 & 0 & 6 \end{bmatrix} \hat{\boldsymbol{\xi}}_1 = \begin{bmatrix} 17 \\ 12 \\ 29 \\ 25 \\ 22 \\ 34 \end{bmatrix},$$

which leads to

$$\hat{\xi}_1 = (2.8333, 4.0000, 7.2500, 5.0000, 5.5000, 5.6667)'.$$

Then

$$\mathrm{SS}(\mu, \alpha) = \hat{\xi}_1' \mathbf{G}_1' \mathbf{Y} = 745.0834,$$

with $a = 6$ degrees of freedom, and

$$\mathrm{SS}(\mu) = N\bar{Y}^2_{\cdots\cdot} = \frac{(139)^2}{28} = 690.0357,$$

with 1 degree of freedom, which leads to

$$\mathrm{SS}(\alpha|\mu) = \mathrm{SS}(\alpha, \mu) - \mathrm{SS}(\mu) = 745.0834 - 690.0357 = 55.0477$$

with $a-1 = 6-1 = 5$ degrees of freedom.

To obtain $\mathrm{SS}(\beta|\mu, \alpha)$ we consider the second normal equations

$$(\mathbf{G}_2'\mathbf{G}_2)\hat{\xi}_2 = \mathbf{G}_2'\mathbf{Y},$$

where $\mathbf{G}_2 = (\mathbf{I} - \mathbf{P}_1)\mathbf{B}\boldsymbol{\Gamma}$, $\mathbf{P}_1 = \mathbf{G}_1(\mathbf{G}_1'\mathbf{G}_1)^{-1}\mathbf{G}_1'$, and $\mathbf{G}_1 = \mathbf{A}$. So we have

$$\boldsymbol{\Gamma}'[\mathbf{B}'\mathbf{B} - \mathbf{B}'\mathbf{A}(\mathbf{A}'\mathbf{A})^{-1}\mathbf{A}'\mathbf{B}]\boldsymbol{\Gamma}\hat{\xi}_2 = \boldsymbol{\Gamma}'[\mathbf{B}'\mathbf{Y} - \mathbf{B}'\mathbf{A}(\mathbf{A}'\mathbf{A})^{-1}\mathbf{A}'\mathbf{Y}],$$

or

$$\boldsymbol{\Gamma}'[\mathbf{B}'\mathbf{B} - \mathbf{B}'\mathbf{A}(\mathbf{A}'\mathbf{A})^{-1}\mathbf{A}'\mathbf{B}]\boldsymbol{\Gamma}\hat{\xi}_2 = \boldsymbol{\Gamma}'(\mathbf{B}'\mathbf{Y} - \mathbf{B}'\mathbf{A}\hat{\xi}_1).$$

The matrices $\mathbf{A}'\mathbf{B}$ (and its transpose $\mathbf{B}'\mathbf{A}$), $\mathbf{B}'\mathbf{B}$, and $\mathbf{B}'\mathbf{Y}$ can also be found in column totals. Here $\mathbf{B}'\mathbf{B} = \mathrm{diag}(n_{.1..}, n_{.2..}, \ldots, n_{.6..}) = \mathrm{diag}(6, 3, 4, 7, 5, 6)$ and the vector $\mathbf{B}'\mathbf{Y} = (Y_{.1..}, Y_{.2..}, \ldots, Y_{.6..})' = (34, 15, 25, 22, 23, 20)'$ are both provided in the column totals.

It remains to find $\bar{f}_\beta$ and $\boldsymbol{\Gamma}'$. This is easily accomplished by supposing γ effects are not present in the model and just considering a two-way model

$$E(Y_{ijh}) = \mu + \alpha_i + \beta_j \tag{6.20}$$

where $i, j = 1, \ldots, 6$ and $h = 1, \ldots, n_{ij}$.

The incidence matrix corresponding to the two-way model with just α and β effect is

$$\mathbf{A}'\mathbf{B} = (n_{ij.}) = \begin{bmatrix} 0 & 1 & 0 & 3 & 1 & 1 \\ 2 & 0 & 1 & 0 & 0 & 0 \\ 0 & 2 & 2 & 0 & 0 & 0 \\ 0 & 0 & 0 & 2 & 2 & 1 \\ 1 & 0 & 1 & 0 & 2 & 0 \\ 3 & 0 & 0 & 2 & 0 & 1 \end{bmatrix}.$$

For the estimability considerations relative to this two-way model we need only suppose the nonzero elements of $\mathbf{A}'\mathbf{B}$ are ones, and this would be exactly the same matrix obtained by changing all nonzero entries in $\mathbf{A}'\mathbf{B}$ to ones. Let $\mathbf{N}_{\alpha\beta}$ denote such a matrix. So we have:

$$\mathbf{N}_{\alpha\beta}=\begin{bmatrix}0&1&0&1&1&1\\1&0&1&0&0&0\\0&1&1&0&0&0\\0&0&0&1&1&1\\1&0&1&0&1&0\\1&0&0&1&0&1\end{bmatrix}.$$

The R process applied to $\mathbf{N}_{\alpha\beta}$ results in a final matrix $\bar{\mathbf{M}}_{\alpha\beta}$ consisting of all ones. Doing this and applying the methods of Chapter 5 we obtain the counter triangular (subdiagonal portion of $\mathbf{M}'_{\alpha\beta}\mathbf{M}_{\alpha\beta}$) for the β differences to be

$$\mathbf{C}_\beta=\begin{array}{c|ccccc} & 1 & 2 & 3 & 4 & 5\\ \hline 2 & 1 \\ 3 & 1 & 1 \\ 4 & 1 & 1 & 1 \\ 5 & 1 & 1 & 1 & 1 \\ 6 & 1 & 1 & 1 & 1 & 1\end{array}$$

From $\mathbf{C}_\beta$ we see that all β differences are estimable, and hence $\bar{f}_\beta=5$, and one may choose

$$\boldsymbol{\Gamma}'=\begin{bmatrix}1&-1&0&0&0&0\\0&1&-1&0&0&0\\0&0&1&-1&0&0\\0&0&0&1&-1&0\\0&0&0&0&1&-1\end{bmatrix}.$$

Hence the two-way model matrix $(\mathbf{1}, \mathbf{A}, \mathbf{B})$ is of maximal rank, $r(\mathbf{A}'\ \mathbf{B})=a+\bar{f}_\beta=6+5=11$, and we have:

$$\begin{bmatrix}1&-1&0&0&0&0\\0&1&-1&0&0&0\\0&0&1&-1&0&0\\0&0&0&1&-1&0\\0&0&0&0&1&-1\end{bmatrix}\left(\begin{bmatrix}6&0&0&0&0&0\\0&3&0&0&0&0\\0&0&4&0&0&0\\0&0&0&7&0&0\\0&0&0&0&5&0\\0&0&0&0&0&3\end{bmatrix}\right.$$

$$-\begin{bmatrix} 0 & 2 & 0 & 0 & 1 & 3 \\ 1 & 0 & 2 & 0 & 0 & 0 \\ 0 & 1 & 2 & 0 & 1 & 0 \\ 3 & 0 & 0 & 2 & 0 & 2 \\ 1 & 0 & 0 & 2 & 2 & 0 \\ 1 & 0 & 0 & 1 & 0 & 1 \end{bmatrix} \cdot \begin{bmatrix} 6 & 0 & 0 & 0 & 0 & 0 \\ 0 & 3 & 0 & 0 & 0 & 0 \\ 0 & 0 & 4 & 0 & 0 & 0 \\ 0 & 0 & 0 & 5 & 0 & 0 \\ 0 & 0 & 0 & 0 & 4 & 0 \\ 0 & 0 & 0 & 0 & 0 & 6 \end{bmatrix}^{-1} \cdot \begin{bmatrix} 0 & 1 & 0 & 3 & 1 & 1 \\ 2 & 0 & 1 & 0 & 0 & 0 \\ 0 & 2 & 2 & 0 & 0 & 0 \\ 0 & 0 & 0 & 2 & 2 & 1 \\ 1 & 0 & 1 & 0 & 2 & 0 \\ 3 & 0 & 0 & 2 & 0 & 1 \end{bmatrix} \Bigg)$$

$$\times \begin{bmatrix} 1 & 0 & 0 & 0 & 0 \\ -1 & 1 & 0 & 0 & 0 \\ 0 & -1 & 1 & 0 & 0 \\ 0 & 0 & -1 & 1 & 0 \\ 0 & 0 & 0 & -1 & 1 \\ 0 & 0 & 0 & 0 & -1 \end{bmatrix} \hat{\xi}_2 = \begin{bmatrix} 1 & -1 & 0 & 0 & 0 & 0 \\ 0 & 1 & -1 & 0 & 0 & 0 \\ 0 & 0 & 1 & -1 & 0 & 0 \\ 0 & 0 & 0 & 1 & -1 & 0 \\ 0 & 0 & 0 & 0 & 1 & -1 \end{bmatrix}$$

$$\times \left(\begin{bmatrix} 34 \\ 15 \\ 25 \\ 22 \\ 23 \\ 20 \end{bmatrix} - \begin{bmatrix} 0 & 2 & 0 & 0 & 1 & 3 \\ 1 & 0 & 2 & 0 & 0 & 0 \\ 0 & 1 & 2 & 0 & 1 & 0 \\ 3 & 0 & 0 & 2 & 0 & 2 \\ 1 & 0 & 0 & 2 & 2 & 0 \\ 1 & 0 & 0 & 1 & 0 & 1 \end{bmatrix} \begin{bmatrix} 2.8333 \\ 4.0000 \\ 7.2500 \\ 5.0000 \\ 5.5000 \\ 5.6667 \end{bmatrix} \right).$$

So we have

$$\begin{bmatrix} 4.7500 & -1.9167 & 0.5833 & -0.1667 & 0.0000 \\ -1.9167 & 6.2500 & -2.9167 & -0.8333 & 0.5000 \\ 0.5833 & -2.9167 & 6.4500 & -4.8333 & -0.4300 \\ -0.1667 & -0.8333 & -4.8333 & 9.6667 & -3.6667 \\ 0.0000 & 0.5000 & -0.4333 & -3.6667 & 6.6333 \end{bmatrix} \hat{\xi}_2 = \begin{bmatrix} 5.8332 \\ -3.3333 \\ 8.8333 \\ -7.0000 \\ -7.3333 \end{bmatrix}$$

which gives

$$\hat{\xi}_2 = \begin{bmatrix} 1.0491 \\ -0.3734 \\ -0.2281 \\ -1.6026 \\ -1.9781 \end{bmatrix}$$

and

$$SS(\beta \mid \mu, \alpha) = \hat{\xi}_2' \mathbf{G}_2' \mathbf{Y} = 31.0745.$$

To obtain $f_\gamma = \dim \bar{\mathscr{G}}$ we use the methods developed in Section 6.4. Since there are no direct γ differences $\bar{\mathscr{D}}_\gamma = \{0\}$ and hence $\bar{\mathscr{G}} = \bar{\mathscr{S}}_\gamma$. Hence **M** is obtained from $\mathbf{N}$ by changing each nonzero element to a 1. The matrix $\bar{\mathbf{M}}_\gamma = 1\mathbf{M}_1 + 2\mathbf{M}_2 + 3\mathbf{M}_3 + 4\mathbf{M}_4 + 5\mathbf{M}_5$ is also available in Table 6.3. Now

$$\bar{\mathbf{M}}_\gamma = \begin{bmatrix} 0 & 3 & 0 & 1 & 2 & 1 \\ 1 & 0 & 2 & 0 & 0 & 0 \\ 0 & 2 & 4 & 0 & 0 & 0 \\ 0 & 0 & 0 & 5 & 1 & 5 \\ 5 & 0 & 1 & 0 & 3 & 0 \\ 3 & 0 & 0 & 2 & 0 & 4 \end{bmatrix}.$$

By applying the Q process to this two-dimensional matrix, we see that the a spanning set for the vector space $\bar{\mathscr{G}}$ of the estimable γ contrasts is $\{\gamma_2 - \gamma_4,\ -2\gamma_1 + \gamma_2 + \gamma_5,\ 0,\ -2\gamma_1 + \gamma_2 + \gamma_5,\ -\gamma_1 + 2\gamma_2 - 2\gamma_3 + \gamma_5\}$. A basis for $\bar{\mathscr{G}}$ is therefore $\{\gamma_2 - \gamma_4,\ -2\gamma_1 + \gamma_2 + \gamma_5,\ -\gamma_1 + 2\gamma_2 - 2\gamma_3 + \gamma_5\}$ and hence $f_\gamma = 3$ and for the analysis one would use

$$\Lambda' = \begin{bmatrix} 0 & 1 & 0 & -1 & 0 \\ -2 & 1 & 0 & 0 & 1 \\ -1 & 2 & -2 & 0 & 1 \end{bmatrix}.$$

Thus,

$$r(\mathbf{X}) = r(\mathbf{A}, \mathbf{B}) + f_\gamma = 11 + 3 = 14,$$

and hence $\mathbf{X}$ is *not* of maximal rank, that is, not all cell expectations are estimable.

Observe that from Table 6.3 we would also obtain $SS(\mu, \beta)$, $SS(\alpha \mid \mu, \beta)$, and their associated degrees of freedoms. However, we would prefer to continue the analysis step by step (providing two additional tables) in order to reduce all ambiguity!

For obtaining a basis for $\bar{\mathscr{A}}$ we need to construct $\bar{\mathbf{M}}_\alpha$ from the α levels of $\mathbf{N}$. So we construct a new table, Table 6.4, in which the rows correspond to β and the columns correspond to γ along n_{ijkl}'s and Y_{ijkl}'s (similar to Table 6.3).

From the above table we can find:

1 $SS(\mu, \beta)$ and its associated degrees of freedom, which is b.
2 $SS(\gamma \mid \mu, \beta)$ and its associated degrees of freedom $\bar{f}_\gamma$.
3 The degrees of freedom associated with the α effects, f_α.

To obtain $SS(\mu, \beta)$ we consider the first set of normal equations,

$$(\mathbf{G}_1'\mathbf{G}_1)\hat{\boldsymbol{\xi}}_1 = \mathbf{G}_1'\mathbf{Y}.$$

Table 6.4 The $\mathbf{B'C}$ *Matrix Superimposed over* $\bar{\mathbf{M}}_\alpha$ *(with* α *Levels Being Present),* n_{ijkl}*'s, and the Observations,* Y_{ijkl}*'s.*

	γ_1	γ_2	γ_3	γ_4	γ_5	Totals
β_1	α_2 $l=2$ $Y_{1121}=3$ $Y_{1122}=2$		α_6 3 5, 6, 7		α_5 1 11	$n_{.1..}=6$ $Y_{.1..}=34$
β_2		α_3 2 5, 7	α_1 1 3			3 15
β_3	α_5 1 1	α_2 1 7		α_3 2 8, 9		4 25
β_4	α_1 3 3, 1, 2	α_6 2 4, 3			α_4 2 4, 5	7 22
β_5	α_4 2 3, 4	α_1 1 6	α_5 2 5, 5			5 23
β_6	α_1 1 2			α_6 1 9	α_4 1 9	3 20
Totals	$n_{..1.}=9$ $Y_{..1.}=21$	6 32	6 31	3 26	4 29	$n_{....}=28$ $Y_{....}=139$

Here $\mathbf{G}_1=\mathbf{B}$ and so we have

$$(\mathbf{B}'\mathbf{B})\hat{\xi}_1=\mathbf{B}'\mathbf{Y}.$$

From the row totals we obtain

$$\begin{bmatrix} 6 & 0 & 0 & 0 & 0 & 0 \\ 0 & 3 & 0 & 0 & 0 & 0 \\ 0 & 0 & 4 & 0 & 0 & 0 \\ 0 & 0 & 0 & 7 & 0 & 0 \\ 0 & 0 & 0 & 0 & 5 & 0 \\ 0 & 0 & 0 & 0 & 0 & 3 \end{bmatrix} \hat{\boldsymbol{\xi}}_1 = \begin{bmatrix} 34 \\ 15 \\ 25 \\ 22 \\ 23 \\ 20 \end{bmatrix}.$$

We find

$$\hat{\boldsymbol{\xi}}_1 = (5.6667, 5.0000, 6.2500, 3.1429, 4.6000, 6.6667)'$$

and

$$SS(\mu, \beta) = \hat{\boldsymbol{\xi}}_1' \mathbf{G}_1' \mathbf{Y} = 732.1929$$

with $b = 6$ degrees of freedom, and that

$$\begin{aligned} SS(\beta \mid \mu) &= SS(\mu, \beta) - SS(\mu) \\ &= 732.1929 - 690.0357 = 42.1572 \end{aligned}$$

with $b - 1 = 6 - 1 = 5$ degrees of freedom.

To obtain $SS(\gamma \mid \mu, \beta)$ we consider the second normal equations,

$$(\mathbf{G}_2'\mathbf{G}_2)\hat{\boldsymbol{\xi}}_2 = \mathbf{G}_2'\mathbf{Y},$$

where

$$\mathbf{G}_2 = (\mathbf{I} - \mathbf{P}_1)\mathbf{C}\boldsymbol{\Gamma}, \qquad \mathbf{P}_1 = \mathbf{G}_1(\mathbf{G}_1'\mathbf{G}_1)^{-1}\mathbf{G}_1', \qquad \text{and} \qquad \mathbf{G}_1 = \mathbf{B}.$$

So we have

$$\boldsymbol{\Gamma}'[\mathbf{C}'\mathbf{C} - \mathbf{C}'\mathbf{B}(\mathbf{B}'\mathbf{B})^{-1}\mathbf{B}'\mathbf{C}]\boldsymbol{\Gamma}\hat{\boldsymbol{\xi}}_2 = \boldsymbol{\Gamma}'[\mathbf{C}'\mathbf{Y} - \mathbf{C}'\mathbf{B}(\mathbf{B}'\mathbf{B})^{-1}\mathbf{B}'\mathbf{Y}],$$

or

$$\boldsymbol{\Gamma}'[\mathbf{C}'\mathbf{C} - \mathbf{C}'\mathbf{B}(\mathbf{B}'\mathbf{B})^{-1}\mathbf{B}'\mathbf{C}]\boldsymbol{\Gamma}\hat{\boldsymbol{\xi}}_2 = \boldsymbol{\Gamma}'(\mathbf{C}'\mathbf{Y} - \mathbf{C}'\mathbf{B}\hat{\boldsymbol{\xi}}_1).$$

The matrices $\mathbf{C}'\mathbf{C}$, $\mathbf{C}'\mathbf{B}$, $\mathbf{B}'\mathbf{B}$, and $\mathbf{C}'\mathbf{Y}$ are all available from Table 6.4. We only need to find $\boldsymbol{\Gamma}'$. This is easily accomplished by supposing α effects are not present in the model and just considering a two-way model

$$E(Y_{jkh}) = \mu + \beta_j + \gamma_k \tag{6.21}$$

where $j = 1, \ldots, 6$, $k = 1, \ldots, 5$, and $h = 1, \ldots, n_{ij}$.

The incidence matrix corresponding to the two-way model with just the β and γ effects is

$$\mathbf{B}'\mathbf{C} = (n_{.ij}) = \begin{bmatrix} 2 & 0 & 3 & 0 & 1 \\ 0 & 2 & 1 & 0 & 0 \\ 1 & 1 & 0 & 2 & 0 \\ 3 & 2 & 0 & 0 & 2 \\ 2 & 1 & 2 & 0 & 0 \\ 1 & 0 & 0 & 1 & 1 \end{bmatrix}.$$

For estimability considerations relative to this model we may suppose the incidence matrix $\mathbf{B}'\mathbf{C}$ with all nonzero entries changed to ones. Let $\mathbf{N}_{\beta\gamma}$ be such a matrix. Then we have

$$\mathbf{N}_{\beta\gamma}=\begin{bmatrix}1&0&1&0&1\\0&1&1&0&0\\1&1&0&1&0\\1&1&0&0&1\\1&1&1&0&0\\1&0&0&1&1\end{bmatrix}.$$

By applying the R process to $\mathbf{N}_{\beta\gamma}$ results in a final matrix $\bar{\mathbf{M}}_{\beta\gamma}$ consisting of all ones. Doing this and using the methods developed in Chapter 5, we obtain $\gamma_1-\gamma_2, \gamma_1-\gamma_3, \gamma_1-\gamma_4$, and $\gamma_1-\gamma_5$ as a basis for the γ contrasts relative to the model in (6.21). Thus, we may choose

$$\boldsymbol{\Gamma}'=\begin{bmatrix}1&-1&0&0&0\\1&0&-1&0&0\\1&0&0&-1&0\\1&0&0&0&-1\end{bmatrix},$$

and $f_\gamma=5$. The vector $\mathbf{C}'\mathbf{Y}=(Y_{..1.}, Y_{..2.},\ldots, Y_{..5.})'=(21, 32, 31, 26, 29)'$ is also available in Table 6.4.

By substituting these matrices into the second normal equations, we obtain

$$\begin{bmatrix}1&-1&0&0&0\\1&0&-1&0&0\\1&0&0&-1&0\\1&0&0&0&-1\end{bmatrix}\left(\begin{bmatrix}9&0&0&0&0\\0&6&0&0&0\\0&0&6&0&0\\0&0&0&3&0\\0&0&0&0&4\end{bmatrix}-\begin{bmatrix}2&0&1&3&2&1\\0&2&1&2&1&0\\3&1&0&0&2&0\\0&0&2&0&0&1\\1&0&0&2&0&1\end{bmatrix}\right.$$

$$\left.\times\begin{bmatrix}6&0&0&0&0&0\\0&3&0&0&0&0\\0&0&4&0&0&0\\0&0&0&7&0&0\\0&0&0&0&5&0\\0&0&0&0&0&3\end{bmatrix}^{-1}\cdot\begin{bmatrix}2&0&3&0&1\\0&2&1&0&0\\1&1&0&2&0\\3&2&0&0&2\\2&1&2&0&0\\1&0&0&1&1\end{bmatrix}\right)\begin{bmatrix}1&1&1&1\\-1&0&0&0\\0&-1&0&0\\0&0&-1&0\\0&0&0&-1\end{bmatrix}\times$$

$$\hat{\boldsymbol{\xi}}_2=\begin{bmatrix}1&-1&0&0&0\\1&0&-1&0&0\\1&0&0&-1&0\\1&0&0&0&-1\end{bmatrix}\left(\begin{bmatrix}21\\32\\31\\26\\29\end{bmatrix}-\begin{bmatrix}2&0&1&3&2&1\\0&2&1&2&1&0\\3&1&0&0&2&0\\0&0&2&0&0&1\\1&0&0&2&0&1\end{bmatrix}\begin{bmatrix}5.6667\\5.0000\\6.2500\\3.1429\\4.6000\\6.6667\end{bmatrix}\right),$$

so

$$\begin{bmatrix}12.3238&7.9048&7.5048&8.1238\\7.9048&12.6310&8.2976&8.4881\\7.5048&8.2976&8.9976&7.6881\\8.1238&8.4881&7.6881&11.6405\end{bmatrix}\hat{\boldsymbol{\xi}}_2=\begin{bmatrix}-26.75\\-21.67\\-28.71\\-32.26\end{bmatrix}$$

which gives

$$\hat{\boldsymbol{\xi}}_2=\begin{bmatrix}-0.1768\\1.5283\\-2.8421\\-1.8852\end{bmatrix}$$

and

$$\text{SS}(\gamma|\mu,\beta)=\hat{\boldsymbol{\xi}}_2\mathbf{G}_2'\mathbf{Y}=114.0166.$$

To obtain $f_\alpha=\dim\bar{\mathscr{A}}$ we use the methods developed in Section 6.4. Since there are no direct α differences $\bar{\mathscr{A}}=\mathscr{S}_\alpha$. The matrix $\bar{\mathbf{M}}_\alpha$ is already available in Table 6.4.

$$\bar{\mathbf{M}}_\alpha=\begin{bmatrix}2&0&6&0&5\\0&3&1&0&0\\5&2&0&3&0\\1&6&0&0&4\\4&1&5&0&0\\1&0&0&6&4\end{bmatrix}.$$

By applying the Q process to this two-dimensional matrix, we obtain a basis for the estimable α contrasts as $\{\alpha_2-\alpha_3,\ \alpha_1+\alpha_2-\alpha_5-\alpha_6,\ \alpha_1-\alpha_2-\alpha_4+\alpha_5,\ \alpha_2-\alpha_4+\alpha_5-\alpha_6\}$, and $f_\alpha=4$.

To obtain a basis for $\bar{\mathscr{B}}$ we need to construct $\bar{\mathbf{M}}_\beta$ from the β levels of $\mathbf{N}$. We form a new table, Table 6.5, in which the rows correspond to α and the columns correspond to γ. In this way we obtain Table 6.5 below:

Table 6.5 The C′A Matrix Superimposed over $\bar{\mathbf{M}}_\beta$ (with β Levels Being present), n_{ijkl}'s, and the Observations, Y_{ijkl}'s.

	α_1	α_2	α_3	α_4	α_5	α_6	Totals
γ_1	β_4: 3; 3, 1, 2 / β_6: 1; 2	β_1: 2; 3, 2		β_5: 2; 3, 4	β_3: 1; 1		$n_{..1.}=9$, $Y_{..1.}=21$
γ_2	β_5: 1; 6	β_3: 1; 7	β_2: 2; 5, 7			β_4: 2; 4, 3	6, 32
γ_3	β_2: 1; 3				β_5: 2; 5, 5	β_1: 3; 5, 6, 7	6, 31
γ_4			β_3: 2; 8, 9			β_6: 1; 9	3, 26
γ_5				β_4: 2; 4, 5 / β_6: 1; 9	β_1: 1; 11		4, 29
Totals	$n_{1...}=6$, $Y_{1...}=17$	3, 12	4, 29	5, 25	4, 22	6, 34	$n_{....}=28$, $Y_{....}=139$

From Table 6.5 we can find:

1 SS($\mu|\gamma$) and its associated degrees of freedom, which is c.
2 SS($\alpha|\mu, \gamma$) and its associated degrees of freedom, $\bar{f}_\alpha$.
3 The degrees of freedom for the β effect, f_β.

To obtain SS(μ, γ) we consider again the first set of normal equations,

$$(\mathbf{G}_1'\mathbf{G}_1)\hat{\boldsymbol{\xi}}_1 = \mathbf{G}_1'\mathbf{Y},$$

with $\mathbf{G}_1 = \mathbf{C}$. So we obtain

$$(\mathbf{C}'\mathbf{C})\hat{\boldsymbol{\xi}}_1 = \mathbf{C}'\mathbf{Y}.$$

From the row totals of Table 6.5 we obtain

$$\begin{bmatrix} 9 & 0 & 0 & 0 & 0 \\ 0 & 6 & 0 & 0 & 0 \\ 0 & 0 & 6 & 0 & 0 \\ 0 & 0 & 0 & 3 & 0 \\ 0 & 0 & 0 & 0 & 4 \end{bmatrix} \hat{\boldsymbol{\xi}}_1 = \begin{bmatrix} 21 \\ 32 \\ 31 \\ 26 \\ 29 \end{bmatrix},$$

and thus

$$\hat{\boldsymbol{\xi}}_1' = (2.3333, 5.3333, 5.1667, 8.6667, 7.2500)'$$

and

$$SS(\mu, \gamma) = \hat{\boldsymbol{\xi}}_1' \mathbf{G}_1' \mathbf{Y} = 815.4167$$

with 5 degrees of freedom, and that

$$SS(\gamma \mid \mu) = SS(\mu, \gamma) - SS(\mu) = 815.4167 - 690.0357 = 125.3810$$

with $5 - 1 = 4$ degrees of freedom.

To obtain $SS(\alpha \mid \mu, \gamma)$ we use the second normal equations

$$(\mathbf{G}_2'\mathbf{G}_2)\hat{\boldsymbol{\xi}}_2 = \mathbf{G}_2'\mathbf{Y}$$

where $\mathbf{G}_2 = (\mathbf{I} - \mathbf{P}_1)\mathbf{A}\boldsymbol{\Gamma}$, $\mathbf{P}_1 = \mathbf{G}_1(\mathbf{G}_1'\mathbf{G}_1)^{-1}\mathbf{G}_1'$, and $\mathbf{G}_1 = \mathbf{C}$. So we have

$$\boldsymbol{\Gamma}'[\mathbf{A}'\mathbf{A} - \mathbf{A}'\mathbf{C}(\mathbf{C}'\mathbf{C})^{-1}\mathbf{C}'\mathbf{A}]\boldsymbol{\Gamma}\hat{\boldsymbol{\xi}}_2 = \boldsymbol{\Gamma}'[\mathbf{A}'\mathbf{Y} - \mathbf{A}'\mathbf{C}(\mathbf{C}'\mathbf{C})^{-1}\mathbf{C}'\mathbf{Y}]$$

or

$$\boldsymbol{\Gamma}'[\mathbf{A}'\mathbf{A} - \mathbf{A}'\mathbf{C}(\mathbf{C}'\mathbf{C})^{-1}\mathbf{C}'\mathbf{A}]\boldsymbol{\Gamma}\hat{\boldsymbol{\xi}}_2 = \boldsymbol{\Gamma}'(\mathbf{A}'\mathbf{Y} - \mathbf{A}'\mathbf{C}\hat{\boldsymbol{\xi}}_1).$$

The matrices $\mathbf{A}'\mathbf{A}$, $\mathbf{A}'\mathbf{C}$, $\mathbf{C}'\mathbf{C}$, and $\mathbf{A}'\mathbf{Y}$ are available in Table 6.5. To obtain $\boldsymbol{\Gamma}'$ we suppose β effects are not present in the model. So we have

$$E(Y_{kih}) = \mu + \gamma_k + \alpha_i \tag{6.22}$$

where $k = 1, \ldots, 5$, $i = 1, \ldots, 6$, and $h = 1, \ldots, n_{ij}$. The incidence matrix corresponding to the two-way model (6.22) is

$$\mathbf{C}'\mathbf{A} = (n_{i.k}) = \begin{bmatrix} 4 & 2 & 0 & 2 & 1 & 0 \\ 1 & 1 & 2 & 0 & 0 & 2 \\ 1 & 0 & 0 & 0 & 2 & 3 \\ 0 & 0 & 2 & 0 & 0 & 1 \\ 0 & 0 & 0 & 3 & 1 & 0 \end{bmatrix}.$$

Again for estimability considerations relative to this model we may suppose that the incidence matrix is $\mathbf{C}'\mathbf{A}$ with all nonzero entries changed to ones. Let $\mathbf{N}_{\gamma\alpha}$ denote such a matrix. Then we have

$$\mathbf{N}_{\gamma\alpha}=\begin{bmatrix}1&1&0&1&1&0\\1&1&1&0&0&1\\1&0&0&0&1&1\\0&0&1&0&0&1\\0&0&0&1&0&1\end{bmatrix}.$$

Applying the R process to this two-dimensional matrix $\mathbf{N}_{\gamma\alpha}$ results in a final matrix $\bar{\mathbf{M}}_{\gamma\alpha}$ consisting of all ones. So all γ and α differences are estimable and we may choose $\mathbf{\Gamma}'$ to be

$$\mathbf{\Gamma}'=\begin{bmatrix}1&-1&0&0&0&0\\1&0&-1&0&0&0\\1&0&0&-1&0&0\\1&0&0&0&-1&0\\1&0&0&0&0&-1\end{bmatrix},$$

and $\bar{f}_\alpha=5$. The vector $\mathbf{A}'\mathbf{Y}=(17, 12, 29, 25, 22, 34)'$ is also available in Table 6.5.

By substituting these matrices into the second normal equations we obtain

$$\begin{bmatrix}8.3889&4.9444&5.3889&5.5000&5.4444\\4.9444&6.5556&5.1111&5.0000&3.7222\\5.3889&5.1111&7.9722&4.5833&5.6111\\5.5000&5.0000&4.5833&8.4167&4.5000\\5.4444&3.7222&5.6111&4.5000&9.0556\end{bmatrix}\hat{\boldsymbol{\xi}}_2=\begin{bmatrix}-4.8333\\-3.8333\\-1.4167\\-4.9167\\-2.0000\end{bmatrix}$$

which leads to

$$\hat{\boldsymbol{\xi}}_2=(-0.5277,\ -0.4341,\ 0.6083,\ -0.3514,\ 0.0726)'$$

To obtain $f_\beta=\dim\bar{\mathscr{B}}$ we use the methods developed in Section 6.4. We know that $\bar{\mathscr{B}}=\bar{\mathscr{D}}_\beta+\bar{\mathscr{S}}_\beta$. From Table 6.5 we see that in cells (γ_1, α_1) and (γ_5, α_4) we have both β_4 and β_6. This implies that $\beta_4-\beta_6$ is estimable. So $D_\beta=\{\beta_4-\beta_6\}$. For estimability considerations we need to keep either β_4 or β_6. We eliminate β_6 and keep β_4 in $\bar{\mathbf{M}}_\beta$. Doing this we obtain

$$\bar{\mathbf{M}}_\beta=\begin{bmatrix}4&1&0&5&3&0\\5&3&2&0&0&4\\2&0&0&0&5&1\\0&0&3&0&0&0\\0&0&0&4&1&0\end{bmatrix}.$$

We apply the Q process to $\bar{\mathbf{M}}_\beta$ and obtain a spanning set $\{-\beta_1+\beta_3+\beta_4-\beta_5, \beta_1-\beta_3-\beta_4+\beta_5, -\beta_1+\beta_2-\beta_4+\beta_5, -2\beta_3+\beta_4+\beta_5\}$ for $\bar{\mathscr{S}}_\beta$. A basis for $\mathscr{B}$, the space of estimable β contrasts, is $\{\beta_2-\beta_3, \beta_4-\beta_6, \beta_1-\beta_2-\beta_4+\beta_5, 2\beta_2-\beta_4-\beta_5\}$, and $f_\beta=4$.

It remains to calculate SS($\alpha|\mu, \alpha, \beta$), SS($\beta|\mu, \alpha, \gamma$), SS($\gamma|\mu, \alpha, \beta$), $\hat{R}$, and the total sums of squares to complete the analysis of variance table.

To obtain SS($\alpha|\mu, \alpha, \beta$), SS($\beta|\mu, \alpha, \gamma$), or SS($\gamma|\mu, \alpha, \beta$) we use the third normal equations

$$(\mathbf{G}_3'\mathbf{G}_3)\hat{\boldsymbol{\xi}}_3=\mathbf{G}_3'\mathbf{Y}.$$

Let us begin by finding SS($\gamma|\mu, \alpha, \beta$). In this case $\mathbf{G}_3=(\mathbf{I}-\mathbf{P}_{12})\mathbf{C}\boldsymbol{\Lambda}$, $\mathbf{P}_{12}=\mathbf{P}_1+\mathbf{P}_2$ where $\mathbf{P}_2=\mathbf{G}_2(\mathbf{G}_2'\mathbf{G}_2)^{-1}\mathbf{G}_2'$. Then

$$\mathbf{G}_3'\mathbf{G}_3=\boldsymbol{\Lambda}'[\mathbf{C}'\mathbf{C}-\mathbf{C}'\mathbf{A}(\mathbf{A}'\mathbf{A})^{-1}\mathbf{A}'\mathbf{C}]\boldsymbol{\Lambda}-\boldsymbol{\Lambda}'\mathbf{C}'\mathbf{G}_2(\mathbf{G}_2'\mathbf{G}_2)^{-1}\mathbf{G}_2'\mathbf{C}\boldsymbol{\Lambda},$$

$$\mathbf{G}_3'\mathbf{Y}=\boldsymbol{\Lambda}'(\mathbf{C}'\mathbf{Y}-\mathbf{C}'\mathbf{A}\hat{\boldsymbol{\xi}}_1-\mathbf{C}'\mathbf{G}_2\hat{\boldsymbol{\xi}}_2),$$

and

$$\mathbf{C}'\mathbf{G}_2=\mathbf{C}'\mathbf{B}\boldsymbol{\Gamma}-\mathbf{C}'\mathbf{A}(\mathbf{A}'\mathbf{A})^{-1}\mathbf{A}'\mathbf{B}\boldsymbol{\Gamma}.$$

From Table 6.3 we found

$$\boldsymbol{\Gamma}'=\begin{bmatrix}1&-1&0&0&0&0\\0&1&-1&0&0&0\\0&0&1&-1&0&0\\0&0&0&1&-1&0\\0&0&0&0&1&-1\end{bmatrix}\quad\text{and}\quad\boldsymbol{\Lambda}'=\begin{bmatrix}0&1&0&-1&0\\-2&1&0&0&1\\-1&2&-2&0&1\end{bmatrix}$$

and

$$\hat{\boldsymbol{\xi}}_1=(2.8333, 4.0000, 7.2500, 5.0000, 5.5000, 5.6667)',$$

$$\hat{\boldsymbol{\xi}}_2=(1.0491, -0.3734, -0.2281, -1.6026, -1.9781)'.$$

Other matrices such as $(\mathbf{G}_2'\mathbf{G}_2)^{-1}$, $\mathbf{C}'\mathbf{C}$, $\mathbf{C}'\mathbf{Y}$, $\mathbf{A}'\mathbf{B}$, $\mathbf{A}'\mathbf{C}$, and $\mathbf{B}'\mathbf{C}$, and $(\mathbf{A}'\mathbf{A})^{-1}$ are all known. By substituting these quantities into the third normal equation we obtain

$$\begin{bmatrix}4.7868&4.8446&9.6423\\4.8446&28.5500&21.7814\\9.6423&21.7814&35.3931\end{bmatrix}\hat{\boldsymbol{\xi}}_3=\begin{bmatrix}2.0801\\41.8062\\31.8666\end{bmatrix},$$

so

$$\hat{\boldsymbol{\xi}}_3=(-2.4044, 1.2925, 0.7599)'$$

then

$$SS(\gamma\mid\mu,\alpha,\beta)=\hat{\boldsymbol{\xi}}_3\mathbf{G}_3'\mathbf{Y}=73.2516,$$

with $f_\gamma=3$ degrees of freedom.

The regression sum of squares $SS(\theta)$ can now be obtained as

$$\begin{aligned}SS(\theta)&=SS(\mu,\alpha)+SS(\beta\mid\mu,\alpha)+SS(\gamma\mid\mu,\alpha,\beta)=745.0834+31.0745+73.2516\\&=849.4094\end{aligned}$$

with $r=14$ degrees of freedom.

Total sum of squares and the residual sum of squares are easily obtained by

$$\begin{aligned}SS_{\text{tot}}=\mathbf{Y}'\mathbf{Y}&=\sum_{i=1}^{6}\sum_{j=1}^{6}\sum_{k=1}^{5}\sum_{l=1}^{n_{ijk}}Y_{ijkl}^2\\&=3^2+3^2+1^2+\cdots+9^2\\&=879.0,\end{aligned}$$

$$\hat{R}=\mathbf{Y}'\mathbf{Y}-SS(\theta)=879.0-849.4094=29.5906.$$

In a similar way we can calculate $SS(\alpha\mid\mu,\beta,\gamma)$ by using again the third normal equation

$$(\mathbf{G}_3'\mathbf{G}_3)\hat{\boldsymbol{\xi}}_3=\mathbf{G}_3'\mathbf{Y},$$

where in this case

$$\begin{aligned}\mathbf{G}_3'\mathbf{G}_3&=\boldsymbol{\Lambda}'[\mathbf{A}'\mathbf{A}-\mathbf{A}'\mathbf{B}(\mathbf{B}'\mathbf{B})^{-1}\mathbf{B}'\mathbf{A}]\boldsymbol{\Lambda}-\boldsymbol{\Lambda}'\mathbf{A}'\mathbf{G}_2(\mathbf{G}_2'\mathbf{G}_2)^{-1}\mathbf{G}_2'\mathbf{A}\boldsymbol{\Lambda},\\\mathbf{G}_3'\mathbf{Y}&=\boldsymbol{\Lambda}'(\mathbf{A}'\mathbf{Y}-\mathbf{A}'\mathbf{B}\hat{\boldsymbol{\xi}}_1-\mathbf{A}'\mathbf{G}_2'\hat{\boldsymbol{\xi}}_2),\end{aligned}$$

and

$$\mathbf{A}'\mathbf{G}_2=\mathbf{A}'\mathbf{C}\boldsymbol{\Gamma}-\mathbf{A}'\mathbf{B}(\mathbf{B}'\mathbf{B})^{-1}\mathbf{B}'\mathbf{C}\boldsymbol{\Gamma}.$$

Note that the matrix **A** is changed to **B**, **B** changed to **C**, and **C** changed to **A** in the third normal equation. From Table 6.4 we obtained

$$\boldsymbol{\Gamma}'=\begin{bmatrix}1&-1&0&0&0&0\\1&0&-1&0&0&0\\1&0&0&-1&0&0\\1&0&0&0&-1&0\\1&0&0&0&0&-1\end{bmatrix},\qquad\boldsymbol{\Lambda}'=\begin{bmatrix}1&1&0&0&-1&-1\\0&1&-1&0&0&0\\1&-1&0&-1&1&0\\0&1&0&-1&1&-1\end{bmatrix},$$

and that

$$\hat{\boldsymbol{\xi}}_1=(5.6667,\ 5.0000,\ 6.2500,\ 3.1429,\ 4.6000,\ 6.6667)',$$

$$\hat{\boldsymbol{\xi}}_2=(-0.1768,\ 1.5283,\ -2.8422,\ -1.8852)',$$

so we have

$$\begin{bmatrix} 11.5266 & 2.6059 & -0.8355 & 2.4973 \\ 2.6059 & 3.1891 & -2.1918 & 1.9770 \\ -0.8335 & -2.1918 & 13.6255 & 6.6841 \\ 2.4973 & 1.9770 & 6.6841 & 11.3451 \end{bmatrix} \hat{\boldsymbol{\xi}}_3 = \begin{bmatrix} 0.1115 \\ -2.4591 \\ 5.0548 \\ 0.8139 \end{bmatrix}$$

which gives

$$\hat{\boldsymbol{\xi}}_3 = (0.1987,\ -0.7661,\ 0.2541,\ 0.0118)'.$$

So we obtain

$$\mathrm{SS}(\alpha \mid \mu, \beta, \gamma) = \hat{\boldsymbol{\xi}}_3' \mathbf{G}_3' \mathbf{Y} = 3.2003.$$

The regression sum of squares can also be calculated by

$$\begin{aligned} \mathrm{SS}(\theta) &= \mathrm{SS}(\mu, \beta) + \mathrm{SS}(\gamma \mid \mu, \beta) + \mathrm{SS}(\alpha \mid \mu, \beta, \gamma) \\ &= 732.1929 + 114.0166 + 3.2003 = 849.4098 \end{aligned}$$

with naturally $r = 14$ degrees of freedom.

To obtain $\mathrm{SS}(\beta \mid \mu, \alpha, \gamma)$ we again use the third normal equation

$$(\mathbf{G}_3'\mathbf{G}_3)\hat{\boldsymbol{\xi}}_3 = \mathbf{G}_3'\mathbf{Y},$$

with

$$\begin{aligned} \mathbf{G}_3'\mathbf{G} &= \boldsymbol{\Lambda}'[\mathbf{B}'\mathbf{B} - \mathbf{B}'\mathbf{C}(\mathbf{C}'\mathbf{C})^{-1}\mathbf{C}'\mathbf{B}]\boldsymbol{\Lambda} - \boldsymbol{\Lambda}'\mathbf{B}'\mathbf{G}_2(\mathbf{G}_2'\mathbf{G}_2)^{-1}\mathbf{G}_2'\mathbf{B}\boldsymbol{\Lambda}, \\ \mathbf{G}_3'\mathbf{Y} &= \boldsymbol{\Lambda}'(\mathbf{B}'\mathbf{Y} - \mathbf{B}'\mathbf{C}\hat{\boldsymbol{\xi}}_1 - \mathbf{B}'\mathbf{G}_2\hat{\boldsymbol{\xi}}_2), \\ \mathbf{B}'\mathbf{G}_2 &= \mathbf{B}'\mathbf{A}\boldsymbol{\Gamma} - \mathbf{B}'\mathbf{C}(\mathbf{C}'\mathbf{C})^{-1}\mathbf{C}'\mathbf{A}\boldsymbol{\Gamma}. \end{aligned}$$

From Table 6.5 we had

$$\boldsymbol{\Gamma}' = \begin{bmatrix} 1 & -1 & 0 & 0 & 0 & 0 \\ 1 & 0 & -1 & 0 & 0 & 0 \\ 1 & 0 & 0 & -1 & 0 & 0 \\ 1 & 0 & 0 & 0 & -1 & 0 \\ 1 & 0 & 0 & 0 & 1 & -1 \end{bmatrix}, \qquad \boldsymbol{\Lambda}' = \begin{bmatrix} 1 & -1 & 0 & -1 & 1 & 0 \\ 0 & 1 & -1 & 0 & 0 & 0 \\ 0 & 0 & 0 & 1 & 0 & -1 \\ 0 & 2 & 0 & -1 & -1 & 0 \end{bmatrix}.$$

$$\begin{aligned} \hat{\boldsymbol{\xi}}_1 &= (2.3333,\ 5.3333,\ 5.1667,\ 8.6667,\ 7.2500)', \\ \hat{\boldsymbol{\xi}}_2 &= (-0.5277,\ -0.4341,\ 0.6083,\ -0.3514,\ 0.0726)'. \end{aligned}$$

So by substituting these quantities in the above normal equation, we obtain

Table 6.6 Analysis of Variance Table for Example 6.7

Source	Degrees of Freedom	Sum of Squares
Regression	14	849.409
Fitting overall mean	1	690.035
Factor A adjusted for mean	5	55.048
Factor B adjusted for mean and A	5	31.074
Factor C adjusted for mean, A, and B	3	73.252
Fitting overall mean	1	690.035
Factor B adjusted for mean	5	42.157
Factor C adjusted for mean and B	4	114.017
Factor A adjusted for mean, B, and C	4	3.200
Fitting overall mean	1	690.035
Factor C adjusted for mean	4	125.381
Factor A adjusted for mean and C	5	4.936
Factor B adjusted for mean, A, and C	4	29.057
Residual	14	29.591
Total	28	879.000

$$\begin{bmatrix} 10.8356 & -0.5331 & -4.5123 & 0.7846 \\ -0.5331 & 3.7907 & -1.3636 & 2.9095 \\ -4.5123 & -1.3636 & 7.7829 & -2.5890 \\ 0.7846 & 2.9095 & -2.5890 & 6.8351 \end{bmatrix} \hat{\xi}_3 = \begin{bmatrix} 16.8848 \\ 0.4863 \\ -10.9182 \\ 3.7424 \end{bmatrix},$$

so

$$\hat{\xi}_3 = (1.2922,\ -0.0559,\ -0.5981,\ 0.1965)',$$

and

$$SS(\beta \mid u, \alpha, \gamma) = \hat{\xi}_3' G_2' Y = 29.0571.$$

We are now in a position to provide the complete analysis of variance table, similar to that of Table 6.2, as in Table 6.6.

Compare the above results with computer outputs at the end of this chapter for the same example.

PROGRAM THREEWAY

Introduction

It performs the calculations required for analyzing a set of data following an additive three-way classification model.

The rows contain the levels of factor A, the columns contain the levels of factor B, and the third dimension rows contain the levels of the factor C.

The only restriction on the number of observations per cell is that each row and each column, as well as each third dimension row, must have at least one observation.

Currently the program is limited by $a, b, c = 50$, although this could easily be changed. If higher precision is wanted, replace inverse subroutines (minv.for and det.for) by the appropriate one.

1 The input for the program includes:

- **(a)** Dimensions and number of incidence matrix (a, b, c).
- **(b)** Number of observations per cell, that is, the incidence matrix.
- **(c)** Vector of observations.

2 The output of this program includes:

- **(a)** Three tables of sums of squares and degrees of freedom.
- **(b)** Normal equation solutions.

Input Order and Format Specifications

The input is in a file which name is chosen by the user, the file extension has to be ".dat", and the file name is asked by the program:

1. The dimensions of the incidence matrix, that is, a, b, c. The format is free.
2. The rows of the incidence matrix. The format is free.
3. The vectors of observations, one per line, and ordered in row-wise fashion (i.e., the observations for cell (1, 1) are in row 1, those for cell (1, 2) are in row 2, etc.). The format is free.

Output Description

The output is in a file which name is the same as the input file's but the extension is ".IMP":

1 The first table has the following structure:

Source of variation	Degrees of Freedom		Sums of Squares	
Regression	r		SS(θ)	
Fitting overall mean		1		SS(μ)
Factor A unadjusted		$a-1$		SS($\alpha \mid \mu$)
Factor B adjusted for A		$\bar{f}_\beta$		SS($\beta \mid \mu, \alpha$)
Factor C adjusted for A and B		f_γ		SS($\gamma \mid \mu, \alpha, \beta$)
Residual	$N-r$		Subtraction	
Total	N		$\mathbf{Y}'\mathbf{Y}$	

The program provides two other tables. In this way we will have all the elements of Table 6.2.

2 A solution $\hat{\boldsymbol{\theta}}$ to the normal equations.

BIBLIOGRAPHICAL NOTES

6.2 For detail application of Latin squares see Denes and Keedwell (1974). For the construction of orthogonal Latin squares using mathematical programming approach see Arthanari and Dodge (1981). See also Scheffé (1959) Chapter 5 on Latin squares, and incomplete blocks.

6.3–6.5 Birkes, Dodge, and Seely (1972) provided results on estimability for an additive three-way classification model with arbitrary incidence. They introduced the *R*3 *process* which provides a sufficient condition for a cell expectation to be estimable. They also gave an algorithm for obtaining a spanning set for the estimable contrasts involving only a single effect. The main part of the algorithm is called the *Q process*. See also Birkes, Dodge, and Seely (1976). Seely and Birkes (1980) presented some estimability facts for partitioned linear models with constraints. In particular, they presented a theorem (Theorem 3.11) which is very useful in an additive three-way classification when all contrasts involving one effect are estimable. In such a case the problem of estimability reduces to a two-way classification model. Many examples of such are provided also in Chapter 7. See also Dodge (1974), Eccleston and Russell (1975), and Raghovarao and Federer (1975) and Shah and Dodge (1977). Wynn (1976) has also given results for an unconstrained model that says if a subset of parameters is estimable, then it can be dropped from the model for further estimability considerations. See also Butz (1982) for connectivity in multifactor designs.

REFERENCES

Arthanari, T. S. and Dodge, Y. (1981). *Mathematical Programming in Statistics.* New York: Wiley.

Birkes, D., Dodge, Y., and Seely, J. (1972). Estimability and analysis for additive three-way classification models with missing observations. Technical Report No. 30. Department of Statistics, Oregon State University.

Birkes, D., Dodge, Y., and Seely, J. (1976). Spanning sets in linear models. *Annals of Statist.* **4**, 86–107.

Butz, L. (1982). *Connectivity in Multi factor Designs: A combinatorial Approach.* Berlin: Heldermann Verlag.

Denes, J. and Keedwell, A. D. (1974). *Latin Squares and Their Applications.* New York: Academic Press.

Dodge, Y. (1974). Estimability considerations of N-way classification experimental arrangements with missing observations. Ph.D. thesis, Oregon State University, Corvallis.

Eccleston, J, and Russell, K. (1975). Connectedness and orthogonality in multi-factor designs. *Biometrika* **62**, 341–345.

Raghovarao, D. and Federer, W. T. (1975). On connectedness in two-way elimination of heterogeneity designs. *Ann. Statist.* **3**, 730–735.

Seely, J. and Birkes, D. (1980). Estimability in partitioned linear models. *Ann. Statist.* **8**, 399–406.

Shah, K. R. and Dodge, Y. (1977). On connectedness of designs. *Sankhya* **39**, 284.

Wynn, H. P. (1976). A note on estimability in multi-way layouts. Unpublished manuscript.

PROGRAM THREEWAY

```
C
C         THIS INTERACTIVE PROGRAM PERFORMS THE CALCULATIONS
C         REQUIRED FOR ANALYZING A SET OF DATA FOLLOWING AN ADDITIVE
C         THREE-WAY CLASSIFICATION MODEL WITH MISSING OBSERVATIONS,
C         USING THE INCIDENCE MATRIX N.
C         RESTRICTIONS:
C         GIVEN A THREE-WAY INCIDENCE MATRIX N, EACH ROW AS WELL AS
C         EACH COLUMN MUST HAVE AT LEAST ONE OBSERVATION.
C
C         INPUT:
C         THE FIRST INPUT LINE CONSISTS OF DIMENSION OF THE INCIDENCE
C         MATRIX N (NAxNBxNC). NEXT INPUT LINES ARE THE ENTRIES OF THE
C         INCIDENCE MATRIX. THERE ARE IN FACT NC INCIDENCE MATRICES
C         WHICH ARE READ IN ONE AFTER ANOTHER, ROW BY ROW. THIS FOLLOWS
C         BY THE ACTUAL OBSERVATIONS. FOR EACH NON EMPTY CELL, THE
C         ACTUAL OBSERVATIONS ARE READ IN SO THAT EACH DATA TAKES
C         ONE LINE. THE NON EMPTY CELL ARE READ IN FROM LEFT TO RIGHT,
C         ROW WISE.
C
C
C         STRUCTURE
C         VARIABLES
C         ESTMU    OUTPUT: ESTIMATOR OF MEAN
C         FOM      OUTPUT: SUM OF SQUARES OF FITTING OVERALL MEAN
C         IDFA     OUTPUT: DEGREES OF FREEDOM OF FACTOR A
C         IDFB     OUTPUT: DEGREES OF FREEDOM OF FACTOR B
C         IDFG     OUTPUT: DEGREES OF FREEDOM OF FACTOR C
C         IDFREG   OUTPUT: DEGREES OF FREEDOM OF REGRESSION
C         IFAUX    INPUT:  ERROR TEST
C         IQUOI    OUTPUT: CURRENT ITERATION
C         NA       INPUT:  NUMBER OF ENTRIES OF FACTOR A
C         NB       INPUT:  NUMBER OF ENTRIES OF FACTOR B
C         NC       INPUT:  NUMBER OF ENTRIES OF FACTOR C
C         NTOT     INPUT:  TOTAL DATA NUMBER
C         SSALP    OUTPUT: SUM OF SQUARES OF FACTOR A
C         SSBET    OUTPUT: SUM OF SQUARES OF FACTOR B
C         SSGAM    OUTPUT: SIM OF SQUARES OF FACTOR C
C         SSREG    OUTPUT: SUM OF SQUARES OF REGRESSION
C         SSTOT    OUTPUT: TOTAL SUM OF SQUARES
C         WEB      INPUT:  NAME OF INPUT FILE
C         YSUM     INPUT:  SUM OF EACH CELL
C
C         ARRAYS
C         BASE     INPUT:  BASIS FOR COMPUTING SUMS OF SQUARES
C         ESTALP   OUTPUT: ESTIMATORS OF FACTOR A
C         ESTBET   OUTPUT: ESTIMATORS OF FACTOR B
C         ESTGAM   OUTPUT: ESTIMATORS OF FACTOR C
C         N        INPUT:  INCIDENCE MATRIX
C         Y        INPUT:  DATA
C
C         SUBROUTINES
C         CHANGE      COMPUTES PERMUTATION OF INCIDENCE MATRIX AND DATA
C         COMPUTE     COMPUTES EVERY SUMS OF SQUARES, ESTIMATORS
C         DEGRE       COMPUTES EVERY DEGREES OF FREEDOM
C         OUTPUT123   DRAWS TABLEAU ONE TO THREE
C         XPROCESS    COMPUTES THE X-PROCESS
C
```

```
      PROGRAM THREEWAY
      PARAMETER IA=50,IB=IA**2
      DIMENSION Y(IA,IA,IA),BASE(IA,IA),N(IA,IA,IA),ESTGAM(IB),JP(3),
     1 ESTBET(IB),ESTALP(IB)
      CHARACTER WEB*7,FACT(3,3)
      COMMON /BLOCK1/ESTMU,ESTALP,ESTBET,ESTGAM,SSALP,SSBET,SSGAM
C
C     INITIALIZE FACT WHICH APPEARS IN THE OUTPUT SUBROUTINE
C
      FACT(1,1)='A'
      FACT(1,2)='B'
      FACT(1,3)='C'
      FACT(2,1)='C'
      FACT(2,2)='A'
      FACT(2,3)='B'
      FACT(3,1)='B'
      FACT(3,2)='C'
      FACT(3,3)='A'
C
C     A FILE NAME IS EXPECTED ( WITHOUT FILE EXTENTION WHICH
C     MUST BE ".DAT")
C
      TYPE 1
1     FORMAT(/,1X,'NAME OF INPUT FILE ? ',$)
      ACCEPT 2,WEB
2     FORMAT(A)
      OPEN(UNIT=1,FILE=WEB//'.DAT',STATUS='OLD')
      OPEN(UNIT=2,FILE=WEB//'.IMP',STATUS='NEW')
C
C     READ IN DIMENSION OF INCIDENCE MATRIX AND NUMBER
C     OF INCIDENCE MATRIX TO BE READ IN
C
      READ(1,*) NA,NB,NC
C
C     INITIALIZE JP WHICH IS USED IN OUTPUT SUBROUTINE OUTPUT123
C
      JP(1)=NA-1
      JP(2)=NC-1
      JP(3)=NB-1
      IF(NA.GT.IA.OR.NB.GT.IA.OR.NC.GT.IA) THEN
      TYPE 8
8     FORMAT(//,1X,'THE INCIDENCE MATRIX IS TOO BIG')
      STOP
      END IF
C
C     READ IN SUCCESIVELY EVERY INCIDENCE MATRIX ROW BY ROW
C
      DO I=1,NC
              DO K=1,NA
                      READ(1,*) (N(K,L,I),L=1,NB)
              END DO
      END DO
      IND=0
      SSTOT=0.
      NTOT=0
      FOM=0.
      IQUOI=0
```

```
C
C       READ IN OBSERVATIONS, BEGINNNING WITH THE FIRST INCIDENCE
C       MATRIX AND READING THE CELLS FROM LEFT TO RIGHT, ROW BY
C       ROW UNTIL LAST MATRIX
C       ONE LINE CONTAINS ONE OBSERVATION.
C
        DO I=1,NC
              DO K=1,NA
                    DO 3 L=1,NB
                          Y(K,L,I)=0.
                          NTOT=NTOT+N(K,L,I)
                          IF(N(K,L,I).EQ.0) GOTO 3
                          DO I2=1,N(K,L,I)
                                READ(1,*) YX

                                SSTOT=SSTOT+YX**2

                                Y(K,L,I)=Y(K,L,I)+YX

                          END DO
                          FOM=FOM+Y(K,L,I)
3                   CONTINUE
              END DO
        END DO
        CLOSE(UNIT=1)
        FOM=FOM**2/FLOAT(NTOT)
4       IQUOI=IQUOI+1
C
C       SUBROUTINE DEGRE COMPUTES ALL DEGREES OF FREEDOM
C
        CALL DEGRE(N,NA,NB,NC,IDFA,IDFB,IDFG)
C
C       TEST TO FIND WHETHER IT IS A TWOWAY ANALYSIS OR NOT
C
        IF(IDFA.EQ.0.OR.IDFB.EQ.0.OR.IDFG.EQ.0) THEN
              WRITE(2,5)
5       FORMAT(///,1X,'IT IS A TWOWAY ANALYSIS')
              STOP
        END IF
C
C       SUBROUTINE COMPUTES SUMS OF SQUARES AND ESTIMATORS FOR
C       FACTOR A, B ,C AND MEAN ESTIMATOR
C
        CALL COMPUTE(N,Y,NA,NB,NC,IDFB,IDFG,IFAUX)
        SSREG=SSBET+SSGAM+SSALP
        IDFREG=JP(IQUOI)+1+IDFB+IDFG
C
C       ERROR TEST
C
        IF(IFAUX.EQ.1) THEN
              IFAUX=0
              WRITE(2,6) IQUOI
6       FORMAT('1',///,1X,'ERROR IN COMPUTING TABLEAU ',I1)
              GOTO 7
        END IF
C
C       SUBROUTINE OUTPUT123 DRAWS THE THREE TABLEAU
C
        CALL OUTPUT123(NA,NB,NC,IQUOI,IDFB,IDFG,FOM,NTOT,SSTOT,JP,FACT)
```

```
7         IF(IQUOI.LT.3) THEN
          IF(IQUOI.EQ.1) THEN
C
C         SUBROUTINE CHANGE MAKES APPROPRIATE CHANGES IN THE INCIDENCE AND
C         OBSERVATION MATRICES IN ORDER TO FIND THE NECESSARY INFORMATION
C         REGARDING A SPECIFIC FACTOR
C
                  CALL CHANGE(N,Y,NA,NB,NC,1,WEB)
                  ELSE
                  CALL CHANGE(N,Y,NA,NB,NC,2,WEB)
          END IF
          GOTO 4
          END IF
          CLOSE(UNIT=2)
          STOP
          END
C
C
C
          SUBROUTINE DEGRE(N,NA,NB,NC,IDFA,IDFB,IDFG)
C         ARRAYS
C         M       INPUT: SUM OF ALL INCIDENCE MATRIX
C         M2      INPUT: COPY OF M MATRIX
C         N2      INPUT: COPY OF INCIDENCE MATRIX N
C
C         SUBROUTINES
C         BETAT   COMPUTES DEGREES OF FREEDOM OF FACTOR A AND B
C         FIND    FINDS THE LOOPS NEEDED TO COMPUTE DEGREE OF FREEDOM
C                 OF ADJUSTED FACTOR C
C         UNION   FINDS THE DIRECT DIFFERENCES BETWEEN EACH Mi MATRICES
C
          PARAMETER IA=50
          DIMENSION N(IA,IA,IA),BASE(IA,IA),M(IA,IA),N2(IA,IA,IA),M2(IA,IA)
          DO J1=1,IA
          DO J2=1,IA
                  BASE(J1,J2)=0.
                  M(J1,J2)=0
                  M2(J1,J2)=0
          END DO
          END DO
          DO I=1,NA
                  DO J=1,NB
                          DO K=1,NC
                                  N2(I,J,K)=N(I,J,K)
                          END DO
                  END DO
          END DO
          I1=0
C
C         REPLACE NON-ZERO VALUES BY APPROPRIATE VALUE
C         TEST OF DIRECT DIFFERENCES
C
          DO II=1,NC-1
          DO 1 I=II+1,NC
                  DO K=1,NB
```

```
                    DO L=1,NA
                            IF(N2(L,K,I).GT.0.AND.N2(L,K,II).GT.0)
     1 THEN
                            CALL UNION(BASE,I1,II,N2,I,NA,NB,IA)
                            GOTO 1
                            END IF
                    END DO
            END DO
1    CONTINUE
     END DO
     IVON=I1+1
C    CONSTRUCT M
     DO K=1,NB
            DO L=1,NA
                    DO I=1,NC
                            IF(N2(L,K,I).GT.0) THEN
                            M(L,K)=I
                            M2(L,K)=I
                            END IF
                    END DO
            END DO
     END DO
     IF(I1.EQ.NC-1) THEN
C
C    DEGREE OF FREEDOM FOR ADJUSTED FACTOR C IS MAXIMUM
C
            CALL BETAT(M,NA,NB,IDFB,IDFA)
            IDFG=NC-1
            RETURN
     END IF
C
C    DEGREE OF FREEDOM FOR ADJUSTED FACTOR C IS NOT MAXIMUM
C    THEREFORE A SPECIAL CALCULATION MUST BE MADE IN
C    SUBROUTINE FIND
C
     CALL FIND(M,NA,NB,NC,BASE,I1,IDFG,IVON)
     CALL BETAT(M2,NA,NB,IDFB,IDFA)
     RETURN
     END
C
C
C
     SUBROUTINE FIND(M,NA,NB,NC,BASE,I1,IDFG,IVON)
C    SUBROUTINES
C    LOOP    FINDS THE LOOPS WHICH FORM A BASIS
C    RANG    FINDS THE RANK OF MATRIX BASE
C    XPROCESS  COMPUTES X-PROCESS
C
     PARAMETER IA=50
     DIMENSION M(IA,IA),BASE(IA,IA),ZIND(IA)
C
C    FIND SIMPLE LOOPS
C
     DO I=1,NB-1
     DO 3 IL=1,I
            K=0
```

```
2               K=K+1
                IF(K.GE.NA) GOTO 3
                L=K
                IF(M(K,IL).EQ.0.OR.M(K,I+1).EQ.0) GOTO 2
4               L=L+1
                IF(L.GT.NA) GOTO 3
                IF(M(L,IL).EQ.0.OR.M(L,I+1).EQ.0) GOTO 4
                I1=I1+1
                I2=0
                J1=K
5               CONTINUE
                I2=I2+1
                BASE(I1,I2)=FLOAT(M(J1,IL))
                I2=I2+1
                BASE(I1,I2)=FLOAT(M(J1,I+1))
                IF(J1.EQ.K) THEN
                        J1=L
                        GOTO 5
                END IF
                M(K,IL)=0
                GOTO 2
3         CONTINUE
          END DO
C
C         SORT THE DATA AND FORM THE BASIS
C
          DO I=IVON,I1
                DO K=1,IA
                        ZIND(K)=0
                END DO
                DO K=1,NC
                        MN=0
                        DO L=1,4
                                MN=MN+1
                                IF(L.GT.2) THEN
                                IO=1
                                ELSE
                                IO=0
                                END IF
                                IF(BASE(I,L).EQ.K) ZIND(K)=ZIND(K)+
         1 (-1.)**(MN+1+IO)
                        END DO
                END DO
                DO K=1,NC
                        BASE(I,K)=ZIND(K)
                END DO
          END DO
C
C         SUBROUTINE XPROCESS COMPUTES X-PROCESS
C
          CALL XPROCESS(M,NA,NB,IA)
C
C         SUBROUTINE LOOP FINDS THE COMPLEX LOOPS
C
          CALL LOOP(M,BASE,NA,NB,I1,NC)
```

```
C
C       SUBROUTINE RANG COMPUTES RANK OF MATRIX BASE
C
        CALL RANG(BASE,I1,NC,IDFG,IA)
        RETURN
        END
C
C
C
        SUBROUTINE UNION(BASE,I1,II,N,I,NA,NB,IA)
        DIMENSION N(IA,IA,IA),BASE(IA,IA)
        I1=I1+1
        BASE(I1,II)=1.
        BASE(I1,I)=-1.
        DO L=1,NA
                DO K=1,NB
                        IF(N(L,K,II).EQ.0.AND.N(L,K,I).GT.0) N(L,K,II)=
        1 N(L,K,I)
                        N(L,K,I)=0
                END DO
        END DO
        RETURN
        END
C
C
C
        SUBROUTINE LOOP(M,BASE,NA,NB,I1,NC)
        PARAMETER IB=50
        DIMENSION M(IB,IB),IOU(IB),BASE(IB,IB)
        L1=0
C
C       SEARCH FIRST NON-ZERO ELEMENT
C
1       L1=L1+1
        L2=0
        IF(L1.GT.NA) GOTO 2
3       L2=L2+1
        IF(L2.GT.NB) GOTO 1
        IF(M(L1,L2).EQ.0) GOTO 3
C
C       TEST IN ORDER TO DETECT A ROW WITH ONLY ONE NON-ZERO ELEMENT
C
        ITEST=0
        DO I=1,NB
                IF(M(L1,I).GT.0) ITEST=ITEST+1
        END DO
        IF(ITEST.LE.1) GOTO 1
        ID=0
        IOU(1)=0
        GOTO 4
5       IF(IOU(2).EQ.L1.AND.IOU(3).EQ.L2.AND.IOU(1).GE.3.AND.
        1 ID.NE.MEM.AND.ID.NE.MEM+2) THEN
C
C       TEST TO DETECT A LOOP
C
        ISI=1
        I1=I1+1
```

```
      DO J2=1,NC
              BASE(I1,J2)=0.
      END DO
C
C     RECORD THE LOOP
C
      DO J1=1,IOU(1)
              N=2*J1
              L=N+1
              ISI=ISI+1
              IMS=ABS(M(IOU(N),IOU(L)))
              BASE(I1,IMS)=BASE(I1,IMS)+(-1.)**ISI
      END DO
      M(IOU(2),IOU(3))=0
C
C     COMPUTE X-PROCESS
C
      CALL XPROCESS(M,NA,NB,IB)
6     L1=IOU(2)
      L2=IOU(3)
      GOTO 3
      END IF
C
C     TEST TO DETECT IF THE ALGORITHM ALREADY PASSED THIS POINT
C
      DO K=1,IOU(1)
              J=2*K
              L=J+1
              IF(IOU(J).EQ.L1.AND.IOU(L).EQ.L2) THEN
              IF(ID.EQ.1.OR.ID.EQ.2.OR.ID.EQ.3) GOTO 7
8             IOU(1)=IOU(1)-1
              IF(IOU(1).EQ.0) GOTO 6
              J=2*IOU(1)
              L=J+1
              L1=IOU(J)
              L2=IOU(L)
              J=2*(IOU(1)+1)
              L=J+1
              ID=0
              IF(IOU(1)-1.GT.0) THEN
              IJ=2*(IOU(1)-1)
              IL=IJ+1
              IF(L1.LT.IOU(IJ)) IK=1
              IF(L1.GT.IOU(IJ)) IK=3
              IF(L2.GT.IOU(IL)) IK=2
              IF(L2.LT.IOU(IL)) IK=4
              END IF
              IF(L1.LT.IOU(J)) GOTO 12
              IF(L1.GT.IOU(J)) GOTO 10
              IF(L2.GT.IOU(L)) GOTO 8
              IF(L2.LT.IOU(L)) GOTO 11
              END IF
      END DO
4     IOU(1)=IOU(1)+1
```

```
C
C          RECORD THE POINT
C
           J=2*IOU(1)
           L=J+1
           IOU(J)=L1
           IOU(L)=L2
           IF(IOU(1).LE.1) THEN
           IK=0
           GOTO 9
           END IF
           J=2*(IOU(1)-1)
           L=J+1
C
C          TEST IN ORDER TO CHOOSE THE RIGHT DIRECTION
C
           IF(IOU(J).LT.L1) IK=3
           IF(IOU(J).GT.L1) IK=1
           IF(IOU(L).LT.L2) IK=2
           IF(IOU(L).GT.L2) IK=4
           IF(IOU(1).EQ.2) THEN
           IF(IK.EQ.2.OR.IK.EQ.4) MEM=2
           IF(IK.EQ.1.OR.IK.EQ.3) MEM=1
           END IF
C
C          DIRECTION 1: UP
C
9          IF(ID.EQ.3.OR.IK.EQ.1.OR.IK.EQ.3) GOTO 10
           IF(L1-1.LE.0.AND.M(L1,L2).EQ.0) GOTO 7
           IF(L1-1.LE.0) GOTO 10
           L1=L1-1
           ID=1
           IF(M(L1,L2).GT.0) GOTO 5
           GOTO 9
C
C          DIRECTION 2: RIGHT
C
10         IF(ID.EQ.4.OR.IK.EQ.2.OR.IK.EQ.4) GOTO 11
           IF(L2+1.GT.NB.AND.M(L1,L2).EQ.0) GOTO 7
           IF(L2+1.GT.NB) GOTO 11
           L2=L2+1
           ID=2
           IF(M(L1,L2).GT.0) GOTO 5
           GOTO 10
C
C          DIRECTION 3: DOWN
C
11         IF(ID.EQ.1.OR.IK.EQ.1.OR.IK.EQ.3) GOTO 12
           IF(L1+1.GT.NA.AND.M(L1,L2).EQ.0) GOTO 7
           IF(L1+1.GT.NA) GOTO 12
           L1=L1+1
           ID=3
           IF(M(L1,L2).GT.0) GOTO 5
           GOTO 11
```

```
C
C         DIRECTION 4: LEFT
C
12        IF(ID.EQ.2.OR.IK.EQ.2.OR.IK.EQ.4) GOTO 8
          IF(L2-1.LE.0.AND.M(L1,L2).EQ.0) GOTO 7
          IF(L2-1.LE.0) GOTO 8
          L2=L2-1
          ID=4
          IF(M(L1,L2).GT.0) GOTO 5
          GOTO 12
7         CONTINUE
          L1B=L1
          L2B=L2
          J=2*IOU(1)
          L=J+1
          L1=IOU(J)
          L2=IOU(L)
          J=2*(IOU(1)+1)
          L=J+1
          ID=0
          IF(L1.LT.L1B) GOTO 12
          IF(L1.GT.L1B) GOTO 10
          IF(L2.GT.L2B) GOTO 8
          IF(L2.LT.L2B) GOTO 11
2         RETURN
          END
C
C
C
          SUBROUTINE BASIS(BASE,NA,NB)
C
C         SUBROUTINE BASIS COMPUTES THE BASIS MATRICES USED FOR
C         THE SUMS OF SQUARES CALCULATIONS
C
          PARAMETER IA=50,IB=IA**2
          DIMENSION BASE(IB),Z(IA,IA)
          ID=1
          DO I1=1,NA
                  DO I2=1,NB
                          Z(I1,I2)=0.
                  END DO
                  Z(I1,1)=1.
                  ID=ID+1
                  Z(I1,ID)=-1.
          END DO
          CALL ARRAY(Z,BASE,NA,NB,1,IA)
          RETURN
          END
C
C
C
```

```
      SUBROUTINE BETAT(M,NA,NB,IDFB,IDFA)
C     SUBROUTINE
C     MPM     MULTIPLICATES TRANSPOSED M MATRIX BY M MATRIX
C
      PARAMETER IA=50
      DIMENSION M(IA,IA),M2(IA,IA),MT(IA,IA)
C
C     INITIALIZE MATRIX M2
C
      DO I=1,NA
              DO J=1,NB
                      M2(I,J)=0.
                      IF(M(I,J).GT.0) M2(I,J)=1
              END DO
      END DO
C
C     COMPUTE R-PROCESS
C
1     DO I=1,NA
              DO J=1,NB
              IL=0
              DO K=1,NA
                      DO L=1,NB
                              IL=IL+M2(I,L)*M2(K,L)*M2(K,J)
                      END DO
              END DO
              IF(IL.GT.0) MT(I,J)=1
              END DO
      END DO
      I2=0
      IT=0
      DO I=1,NA
              DO J=1,NB
                      I2=I2+M2(I,J)
                      IT=IT+MT(I,J)
              END DO
      END DO
      IF(I2.EQ.IT) GOTO 2
      DO I=1,NA
              DO J=1,NB
                      M2(I,J)=MT(I,J)
              END DO
      END DO
      GOTO 1
2     CALL MPM(MT,NA,NB,IDFB,IDFA)
      RETURN
      END
C
C
C
      SUBROUTINE ARRAY(A,B,N,M,ICODE,IA)
C
C     SUBROUTINE ARRAY TRANSFORMS A MATRIX INTO A VECTOR BY
C     USING ICODE 1 AND TRANSFORMS A VECTOR INTO A MATRIX BY
C     USING ICODE 2
C
```

```
      DIMENSION A(IA,IA),B(IA**2)
      IF(ICODE.EQ.2) GOTO 1
      K=1
      DO I=1,N
      DO L=1,M
              B(K)=A(I,L)
              K=K+1
      END DO
      END DO
      GOTO 2
1     I=1
      L=1
      DO K=1,N*M
              A(I,L)=B(K)
              IF(L.LT.M) GOTO 3
              L=0
              I=I+1
3             L=L+1
      END DO
2     RETURN
      END
C
C
C
      SUBROUTINE MADD(A,B,C,N,IA)
C
C     SUBROUTINE MADD ADDS A DIAGONAL MATRIX A TO A NORMAL MATRIX B
C
      DIMENSION A(IA),B(IA**2),C(IA**2)
      K=0
      DO 2 I=1,N**2
              IF(I.EQ.1+K*(N+1)) GOTO 1
              C(I)=B(I)
              GOTO 2
1             C(I)=A(K+1)+B(I)
              K=K+1
2     CONTINUE
      RETURN
      END
C
C
C
      SUBROUTINE MINV(A,M,IFAUX)
C
C     SUBROUTINE MINV INVERSES A MATRIX A
C     FOR HIGHER PRECISION USE SYSTEM SUBROUTINES
C     ERROR TESTS ARE PROVIDED
C
      PARAMETER IA=50
      DIMENSION A(IA,IA),B(IA,IA),C(IA,IA)
      IFAUX=0
      IF(M.EQ.0.OR.(M.EQ.1.AND.A(1,1).EQ.0.)) THEN
              IFAUX=1
              WRITE(2,*)'ERROR IN COMPUTING INVERSE'
              RETURN
      END IF
```

```
      IF(M.EQ.1) THEN
              A(1,1)=1./A(1,1)
              RETURN
      END IF
      DO I=1,M
      DO K=1,M
              C(I,K)=A(I,K)
      END DO
      END DO
      CALL DET(C,M,DETER,IFAUX)
      IF(DETER.EQ.0.OR.IFAUX.EQ.1) THEN
              IFAUX=1
              WRITE(2,*)'ERROR IN COMPUTING INVERSE'
              RETURN
      END IF
      IF(M.EQ.2) GOTO 6
      DO I=1,M
      DO 5 K=1,M
              J1=1
              J2=0
              DO 1 J=1,M
              DO 2 L=1,M
                      IF(J.EQ.I) GOTO 1
                      IF(L.EQ.K) GOTO 2
                      J2=J2+1
                      IF(J2.LE.M-1) GOTO 3
                      J2=1
                      J1=J1+1
3                     C(J1,J2)=A(J,L)
2             CONTINUE
1             CONTINUE
              IF(M.NE.2) GOTO 7
              DE=C(1,1)
              GOTO 8
7             CALL DET(C,M-1,DE,IFAUX)
              IF(IFAUX.EQ.1) RETURN
8             CONTINUE
              B(K,I)=DE/DETER*(-1.)**(I+K)
5     CONTINUE
      END DO
      DO I=1,M
      DO J=1,M
              A(I,J)=B(I,J)
      END DO
      END DO
      GOTO 9
6     B1=A(2,2)/DETER
      B2=-A(1,2)/DETER
      B3=-A(2,1)/DETER
      B4=A(1,1)/DETER
      A(1,1)=B1
      A(1,2)=B2
      A(2,1)=B3
      A(2,2)=B4
9     RETURN
      END
C
```

```
C
C
      SUBROUTINE DET(C,N,TOTAL,IFAUX)
C
C     SUBROUTINE DET COMPUTES DETERMINANT OF MATRIX C
C     IT COMPUTES IN DOUBLE PRECISION
C     ERROR TESTS ARE PROVIDED
C
      PARAMETER IB=50
      DIMENSION C(IB,IB)
      DOUBLE PRECISION A(IB,IB),RAP
      IFAUX=0
      DO I=1,N
            DO K=1,N
                  A(I,K)=C(I,K)
            END DO
      END DO
      RAP=0.
      SIGNE=1.
      TOTAL=1.
      DO I=N,2,-1
      DO 1 L=I-1,1,-1
            IF(A(L,I).EQ.0.) GOTO 1
            IF(A(I,I).NE.0.) GOTO 4
            DO I1=L,1,-1
                  IND=I1
                  IF(A(I1,I).NE.0.) GOTO 3
            END DO
            IFAUX=1
            WRITE(2,*)'ERROR IN COMPUTING INVERSE'
            RETURN
3           DO I1=1,N
                  B=A(IND,I1)
                  A(IND,I1)=A(I,I1)
                  A(I,I1)=B
            END DO
            SIGNE=(-1.)**(IND+I)*SIGNE
4           RAP=A(L,I)/A(I,I)
            DO K=I-1,1,-1
                  A(L,K)=A(L,K)-RAP*A(I,K)
            END DO
1     CONTINUE
      END DO
      DO I=1,N
            TOTAL=TOTAL*A(I,I)
      END DO
      TOTAL=TOTAL*SIGNE
      RETURN
      END
C
C
C
```

```
      SUBROUTINE MPRO(A,B,C,N,II,M,ICODE,ICA,ICB,ICC)
      DIMENSION A(ICA),B(ICB),C(ICC)
C
C     SUBROUTINE MPRO MULTIPLICATES A MATRIX A BY A MATRIX B.
C     FOUR CASES MAY APPEAR:
C     ICODE 1: TWO NORMAL MATRICES ARE INVOLVED
C     ICODE 2: A NORMAL AND A DIAGONAL MATRIX ARE INVOLVED
C     ICODE 3: A DIAGONAL AND A NORMAL MATRIX ARE INVOLVED
C     ICODE 4: TWO VECTORS ARE MULTIPLICATED IN ORDER TO FORM
C              A MATRIX
C
      IF(ICODE.NE.1) GOTO 1
      IPAS=1
      DO K=1,N
      DO J=1,M
              L=0
              C(IPAS)=0.
              DO I=1+II*(K-1),II*K
                      C(IPAS)=A(I)*B(L*M+J)+C(IPAS)
                      L=L+1
              END DO
              IPAS=IPAS+1
      END DO
      END DO
      GOTO 2
1     IF(ICODE.NE.2) GOTO 3
      K=1
      DO I=1,N*II
              C(I)=A(I)*B(K)
              K=K+1
              IF(K.GT.II) K=1
      END DO
      GOTO 2
3     IF(ICODE.NE.3) GOTO 5
      K=1
      J=0
      DO 4 I=1,N*M
              C(I)=A(K)*B(I)
              J=J+1
              IF(J.LT.M) GOTO 4
              K=K+1
              J=0
4     CONTINUE
      GOTO 2
5     IPAS=1
      DO I=1,M
      DO K=1,N
              C(IPAS)=A(K)*B(I)
              IPAS=IPAS+1
      END DO
      END DO
2     RETURN
      END
C
C
```

```
C
      SUBROUTINE MSUB(A,B,C,N,M,ICODE,ICA,IA,IB)
C
C     SUBROUTINE MSUB SUBTRACTS A MATRIX B FROM A MATRIX A
C     ICODE 1: TWO NORMAL MATRICES ARE INVOLVED
C     ICODE 2: A DIAGONAL AND A NORMAL MATRIX ARE INVOLVED
C
      DIMENSION A(ICA),B(IA),C(IB)
      K=0
      IF(ICODE.NE.1) GOTO 3
      DO I=1,N*M
              C(I)=A(I)-B(I)
      END DO
      GOTO 4
3     DO 2 I=1,N*M
              IF(I.EQ.1+K*(M+1)) GOTO 1
              C(I)=-B(I)
              GOTO 2
1             C(I)=A(K+1)-B(I)
              K=K+1
2     CONTINUE
4     RETURN
      END
C
C
C
      SUBROUTINE MTRA(A,B,N,M,IC,ID)
C
C     SUBROUTINE MTRA TRANSPOSES A MATRIX A INTO A MATRIX B
C
      DIMENSION A(IC),B(ID)
      K=0
      IPAS=1
      DO 1 I=1,M*N
              B(I)=A(IPAS+K*M)
              K=K+1
              IF(K.LT.N) GOTO 1
              IPAS=IPAS+1
              K=0
1     CONTINUE
      RETURN
      END
C
C
C
      SUBROUTINE RANG(BASE,NA,NB,IT,IA)
C     SUBROUTINES
C     WH01    MAKES APPROPRIATE PERMUTATIONS TO BRING A 1 TO POSITION
C             i,i
C     WH02    MAKES APPROPRIATE PERMUTATIONS TO BRING AN NON-ZERO
C             ELEMENT IN POSITION i,i
C
      DIMENSION BASE(IA,IA)
      IT=0
      I=1
      DO K1=1,IA
              DO K2=NB+1,IA
```

```
                        BASE(K1,K2)=0.
                END DO
        END DO
1       CONTINUE
        CALL WHO1(BASE,NA,NB,I,IREP,IA)
        IF(IREP.EQ.0) THEN
                CALL WHO2(BASE,NA,NB,I,IREP,IA)
                IF(IREP.EQ.0) GOTO 2
                DO I1=I+1,NB
                        BASE(I,I1)=BASE(I,I1)/BASE(I,I)
                END DO
                BASE(I,I)=1.
        END IF
        IF(I+1.GT.NA) GOTO 2
        DO K=I+1,NA
                RAP=BASE(K,I)/BASE(I,I)
                DO L=I,NB
                        BASE(K,L)=BASE(K,L)-RAP*BASE(I,L)
                END DO
        END DO
        IF(I+1.GT.NB) GOTO 2
        DO K=I+1,NB
                RAP=BASE(I,K)/BASE(I,I)
                DO L=I,NA
                        BASE(L,K)=BASE(L,K)-RAP*BASE(L,I)
                END DO
        END DO
        I=I+1
        GOTO 1
2       NC=NA
        IF(NB.GT.NA) NC=NB
        DO I1=1,NC
                IF(BASE(I1,I1).GT.0.0001.OR.BASE(I1,I1).LT.-0.0001) IT=IT+1
        END DO
        RETURN
        END
C
C
C
        SUBROUTINE WHO1(BASE,NA,NB,I,IREP,IA)
C
C       SUBROUTINE WHO1 FINDS A ELEMENT EQUAL TO 1 OR -1 AND
C       BRINGS IT INTO POSITION i,i
C
        DIMENSION BASE(IA,IA)
        IREP=0
        DO K=I,NA
                DO L=I,NB
                        IF(BASE(K,L).GT.0.99999.AND.BASE(K,L).LT.1.00001)
        1 GOTO 1
                        IF(BASE(K,L).GT.-1.00001.AND.BASE(K,L).LT.-0.99999)
        1 THEN
                        CALL MINUS(BASE,K,NB,IA)
1                       IF(K.NE.I) CALL PERLI(BASE,I,K,NB,IA)
                        IF(L.NE.I) CALL PERCO(BASE,I,L,NA,IA)
                        IREP=1
                        GOTO 2
```

```
                END IF
                END DO
        END DO
2       RETURN
        END
C
C
C
        SUBROUTINE MINUS(BASE,I1,NB,IA)
        DIMENSION BASE(IA,IA)
        DO I=1,NB
                BASE(I1,I)=-BASE(I1,I)
        END DO
        RETURN
        END
C
C
C
        SUBROUTINE PERLI(BASE,I,I1,NB,IA)
C
C       SUBROUTINE PERLI MAKES APPROPRIATE LINE PERMUTATIONS
C
        DIMENSION BASE(IA,IA)
        DO K=1,NB
                A=BASE(I,K)
                BASE(I,K)=BASE(I1,K)
                BASE(I1,K)=A
        END DO
        RETURN
        END
C
C
C
        SUBROUTINE PERCO(BASE,I,I2,NA,IA)
C
C       SUBROUTINE PERCO MAKES APPROPRIATE COLUMN PERMUTATION
C
        DIMENSION BASE(IA,IA)
        DO K=1,NA
                A=BASE(K,I)
                BASE(K,I)=BASE(K,I2)
                BASE(K,I2)=A
        END DO
        RETURN
        END
C
C
C
        SUBROUTINE WHO2(BASE,NA,NB,I,IREP,IA)
C
C       SUBROUTINE WHO2 FINDS A NON-ZERO ELEMENT ABD BRINGS
C       IT TO POSITION i,i
C
```

```
      DIMENSION BASE(IA,IA)
      IREP=0
      DO K=I,NA
              DO L=I,NB
                      IF(BASE(K,L).LT.-0.0001.OR.BASE(K,L).GT.0.0001)
      1 THEN
                      IF(K.NE.I) CALL PERLI(BASE,I,K,NB,IA)
                      IF(L.NE.I) CALL PERCO(BASE,I,L,NA,IA)
                      IREP=1
                      GOTO 2
                      END IF
              END DO
      END DO
2     RETURN
      END
C
C
C
      SUBROUTINE MPM(MT,NA,NB,IDFB,IDFA)
      PARAMETER IA=50
      DIMENSION M2(IA,IA),MT(IA,IA)
      DO I=1,NB
              DO J=1,NB
                      M2(I,J)=0
                      DO K=1,NA
                              M2(I,J)=M2(I,J)+MT(K,I)*MT(K,J)
                      END DO
              END DO
      END DO
      IDFB=0
      DO I=2,NB
              IL=0
              DO J=1,I-1
                      IF(M2(I,J).GT.0) IL=IL+1
              END DO
              IF(IL.GT.0) IDFB=IDFB+1
      END DO
      DO I=1,NA
              DO J=1,NA
                      M2(I,J)=0
                      DO K=1,NB
                              M2(I,J)=M2(I;J)+MT(I,K)*MT(J,K)
                      END DO
              END DO
      END DO
      IDFA=0
      DO I=2,NA
              IL=0
              DO J=1,I-1
                      IF(M2(I,J).GT.0) IL=IL+1
              END DO
              IF(IL.GT.0) IDFA=IDFA+1
      END DO
      RETURN
      END
C
C
```

```
C
      SUBROUTINE OUTPUT123(NA,NB,NC,IZ,IDFB,IDFG,FOM,NTOT,SSTOT,JP,
     1 FACT)
      PARAMETER IA=50,IB=IA**2
      DIMENSION JP(3),ESTALP(IB),ESTBET(IB),ESTGAM(IB)
      CHARACTER FACT(3,3)
      COMMON /BLOCK1/ESTMU,ESTALP,ESTBET,ESTGAM,SSALP,SSBET,SSGAM
      IDT=JP(IZ)+1+IDFB+IDFG
      SSALU=SSALP-FOM
      SSREG=SSBET+SSGAM+SSALP
      WRITE(2,2) IDT,SSREG,FOM
2     FORMAT('1',//,1X,72('*'),/,1X,35X,'*',12X,'*',/,3X,'SOURCE OF ',
     1 'VARIATION',14X,'*',4X,'D.F.',4X,'*',3X,'SUMS OF SQUARES',/,
     2 36X,'*',12X,'*',/,1X,72('*'),/,36X,'*',12X,'*',/,2X,
     3 'REGRESSION',24X,'*',2X,I3,7X,'*',2X,E15.7,/,36X,'*',12X,'*',
     4 /,36X,'*',12X,'*',/,6X,'FITTING OVERALL MEAN',10X,'*',8X,
     5 '1',3X,'*',6X,E15.7,/,36X,'*',12X,'*')
      WRITE(2,3) FACT(IZ,1),JP(IZ),SSALU,FACT(IZ,2),FACT(IZ,1),IDFB,
     1 SSBET,FACT(IZ,3),FACT(IZ,1),FACT(IZ,2),IDFG,SSGAM
3     FORMAT(6X,'FACTOR ',A1,' UNADJUSTED',11X,'*',6X,I3,3X,'*',6X,
     1 E15.7,/,36X,'*',12X,'*',/,6X,'FACTOR ',A1,' ADJUSTED FOR ',
     2 A1,7X,'*',6X,I3,3X,'*',6X,E15.7,/,36X,'*',12X,'*',/,
     3 6X,'FACTOR ',A1,' ADJUSTED FOR ',A1,' AND ',A1,1X,'*',6X,I3,
     4 3X,'*',6X,E15.7,/,36X,'*',12X,'*',/,36X,'*',12X,'*')
      WRITE(2,4) NTOT-IDT,SSTOT-SSREG,NTOT,SSTOT,ESTMU
4     FORMAT( 2X,'RESIDUAL',26X,'*',2X,I3,7X,'*',2X,
     1 E15.7,/,36X,'*',12X,'*',/,1X,72('*'),/,36X,'*',12X,'*',/,
     2 2X,'TOTAL',29X,'*',2X,I3,7X,'*',2X,E15.7,/,'1',///,2X,'ESTIMATE OI
     3 ' MEAN  ',E15.7)
      WRITE(2,5) FACT(IZ,1)
5     FORMAT(///,2X,'ESTIMATES OF ',A1,' EFFECTS ',/)
      DO I=1,NA
              WRITE(2,6) ESTALP(I)
6             FORMAT(20X,E15.7)
      END DO
      WRITE(2,7) FACT(IZ,2)
7     FORMAT(///,3X,'ESTIMATES OF ',A1,' EFFECTS',/)
      DO I=1,NB
              WRITE(2,6) ESTBET(I)
      END DO
      WRITE(2,8) FACT(IZ,3)
8     FORMAT(///,2X,'ESTIMATES OF ',A1,' EFFECTS',/)
      DO I=1,NC
              WRITE(2,6) ESTGAM(I)
      END DO
      RETURN
      END
C
C
C
      SUBROUTINE CHANGE(N,Y,NA,NB,NC,IQUOI,WEB)
      PARAMETER IA=50,IB=IA**2
      DIMENSION N(IA,IA,IA),Y(IA,IA,IA),N2(IA,IA,IA),Y2(IA,IA,IA)
      CHARACTER WEB*7
      IF(IQUOI.EQ.1) GOTO 1
```

```
C
C         DATA MUST BE READ IN ONE MORE TIME BECAUSE THEY HAVE
C         BEEN AFFECTED BY THE PREVIOUS CHANGE
C
          OPEN(UNIT=1,FILE=WEB//'.DAT',STATUS='OLD')
          READ(1,*) NA,NB,NC
          DO I=1,NC
                  DO K=1,NA
                          READ(1,*) (N(K,L,I),L=1,NB)
                  END DO
          END DO
          DO L=1,NC
          DO I=1,NA
          DO K=1,NB
                  Y(I,K,L)=0.
                  IF(N(I,K,L).GT.0) THEN
                          DO J=1,N(I,K,L)
                                  READ(1,*) YX
                                  Y(I,K,L)=Y(I,K,L)+YX
                          END DO
                  END IF
          END DO
          END DO
          END DO
          DO I=1,NC
          DO J=1,NA
          DO K=1,NB
                  N2(K,I,J)=N(J,K,I)
                  Y2(K,I,J)=Y(J,K,I)
          END DO
          END DO
          END DO
          NO=NC
          NC=NA
          NA=NB
          NB=NO
          GOTO 2
1         CONTINUE
          DO I=1,NC
          DO J=1,NA
          DO K=1,NB
                  N2(I,J,K)=N(J,K,I)
                  Y2(I,J,K)=Y(J,K,I)
          END DO
          END DO
          END DO
          NO=NB
          NB=NA
          NA=NC
          NC=NO
2         CONTINUE
          DO K=1,NC
          DO I=1,NA
          DO J=1,NB
                  N(I,J,K)=N2(I,J,K)
                  Y(I,J,K)=Y2(I,J,K)
          END DO
```

```
      END DO
      END DO
      RETURN
      END
C
C
C
      SUBROUTINE XPROCESS(M,NA,NB,IA)
      DIMENSION M(IA,IA)
      DO I=1,NA
              ISOM=0
              DO K=1,NB
                      IF(M(I,K).EQ.0) ISOM=ISOM+1
              END DO
              IF(ISOM.EQ.NB-1) THEN
                      DO K=1,NB
                              M(I,K)=0
                      END DO
              END IF
      END DO
      DO I=1,NB
              ISOM=0
              DO K=1,NA
                      IF(M(K,I).EQ.0) ISOM=ISOM+1
              END DO
              IF(ISOM.EQ.NA-1) THEN
                      DO K=1,NA
                              M(K,I)=0
                      END DO
              END IF
      END DO
      RETURN
      END
C
C
C
C
      SUBROUTINE COMPUTE(N,Y,NA,NB,NC,IDFB,IDFG,IFAUX)
C
C     SUBROUTINES
C     ARRAY   CHANGES VECTORS TO MATRIX AND MATRIX TO VECTORS
C     BASIS   COMPUTES BASIS OF FACTOR B AND C
C     DO_ALL  COMPUTES APAI,APY,BPB,BPBI,BPY,CPC,CPY,CPA,BPC
C     MADD    ADDS TO MATRIX
C     MINV    COMPUTES INVERSE MATRIX
C     MPRO    MULTIPLICATES TWO MATRIX
C     MSUB    SUBTRACTS TWO MATRIX
C     MTRA    TRANPOSES A MATRIX
C
      PARAMETER IA=50,IB=IA**2
      DIMENSION APAI(IA),APY(IB),BPB(IA),BPY(IB),APB(IB),
      1 CPC(IA),CPA(IB),CPY(IB),BPC(IB),P1(IB),P2(IB),P3(IB),P4(IB),
      2 P5(IB),P6(IB),P7(IB),ESTALP(IB),ESTBET(IB),ESTGAM(IB),
      3 Y(IA,IA,IA),ATA(IA,IA),N(IA,IA,IA),P8(IB)
      COMMON /BLOCK1/ESTMU,ESTALP,ESTBET,ESTGAM,SSALP,SSBET,SSGAM
      COMMON /BLOCK2/ APAI,APY,BPB,BPY,APB,CPC,CPA,CPY,BPC
```

```
C
C       COMPUTE SUMS OF SQUARES OF ALPHA EFFECT
C
        CALL DO_ALL(N,Y,NA,NB,NC)
        CALL MPRO(APAI,APY,P1,NA,NA,1,3,IA,IB,IB)
        CALL MPRO(P1,APY,P2,1,NA,1,1,IB,IB,IB)
        SSALP=P2(1)
        CALL MPRO(APAI,APB,P2,NA,NA,NB,3,IA,IB,IB)
C
C       COMPUTE SUMS OF SQUARES OF BETA EFFECT
C
        CALL MTRA(APB,P3,NA,NB,IB,IB)
        CALL MPRO(P3,P2,P1,NB,NA,NB,1,IB,IB,IB)
        CALL MSUB(BPB,P1,P2,NB,NB,2,IA,IB,IB)
        CALL BASIS(P1,IDFB,NB)
        CALL MPRO(P1,P2,P3,IDFB,NB,NB,1,IB,IB,IB)
        CALL MTRA(P1,P2,IDFB,NB,IB,IB)
        CALL MPRO(P3,P2,P1,IDFB,NB,IDFB,1,IB,IB,IB)
        CALL ARRAY(ATA,P1,IDFB,IDFB,2,IA)
        CALL MINV(ATA,IDFB,IFAUX)
        IF(IFAUX.EQ.1) THEN
                WRITE(2,*) 'ERROR IN COMPUTING INVERSE'
                RETURN
        END IF
        CALL ARRAY(ATA,P1,IDFB,IDFB,1,IA)
        CALL MTRA(APB,P3,NA,NB,IB,IB)
        CALL MPRO(P3,APAI,P2,NB,NA,NA,2,IB,IA,IB)
        CALL MPRO(P2,APY,P3,NB,NA,1,1,IB,IB,IB)
        CALL MSUB(BPY,P3,P2,NB,1,1,IB,IB,IB)
        CALL BASIS(P3,IDFB,NB)
        CALL MPRO(P3,P2,P4,IDFB,NB,1,1,IB,IB,IB)
        CALL MPRO(P1,P4,P2,IDFB,IDFB,1,1,IB,IB,IB)
        CALL MPRO(P2,P4,P3,1,IDFB,1,1,IB,IB,IB)
        SSBET=P3(1)
C
C       COMPUTE SUMS OF SQUARES OF GAMMA EFFECT
C
C       COMPUTE G3'G3
        CALL MPRO(CPA,APAI,P2,NC,NA,NA,2,IB,IA,IB)
        CALL MTRA(CPA,P3,NC,NA,IB,IB)
        CALL MPRO(P2,P3,P5,NC,NA,NC,1,IB,IB,IB)
        CALL MSUB(CPC,P5,P2,NC,NC,2,IA,IB,IB)
        CALL BASIS(P3,IDFG,NC)
        CALL MPRO(P3,P2,P5,IDFG,NC,NC,1,IB,IB,IB)
        CALL MTRA(P3,P6,IDFG,NC,IB,IB)
        CALL MPRO(P5,P6,P3,IDFG,NC,IDFG,1,IB,IB,IB)
C
C       COMPUTE C'G2
C
        CALL MPRO(CPA,APAI,P2,NC,NA,NA,2,IB,IA,IB)
        CALL MPRO(P2,APB,P5,NC,NA,NB,1,IB,IB,IB)
        CALL MTRA(BPC,P6,NB,NC,IB,IB)
        CALL MSUB(P6,P5,P7,NC,NB,1,IB,IB,IB)
        CALL BASIS(P2,IDFB,NB)
        CALL MTRA(P2,P5,IDFB,NB,IB,IB)
        CALL MPRO(P7,P5,P2,NC,NB,IDFB,1,IB,IB,IB)
        CALL MPRO(P2,P1,P7,NC,IDFB,IDFB,1,IB,IB,IB)
```

```
      CALL MTRA(P2,P5,NC,IDFB,IB,IB)
      CALL MPRO(P7,P5,P6,NC,IDFB,NC,1,IB,IB,IB)
      CALL BASIS(P5,IDFG,NC)
      CALL MPRO(P5,P6,P7,IDFG,NC,NC,1,IB,IB,IB)
      CALL MTRA(P5,P6,IDFG,NC,IB,IB)
      CALL MPRO(P7,P6,P5,IDFG,NC,IDFG,1,IB,IB,IB)
      CALL MSUB(P3,P5,P7,IDFG,IDFG,1,IB,IB,IB)
      CALL ARRAY(ATA,P7,IDFG,IDFG,2,IA)
      CALL MINV(ATA,IDFG,IFAUX)
      IF(IFAUX.EQ.1) THEN
              WRITE(2,*) 'ERROR IN COMPUTING INVERSE'
              RETURN
      END IF
      CALL ARRAY(ATA,P7,IDFG,IDFG,1,IA)
      CALL MPRO(P2,P1,P3,NC,IDFB,IDFB,1,IB,IB,IB)
      CALL MPRO(P3,P4,P6,NC,IDFB,1,1,IB,IB,IB)
      CALL MSUB(CPY,P6,P8,NC,1,1,IB,IB,IB)
      CALL MPRO(CPA,APAI,P6,NC,NA,NA,2,IB,IA,IB)
      CALL MPRO(P6,APY,P3,NC,NA,1,1,IB,IB,IB)
      CALL MSUB(P8,P3,P6,NC,1,1,IB,IB,IB)
      CALL BASIS(P5,IDFG,NC)
C
C     COMPUTE G3'Y
C
      CALL MPRO(P5,P6,P3,IDFG,NC,1,1,IB,IB,IB)
C
C     COMPUTE EPSILON3
C
      CALL MPRO(P7,P3,P6,IDFG,IDFG,1,1,IB,IB,IB)
      CALL MTRA(P5,P8,IDFG,NC,IB,IB)
      CALL MPRO(P8,P6,ESTGAM,NC,IDFG,1,1,IB,IB,IB)
C
C     COMPUTE SS(EPSILON3)
C
      CALL MPRO(P6,P3,P8,1,IDFG,1,1,IB,IB,IB)
      SSGAM=P8(1)
C
C     COMPUTE GAMMA ESTIMATORS
C
      CALL MTRA(P2,P6,NC,IDFB,IB,IB)
      CALL MPRO(P1,P6,P2,IDFB,IDFB,NC,1,IB,IB,IB)
      CALL MPRO(P2,ESTGAM,P8,IDFB,NC,1,1,IB,IB,IB)
C
C     COMPUTE BETA ESTIMATORS
C
      CALL MPRO(P1,P4,P3,IDFB,IDFB,1,1,IB,IB,IB)
      CALL MSUB(P3,P8,P6,IDFB,1,1,IB,IB,IB)
      CALL BASIS(P8,IDFB,NB)
      CALL MTRA(P8,P4,IDFB,NB,IB,IB)
      CALL MPRO(P4,P6,ESTBET,NB,IDFB,1,1,IB,IB,IB)
C
C     COMPUTE MEAN ESTIMATOR
C
      CALL MPRO(APB,ESTBET,P8,NA,NB,1,1,IB,IB,IB)
      CALL MTRA(CPA,P6,NC,NA,IB,IB)
      CALL MPRO(P6,ESTGAM,P3,NA,NC,1,1,IB,IB,IB)
```

```
      DO I=1,NA
             P1(I)=P3(I)+P8(I)
      END DO
      CALL MSUB(APY,P1,P2,NA,1,1,IB,IB,IB)
      CALL MPRO(APAI,P2,P3,NA,NA,1,3,IA,IB,IB)
      ESTMU=0.
      DO I=1,NA
             ESTMU=ESTMU+P3(I)
      END DO
      ESTMU=ESTMU/FLOAT(NA)
C
C     COMPUTE ALPHA ESTIMATORS
C
      DO I=1,NA
             ESTALP(I)=P3(I)-ESTMU
      END DO
      RETURN
      END
C
C
C
      SUBROUTINE DO_ALL(N,Y,NA,NB,NC)
      PARAMETER IA=50,IB=IA**2
      DIMENSION APAI(IA),APY(IB),BPB(IA),BPY(IB),APB(IB),CPC(IA)
     1 ,CPA(IB),CPY(IB),BPC(IB),ATA(IA,IA),Y(IA,IA,IA),N(IA,IA,IA)
      COMMON /BLOCK2/ APAI,APY,BPB,BPY,APB,CPC,CPA,CPY,BPC
      DO I=1,NA
             DO J=1,NB
                    ATA(I,J)=0.
                    DO K=1,NC
                           ATA(I,J)=ATA(I,J)+Y(I,J,K)
                    END DO
             END DO
      END DO
      DO I=1,NA
             APY(I)=0.
             DO K=1,NB
                    APY(I)=APY(I)+ATA(I,K)
             END DO
      END DO
      DO I=1,NB
             BPY(I)=0.
             DO K=1,NA
                    BPY(I)=BPY(I)+ATA(K,I)
             END DO
      END DO
C
C     CONSTRUCT APB
C
      DO I=1,NA
             DO J=1,NB
                    ATA(I,J)=0.
                    DO K=1,NC
                           ATA(I,J)=ATA(I,J)+FLOAT(N(I,J,K))
                    END DO
             END DO
      END DO
```

```
      CALL ARRAY(ATA,APB,NA,NB,1,IA)
      DO I=1,NA
              APAI(I)=0.
              DO K=1,NB
                      APAI(I)=APAI(I)+ATA(I,K)
              END DO
              APAI(I)=1./APAI(I)
      END DO
      DO I=1,NB
              BPB(I)=0.
              DO K=1,NA
                      BPB(I)=BPB(I)+ATA(K,I)
              END DO
      END DO
C
C     COMPUTE BPC
C
      DO I=1,NC
              DO K=1,NB
                      ATA(K,I)=0.
                      DO L=1,NA
                              ATA(K,I)=ATA(K,I)+FLOAT(N(L,K,I))
                      END DO
              END DO
      END DO
C
C     COMPUTE CPC
C
      DO I=1,NC
              CPC(I)=0.
              DO K=1,NB
                      CPC(I)=CPC(I)+ATA(K,I)
              END DO
      END DO
      CALL ARRAY(ATA,BPC,NB,NC,1,IA)
C
C     COMPUTE CPA
C
      DO I=1,NC
              DO J=1,NA
                      ATA(I,J)=0.
                      DO K=1,NB
                              ATA(I,J)=ATA(I,J)+FLOAT(N(J,K,I))
                      END DO
              END DO
      END DO
      CALL ARRAY(ATA,CPA,NC,NA,1,IA)
C
C     COMPUTE CPY
C
      DO I=1,NC
              CPY(I)=0.
              DO J=1,NA
                      DO K=1,NB
                              IF(N(J,K,I).GT.0) CPY(I)=CPY(I)+Y(J,K,I)
                      END DO
              END DO



      END DO
      RETURN
      END
C
C
C
```

INPUT DATA

```
6 6 5
0 0 0 3 0 1
2 0 0 0 0 0
0 0 0 0 0 0
0 0 0 0 2 0
0 0 1 0 0 0
0 0 0 0 0 0
0 0 0 0 1 0
0 0 1 0 0 0
0 2 0 0 0 0
0 0 0 0 0 0
0 0 0 0 0 0
0 0 0 2 0 0
0 1 0 0 0 0
0 0 0 0 0 0
0 0 0 0 0 0
0 0 0 0 0 0
0 0 0 0 2 0
3 0 0 0 0 0
0 0 0 0 0 0
0 0 0 0 0 0
0 0 2 0 0 0
0 0 0 0 0 0
0 0 0 0 0 0
0 0 0 0 0 1
0 0 0 0 0 0
0 0 0 0 0 0
0 0 0 0 0 0
0 0 0 2 0 1
1 0 0 0 0 0
0 0 0 0 0 0
3
1
2
2
3
2
3
4
1
6
7
5
7
4
3
3
5
5
5
6
7
8
9
9
4
5
9
11
```

SOURCE OF VARIATION	D.F.	SUMS OF SQUARES
REGRESSION	14	0.8494099E+03
FITTING OVERALL MEAN	1	0.6900357E+03
FACTOR A UNADJUSTED	5	0.5504767E+02
FACTOR B ADJUSTED FOR A	5	0.3107491E+02
FACTOR C ADJUSTED FOR A AND B	3	0.7325164E+02
RESIDUAL	14	0.2959009E+02
TOTAL	28	0.8790000E+03

ESTIMATE OF MEAN 0.5663054E+01

ESTIMATES OF A EFFECTS

0.5848269E+00
-0.1469856E+01
0.1060071E+00
0.1450385E+01
0.9913754E+00
-0.1662739E+01

ESTIMATES OF B EFFECTS

0.3034693E+01
-0.1536084E+01
-0.5248966E+00
-0.1505948E+01
-0.4502255E+00
0.9824616E+00

ESTIMATES OF C EFFECTS

-0.3818653E+01
0.1513226E+01
-0.1204205E+01
0.3509632E+01
0.0000000E+00

SOURCE OF VARIATION	D.F.	SUMS OF SQUARES
REGRESSION	14	0.8494099E+03
FITTING OVERALL MEAN	1	0.6900357E+03
FACTOR C UNADJUSTED	4	0.1253810E+03
FACTOR A ADJUSTED FOR C	5	0.4936049E+01
FACTOR B ADJUSTED FOR C AND A	4	0.2905714E+02
RESIDUAL	14	0.2959009E+02
TOTAL	28	0.8790000E+03

ESTIMATE OF MEAN 0.5794048E+01

ESTIMATES OF C EFFECTS

-0.2639706E+01
-0.1237653E+01
-0.1990172E+01
0.7587566E+00
0.5108774E+01

ESTIMATES OF A EFFECTS

0.2573433E+00
0.1675721E+00
0.1743431E+01
-0.2806926E+01
-0.1301023E+01
0.1939603E+01

ESTIMATES OF B EFFECTS

0.8732319E-01
-0.5536230E+00
0.4575604E+00
-0.2488406E+01
0.2497146E+01
0.0000000E+00

SOURCE OF VARIATION	D.F.	SUMS OF SQUARES
REGRESSION	14	0.8494100E+03
FITTING OVERALL MEAN	1	0.6900357E+03
FACTOR B UNADJUSTED	5	0.4215723E+02
FACTOR C ADJUSTED FOR B	4	0.1140168E+03
FACTOR A ADJUSTED FOR B AND C	4	0.3200249E+01
RESIDUAL	14	0.2959003E+02
TOTAL	28	0.8790000E+03

ESTIMATE OF MEAN 0.5723517E+01

ESTIMATES OF B EFFECTS

0.1674268E+01
-0.1082611E+01
-0.7142162E-01
-0.1959423E+01
0.9101968E+00
0.5289893E+00

ESTIMATES OF C EFFECTS

-0.3274484E+01
0.2434985E+00
-0.1566982E+01
0.2239901E+01
0.2358066E+01

ESTIMATES OF A EFFECTS

0.4336679E+00
-0.7140656E+00
0.8617994E+00
-0.5146704E+00
-0.6673142E-01
0.0000000E+00

CHAPTER 7

n-Way Classifications with Missing Observations: Estimability and Analysis

Ho, saki, haste, the beaker bring,
Fill up, and pass it round the ring;
Love seemed at first an easy thing—
But ah! the hard awakening.

HĀFEZ SHIRAZI: *persian poet (1348–1398)*

7.1 INTRODUCTION

In this chapter we apply the methods and results that we have obtained for the additive two-way and three-way classifications to the additive n-way classification, with $n \geqslant 4$ factors. In this way, we can directly obtain spanning sets for the vector spaces of estimable α contrasts, estimable β contrasts, and estimable parametric functions not involving the μ, α, and β parameters. We shall see through examples that sometimes the structure of the incidence matrix is such that spanning sets for all vector spaces consisting of estimable contrasts involving single parameter types can be obtained with little additional effort. Regardless, we can obtain spanning sets for these vector spaces by relabeling the parameters and applying our general method to the resulting models.

We shall consider estimability in the additive four-way classification first. This will illustrate how to apply the two-way and three-way methods to classifications with more than three factors. We also illustrate some shortcuts and how the remaining contrasts can sometimes be separated with little additional work. We then present a parametrization for the additive n-way classification and apply our methods for the estimability of parametric functions to this model. Some general results on estimation of parameters in

Latin and Graeco-Latin squares with missing observations are also given. We conclude this chapter by estimation and hypotheses testing in two-way classification model with missing observations when interactions are present.

7.2 ESTIMABILITY CONSIDERATIONS FOR THE FOUR-WAY MODEL

Let $\{Y_{ijkte}\}$ be a collection of N independently distributed random variables each having a common unknown variance σ^2 and each having an expectation of the form

$$E(Y_{ijkte}) = \mu + \alpha_i + \beta_j + \gamma_k + \delta_t, \tag{7.1}$$

where i, j, k, t range from 1 to a, b, c, d, respectively, and for a given i, j, k, t the index e ranges from 1 to n_{ijkt}. As usual, $n_{ijkt} = 0$ means that no random variables with the first four subscripts i, j, k, t occur in the collection. We assume that

$$\begin{aligned}
n_{i\ldots} &= \sum_{jkt} n_{ijkt} \neq 0 \quad \text{for } i = 1, \ldots, a,\\
n_{.j..} &= \sum_{ikt} n_{ijkt} \neq 0 \quad \text{for } j = 1, \ldots, b,\\
n_{..k.} &= \sum_{ijt} n_{ijkt} \neq 0 \quad \text{for } k = 1, \ldots, c,\\
n_{\ldots t} &= \sum_{ijk} n_{ijkt} \neq 0 \quad \text{for } t = 1, \ldots, d.
\end{aligned} \tag{7.2}$$

Thus, we are assuming an additive fixed effects four-way classification model with no restrictions on the unknown parameters occurring in the above expectations.

As before, an estimable linear parametric function is any linear combination of the parameters which can be expressed as a linear combination of the cell expectations $\mu + \alpha_i + \beta_j + \gamma_k + \delta_t$ for which $n_{ijkt} \geq 1$. Theorem 5.2 applies here as well, and the vector space $\bar{\mathscr{A}}$ of estimable linear parametric functions involving only the α parameters is the vector space of estimable α contrasts. Similarly for the vector spaces $\bar{\mathscr{B}}$, $\bar{\ell}$, and $\bar{\mathscr{H}}$, the vector spaces of estimable parametric functions involve only the β, the γ, and the δ parameters, respectively.

An estimable linear parametric function not involving μ must necessarily be a sum of α, β, γ, and δ contrasts. To see this let f be an estimable linear parametric function not involving μ. Then we can write

$$f = \sum_i \sum_j \sum_k \sum_t c_{ijkt}(\mu + \alpha_i + \beta_j + \gamma_k + \delta_t).$$

Using the usual dot notation to denote summation over the suppressed subscripts we can write

$$f=c_{....}\mu+\sum_i c_{i...}\alpha_i+\sum_j c_{.j..}\beta_j+\sum_k c_{..k.}\gamma_k+\sum_t c_{...t}\delta_t.$$

Since f does not involve μ it follows that $c_{....}=0$. Then $\sum_{i=1}^{a} c_{i...}\alpha_i$ is an α contrast because $\sum_{i=1}^{a} c_{i...}=c_{....}=0$. Similarly the other terms of f are seen to be β, γ, and δ contrasts, respectively.

$\bar{\mathscr{A}}$, $\bar{\mathscr{B}}$, $\bar{\ell}$, and $\bar{\mathscr{H}}$ are linearly independent subspaces of $\bar{\Theta}$, the space of estimable linear parametric functions. $\bar{\Theta}$ can be expressed as the direct sum

$$\bar{\Theta}=\bar{\mathscr{A}}\oplus\bar{\mathscr{B}}\oplus\bar{\ell}\oplus\bar{\mathscr{H}}\oplus\bar{\mathscr{C}} \tag{7.3}$$

where, as before, the subspace $\bar{\mathscr{C}}$ is not uniquely determined. It is our goal to present procedures that provide bases for these subspaces.

If we could determine that all cell expectations were estimable, then we would have

$$\begin{aligned}
\bar{\mathscr{A}}&=\text{span}\{\alpha_1-\alpha_2,\alpha_1-\alpha_3,\ldots,\alpha_1-\alpha_a\},\\
\bar{\mathscr{B}}&=\text{span}\{\beta_1-\beta_2,\beta_1-\beta_3,\ldots,\beta_1-\beta_b\},\\
\bar{\ell}&=\text{span}\{\gamma_1-\gamma_2,\gamma_1-\gamma_3,\ldots,\gamma_1-\gamma_c\},\\
\bar{\mathscr{H}}&=\text{span}\{\delta_1-\delta_2,\delta_1-\delta_3,\ldots,\delta_1-\delta_d\}\\
\bar{\mathscr{C}}&=\text{span}\{\mu+\alpha_1+\beta_1+\gamma_1+\delta_1\}.
\end{aligned} \tag{7.4}$$

These spanning sets would then be bases, and the dimensions of the subspaces would be $f_\alpha=a-1$, $f_\beta=b-1$, $f_\gamma=c-1$, $f_\delta=d-1$, and $f_c=1$. Note that any cell expectation could be used as a basis for a subspace $\bar{\mathscr{C}}$ satisfying (7.3).

In this section we develop a procedure for obtaining a spanning set for the vector space of estimable parametric functions involving only one of the four classification effects. For convenience we concentrate on finding a spanning set for $\bar{\mathscr{H}}$, the vector space of all estimable δ contrasts. Once such a spanning set is obtained a basis for $\bar{\mathscr{H}}$ can then be extracted by standard methods.

In order to obtain a spanning set for $\bar{\mathscr{H}}$ we do the following steps:

1 Direct δ Differences First apply the R process described in Chapter 6 to a special two-dimensional matrix. By doing this we generally find more estimable cell expectations. That is, even though a particular n_{ijkt} may be zero, it is possible that the cell expectation $\mu+\alpha_i+\beta_j+\gamma_k+\delta_t$ is estimable. After applying the R process, some δ contrasts may be "directly" seen to be estimable. For example, if $\mu+\alpha_1+\beta_2+\gamma_3+\delta_1$ and $\mu+\alpha_1+\beta_2+\gamma_3+\delta_3$ are both estimable, then their difference $\delta_1-\delta_3$ is estimable. We collect all the "direct δ-differences" that can be obtained in this way. These direct δ differences form a set D.

2 Direct ω Differences We form a new two-dimensional matrix and apply the R process again in order to find more estimable cell expectations. This time we find some estimable contrasts, called "direct ω differences," involving γ and δ effects. The reason that we bring γ effects into consideration is that our procedure makes use of the Q process described in Chapter 6 which eliminates only μ, α, and β effects. We collect all "direct ω differences" that can be obtained in this way. These direct ω differences form a set E.

3 Contrasts from the Q Process At this step we form a special matrix $\bar{\mathbf{M}}$. By applying the Q process we obtain more estimable ω contrasts. These contrasts form the set F.

4 Separation of δ Contrasts Since the set E and F found at steps 2 and 3 contain contrasts involving γ effects as well as δ effects, we must take linear combinations of these ω contrasts to obtain δ contrasts.

Since the maximum dimension of the vector space is $d-1$, if at any point we find $d-1$ linearly independent estimable functions involving δ effects we stop the procedure.

We now describe in detail the first step in obtaining a spanning set for $\bar{\mathscr{H}}$. Transform the four-dimensional $a \times b \times c \times d$ matrix $\mathbf{N}=(n_{ijkt})$ into a special two-dimensional $abc \times d$ matrix as shown in the diagram below, where the rows are identified by triples.

$$
\alpha\beta\gamma
\begin{array}{c}
(1,1,1) \\
(2,1,1) \\
\vdots \\
(i,j,k) \\
\vdots \\
(a,b,c)
\end{array}
\overset{\displaystyle \overset{\delta}{1 \quad 2 \quad \cdots \quad t \quad \cdots \quad d}}{
\left[
\begin{array}{ccc}
& & \\
& & \\
\hline
& n_{ijkt} & \\
\hline
& & \\
\end{array}
\right]}
$$

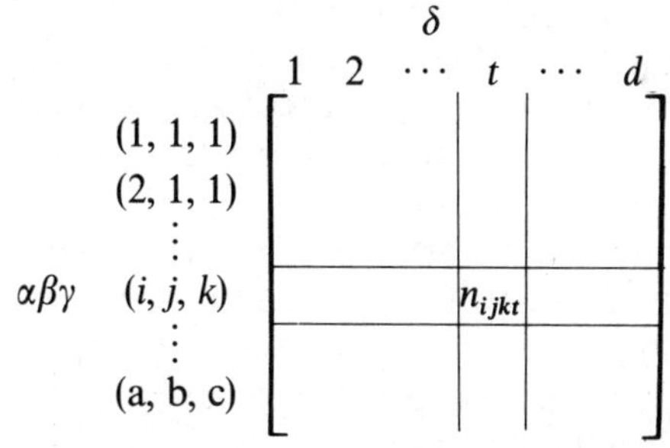

Apply the R process of Chapter 6 to the above two-dimensional matrix to obtain a final matrix $\mathbf{Z}$. Now transform this two-dimensional matrix $\mathbf{Z}$ into a four-dimensional matrix $\mathbf{M}$ with entries $m_{ijkt}=z_{(i,j,k),t}$.

By relabeling parameters and subscripts, we see that Theorem 6.1 applies here, giving us the following:

COROLLARY 7.1 If $m_{ijkt}=1$, then the parametric function $\mu+\alpha_i+\beta_j+\gamma_k+\delta_t$ is estimable.

Although this corollary provides a sufficient condition for a cell expectation to be estimable, it does not in general provide a necessary and sufficient condition. The matrix **M** can, however, for estimability considerations be viewed as the incidence matrix for the original pattern. This follows from the above corollary and from the fact that $n_{ijkt} \neq 0$ in the original incidence matrix **N** implies that $m_{ijkt} \neq 0$ in the matrix **M**. We continue now with step 1 essentially treating **M** as the incidence matrix for the data pattern.

For $t=1,\ldots,d$ let $\mathbf{M}_t$ denote the $a \times b \times c$ matrix having entries m_{ijkt}; we call these matrices the δ levels of **M**. Define an equivalence relation on $\{1,2,\ldots,d\}$ by $t \sim h$ if $\mathbf{M}_t = \mathbf{M}_h$. Suppose there are s equivalence classes; by relabeling, we can assume $\{1,2,\ldots,s\}$ is a complete set of representatives for these equivalence classes. If $t \sim h$, then $\delta_t - \delta_h$ is estimable by Theorem 6.2. We refer to these estimable contrasts as *direct δ differences.*

Let $D = \{\delta_t - \delta_h\colon 1 \leqslant t \leqslant s,\ s+1 \leqslant h \leqslant d,\ t \sim h\}$, and let $\bar{\mathscr{D}}$ be the vector space spanned by D. As explained in Chapter 6, once we have D we can reduce estimability problems to a model with fewer δ effects by keeping δ_t for only one t in each equivalence class in $\{1,2,\ldots,d\}$. That is, in order to find more estimable δ contrasts we need to consider only the estimable contrasts of the form

$$\sum_{i=1}^{a} \sum_{j=1}^{b} \sum_{k=1}^{c} \sum_{t=1}^{s} c_{ijkt}(\mu + \alpha_i + \beta_j + \gamma_k + \delta_t), \tag{7.5}$$

where $c_{ijkt} = 0$ if $m_{ijkt} = 0$.

Similar considerations could also be made with regard to α, β, and γ effects. For instance, by applying the R process to a special two-dimensional matrix having the columns identified by the levels of γ effects, we may find a set, say K, which consists of direct γ differences. In this case, if it happens that we find $c-z$ direct γ differences then we only need to consider contrasts involving z of the γ effects.

It is not actually necessary to do step 1 for our algorithm to work. That is, one could start immediately with step 2 described below by simply setting the matrix $\mathbf{M}'$ equals to **N**. But in many cases doing step 1 and finding the set D will facilitate the search for a spanning set for $\bar{\mathscr{H}}$. The second algorithm of the next section bypasses the set D.

We now start the second step toward finding a spanning set for $\bar{\mathscr{H}}$ by working with the $a \times b \times c \times s$ submatrix of **M** consisting of the first s levels $\mathbf{M}_1, \ldots, \mathbf{M}_s$; denote this matrix by $\mathbf{M}'$.

Define $\boldsymbol{\omega}$ to be the column vector whose transpose is $(\gamma_1, \ldots, \gamma_c, \delta_1, \ldots, \delta_s)$. The reason that we bring γ effects into consideration is that we want to make use of the Q process of Chapter 6 which eliminates only μ, α, and β effects.

We now form the two-dimensional $ab \times cs$ matrix, as shown in the diagram below, where rows and columns are identified by pairs:

$$\begin{array}{cc} & \gamma\delta \\ & \begin{array}{ccccc}(1,1) & \cdots & (k,t) & \cdots & (c,s)\end{array} \\ \alpha\beta \begin{array}{c}(1,1)\\ \vdots \\ (i,j) \\ \vdots \\ (a,b)\end{array} & \left[\begin{array}{ccccc} & & & & \\ & & \vdots & & \\ & \cdots & m_{ijkt} & \cdots & \\ & & \vdots & & \\ & & & & \end{array}\right]. \end{array}$$

Apply the R process to the above two-dimensional matrix to obtain a final matrix $\mathbf{G}$. Now we transform this two-dimensional matrix into the four-dimensional $a \times b \times c \times s$ matrix $\mathbf{M}''$ with entries $m''_{ijkt} = g_{(i,j),(k,t)}$ (note that the rows and columns of $\mathbf{G}$ are identified by pairs).

For each pair of indices (k, t) where $1 \leqslant k \leqslant c$ and $1 \leqslant t \leqslant s$ let $\mathbf{M}''_{(k,t)}$ denote the $a \times b$ matrix with entries m''_{ijkt} and let $\omega_{(k,t)} = \gamma_k + \delta_t$. Because the R process has been applied, either $\mathbf{M}''_{(g,h)} = \mathbf{M}''_{(u,v)}$ or else there are no i, j such that $m''_{ijgh} = 1$ and $m''_{ijuv} = 1$. If $\mathbf{M}''_{(g,h)} = \mathbf{M}''_{(u,v)}$, and $m''_{qrgh} \neq 0$ for some q, r, then $\omega_{(g,h)} - \omega_{(u,v)}$ is directly estimable, and moreover if $m''_{ijuv} = 1$, then

$$\mu + \alpha_i + \beta_j + \gamma_g + \delta_h = (\mu + \alpha_i + \beta_j + \gamma_u + \delta_v) + (\gamma_g + \delta_h - \gamma_u - \delta_v)$$

is estimable. Define an equivalence relation on $C = \{(1, 1), \ldots, (c, s)\}$ by $(g, h) \sim (u, v)$ if $\mathbf{M}''_{(g,h)} = \mathbf{M}''_{(u,v)}$ and $m''_{ijgh} \neq 0$ for some i, j.

Let S be a complete set of representatives for these equivalence classes. For each i, j, there is at most one pair (k, t), $(k, t) \in S$, such that $m_{ijkt} \neq 0$. Let $E = \{\omega_{(g,h)} - \omega_{(k,t)} | (g, h) \in S, (g, h) \sim (k, t)\}$. Let $\mathscr{E}$ be the vector space spanned by E. Then it follows that in order to find more estimable δ contrasts we need to consider only the ω contrasts of the form

$$\lambda'\omega = \sum_{i=1}^{a} \sum_{j=1}^{b} \sum_{(k,t)\in S} d_{ijkt}(\mu + \alpha_i + \beta_j + \gamma_k + \delta_t) \tag{7.6}$$

where $d_{ijkt} = 0$ if $m''_{ijkt} = 0$. Now we start the third step toward finding a spanning set for $\mathscr{H}$.

Let $\mathscr{W}$ be the vector space of estimable ω contrasts of the form $\sum\sum\sum_{(k,t)\in S} d_{ijkt}(\mu + \alpha_i + \beta_j + \gamma_k + \delta_t)$.

Define the matrix $\bar{\mathbf{M}}$ such that

$$\bar{m}_{ij} = \begin{cases} (k, t) & \text{if } m''_{ijkt} = 1, \quad (k, t) \in S \\ 0 & \text{otherwise.} \end{cases}$$

By applying Theorem 6.3, we obtain the following:

COROLLARY 7.2 The vector space $\mathscr{W}$ of estimable linear parametric functions involving $\omega_{(k,t)}$, $(k, t) \in S$, is spanned by contrasts

$$(\gamma_{k_1}+\delta_{t_1})-(\gamma_{k_2}+\delta_{t_2})+\cdots-(\gamma_{k_u}+\delta_{t_u})$$

where $(k_r, t_r)=\bar{m}_{i_r j_r}$ and $(i_1, j_1), (i_2, j_2), \ldots, (i_u, j_u)$ is a loop in $\bar{\mathbf{M}}$.

We shall denote by F the set of contrasts obtained from the loops in $\bar{\mathbf{M}}$. Then Corollary 7.2 says that $\bar{\mathscr{W}}=\text{span } F$.

THEOREM 7.1 The vector space $\bar{\mathscr{G}}$ of estimable ω contrasts may be written as

$$\bar{\mathscr{G}}=\bar{\mathscr{D}}\oplus\{\bar{\mathscr{E}}+\bar{\mathscr{W}}\}.$$

The direct sum in Theorem 7.1 follows from the fact that after finding a direct δ difference $\delta_t-\delta_h$ for the set D, we can eliminate δ_h, $s+1\leqslant h\leqslant d$, and only keep δ_t, $1\leqslant t\leqslant s$, in the model for the remainder of the process. Therefore no linear combinations of the elements in $\bar{\mathscr{D}}$ will occur in $\{\bar{\mathscr{E}}+\bar{\mathscr{W}}\}$, that is

$$\bar{\mathscr{D}}\cap\{\bar{\mathscr{E}}+\bar{\mathscr{W}}\}=\{\emptyset\}.$$

Recall that $\bar{\mathscr{H}}$ is the vector space of all estimable δ contrasts. Since $\bar{\mathscr{D}}\subset\bar{\mathscr{H}}\subset\bar{\mathscr{G}}$, we see that

$$\bar{\mathscr{H}}=\bar{\mathscr{D}}\oplus\bar{\mathscr{F}},$$

where $\bar{\mathscr{F}}=\bar{\mathscr{H}}\cap\{\bar{\mathscr{E}}+\bar{\mathscr{W}}\}$. The fourth and final step for finding a spanning set for $\bar{\mathscr{H}}$ is to take linear combinations of the ω contrasts in E and F to obtain δ contrasts spanning $\bar{\mathscr{F}}$.

The δ differences found directly after applying the R process at the first step yield a linearly independent set D of $d-s$ differences. To obtain a basis for $\bar{\mathscr{H}}$, the problem is then to obtain a basis for estimable δ contrasts from the sets E and F. Fortunately, for small experiments such a basis is often easily obtained by hand.

7.3 TWO ALGORITHMS FOR FINDING A SPANNING SET FOR $\bar{\mathscr{G}}$

In this section we give two algorithms for finding a spanning set for ω contrasts. We use the same model, assumptions, and notation introduced previously.

The first algorithm follows approximately the proof of Corollary 7.1. The second algorithm is different from the first one in that the R process is only implicitly used and an earlier remark is taken into account. It will be seen that each of these two algorithms has some advantages over the other depending upon the structure of the incidence matrix. We will demonstrate both these algorithms with some examples.

Algorithm 7.1

Step 1 Transform the four-dimensional $a \times b \times c \times d$ matrix $\mathbf{N} = (n_{ijkt})$ into a two-dimensional $abc \times d$ matrix whose columns correspond to δ effects. (See the previous section for the exact transformation.)

Step 2 Apply the R process to this two-dimensional matrix to obtain a final matrix.

Step 3 Compare the columns of the final matrix. If two columns, say t and h, have nonzero entries in the same row, then $\delta_t - \delta_h$ is estimable. Keep one column and eliminate the other one. Collect all the direct δ differences that can be obtained in this way. These direct δ differences form the set D.

Step 4 Suppose we are left with s of these columns. Transform this $abc \times s$ matrix into an $a \times b \times c \times s$ matrix $\mathbf{M}'$. (Note that this is a submatrix of the matrix $\mathbf{M}$ in the previous section.)

Step 5 Form a two-dimensional $ab \times cs$ matrix whose columns corresponded to pairs (γ_k, δ_t).

Step 6 Apply the R process to this two-dimensional matrix to obtain a final matrix.

Step 7 Transform this final matrix into a four-dimensional matrix $\mathbf{M}''$.

Step 8 For $(g, h) \neq (u, v)$ if $\mathbf{M}''_{(g,h)} = \mathbf{M}''_{(u,v)}$, and $m''_{ijgh} \neq 0$ for some i, j then $(\gamma_g + \delta_h) - (\gamma_u + \delta_v)$ is estimable. Keep either $\mathbf{M}''_{(g,h)}$ or $\mathbf{M}''_{(u,v)}$ and ignore the other one. Collect all direct ω differences obtained at this step. These direct ω differences form the set E.

Step 9 Form a matrix $\bar{\mathbf{M}}$ as described in the previous section.

Step 10 Apply the Q process to this matrix. Collect the ω contrasts obtained at this step into our spanning set F for $\bar{\mathscr{W}}$ and stop.

EXAMPLE 7.1 For an $a \times b \times c \times d$ incidence matrix $\mathbf{N}$ let $\mathbf{N}_1, \ldots, \mathbf{N}_d$ denote the δ levels, that is, $\mathbf{N}_t$ is an $a \times b \times c$ matrix with entries n_{ijkt}. Consider the following $4 \times 3 \times 2 \times 5$ incidence matrix $\mathbf{N}$:

$\mathbf{N}_1 =$ (δ_1)

	γ_1: β_1	β_2	β_3	γ_2: β_1	β_2	β_3
α_1	3				1	
α_2			2			
α_3						
α_4						

,

$\mathbf{N}_2 =$ (δ_2)

	γ_1: β_1	β_2	β_3	γ_2: β_1	β_2	β_3
α_1						
α_2						
α_3						
α_4	1					1

,

$N_3 =$

	δ_3					
	γ_1			γ_2		
	β_1	β_2	β_3	β_1	β_2	β_3
α_1					1	
α_2						
α_3						
α_4		4			2	

,

$N_4 =$

	δ_4					
	γ_1			γ_2		
	β_1	β_2	β_3	β_1	β_2	β_3
α_1						
α_2	1					
α_3					6	
α_4			1			

,

$N_5 =$

	δ_5					
	γ_1			γ_2		
	β_1	β_2	β_3	β_1	β_2	β_3
α_1	1					
α_2						
α_3				1		
α_4						

.

Steps 1–2 Transform the four-dimensional $4 \times 3 \times 2 \times 5$ matrix $\mathbf{N} = (n_{ijkt})$ into a two-dimensional 24×5 matrix. The final matrix $\mathbf{Z}$ obtained from $\mathbf{N}$ by the R process is

$$
\mathbf{Z} = \begin{array}{c} \alpha\beta\gamma \\ (1,1,1) \\ (2,1,1) \\ (3,1,1) \\ (4,1,1) \\ (1,2,1) \\ (2,2,1) \\ (3,2,1) \\ (4,2,1) \\ (1,3,1) \\ (2,3,1) \\ (3,3,1) \\ (4,3,1) \\ (1,1,2) \\ (2,1,2) \\ (3,1,2) \\ (4,1,2) \\ (1,2,2) \\ (2,2,2) \\ (3,2,2) \\ (4,2,2) \\ (1,3,2) \\ (2,3,2) \\ (3,3,2) \\ (4,3,2) \end{array}
\begin{array}{c} \begin{array}{ccccc} \delta_1 & \delta_2 & \delta_3 & \delta_4 & \delta_5 \end{array} \\
\left[\begin{array}{ccccc}
1 & 0 & \mathrm{x} & 0 & 1 \\
0 & 0 & 0 & 1 & 0 \\
0 & 0 & 0 & 0 & 0 \\
0 & 1 & 0 & 0 & 0 \\
0 & 0 & 0 & 0 & 0 \\
0 & 0 & 0 & 0 & 0 \\
0 & 0 & 0 & 0 & 0 \\
\mathrm{x} & 0 & 1 & 0 & \mathrm{x} \\
0 & 0 & 0 & 0 & 0 \\
1 & 0 & \mathrm{x} & 0 & \mathrm{x} \\
0 & 0 & 0 & 0 & 0 \\
0 & 0 & 0 & 1 & 0 \\
0 & 0 & 0 & 0 & 0 \\
0 & 0 & 0 & 0 & 0 \\
\mathrm{x} & 0 & \mathrm{x} & 0 & 1 \\
0 & 0 & 0 & 0 & 0 \\
1 & 0 & 1 & 0 & \mathrm{x} \\
0 & 0 & 0 & 0 & 0 \\
0 & 0 & 0 & 1 & 0 \\
\mathrm{x} & 0 & 1 & 0 & \mathrm{x} \\
0 & 0 & 0 & 0 & 0 \\
0 & 0 & 0 & 0 & 0 \\
0 & 0 & 0 & 0 & 0 \\
0 & 1 & 0 & 0 & 0
\end{array}\right] \end{array},
$$

where x indicates a cell that has been filled by applying the R process.

Step 3 We see that columns 1, 3, and 5 have nonzero entries in the first row. Thus $\delta_1 - \delta_3$ and $\delta_1 - \delta_5$ are estimable, and they form the set D. Keep columns 1, 2, and 4 and eliminate columns 3 and 5. Thus we are left with a 24×3 matrix.

Step 4 Transform the 24×3 matrix into the four-dimensional $4 \times 3 \times 2 \times 3$ matrix $\mathbf{M}'$ with entries $m'_{ijkt} = z_{(i,j,k),t}$. Thus we have

$\mathbf{M}'_1 =$

1	0	0		0	1	0
0	0	1		0	0	0
0	0	0		1	0	0
0	1	0		0	1	0

,

$\mathbf{M}'_2 =$

0	0	0		0	0	0
0	0	0		0	0	0
0	0	0		0	0	0
1	0	0		0	0	1

,

$\mathbf{M}'_4 =$

0	0	0		0	0	0
1	0	0		0	0	0
0	0	0		0	1	0
0	0	1		0	0	0

.

Steps 5–6 Form a two-dimensional 12×6 matrix with rows and columns identified by pairs, and apply the R process to this two-dimensional matrix to obtain a final matrix $\mathbf{G}$. After applying the R process we have the following matrix where, as before, $\mathbf{x}$ indicates a cell filled in by the R process.

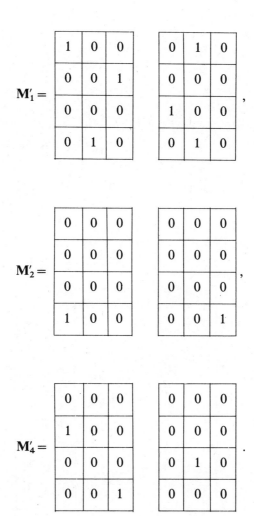

$$
\mathbf{G} = \begin{array}{cc} & (\gamma,\delta) \\ & \begin{array}{cccccc} (1,1) & (2,1) & (1,2) & (2,2) & (1,4) & (2,4) \end{array} \\ (\alpha,\beta) \begin{array}{c} (1,1) \\ (2,1) \\ (3,1) \\ (4,1) \\ (1,2) \\ (2,2) \\ (3,2) \\ (4,2) \\ (1,3) \\ (2,3) \\ (3,3) \\ (4,3) \end{array} & \begin{bmatrix} 1 & \mathrm{x} & 0 & 0 & 0 & 0 \\ 0 & 0 & 0 & \mathrm{x} & 1 & 0 \\ \mathrm{x} & 1 & 0 & 0 & 0 & 0 \\ 0 & 0 & 1 & 0 & 0 & 0 \\ \mathrm{x} & 1 & 0 & 0 & 0 & 0 \\ 0 & 0 & 0 & 0 & 0 & 0 \\ 0 & 0 & 0 & 0 & 0 & 1 \\ 1 & 1 & 0 & 0 & 0 & 0 \\ 0 & 0 & 0 & 0 & 0 & 0 \\ 1 & \mathrm{x} & 0 & 0 & 0 & 0 \\ 0 & 0 & 0 & 0 & 0 & 0 \\ 0 & 0 & 0 & 1 & 1 & 0 \end{bmatrix} \end{array}.
$$

***Step* 7** Transforming this two-dimensional matrix into the four-dimensional $4 \times 3 \times 2 \times 3$ matrix $\mathbf{M}''$ with entries $m''_{ijkt} = g_{(i,j),(k,t)}$. We have

$$
\mathbf{M}''_{(1,1)} = \begin{bmatrix} 1 & 1 & 0 \\ 0 & 0 & 1 \\ 1 & 0 & 0 \\ 0 & 1 & 0 \end{bmatrix}, \qquad \mathbf{M}''_{(2,1)} = \begin{bmatrix} 1 & 1 & 0 \\ 0 & 0 & 1 \\ 1 & 0 & 0 \\ 0 & 1 & 0 \end{bmatrix},
$$

$$
\mathbf{M}''_{(1,2)} = \begin{bmatrix} 0 & 0 & 0 \\ 0 & 0 & 0 \\ 0 & 0 & 0 \\ 1 & 0 & 0 \end{bmatrix}, \qquad \mathbf{M}''_{(2,2)} = \begin{bmatrix} 0 & 0 & 0 \\ 1 & 0 & 0 \\ 0 & 0 & 0 \\ 0 & 0 & 1 \end{bmatrix},
$$

$$
\mathbf{M}''_{(1,4)} = \begin{bmatrix} 0 & 0 & 0 \\ 1 & 0 & 0 \\ 0 & 0 & 0 \\ 0 & 0 & 1 \end{bmatrix}, \qquad \mathbf{M}''_{(2,4)} = \begin{bmatrix} 0 & 0 & 0 \\ 0 & 0 & 0 \\ 0 & 1 & 0 \\ 0 & 0 & 0 \end{bmatrix}.
$$

***Step* 8** We see that $\mathbf{M}''_{(1,1)} = \mathbf{M}''_{(2,1)}$ and $\mathbf{M}''_{(2,2)} = \mathbf{M}''_{(1,4)}$ which implies $(\gamma_1 + \delta_1) - (\gamma_2 + \delta_1)$ and $(\gamma_2 + \delta_2) - (\gamma_1 + \delta_4)$ are estimable. Thus $E = \{\gamma_1 - \gamma_2, -\gamma_1 + \gamma_2 + \delta_2 - \delta_4\}$. Keep $\mathbf{M}''_{(1,1)}$, $\mathbf{M}''_{(1,2)}$, $\mathbf{M}''_{(2,2)}$, and $\mathbf{M}''_{(2,4)}$.

***Step* 9** Form a matrix $\bar{\mathbf{M}}$:

$\bar{\mathbf{M}} =$

(1, 1)	(1, 1)	—
(2, 2)	—	(1, 1)
(1, 1)	(2, 4)	—
(1, 2)	(1, 1)	(2, 2)

.

Step 10 Apply the Q process:

(a)

(1, 1)*	(1, 1)*	—
(2, 2)	—	(1, 1)
*(1, 1)	*(2, 4)	—
(1, 2)	(1, 1)	(2, 2)

leads to $(\gamma_1+\delta_1)-(\gamma_1+\delta_1)+(\gamma_2+\delta_4)-(\gamma_1+\delta_1)$,

(b)

X	(1, 1)	—
(2, 2)*	—	(1, 1)*
(1, 1)	(2, 4)	—
*(1, 2)	(1, 1)	*(2, 2)

and $(\gamma_2+\delta_2)-(\gamma_1+\delta_1)+(\gamma_2+\delta_2)-(\gamma_1+\delta_2)$,

(c)

X	(1, 1)	—
X	—	(1, 1)
(1, 1)*	(2, 4)*	—
*(1, 2)	*(1, 1)	(2, 2)

and $(\gamma_1+\delta_1)-(\gamma_2+\delta_4)+(\gamma_1+\delta_1)-(\gamma_1+\delta_2)$ and

(d)

×	(1, 1)	—
×	—	(1, 1)
×	(2, 4)	—
(1, 2)	(1, 1)	(2, 2)

At this point there are no further loops which can be obtained and hence the Q process and step 10 are concluded. From this step the Q process has given us the following three estimable contrasts:

γ_1	γ_2	δ_1	δ_2	δ_4	
-1	1	-1	0	1	
-2	2	-1	1	0	Set F
1	-1	2	-1	-1	

Note There are many different ways we could have done the Q process above and each way would give a different set of three estimable contrasts. But they would always span the same vector space $\bar{\mathscr{W}}$.

Thus from step 4 we found the set $D=\{\delta_1-\delta_3, \delta_1-\delta_5\}$ and from step 8 we found the set $E=\{\gamma_1-\gamma_2, -\gamma_1+\gamma_2+\delta_2-\delta_4\}$ and from step 10 we found $F=\{-\gamma_1+\gamma_2-\delta_1+\delta_4, -2\gamma_1+2\gamma_2-\delta_1+\delta_2, \gamma_1-\gamma_2+2\delta_1-\delta_2-\delta_4\}$.

By Theorem 7.1 the vector space $\bar{\mathscr{G}}$ of estimable ω contrasts is spanned by D, E, and F. After taking linear combinations, we see that $\bar{\mathscr{G}}$ is spanned by $\{\gamma_1-\gamma_2, \delta_1-\delta_2, \delta_1-\delta_3, \delta_1-\delta_4, \delta_1-\delta_5\}$. Thus dim $\bar{\mathscr{G}}=5$.

Let us point out the difference between Algorithm 3.1 above and Algorithm 3.2 which is presented below. Steps 1–4 of the first algorithm are bypassed. Steps 5–8 are replaced by another method of finding direct ω differences. Steps 9–10 of the first algorithm are the same as steps 5–6 of the second algorithm.

Algorithm 7.2

Step 1 Begin to form the matrix **M** by changing every nonzero entry of **N** to 1.

Step 2 For $(k, t)\neq(u, v)$, if $m_{fgkt}=1$ and $m_{fguv}=1$ for some f, g then

$\omega_{(k,t)} - \omega_{(u,v)}$ is estimable (recall $\omega_{(k,t)} = \gamma_k + \delta_t$) and should be put into the set E' of direct ω differences.

Step 3 For (k, t) and (u, v) as in step 2, redefine the submatrix $\mathbf{M}_{(k,t)}$ and $\mathbf{M}_{(u,v)}$ of $\mathbf{M}$ by

$$m_{ijkt} = \begin{cases} 1 & \text{if } m_{ijkt} = 1 \text{ or } m_{ijuv} = 1 \\ 0 & \text{otherwise.} \end{cases}$$

Eliminate the submatrix $\mathbf{M}_{(u,v)}$.

Step 4 Repeat steps 2 and 3 until no more changes can be made in the matrix $\mathbf{M}$.

Step 5 Form the matrix $\bar{\mathbf{M}}$ as defined in the previous section.

Step 6 Apply the Q process to the matrix $\bar{\mathbf{M}}$. Add the ω contrasts that we can obtain by this step to those found by step 2, and stop. We now have a spanning set for $\bar{\mathscr{G}}$.

A method similar to steps 1–4 of Algorithm 7.2 can be used for finding direct ω differences. This is usually preferable to the method in steps 1–4 of Algorithm 7.1 when calculations are being done by hand. A complete description is given in the next section.

EXAMPLE 7.2 Consider a $4 \times 4 \times 3 \times 3$ factorial experiment with the following incidence matrix:

$\mathbf{N}_1 =$

	δ_1											
	γ_1				γ_2				γ_3			
	β_1	β_2	β_3	β_4								
α_1	2							1				
α_2						1						
α_3			3									
α_4										1		

,

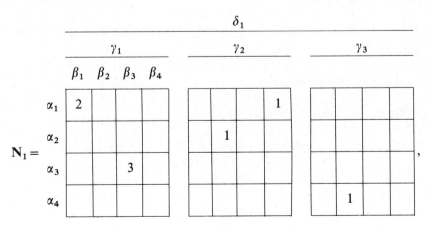

δ_2

$N_2 =$

						1					
			4								2
							3				

,

δ_3

$N_3 =$

	5								3		
					2						

.

Step 1 Change every nonzero entry of **N** to 1. We can assume such a change without drawing the above pattern again.

Step 2 Since $m_{2221} = 1$ and $m_{2213} = 1$. Thus $\omega_{(2,1)} - \omega_{(1,3)}$ is estimable. We put this in the set E'.

Step 3 Eliminate the submatrix **M**(1, 3) and transform its nonzero elements into the submatrix **M**(2, 1). We now have two nonzero elements in the submatrix **M**(2, 1). These elements are m_{1421} and m_{2221}. Note that m_{2213} does not exist any more.

Step 4 In repeating steps 2–3, we find:

(a) From m_{2221} and m_{2233}, we find that $\omega_{(2,1)} - \omega_{(3,3)}$ is estimable. Again we keep one of the submatrices and eliminate the other one. We keep **M**(2, 1).

(b) $m_{3412} = 1$ and $m_{3432} = 1$, which implies $\omega_{(1,2)} - \omega_{(3,2)}$ is estimable. Keep **M**(1, 2) and eliminate **M**(3, 2).

(c) Finally, because $m_{4231} = 1$ and $m_{4223} = 1$, $\omega_{(3,1)} - \omega_{(2,3)}$ is estimable.

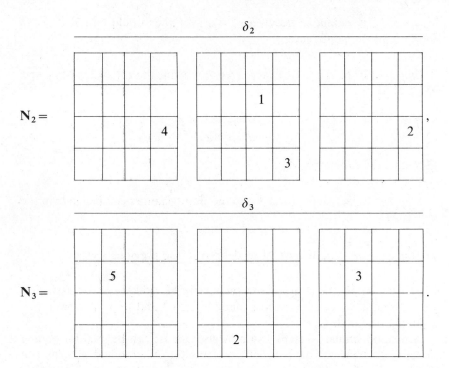

Keep $\mathbf{M}(3, 1)$ and eliminate $\mathbf{M}(2, 3)$. At the end of step 4 we are left with the following submatrices:

$\mathbf{M}(1, 1) =$

1			
		1	

,

$\mathbf{M}(2, 1) =$

			1
	1		

,

$\mathbf{M}(3, 1) =$

	1		

,

$\mathbf{M}(1, 2) =$

			1

.

$\mathbf{M}(2, 2) =$

		1	
			1

,

Step 5 Form the matrix $\bar{\mathbf{M}}$.

$\bar{\mathbf{M}} =$

(1, 1)	—	—	(2, 1)
—	(2, 1)	(2, 2)	—
—	—	(1, 1)	(1, 2)
—	(3, 1)	—	(2, 2)

.

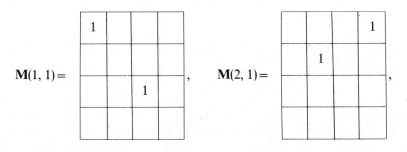

Step 6 By applying the Q process to the matrix $\bar{\mathbf{M}}$, we find that

$$\omega_{(2,1)}-\omega_{(2,2)}+\omega_{(1,1)}-\omega_{(1,2)}+\omega_{(2,2)}-\omega_{(3,1)}$$

is estimable. Thus we get sets

$$E'=\{\omega_{(2,1)}-\omega_{(1,3)},\ \omega_{(2,1)}-\omega_{(3,3)},\ \omega_{(1,2)}-\omega_{(3,2)},\ \omega_{(3,1)}-\omega_{(2,3)}\}$$

from steps 2–4 and

$$F=\{\omega_{(2,1)}+\omega_{(1,1)}-\omega_{(1,2)}-\omega_{(3,1)}\}$$

from step 6.

$\bar{\mathscr{G}}$ is spanned by E' and F. This is true by Theorem 7.1 and the fact that all direct δ differences in D will occur in the set E' if Algorithm 7.2 is used. Thus we have the following five estimable contrasts:

γ_1	γ_2	γ_3	δ_1	δ_2	δ_3
−1	1	0	1	0	−1
0	1	−1	1	0	−1
1	0	−1	0	0	0
0	−1	1	1	0	−1
0	1	−1	1	−1	0

If we row-reduce the above 5×6 matrix we see that $\dim \bar{\mathscr{G}}=4$.

7.4 SOME USEFUL SHORTCUTS

In this section we continue to use the same model, assumptions, and notation as introduced previously. The purposes of this section are: to describe the method for finding direct δ differences when using Algorithm 7.2; to show how estimability problems can be reduced to models with fewer effects and sometimes even fewer factors; and to illustrate some miscellaneous shortcuts.

In the following examples we will use $\mathbf{X}$ to denote the model matrix. We can write $\mathbf{X}=(\mathbf{1}, \mathbf{A}, \mathbf{B}, \mathbf{C}, \mathbf{D})$, where $\mathbf{1}, \mathbf{A}, \mathbf{B}, \mathbf{C}, \mathbf{D}$ are the submatrices corresponding respectively to μ, α, β, γ, and δ effects.

The *degrees of freedom* for any effect is defined to be the dimension of the subspace of estimable linear parametric functions involving that effect.

An Alternative Method for Finding Direct δ Differences

Suppose we have an additive four-way classification model with incidence matrix $\mathbf{N}$. For $t=1,\ldots,d$, let $\mathbf{N}_t$ be the $a\times b\times c$ matrix having entries n_{ijkt}. To find direct δ differences we do the following steps:

Step 1 Begin to form the matrix $\mathbf{M}'$ (the same $\mathbf{M}'$ as in step 4 of Algorithm 7.1) by changing every nonzero entry of $\mathbf{N}$ to 1.

Step 2 For $t \neq h$, if $m'_{efgt} = 1$ and $m'_{efgh} = 1$ for some e, f, g, then $\delta_t - \delta_h$ is estimable and should be put into the set D of direct δ differences.

Step 3 For t and h as in step 2, redefine the submatrix $\mathbf{M}'_t$ of $\mathbf{M}'$ by

$$m'_{ijkt} = \begin{cases} 1 & \text{if } m'_{ijkt} = 1 \text{ or } m'_{ijkh} = 1 \\ 0 & \text{otherwise.} \end{cases}$$

Eliminate the submatrix $\mathbf{M}'_h$.

Step 4 Repeat steps 2 and 3 until no more changes can be made in the matrix $\mathbf{M}'$.

EXAMPLE 7.3 Consider again Example 7.1. After step 1 we have the following pattern:

$\mathbf{M}'_1 =$

	δ_1					
	γ_1			γ_2		
	β_1	β_2	β_3	β_1	β_2	β_2
α_1	1				1	
α_2			1			
α_3						
α_4						

,

$\mathbf{M}'_2 =$

δ_2					
1					1

,

δ_3

$\mathbf{M}'_3 =$

					1		
							,
	1				1		—— m'_{4223}

δ_4

$\mathbf{M}'_4 =$

1							
					1		,
		1					

δ_5

$\mathbf{M}'_5 =$

1							
				1			.

We see that $m'_{1221}=1$ and $m'_{1223}=1$; thus $\delta_1-\delta_3$ is estimable. We now eliminate the submatrix $\mathbf{M}'_3$ and place ones in the appropriate cells in $\mathbf{M}'_1$. Therefore we have the following pattern:

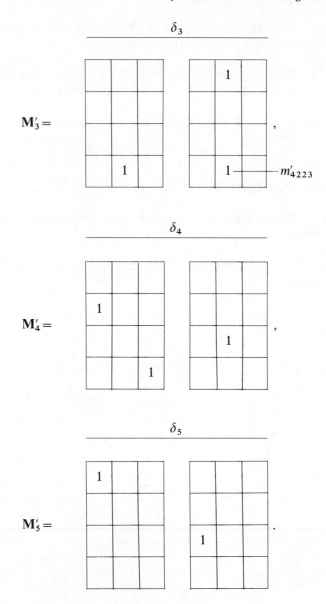

$$\mathbf{M}_1' = \begin{array}{|c|c|c|}\hline 1 & & \\ \hline & & 1 \\ \hline & & \\ \hline & 1 & \\ \hline \end{array} \quad \begin{array}{|c|c|c|}\hline & 1 & \\ \hline & & \\ \hline & & \\ \hline & 1 & \\ \hline \end{array},$$

$$\mathbf{M}_2' = \begin{array}{|c|c|c|}\hline & & \\ \hline & & \\ \hline & & \\ \hline 1 & & \\ \hline \end{array} \quad \begin{array}{|c|c|c|}\hline & & \\ \hline & & \\ \hline & & \\ \hline & & 1 \\ \hline \end{array},$$

$$\mathbf{M}_4' = \begin{array}{|c|c|c|}\hline & & \\ \hline 1 & & \\ \hline & & \\ \hline & & 1 \\ \hline \end{array} \quad \begin{array}{|c|c|c|}\hline & & \\ \hline & & \\ \hline & 1 & \\ \hline & & \\ \hline \end{array},$$

$$\mathbf{M}_5' = \begin{array}{|c|c|c|}\hline 1 & & \\ \hline & & \\ \hline & & \\ \hline & & \\ \hline \end{array} \quad \begin{array}{|c|c|c|}\hline & & \\ \hline & & \\ \hline 1 & & \\ \hline & & \\ \hline \end{array}.$$

We see that $m'_{1111}=1$ and $m'_{1115}=1$; thus $\delta_1-\delta_5$ is estimable. By the same procedure as above we eliminate $\mathbf{M}_5'$ and keep $\mathbf{M}_1'$. This leads to the following pattern:

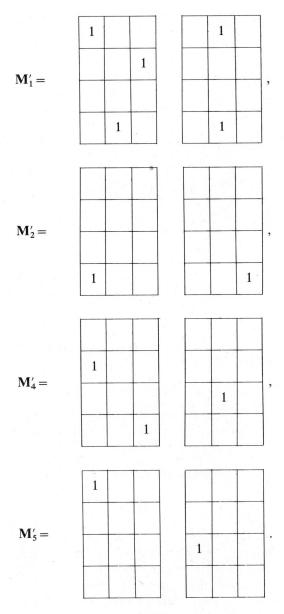

$$
\mathbf{M}_1' = \begin{array}{|c|c|c|}\hline 1 & & \\ \hline & & 1 \\ \hline & & \\ \hline & 1 & \\ \hline \end{array} \quad \begin{array}{|c|c|c|}\hline & 1 & \\ \hline & & \\ \hline 1 & & \\ \hline & 1 & \\ \hline \end{array},
$$

$$
\mathbf{M}_2' = \begin{array}{|c|c|c|}\hline & & \\ \hline & & \\ \hline & & \\ \hline 1 & & \\ \hline \end{array} \quad \begin{array}{|c|c|c|}\hline & & \\ \hline & & \\ \hline & & \\ \hline & & 1 \\ \hline \end{array},
$$

$$
\mathbf{M}_4' = \begin{array}{|c|c|c|}\hline & & \\ \hline 1 & & \\ \hline & & \\ \hline & & 1 \\ \hline \end{array} \quad \begin{array}{|c|c|c|}\hline & & \\ \hline & & \\ \hline & 1 & \\ \hline & & \\ \hline \end{array}.
$$

We see that no more changes can be made in the matrix $\mathbf{M}'$, and the set D is $\{\delta_1-\delta_3, \delta_1-\delta_5\}$.

Reduced Model

In order to show how a problem can be reduced to a smaller one let us consider the $4\times4\times3\times3$ factorial experiment of Example 7.2. Using Algorithm 7.2, we found that $\dim \bar{\mathscr{G}}=4$, which implies that we have full degrees of freedom for γ and δ effects; in other words, all the linear contrasts of γ effects and of δ effects are estimable. In particular we know that $\delta_1-\delta_2$ is estimable. Consider any occupied cell in $\mathbf{N}_2$, such as cell (2, 3, 2, 2). We see that

$$
\mu+\alpha_2+\beta_3+\gamma_2+\delta_1=(\mu+\alpha_2+\beta_3+\gamma_2+\delta_2)+(\delta_1-\delta_2)
$$

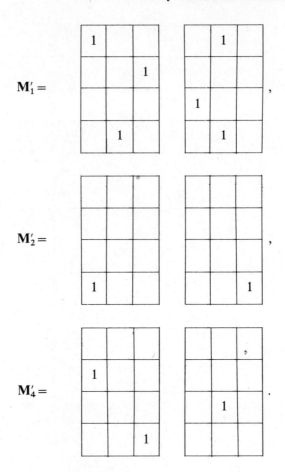

is estimable; we can indicate this by placing a 1 in cell (2, 3, 2, 1). Now we can place a zero in cell (2, 3, 2, 2) without losing any information about estimability, because

$$\mu+\alpha_2+\beta_3+\gamma_2+\delta_2=(\mu+\alpha_2+\beta_3+\gamma_2+\delta_1)-(\delta_1-\delta_2).$$

In general, knowing that $\delta_1-\delta_2$ is estimable, we can change the entries in all occupied cells of $\mathbf{N}_2$ to zero if we place a 1 in the corresponding cells in $\mathbf{N}_1$. Thus we can eliminate $\mathbf{N}_2$ and reduce the problem to a new model with only two δ effects. The new incidence matrix $\mathbf{N}^{(1)}$, obtained from $\mathbf{N}$ as just described, is shown below.

$\mathbf{N}_1^{(1)}=$

	δ_1											
	γ_1				γ_2				γ_3			
	β_1	β_2	β_3	β_4								
α_1	1											
α_2						1	1					
α_3			1	1								1
α_4								1		1		

,

$\mathbf{N}_3^{(1)}=$

	δ_2											
		1								1		
						1						

.

We also know $\delta_1-\delta_3$ is estimable. By the same procedure as above we can eliminate $\mathbf{N}_3^{(1)}$ after placing ones in the appropriate cells in $\mathbf{N}_1^{(1)}$. Now the problem is reduced to a model with incidence matrix $\mathbf{N}^{(2)}$ with only one δ level:

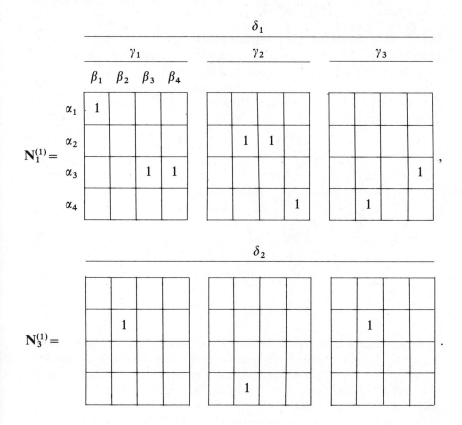

$\mathbf{N}_1^{(2)} =$

	γ_1				γ_2				γ_3			
	β_1	β_2	β_3	β_4	β_1	β_2	β_3	β_4	β_1	β_2	β_3	β_4
α_1	1											
α_2		1				1	1			1		
α_3			1	1								1
α_4						1		1		1		

But of course this is just the incidence matrix of a $4 \times 4 \times 3$ factorial design. Thus, knowing that all δ contrasts are estimable, we have been able to reduce the problem from a four-way model to a three-way model.

Because all the γ contrasts are estimable, by the same procedure the problem reduces from a three-way to a two-way model with the following incidence matrix:

	β_1	β_2	β_3	β_4
α_1	1			
α_2		1	1	
α_3			1	1
α_4		1		1

As we see, it turns out that this incidence matrix has the same pattern as the matrix $\bar{\mathbf{M}}$, that is, they both have nonzero entries in the same positions. The reason for this is that by applying the R process at (any) step 2 and step 6 of Algorithm 7.1 no cell (i, j, h, t) will be filled unless there is already some cell (i, j, u, v) which is filled.

In order to find the degrees of freedom for α and β effects we use the method of Chapter 5.

Apply the R process to the above incidence matrix. We get the following final matrix:

	β_1	β_2	β_3	β_4
α_1	1			
α_2		1	1	1
α_3		1	1	1
α_4		1	1	1

From the final matrix we see that $\alpha_2 - \alpha_3$ and $\alpha_2 - \alpha_4$ form a basis for the space of estimable α contrasts so the degrees of freedom for α effects is 2. We can find $r(\mathbf{A}, \mathbf{B}) = b + (\text{degrees of freedom for } \alpha) = 4 + 2 = 6$, and the degrees of freedom for β effects is $r(\mathbf{A}, \mathbf{B}) - a = 6 - 4 = 2$.

The above procedure for finding the degrees of freedom for α and β effects using a two-way model is valid only because we know there are full degrees of freedom for γ and δ effects. In case there are not full degrees of freedom for γ and δ effects, we can switch the roles of α and β with the roles of γ and δ in order to find the degrees of freedom for α and β effects. However, the above procedure is always valid for finding $r(\mathbf{A}, \mathbf{B})$.

We are now in a position to write the table of degrees of freedom. Note that $r(\mathbf{X}) = r(\mathbf{A}, \mathbf{B}) + \dim \bar{\mathscr{G}} = 6 + 4 = 10$ which implies the model matrix is not of maximal rank (i.e., is not 11).

Source of Variation	Degrees of Freedom
Mean	1
α	2
β	2
γ	2
δ	2
Confounded	1
Residual	$N - r(\mathbf{X}) = 11 - 10 = 1$
Total	11

It may happen that the incidence matrix is such that we can obtain all estimable contrasts by direct differences. In such a case one can use the method developed by Weeks and Williams (1964).

EXAMPLE 7.4 Consider the following $3 \times 3 \times 2 \times 2$ incidence matrix:

$\mathbf{N}_1 =$ (δ_1)

	γ_1: β_1	γ_1: β_2	γ_1: β_3	γ_2: β_1	γ_2: β_2	γ_2: β_3
α_1	1					
α_2		1				
α_3			1			1

,

$\mathbf{N}_2 =$ (δ_2)

	γ_1: β_1	γ_1: β_2	γ_1: β_3	γ_2: β_1	γ_2: β_2	γ_2: β_3
α_1				1		
α_2	1				1	1
α_3			1			

.

Notice that cells (3, 3, 1, 1) and (3, 3, 2, 1) lead to the estimability of $\gamma_1 - \gamma_2$, and cells (3, 3, 1, 1) and (3, 3, 1, 2) lead to the estimability of $\delta_1 - \delta_2$. Now the problem is to find estimable contrasts for α and β effects.

Because $\gamma_1 - \gamma_2$ and $\delta_1 - \delta_2$ are estimable, we can form a two-way table as follows:

	β_1	β_2	β_3
α_1	1		
α_2	1	1	1
α_3			1

.

From cells (1, 1) and (2, 1) we find $\alpha_1 - \alpha_2$ to be estimable, and from cells (2, 3), (3, 3), $\alpha_2 - \alpha_3$ is estimable; similarly from (2, 1), (2, 2) and (2, 2), (2, 3) we find that $\beta_1 - \beta_2$ and $\beta_2 - \beta_3$ are estimable. Note that all estimable functions have been found simply by *direct differences.* The design matrix for the above problem is of maximal rank, that is, $r(\mathbf{X}) = 7$, and table of degrees of freedom is as follows:

Source of Variation	Degrees of Freedom
Mean	1
α	2
β	2
γ	1
δ	1
Residual	2
Total	7

It is important to notice that each $\mathbf{N}_t$ can be considered as the model matrix of a three-way classification model, so it is sometimes wise and efficient to see if we can find all γ contrasts by applying the method of Chapter 6. It it happens that we can obtain all γ contrasts by this method, the problem can be considered as a three-way model by dropping the γ effects.

EXAMPLE 7.5 Consider a 2^4 design with incidence matrix as follows:

δ_1

	γ_1		γ_2	
	β_1	β_2	β_1	β_2
α_1	1			1
α_2		1	1	

δ_2

	γ_1		γ_2	
	β_1	β_2	β_1	β_2
α_1		1		
α_2				

Consider the $\mathbf{N}_1$ matrix corresponding to the first level of δ. $\mathbf{N}_1$ is a $2 \times 2 \times 2$ matrix and applying the Q process, it follows immediately that $\gamma_1 - \gamma_2$ is estimable. We see this by forming the matrix $\bar{\mathbf{N}}_1 = \sum_{k=1}^{c} k\mathbf{N}_1$.

$\bar{\mathbf{N}}_1 =$

1	2
2	1

.

Now applying the Q process to the matrix $\bar{\mathbf{N}}_1$ we get $2\gamma_1 - 2\gamma_2$ to be estimable. Therefore we can now work with the following pattern by combining the γ levels:

$\mathbf{N}_1' =$

	δ_1	
	β_1	β_2
α_1	1	1
α_2	1	1

and

$\mathbf{N}_2' =$

	δ_2	
	β_1	β_2
α_1		1
α_2		

.

Again this problem can be considered as a three-way additive model, from which it immediately follows that all estimable differences of α's, β's, and δ's exist.

In a four-way model with incidence matrix $\mathbf{N}$, if one two-dimensional submatrix $\mathbf{N}_{(k,t)}$ has all its cell expectations estimable (which is equivalent to the condition that all its cells can be filled by applying the R process), then the problem can be considered as a two-way classification model.

EXAMPLE 7.6 Consider a $4 \times 3 \times 2 \times 2$ factorial design with the following pattern:

$N_1 =$

	δ_1					
	γ_1			γ_2		
	β_1	β_2	β_3			
α_1	1	1				
α_2		1	1			
α_3			1		1	
α_4			1			

,

$N_2 =$

	δ_2					
	1					
	1					
		1				

.

After applying the R process to the two-dimensional matrix $N_{(1,1)}$ all cells will be filled. This means that we have full degrees of freedom for α and β effects. The problem of finding degrees of freedom for γ and δ effects reduces to considering a two-way model with the following incidence matrix:

	δ_1	δ_2
γ_1	1	1
γ_2	1	

.

We see that $\gamma_1 - \gamma_2$ and $\delta_1 - \delta_2$ are estimable, and hence the design matrix is of maximal rank.

7.5 A TECHNIQUE FOR SEPARATING γ CONTRASTS FROM δ CONTRASTS

In this section we continue to use the same model, assumptions, and notation as introduced previously. The basic purpose of this section is to present an algorithm, which is essentially a row-reduction algorithm, for obtaining a spanning set for $\bar{\mathscr{H}}$, the vector space of estimable δ contrasts. Recall that several procedures are available for obtaining a spanning set for $\bar{\mathscr{G}}$, the vector space of estimable ω-contrasts, and that our algorithms conveniently provide a decomposition of $\bar{\mathscr{G}}$ of the form

$$\bar{\mathscr{G}} = \bar{\mathscr{D}} \oplus (\bar{\mathscr{E}} + \bar{\mathscr{W}}).$$

Since $\bar{\mathscr{D}} \subset \bar{\mathscr{H}}$ it then follows that

$$\bar{\mathscr{H}} = \bar{\mathscr{D}} \oplus \bar{\mathscr{F}}, \qquad \bar{\mathscr{F}} = \bar{\mathscr{H}} \cap (\bar{\mathscr{E}} + \bar{\mathscr{W}}).$$

And thus we need only concentrate on finding a spanning set for $\bar{\mathscr{F}}$ which is the vector space of all estimable δ contrasts in $\bar{\mathscr{E}} + \bar{\mathscr{W}}$.

In step 3 below if $q_{ijk_1} = 1$ and $q_{ijk_2} = 1$, then we let $P = (k_1, t_1)(k_2, t_2)$, where $t_r (r = 1, 2)$ is the unique index such that $n_{ijk_r t_r} = 1$, and let $\gamma(P) = \gamma_{k_1} - \gamma_{k_2}$ and $\delta(P) = \delta_{t_1} - \delta_{t_2}$. In step 8 below, if $P = (i_1, j_1) \dots (i_u, j_u)$ is a loop in $\bar{Q}$, then we let $\gamma(P) = \gamma_{k_1} - \gamma_{k_2} + \cdots - \gamma_{k_u}$ and $\delta(P) = \delta_{t_1} - \delta_{t_2} + \cdots - \delta_{t_u}$ where $k_r, t_r (r = 1, \dots, u)$ are the unique indices such that $n_{i_r j_r k_r t_r} = 1$.

We assume that the set D of direct δ differences has already been found. Thus we are left with the matrix $\mathbf{M}'$ which is the incidence matrix of a four-way model which has no direct δ differences. Therefore, we can assume that we have an $a \times b \times c \times s$ incidence matrix $\mathbf{N}$ with all entries either zero or 1 and with no direct δ differences.

Our technique involves finding a spanning set for γ contrasts which would be estimable if δ effects were not in the model. We refer to such a γ contrast as a γ *path*.

The technique consists of the following steps:

Step 1 Form the matrix $\bar{\mathbf{N}} = \sum_{t=1}^{s} t\mathbf{N}_t$, where $\mathbf{N}_t$ is an $a \times b \times c$ matrix with entries n_{ijkt}.

Step 2 Form the $a \times b \times c$ matrix $\mathbf{Q}$ with entries

$$q_{ijk} = \begin{cases} 1 & \text{if } \bar{n}_{ijk} \neq 0 \\ 0 & \text{otherwise.} \end{cases}$$

Let $\mathbf{Q}_k$ be the $a \times b$ matrix with entries q_{ijk}.

Step 3 If $q_{ijk} = 1$ and $q_{ijh} = 1$ for some i, j, then we have $\gamma(P_1) = \gamma_k - \gamma_h$

where $P_1=(k, t_k)(h, t_h)$ for some t_k, t_h. We see that $\gamma(P_1)+\delta(P_1)=(\gamma_k+\delta_{t_k})-(\gamma_h+\delta_{t_h})$ is estimable, but that $\gamma(P_1)=\gamma_k-\gamma_h$ is not necessarily estimable in the four-way model. Set either q_{ijk} or q_{ijh} equal to zero and keep the other. Continue in this way until there are no nonzero cells in common between $\mathbf{Q}_k$'s for $k=1,\ldots,c$. Collect all $\gamma(P)$'s that can be obtained by this step, and also keep track of the $\delta(P)$'s using $\bar{\mathbf{N}}$. If $\gamma(P)=0$, keep $\delta(P)$ as an estimable δ-contrast.

Step 4 In the collection of nonzero γ paths, find, if possible, $\gamma_1(P_1),\ldots,\gamma(P_m)$ such that $a_1\gamma(P_1)+\cdots+a_m\gamma(P_m)=0$ for some numbers $a_1,\ldots,a_m$ which are not all zero. If this is not possible go to step 7.

Step 5 Form $a_1\delta(P_1)+\cdots+a_m\delta(P_m)$ and keep it as an estimable δ contrast.

Step 6 For one a_i in step 4, $a_i\neq0$, eliminate $\gamma(P_i)$ from the collection of γ paths. (The reason we can do this is shown in Lemma 7.1 below.) Go to step 4.

Step 7 Form the matrix $\bar{\mathbf{Q}}=\sum_{k=1}^{c}k\mathbf{Q}_k$.

Step 8 Apply the Q process to the matrix $\bar{\mathbf{Q}}$. Add all $\gamma(P)$'s (where P is a loop in $\bar{\mathbf{Q}}$) that can be obtained by this step to the collection of γ paths and also keep track of the corresponding $\delta(P)$'s using $\bar{\mathbf{N}}$.

Step 9 Repeat steps 4–6 with "go to step 7" replaced by "go to step 10."

Step 10 Stop. The δ contrasts found in step 5 form a spanning set for $\bar{\mathscr{F}}$.

If at any step we find $d-1$ linearly independent estimable contrasts we stop.

At step 3 and step 8 sometimes it is efficient whenever we find a γ path to search in the matrix $\bar{\mathbf{N}}$ to see whether a corresponding δ contrast is equal to zero. If it is, then depending on the degrees of freedom for γ effects, the problem could possibly be reduced to a three-way model.

LEMMA 7.1 Consider steps 4–6 above. Suppose we find $a_1\gamma(P_1)+\cdots+a_m\gamma(P_m)=0$, $a_1\neq0$, in step 4 and we eliminate $\gamma(P_1)$ from the collection of γ paths in step 6. If, in addition, $b_1\gamma(P_1)+\cdots+b_m\gamma(P_m)=0$, then we will still be able to tell that $b_1\delta(P_1)+\cdots+b_m\delta(P_m)$ is estimable.

Proof From the algorithm in step 4 we find $a_1\gamma(P_1)+\cdots+a_m\gamma(P_m)=0$.

Then we know $a_1\delta(P_1)+\cdots+a_m\delta(P_m)$ is estimable. Suppose $a_1\neq 0$ and at step 6 we eliminate $\gamma(P_1)$. Thus we cannot find the zero combination $b_1\gamma(P_1)+\cdots+b_m\gamma(P_m)=0$ directly from the algorithm because $\gamma(P_1)$ has been eliminated. From step 4 we can write

$$\gamma(P_1)=\left(\frac{1}{a_1}\right)[-a_2\gamma(P_2)-\cdots-a_m\gamma(P_m)], \qquad a_1\neq 0.$$

By substitution we can write

$$b_1\left\{\left(\frac{1}{a_1}\right)[-a_2\gamma(P_2)-\cdots-a_m\gamma(P_m)]\right\}+b_2\gamma(P_2)+\cdots+b_m\gamma(P_m)=0,$$

or

$$\left(b_2-\frac{b_1}{a_1}a_2\right)\gamma(P_2)+\cdots+\left(b_m-\frac{b_1}{a_1}a_m\right)\gamma(P_m)=0. \tag{7.7}$$

It is possible to get (7.7) directly from the remaining γ paths $\gamma(P_2),\ldots,\gamma(P_m)$. Moreover,

$$\begin{aligned}b_1\gamma(P_1)+\cdots+b_m\gamma(P_m)=&\left(\frac{b_1}{a_1}\right)[a_1\gamma(P_1)+\cdots+a_m\gamma(P_m)]\\&+\left[\left(b_2-\frac{b_1}{a_1}a_2\right)\gamma(P_2)\right.\\&\left.+\cdots+\left(b_m-\frac{b_1}{a_1}a_m\right)\gamma(P_m)\right].\end{aligned}$$

Thus eliminating $\gamma(P_1)$ in step 6 would not lead to losing $b_1\gamma(P_1)+\cdots+b_m\gamma(P_m)=0$. Corresponding to the zero combination of γ paths in (7.7), we know

$$\left(b_2-\frac{b_1}{a_1}a_2\right)\delta(P_2)+\cdots+\left(b_m-\frac{b_1}{a_1}a_m\right)\delta(P_m)$$

is estimable. Thus we can write

$$\begin{aligned}b_1\delta(P_1)+\cdots+b_m\delta(P_m)=&\left(\frac{b_1}{a_1}\right)[a_1\delta(P_1)+\cdots+a_m\delta(P_m)]\\&+\left[\left(b_2-\frac{b_1}{a_1}a_2\right)\delta(P_2)\right.\\&\left.+\cdots+\left(b_m-\frac{b_1}{a_1}a_m\right)\delta(P_m)\right].\end{aligned}$$

Since the right-hand side is estimable, the left-hand side is also estimable. □

EXAMPLE 7.7 Consider the following $3 \times 3 \times 2 \times 2$ incidence matrix $\mathbf{N}$:

δ_1

	γ_1			γ_2		
	β_1	β_2	β_3			
α_1	1					1
$\mathbf{N}_1 =$ α_2		1		1		
α_3			1			

,

δ_2

					1	
$\mathbf{N}_2 =$			1			
	1					1

.

Step 1 Form $\bar{\mathbf{N}} = \mathbf{N}_1 + 2\mathbf{N}_2$

	1				2	1
$\bar{\mathbf{N}} =$		1	2	1		
	2		1			2

.

Step 2 Form the matrix $\mathbf{Q}$ as follows:

	1				1	1
$\mathbf{Q} =$		1	1	1		
	1		1			1

.

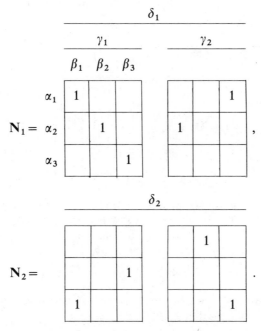

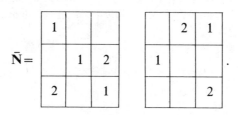

Step 3 From $\mathbf{Q}$ it is seen that we can estimate $\gamma_1 - \gamma_2$ (ignoring δ effects),

because $q_{331}=1$ and $q_{332}=1$. Save the γ path $\gamma(P_1)=\gamma_1-\gamma_2$, the δ path $\delta(P_1)=\delta_1-\delta_2$, and set $q_{331}=0$. This is the only γ path we get from step 3. Thus steps 4–6 will be eliminated.

***Step* 7** Form the matrix $\bar{\mathbf{Q}}$

$\bar{\mathbf{Q}}=\mathbf{Q}_1+2\mathbf{Q}_2=$

1	2	2
2	1	1
1		2

.

***Step* 8** Applying the Q process to the matrix $\bar{\mathbf{Q}}$, we see that $P_2=(1,1)(1,2)(2,2)(2,1)$ forms a loop in $\bar{\mathbf{Q}}$, which leads to the γ path $\gamma(P_2)=\gamma_1-\gamma_2+\gamma_1-\gamma_2=2\gamma_1-2\gamma_2$, and corresponding $\delta(P_2)=\delta_1-\delta_2+\delta_1-\delta_1=\delta_1-\delta_2$. Also, we find $P_3=(1,1)(1,3)(2,3)(2,1)$, $\gamma(P_3)=2\gamma_1-2\gamma_2$, $\delta(P_3)=\delta_2-\delta_1$ and $P_4=(1,1)(1,3)(3,3)(3,1)$, $\gamma(P_4)=0$, $\delta(P_4)=0$.

***Step* 9** From step 3 we have $\gamma(P_1)=\gamma_1-\gamma_2$ and from step 8 we have $\gamma(P_2)=2\gamma_1-2\gamma_2$. We see that $2\gamma(P_1)-\gamma(P_2)=0$.

***Step* 10** We see that $2\delta(P_1)-\delta(P_2)=\delta_1-\delta_2$ is estimable.

We can stop because there can only be $d-1=1$ degree of freedom for δ effects.

EXAMPLE 7.8 Consider the following $4\times5\times2\times4$ incidence matrix $\mathbf{N}$:

$\mathbf{N}_1=$

	δ_1									
	γ_1					γ_2				
	β_1	β_2	β_3	β_4	β_5					
α_1						1				
α_2	1									
α_3										
α_4										

,

$\mathbf{N}_2 =$

δ_2									
	1								
								1	
							1		

,

$\mathbf{N}_3 =$

δ_3									
						1			
	1								
				1					

,

$\mathbf{N}_4 =$

δ_4									
			1						
							1		
1									

.

Step 1 Form $\bar{\mathbf{N}} = \mathbf{N}_1 + 2\mathbf{N}_2 + 3\mathbf{N}_3 + 4\mathbf{N}_4$:

$\bar{\mathbf{N}} =$

γ_1					γ_2				
	2				1				
1			4			3			
	3						4	2	
4				3			2		

.

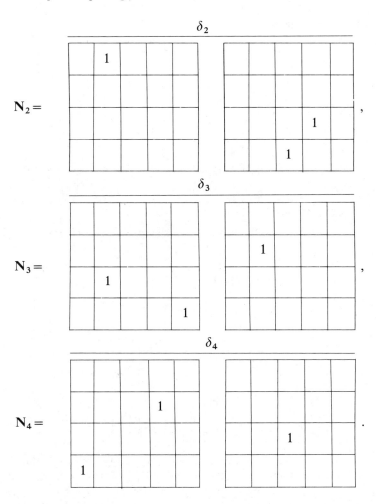

Step 2 Form the matrix $\mathbf{Q}$:

$\mathbf{Q}=$

γ_1					γ_2				
	1				1				
1			1			1			
	1						1	1	
1				1			1		

Step 3 No γ path can be found by this step. Therefore steps 4–6 will be eliminated.

Step 7 Form the matrix $\bar{\mathbf{Q}}=\mathbf{Q}_1+2\mathbf{Q}_2$:

$\bar{\mathbf{Q}}=$

2	1			
1	2		1	
	1	2	2	
1		2		1

Step 8 Apply the Q process to the matrix $\bar{\mathbf{Q}}$. We see that $P_1=(1, 1)(1, 2)(2, 2)(2, 1)$ and $P_2=(3, 3)(3, 2)(1, 2)(1, 1)(4, 1)(4, 3)$ and $P_3=(3, 4)(3, 2)(1, 2)(1, 1)(2, 1)(2, 4)$ are loops in $\bar{\mathbf{Q}}$. The first and the second loops lead to $\gamma(P_1)=-2\gamma_1+2\gamma_2$ and $\gamma(P_2)=\gamma_1-\gamma_2$ and we have $\gamma(P_1)+2\gamma(P_2)=0$. Corresponding to these $\gamma(P)$'s are $\delta(P_1)=\delta_1-\delta_2+\delta_3-\delta_1=-\delta_2+\delta_3$ and $\delta(P_2)=\delta_4-\delta_3+\delta_2-\delta_1+\delta_4-\delta_2=-\delta_1-\delta_3+2\delta_4$. Therefore $\delta(P_1)+2\delta(P_2)=-2\delta_1-\delta_2-\delta_3+4\delta_4$. The third loop leads to $\gamma(P_3)=\gamma_2-\gamma_1+\gamma_1-\gamma_2+\gamma_1-\gamma_1=0$, and corresponding to this γ path is $\delta(P_3)=\delta_4-\delta_3+\delta_2-\delta_1+\delta_1-\delta_4=\delta_2-\delta_3$. Therefore $\{-2\delta_1-\delta_2-\delta_3+4\delta_4, \delta_2-\delta_3\}$ is a basis for $\bar{\mathscr{F}}$.

7.6 ESTIMABILITY CONSIDERATIONS FOR THE n-WAY MODEL

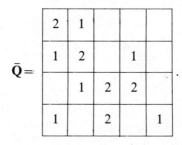

We now extract the essential features of previous sections which are applicable to very general classification models to present a complete and more general solution to the problem of estimability in classification models. Previously, we provided a spanning set for estimable functions involving γ and δ effects. We then introduced an algorithm to separate γ from δ effects within this spanning set. The general model which will be presented here is faced with the same difficulties. That is, it only provides a spanning set for estimable functions not involving μ, α, and β effects. However, as will be seen in several examples, sometimes the structure of the incidence matrix is such that the separation of effects does not require too much effort.

Let $\{Y_{ijk}\}$ be a collection of N independently distributed random variables each having a common unknown variance σ^2 and each having an expectation of the form

$$E(Y_{ijt}) = \mu + \alpha_i + \beta_j + \mathbf{k}_{ijt} \cdot \boldsymbol{\eta} \tag{7.8}$$

where $\boldsymbol{\eta}$ is a column vector of parameters not including μ, $\alpha_1, \ldots, \alpha_a$ and $\beta_1, \ldots, \beta_b$, and $\mathbf{k}_{ijk}$ is a vector of real numbers, so that the dot product $\mathbf{k}_{ijt} \cdot \boldsymbol{\eta}$ is a linear combination of parameters in $\boldsymbol{\eta}$. The indices i and j range from 1 to a and 1 to b respectively, and for each i, j the index t ranges in a set T_{ij} where T_{ij} is a finite (possibly empty) index set.

A linear parametric function is *estimable* if it can be expressed as a linear combination of the expectations $\mu + \alpha_i + \beta_j + \mathbf{k}_{ijt} \cdot \boldsymbol{\eta}$ for $i = 1, \ldots, a, j = 1, \ldots, b$, and $t \in T_{ij}$. An η *contrast* is defined to be any linear parametric function involving only parameters in η.

Let $\bar{\mathscr{G}}$ denote the vector space of all estimable η contrasts. Then our goal is to obtain a spanning set for the vector space $\bar{\mathscr{G}}$ of estimable η contrasts.

Let $K = \{\mathbf{k}_{ijt} | (i, j, t) \in I\}$, where I denotes the index set (i, j, t) such that i, j range from 1 to a, b, respectively, and $t \in T_{ij}$. Note since K is written in set form that K consists of the set of distinct vectors $\mathbf{k}_{ijt}$. Now for each $\mathbf{k} \in K$ define an $a \times b$ matrix $\mathbf{N}(\mathbf{k}) = (n_{ij}(\mathbf{k}))$ such that

$$n_{ij}(\mathbf{k}) = \begin{cases} 1 & \text{if } \mathbf{k} = \mathbf{k}_{ijt} \text{ for some } t \in T_{ij} \\ 0 & \text{otherwise.} \end{cases}$$

LEMMA 7.2 An estimable η contrast can be written in the form $\sum_{\mathbf{k} \in K} c_{\mathbf{k}} (\mathbf{k} \cdot \boldsymbol{\eta})$, where the $c_{\mathbf{k}}$ are real numbers that sum to 0.

The proof of this lemma follows directly from the definition of estimability and the fact that the coefficients of μ and α_i and β_j are zero.

Suppose there are two indices u, v such that T_{uv} has more than one element, and let t_1 and t_2 be two elements of T_{uv}. We see that if we set $\mathbf{k}_1 = \mathbf{k}_{uvt_1}$ and $\mathbf{k}_2 = \mathbf{k}_{uvt_2}$, it follows that

$$(\mathbf{k}_1 - \mathbf{k}_2) \cdot \boldsymbol{\eta} = (\mu + \alpha_u + \beta_v + \mathbf{k}_1 \cdot \boldsymbol{\eta}) - (\mu + \alpha_u + \beta_v + \mathbf{k}_2 \cdot \boldsymbol{\eta})$$

is estimable. This is equivalent to saying that if $\mathbf{N}(\mathbf{k}_1)$ and $\mathbf{N}(\mathbf{k}_2)$ have a nonzero entry in common, that is, for some u, v both $n_{uv}(\mathbf{k}_1) = 1$ and $n_{uv}(\mathbf{k}_2) = 1$, then $(\mathbf{k}_1-\mathbf{k}_2)\cdot\boldsymbol{\eta}$ is directly seen to be estimable. We refer to these direct differences as *direct η differences.* Moreover if $(\mathbf{k}_1-\mathbf{k}_2)\cdot\boldsymbol{\eta}$ is estimable, and $n_{ij}(\mathbf{k}_1)=1$ for some i, j then

$$\mu+\alpha_i+\beta_j+\mathbf{k}_2\cdot\boldsymbol{\eta}=(\mu+\alpha_i+\beta_j+\mathbf{k}_1\cdot\boldsymbol{\eta})-(\mathbf{k}_1-\mathbf{k}_2)\cdot\boldsymbol{\eta}$$

is estimable.

Define a two-dimensional matrix $\mathbf{Z}=(z_{(i,j),k})$ with rows identified by the pairs (i, j) where i, j range from 1 to a, b, respectively, with columns identified by the vectors $\mathbf{k}\in K$ and with entries defined by $z_{(i,j),k}=n_{ij}(\mathbf{k})$. That is, the matrix $\mathbf{N}(\mathbf{k})$ is transformed to column $\mathbf{k}$ of $\mathbf{Z}$. Now apply the R process to the matrix $\mathbf{Z}$ to obtain a final matrix $\mathbf{W}$ with entries $w_{(i,j),k}$. For each $\mathbf{k}\in K$ define an $a\times b$ matrix $\mathbf{M}(\mathbf{k})=(m_{ij}(\mathbf{k}))$ by $m_{ij}(\mathbf{k})=w_{(i,j),k}$, that is, take column $\mathbf{k}$ and put it back into an $a\times b$ matrix.

COROLLARY 7.3 If $m_{ij}(\mathbf{k})=1$, then $\mu+\alpha_i+\beta_j+\mathbf{k}\cdot\boldsymbol{\eta}$ is estimable.

The proof follows exactly as that of Corollary 7.1. If $\mathbf{M}(\mathbf{k}_1)=\mathbf{M}(\mathbf{k}_2)$, then $(\mathbf{k}_1-\mathbf{k}_2)\cdot\boldsymbol{\eta}$ is directly seen to be estimable.

Now define an equivalence relation on K by $\mathbf{k}_1\sim\mathbf{k}_2$ if $\mathbf{M}(\mathbf{k}_1)=\mathbf{M}(\mathbf{k}_2)$. Let S be a complete set of representatives of these equivalence classes in K. For each pair (i, j) there is at most one $\mathbf{s}\in S$ such that $m_{ij}(\mathbf{s})=1$. Let J be the set of triples $(i, j, \mathbf{s})$ such that $m_{ij}(\mathbf{s})=1$, $i=1,\ldots,a$, $j=1,\ldots,b$, $\mathbf{s}\in S$.

Let $D^*=\{(\mathbf{k}-\mathbf{s})\cdot\boldsymbol{\eta}\,|\,\mathbf{k}\in K \text{ and } \mathbf{s}\in S,\ \mathbf{k}\sim\mathbf{s}\}$ and recall from above that the elements of D^* are estimable. Now let $\bar{\mathscr{D}}^*$ be the vector space spanned by D^*. Let $\bar{\mathscr{W}}$ be the vector space of estimable η contrasts of the form $\sum_{(i,j,\mathbf{s})\in J} c_{ij\mathbf{s}}(\mu+\alpha_i+\beta_j+\mathbf{s}\cdot\boldsymbol{\eta})$. Then similar to Theorem 7.1 we get the following:

LEMMA 7.3 $\bar{\mathscr{G}}=\bar{\mathscr{D}}^*+\bar{\mathscr{W}}$.

For each $\mathbf{s}\in S$ select a symbol $\theta_\mathbf{s}$ which is not the numeral 0. Define a matrix $\bar{\mathbf{M}}$ such that

$$\bar{m}_{ij}=\begin{cases}\theta_\mathbf{s} & \text{if } m_{ij}(\mathbf{s})=1 \text{ for some } \mathbf{s}\in S\\ 0 & \text{if } m_{ij}(\mathbf{s})=0 \text{ for all } \mathbf{s}\in S.\end{cases}$$

The symbol $\theta_\mathbf{s}$ is a device which can help us remember which vector $\mathbf{s}$ is associated with which pair (i, j).

Let F denote the set of η contrasts obtained from the loops in $\bar{\mathbf{M}}$. Then by applying Theorem 6.3, we obtain the following:

COROLLARY 7.4 $\bar{\mathscr{W}}=\text{span}(F)$.

Combining Lemma 7.3 and Corollary 7.4, we have the following:

LEMMA 7.4 The vector space $\bar{\mathscr{G}}$ of estimable η contrasts is spanned by $D^* \cup F$.

7.7 EXAMPLES: ESTIMABILITY

We have presented several examples to demonstrate techniques for finding spanning sets for vector spaces of certain estimable parametric functions. We now provide some more examples which illustrate these techniques as well as the general theory of the previous section. With regard to the notation used in the following examples, several comments seem appropriate. In all of the examples we will use η to denote all parameters in the model except $\mu, \alpha_1, \ldots, \alpha_a$ and $\beta_1, \ldots, \beta_b$. Thus, $\bar{\mathscr{G}}$, $\bar{\mathscr{D}}^*$, and $\bar{\mathscr{W}}$ will denote, respectively, the vector space of estimable η contrasts, the vector space spanned by D^*, and the vector space spanned by F, where D^* is the set of direct η difference and F is the set of estimable contrasts obtained from the Q process. For the examples in which we have an additive four-way model we use the notation $\bar{\mathscr{D}}$ to denote the vector space spanned by estimable direct δ differences and $\bar{\mathscr{E}}$ to denote the vector space spanned by estimable ω differences found after obtaining $\bar{\mathscr{D}}$. It should be noted that the notation $\bar{\mathscr{G}}$ and $\bar{\mathscr{W}}$ of the general model is consistent with the notation $\bar{\mathscr{G}}$ and $\bar{\mathscr{W}}$ as used in an additive four-way model. Also note that when the additive four-way model is viewed in the general setting, the vector $\boldsymbol{\eta}$ is the vector ω. We begin with a 2^4 factorial design to illustrate how the results for the general model can be compared with those for the four-way model.

EXAMPLE 7.9 Consider a collection of random variables Y_{ijuve} such that

$$E(Y_{ijuve}) = \mu + \alpha_i + \beta_j + \gamma_u + \delta_v,$$

where i, j, u, and v range from 1 to 2, respectively, and $e = 1, \ldots, n_{ijuv}$. Let n_{ijuv}'s be arranged in incidence matrices $\mathbf{N}_{11}, \mathbf{N}_{21}, \mathbf{N}_{12}$, and $\mathbf{N}_{22}$ with entries (i, j) of $\mathbf{N}_{uv}$ being n_{ijuv}. Suppose that we have the following pattern:

$$\mathbf{N}_{11} = \begin{array}{|c|c|}\hline 3 & \\ \hline & 2 \\ \hline\end{array}, \qquad \mathbf{N}_{21} = \begin{array}{|c|c|}\hline & \\ \hline 1 & \\ \hline\end{array},$$

$$\mathbf{N}_{12} = \begin{array}{|c|c|}\hline & \\ \hline 2 & \\ \hline\end{array}, \qquad \mathbf{N}_{22} = \begin{array}{|c|c|}\hline & 4 \\ \hline & \\ \hline\end{array}.$$

Let $\boldsymbol{\eta}'=(\gamma_1, \gamma_2, \delta_1, \delta_2)$. The set K consists of the vectors $\mathbf{k}=\mathbf{k}_{uv}$ for $u=1, 2$ and $v=1, 2$ where $\mathbf{k}_{uv}\cdot\boldsymbol{\eta}=\gamma_u+\delta_v$. The matrix $\mathbf{N}(\mathbf{k}_{uv})$ is obtained from $\mathbf{N}_{uv}$ by changing all nonzero entries to ones. We see that $\mathbf{k}_{21}\sim\mathbf{k}_{12}$. Hence, let $S=\{\mathbf{k}_{11}, \mathbf{k}_{21}, \mathbf{k}_{22}\}$ so that $D^*=\{\gamma_1-\gamma_2+\delta_2-\delta_1\}$ and let $\theta_{\mathbf{k}_{uv}}=(u, v)$ so that

$$\bar{\mathbf{M}}=\begin{array}{|c|c|}\hline (1, 1) & (2, 2) \\ \hline (2, 1) & (1, 1) \\ \hline \end{array}\,.$$

Apply the Q process to the above matrix $\bar{\mathbf{M}}$. We find that $F=\{2\gamma_1-2\gamma_2+\delta_1-\delta_2\}$. Here the set D^* is the set E for the four-way model.

By Theorem 7.1 $\bar{\mathscr{G}}=\bar{\mathscr{E}}+\bar{\mathscr{W}}$ and by Lemma 7.3 $\bar{\mathscr{G}}=\bar{\mathscr{D}}^*+\bar{\mathscr{W}}$; in other words, by Theorem 7.1 $\bar{\mathscr{G}}$ is spanned by E and F and by Lemma 7.4 $\bar{\mathscr{G}}$ is spanned by $D^*\cup F$. After reduction we see that $\bar{\mathscr{G}}$ is spanned by $\{\gamma_1-\gamma_2, \delta_1-\delta_2\}$.

Since $\gamma_1-\gamma_2$ and $\delta_1-\delta_2$ are estimable, then $\dim\bar{\mathscr{G}}=2$ and $r(\mathbf{X})=r(\mathbf{A}, \mathbf{B})+\dim\bar{\mathscr{G}}=3+2=5$, which implies $\mathbf{X}$ is of maximal rank. This means that the table of degrees of freedom is as follows:

Source	Degrees of Freedom
Mean	1
α	1
β	1
γ	1
δ	1
Residual	7
Total	12

Graeco-Latin Square

Suppose that we take a Latin square and superimpose upon it a second square with the treatments denoted by Greek letters. If the two squares have the property that each Latin letter coincides exactly once with each Greek letter when the squares are superimposed, they then are said to be orthogonal. The combined squares are said to form a *Graeco-Latin square.* Graeco-Latin squares can be used to provide designs for four factors each v levels in v^2 runs. In these designs the treatments are grouped into replicates in three different ways (triple grouping). Each treatment (Latin letters)

	c_1	c_2	c_3	c_4	c_5
r_1	$A\alpha$	$B\beta$	$C\gamma$	$D\delta$	$E\varepsilon$
r_2	$B\gamma$	$C\delta$	$D\varepsilon$	$E\alpha$	$A\beta$
r_3	$C\varepsilon$	$D\alpha$	$E\beta$	$A\gamma$	$B\delta$
r_4	$D\beta$	$E\gamma$	$A\delta$	$B\varepsilon$	$C\alpha$
r_5	$E\delta$	$A\varepsilon$	$B\alpha$	$C\beta$	$D\gamma$

Figure 7.1 A 5×5 Graeco-Latin square design.

appears once in each row and column and once with each Greek letter. A Graeco-Latin square of size five is presented in Figure 7.1.

Adding a third square orthogonal to each of the others gives a *hyper-Graeco–Latin square.* The Graeco-Latin squares and the hyper-Graeco-Latin squares are fractional factorial designs. The Graeco-Latin square is a $1/v^2$ fraction of factorial with four factors, each at v levels.

EXAMPLE 7.10 Consider the following 4×4 Graeco-Latin square with three missing observations:

	c_1	c_2	c_3	c_4
r_1	—	$B\beta$	$C\gamma$	$D\delta$
r_2	$B\gamma$	$A\delta$	—	$C\beta$
r_3	$C\delta$	$D\gamma$	$A\beta$	$B\alpha$
r_4	$D\beta$	—	$B\delta$	$A\gamma$

This design could be viewed as a 4^4 factorial design with a $4 \times 4 \times 4 \times 4$ incidence matrix. This incidence matrix would be presented as sixteen 4×4 submatrices, one corresponding to each pair of levels [e.g., (B, γ)] of the third and fourth factors. There would be no direct ω differences because of the way, in which a Graeco-Latin square is designed. Thus $\bar{\mathscr{G}} = \bar{\mathscr{W}}$. Now apply the Q process.

1.

—	$B\beta$	$C\gamma$	$D\delta$
$B\gamma$	$A\delta$	—	$C\beta$
$C\delta$	$D\gamma$	$A\beta$	$B\alpha$
$D\beta$	—	$B\delta$	$A\gamma$

$B\gamma - A\delta + D\gamma - C\delta.$

2.

—	$B\beta$	$C\gamma$	$D\delta$
	$A\delta$	—	$C\beta$
$C\delta$	$D\gamma$	$A\beta$	$B\alpha$
$D\beta$	—	$B\delta$	$A\gamma$

$B\beta - D\delta + C\beta - A\delta.$

3.

—		$C\gamma$	$D\delta$
	$A\delta$	—	$C\beta$
$C\delta$	$D\gamma$	$A\beta$	$B\alpha$
$D\beta$	—	$B\delta$	$A\gamma$

$A\delta - C\beta + B\alpha - D\gamma.$

4.

		$C\gamma$	$D\delta$
		—	$C\beta$
$C\delta$	$D\gamma$	$A\beta$	$B\alpha$
$D\beta$	—	$B\delta$	$A\gamma$

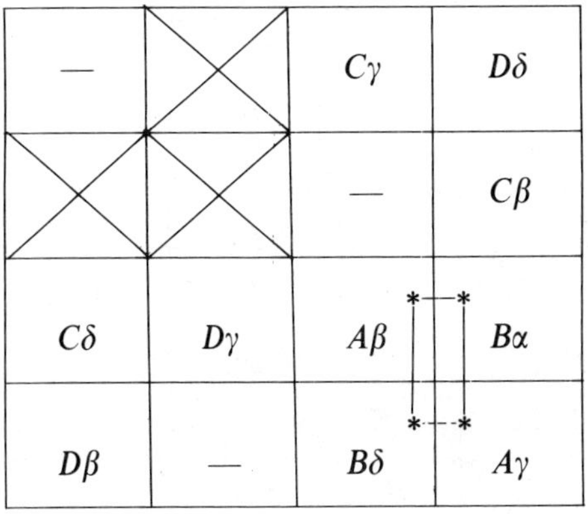

$A\gamma - B\alpha + A\beta - B\delta.$

5.

		$C\gamma$	$D\delta$
		—	$C\beta$
$C\delta$	$D\gamma$	$A\beta$	$B\alpha$
$D\beta$	—	$B\delta$	

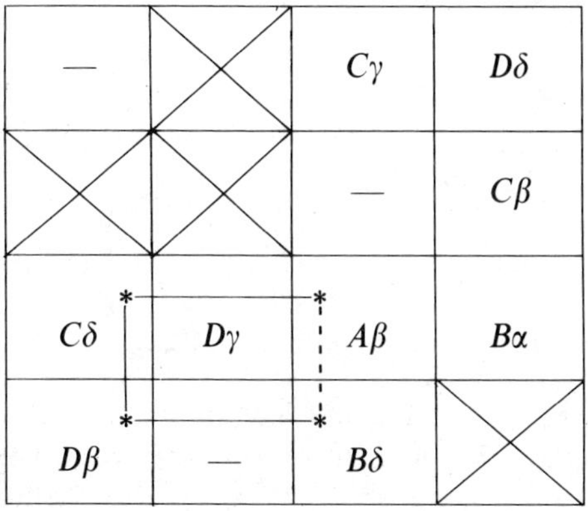

$A\beta - C\delta + D\beta - B\delta.$

6.

		$C\gamma$	$D\delta$
		—	$C\beta$
$C\delta$	$D\gamma$		$B\alpha$
$D\beta$	—	$B\delta$	

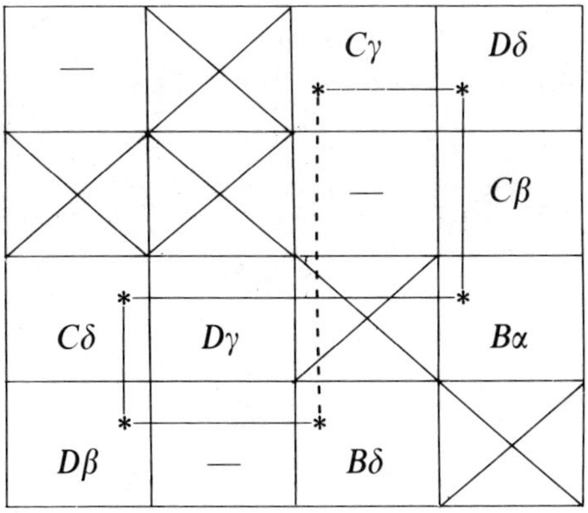

$C\gamma - D\delta + B\alpha - C\delta + D\beta - B\delta.$

We are left with no more loops:

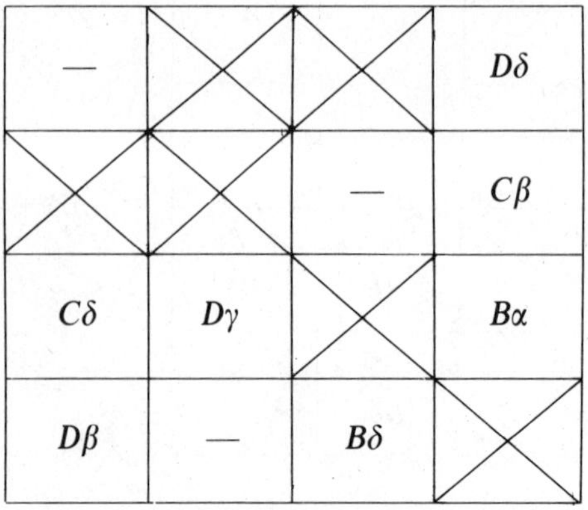

The Q process above has given us the following six estimable contrasts:

	A	B	C	D	α	β	γ	δ
1.	−1	1	−1	1	0	0	2	−2
2.	−1	1	1	−1	0	2	0	−2
3.	1	1	−1	−1	1	−1	−1	1
4.	2	−2	0	0	−1	1	1	−1
5.	1	−1	−1	1	0	2	0	−2
6.	0	0	0	0	1	1	1	−3

Note There are many different ways we could have done the Q process above and each way might give a different set of six estimable contrasts. But they would always span the same vector space $\bar{\mathscr{W}}$. If we row-reduce the above 6×8 matrix we see that $\dim \bar{\mathscr{W}} = 6$. Since $\bar{\mathscr{G}} = \bar{\mathscr{W}}$ it follows that $r(\mathbf{X}) = r(\mathbf{A}, \mathbf{B}) + \dim \bar{\mathscr{W}} = 7 + 6 = 13$, so the design has maximal rank.

We could see that $r(\mathbf{X}) = 13$ by row-reducing the 13×17 matrix $\mathbf{X}$. But clearly the 6×8 matrix is easier to row-reduce. Now we get the following table of degrees of freedom:

Source	Degrees of Freedom
Mean	1
Rows	3
Columns	3
Greek	3
Latin	3
Residual	0
Total	13

EXAMPLE 7.11 Consider another 4×4 Graeco-Latin square, this time with two missing observations:

—	$B\beta$	$C\gamma$	$D\delta$
$B\gamma$	$A\delta$	$D\alpha$	$C\beta$
$C\delta$	—	$A\beta$	$B\alpha$
$D\beta$	$C\alpha$	$B\delta$	$A\gamma$

From the Q process we get seven estimable contrasts involving A, B, C, D, α, β, γ, δ. By row-reducing the corresponding 7×8 matrix we see dim $\bar{\mathscr{G}} = 5$. In fact, we get

Greek letter contrasts: $\beta - \delta$, $\alpha - \beta + \gamma - \delta$
Latin letter contrasts: $B - C$, $A - B - C + D$
Inseparable contrasts: $(D - A) + (\alpha - \gamma)$.

$r(\mathbf{X}) = r(\mathbf{A}, \mathbf{B}) + \dim \bar{\mathscr{G}} = 7 + 5 = 12$, which is not maximal rank. We see that there are 2 degrees of freedom for Greek letter effects and 2 degrees of freedom for Latin letter effects.

If we switch the role of the Greek and Latin letter effects with the role of the row and column effects, we can apply the Q process again to get the corresponding results for row and column effects. The table of degrees of freedom is as follows:

Source	Degrees of Freedom
Mean	1
Rows	2
Columns	2
Greek	2
Latin	2
Confounded	3
Residual	2
Total	14

Note that in the above example we had two missing cells and we found that the design matrix is not of maximal rank, while in Example 7.10 we had three missing cells and we found that the design matrix is of maximal rank. This is due to which cells are missing. Notice that in Example 7.10 all missing cells had α effects in common, but in Example 7.11 the two missing cells had no Greek or Latin letter effects in common.

EXAMPLE 7.12 Consider a 4×4 hyper-Graeco-Latin square with four missing observations according to the following pattern:

	c_1	c_2	c_3	c_4
r_1	—	$B\beta b$	$C\gamma c$	$D\delta d$
r_2	$D\gamma b$	$C\delta a$	$B\alpha d$	—
r_3	$C\beta d$	$D\alpha c$	—	$B\gamma a$
r_4	$B\delta c$	—	$D\beta a$	$C\alpha b$

In order to illustrate how the results for the general model can be applied to the above additive five-way classification model, let T_{11}, T_{24}, T_{33}, and T_{42} be empty and for each occupied (i, j) cell let $T_{ij} = \{t\}$ where t consists of the letter combination capital Latin, Greek, and small Latin in that cell, $\boldsymbol{\eta}' = (A, B, C, D, \alpha, \beta, \gamma, \delta, a, b, c, d)$, and the set K consists of 12 distinct 12×1 vectors which are collections of $\mathbf{k}_{ijt}$'s, where, for example, $\mathbf{k}_{12t} = (0\ 1\ 0\ 0\ 0\ 1\ 0\ 0\ 0\ 1\ 0\ 0)'$. Thus, there are 12 $\mathbf{N}(\mathbf{k})$ matrices. Note that hyper-Graeco-Latin squares are fractional factorial designs, that is, they are $(1/p^3)$ fraction of p^5 design. Therefore, for each $\mathbf{k} = \mathbf{k}_{ijt} \in K$, $\mathbf{N}(\mathbf{k})$ has a 1 in cell (i, j) and zero elsewhere. Thus, as we saw in Example 7.10 there are no direct differences, so that $S = K$, $\mathbf{M}(\mathbf{k}) = \mathbf{N}(\mathbf{k})$ for all $\mathbf{k} \in K$ and the vector space of estimable η contrasts $\mathscr{G}$ is equal to $\bar{\mathscr{W}}$. We now form the matrix $\bar{\mathbf{M}}$. For the symbol θ_s we use t where $\mathbf{s} = \mathbf{k}_{ijt}$. Thus $\bar{\mathbf{M}}$ is the original hyper-Graeco-Latin square having triples symbols [e.g., $(B\alpha c)$] in cell (i, j). Applying the Q process to the above incidence matrix leads to the following five estimable contrasts:

1. $D\gamma b - C\delta a + D\alpha c - C\beta d.$
2. $C\gamma c - B\beta b + C\delta a - B\alpha d.$
3. $C\gamma c - B\beta b + C\delta a - D\gamma b + B\delta c - D\beta a.$
4. $D\delta d - B\beta b + D\alpha c - B\gamma a.$
5. $D\delta d - B\beta b + C\delta a - D\gamma b + B\delta c - C\alpha b.$

The above five estimable contrasts are all linearly independent and thus $\dim \bar{\mathscr{G}} = 5$, $r(\mathbf{X}) = r(\mathbf{A}, \mathbf{B}) + \dim \bar{\mathscr{G}} = 5 + 7 = 12$, so the design matrix is not of maximal rank. Moreover, after row-reduction on the 5×12 matrix which is obtained from the above five estimable contrasts, we find out that there is only 1 degree of freedom for capital letters, 1 degree of freedom for Greek letters, and 3 degrees of freedom are confounded. This fact is rather strange in the sense that in all four missing observations the letter A is common and thus one might expect some degrees of freedom for the small letters.

The above examples bring us to an interesting problem: Is there a relation between the degrees of freedom and the pattern of the design in Latin and Graeco-Latin squares designs?

The above founding leads us to some general results on estimation of parameters in Latin squares (LS) and Graeco-Latin squares (GLS) with missing observations.

We can show that if in an additive model with $p-2$ mutually orthogonal Latin squares (MOLS), if one omits up to $p-1$ observations from the same row, the same column, or which correspond to the same letter in any of the squares, all effects are estimable. On the other hand with only two missing observations not from the same row, the same column, or corresponding to the same letter in any of the squares, 1 degree of freedom is lost for *each* set of effects.

We shall first prove the main theorems and then point out their implications.

THEOREM 7.2 Consider a set of $p-2$ MOLS of side p. In a model with the row effects, the column effects, and the main effects of classification for these $p-2$ LS, all effects are estimable if one omits any set of $p-1$ observations that correspond to a common row, common column, or a common letter in any of the first $p-2$ squares.

Proof We note that the model contains p classifications each at p levels so that each classification gives rise to $p-1$ parameters. When we include the general mean we have p^2-p+1 linearly independent parametric functions. All effects will be said to be estimable if the dimension of the estimation space (of the model with missing observations) is p^2-p+1, that is, the same as that of the estimation space of the model with no missing observations. Since we have p^2-p+1 observations all effects are estimable if and only if there are no zero functions, that is, linear functions of observations with expected value zero.

The system of $p-2$ MOLS can be embedded in a system of $p-1$ MOLS. Let $A_1, A_2, \ldots, A_p$ be the letters for the $(p-1)$th square and let $T_{A_1}, T_{A_2}, \ldots, T_{A_p}$ denote the totals of the observations corresponding to the letters A_1, $A_2, \ldots, A_p$, respectively, when all the p^2 observations are available. In this model any zero function must be a contrast in $T_{A_1}, T_{A_2}, \ldots, T_{A_p}$ and hence

$T_{A_1}-T_{A_p},\ldots,T_{A_{p-1}}-T_{A_p}$ is a basis for the space of zero functions. Let $f_i = T_{A_i}-T_{A_p}$, $i=1,2,\ldots,p-1$.

With the pattern of omitted observations as stated in the theorem no two missing observations can have any A letter in common. This follows from the orthogonality of the complete set of Latin squares. Without loss of generality we can assume that one observation is missing from each of $T_{A_1},\ldots,T_{A_{p-1}}$.

Let f be a zero function. Obviously, f will also be a zero function when all the p^2 observations are available and hence f can be expressed as $a_1 f_1 + a_2 f_2 + \cdots + a_{p-1} f_{p-1}$ for some constants $a_1, a_2, \ldots, a_{p-1}$. Since T_{A_1} is not available when one observation on A_1 is missing, a_1 must be zero. Similarly, it follows that all other a_i's are also zero. Thus, f must be zero identically. This completes the proof of the theorem. □

COROLLARY 7.5 With any t missing observations from the same row, same column, or corresponding to the same letter in any of the $p-2$ squares, all effects are estimable if $t<p-1$.

COROLLARY 7.6 With the pattern of missing observations as given in Theorem 7.2 or Corollary 7.5 all effects are estimable if one omits one or more sets of effects corresponding to rows, columns, or main effects of classification corresponding to any of the $p-2$ squares.

Proof An estimable parametric function continues to be estimable when some parameters not involved in this function are dropped from the model. Thus the result follows. □

COROLLARY 7.7 In a LS of side p which can be embedded in a complete system of $p-1$ MOLS, all effects are estimable if we omit up to $p-1$ observations from any row, any column, or which correspond to the same letter in *any* of the $p-1$ squares.

Proof The proof follows from Corollary 7.6 except when the omitted observations correspond to a common letter in the last square. In such a case one can switch the roles of the first and the last squares and apply Theorem 7.2. □

THEOREM 7.3 In a system of $p-2$ MOLS if one misses any two observations that do not belong to the same row, same column, or correspond to the same letter in any of the squares, each of the row effects, column effects, or square effects carries only $p-2$ degrees of freedom.

Proof We have here p^2-2 observations and p^2-p+1 linearly independent parametric functions. If we have h linearly independent zero functions the rank of the estimation space is at most p^2-h-2. We shall first show that $h=p-2$ so that the rank of the estimation space is at most p^2-p.

Extend the system of $p-2$ MOLS to a complete set of $p-1$ MOLS and let $A_1, A_2, \ldots, A_p$ be the letters for this last square. Since the pair of missing observations does not correspond to a common row, column, or a common letter in any of the first $p-2$ squares they must correspond to a common letter in the last square. Without loss of generality we can assume that they correspond to A_1. It is clear that $f_2, f_3, \ldots, f_{p-1}$ as defined in the proof of Theorem 7.2 are all available and hence the proof follows. □

It is also clear that f_1 is not available. Thus the rank of the estimation space is p^2-p.

We now assert that if we omit any of the classifications corresponding to the rows, columns, or one of the first $p-2$ squares, all the remaining effects are estimable. To show this we note that if from our model we drop this classification and introduce the parameters corresponding to the classification given by the last square by Corollary 7.5 of the theorem, all effects are estimable. If we now drop the newly introduced parameters the truth of the assertion follows.

Thus the rank of the estimation space when any one classification is ignored is $1+(p-1)(p-1)$, while the rank of the full estimation space is p^2-p. This implies that for *any* classification eliminating all others we have $p-2$ degrees of freedom.

COROLLARY 7.8 If in a model with $p-2$ MOLS with t missing observations *any* two missing observations do not belong to the same row or the same column or correspond to the same letter in any of the squares, at least 1 degree of freedom is lost for each classification effect.

REMARK In the design and the model considered in Theorem 7.3 the rank of the estimation space eliminating the general mean is p^2-p-1. On the other hand, for each of the p classifications we have $p-2$ degrees of freedom. Thus there remain $p-1$ estimable parametric functions which involve effects from more than one classification. Since these effects can not be separated, we say that they are confounded with each other.

Latin and Graeco-Latin Squares of Side 3

In this section we shall consider applications of the results of the previous section to Latin and Graeco-Latin squares of side 3 with some missing observations.

By Corollary 7.5 all the effects are estimable if a single observation is missing in LS(3). By Theorem 7.2 it follows that if two observations are missing from the same row, the same column, or with the same treatment all effects are estimable. If the two missing observations do not satisfy this pattern it follows from Theorem 7.3 that only 1 degree of freedom is available for each row, column and treatment effects. This gives the following table of degrees of freedom:

Source	Degrees of Freedom
Rows	1
Columns	1
Treatments	1
Confounded	2
Residual	1
Total	6

We next consider the following GLS(3) with the observation on $A\alpha$ missing.

	c_1	c_2	c_3
r_1	—	$B\beta$	$C\partial$
r_2	$B\partial$	$C\alpha$	$A\beta$
r_3	$C\beta$	$A\partial$	$B\alpha$

Analysis of contrasts orthogonal to row totals and to column totals indicates that only $B-C$ and $\beta-\partial$ effects are estimable. Switching the roles of A, B, C and α, β, ∂ with the rows and the columns shows that r_2-r_3 and c_2-c_3 are estimable. This leaves us with the following three confounded (inseparable) contrasts:

$$\phi_1 = A+B-2C-\alpha+\partial,$$

$$\phi_2 = r_1-r_3-c_1+2c_2-c_3,$$

$$\phi_3 = A-C-c_1-c_2+2c_3.$$

If in a GLS(3), two observations are missing they must have a classification in common. If, for example, they both correspond to letter A only $B-C$ is estimable. No other effects are estimable and hence we must have 5 degrees of freedom for confounded (inseparable) effects.

Latin and Graeco Squares of Side 4

There are only two standard Latin squares of side 4. One of them has no orthogonal mate. For this square we consider patterns of three missing observations all from the same row, same column, or the same treatment. There are 48 such patterns and direct verification shows that for each of these patterns all the effects are estimable. We conjecture that a similar result with $p-1$ missing observations holds for any LS of side p for $p>4$.

The other standard square of side 4 can be extended to a GLS(4) and hence to a complete system of three MOLS. If our model is a LS model and if we have up to three missing observations from the same row or the same column or corresponding to the same letter in any of the three squares by Corollary 7.7, all effects are estimable. In particular, one can omit *any* pair of observations and estimate all the effects.

In a GLS(4) if we have up to three missing observations from the same row, same column, or with the same letter in either square, all the effects are estimable. If on the other hand we have only two missing observations, which do not satisfy the above condition, we must lose 1 degree of freedom for each of the four main effects. Since the rank of the estimation space eliminating the mean is 13 we have the following table of degrees of freedom:

Source	Degrees of Freedom
Rows	2
Columns	2
Greek	2
Latin	2
Confounded	3
Residual	2
Total	13

We close this section by giving an example for the five-way factorial experiment.

EXAMPLE 7.13 Consider a $3^2 \times 2^3$ factorial design with the following incidence pattern:

$\mathbf{N}_1 =$

	τ_1											
	δ_1						δ_2					
	γ_1			γ_2			γ_1			γ_2		
	β_1	β_2	β_3	β_1	β_2	β_3	β_1	β_2	β_3	β_1	β_2	β_3
α_1	1											
α_2					1							1
α_3								1				

,

$\mathbf{N}_2 =$

	τ_2											
	δ_1						δ_2					
	γ_1			γ_2			γ_1			γ_2		
	β_1	β_2	β_3	β_1	β_2	β_3	β_1	β_2	β_3	β_1	β_2	β_3
α_1												1
α_2									1			
α_3			1	1								

.

We see that

$$(\mu+\alpha_2+\beta_3+\gamma_2+\delta_2+\tau_1)-(\mu+\alpha_2+\beta_3+\gamma_1+\delta_2+\tau_2)=\gamma_2-\gamma_1+\tau_1-\tau_2$$

is estimable. This is the set D^*. Now we form the matrix $\bar{\mathbf{M}}$:

$\bar{\mathbf{M}} =$

(1, 1, 1)	—	(2, 2, 2)
—	(2, 1, 1)	(2, 2, 1)
(2, 1, 2)	(1, 2, 1)	(1, 1, 2)

Applying the Q process to the above matrix leads to

(1, 1, 1)	—	(2, 2, 2)	
—	(2, 1, 1)	(2, 2, 1)	$-2\gamma_1+2\gamma_2-\delta_1+\delta_2-\tau_1+\tau_2,$
(2, 1, 2)	(1, 2, 1)	(1, 1, 2)	

and

X	—	(2, 2, 2)	
—	(2, 1, 1)	(2, 2, 1)	$-2\delta_1+2\delta_2+\tau_1-\tau_2.$
(2, 1, 2)	(1, 2, 1)	(1, 1, 2)	

Thus $F=\{-2\gamma_1+2\gamma_2-\delta_1+\delta_2-\tau_1+\tau_2,\ -2\delta_1+2\delta_2+\tau_1-\tau_2\}$, and it is clear that we are left with no loops.

From the Q process and direct η differences we found the following three estimable contrasts:

γ_1	γ_2	δ_1	δ_2	τ_1	τ_2	
-1	1	0	0	1	-1	From the set D^*
-2	2	-1	1	-1	1	From the set F.
0	0	-2	2	1	-1	

If we row-reduce the above 3×6 matrix we see that $\gamma_1-\gamma_2, \delta_1-\delta_2, \tau_1-\tau_2$ are estimable and thus $\dim \bar{\mathscr{G}}=3$. Thus, $r(\mathbf{X})=r(\mathbf{A}, \mathbf{B})+\dim \mathscr{G}=8$, so the design has maximal rank.

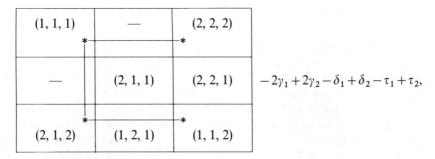

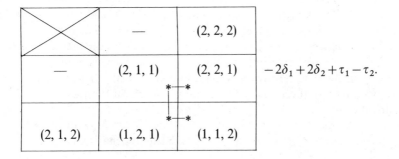

7.8 TWO-WAY CLASSIFICATION WITH INTERACTION

Consider a collection of N independent random variables $\{Y_{ijq}\}$ each having a common variance σ^2 and an expectation of the form

$$E(Y_{ijq})=\mu+\alpha_i+\beta_j+\theta_{ij} \tag{7.9}$$

where i, j range from 1 to a and b, respectively, and for a given i, j the index q ranges from 1 to n_{ij}, with the usual interpretation that when $n_{ij}=0$ there are no random variables with the first two subscripts i, j. The incidence matrix is $\mathbf{N}=(n_{ij})$, which we suppose has no row nor column composed completely of zeros. Thus, we suppose that the Y_{ijq}'s follow a fixed effects two-way classification with interaction, with no restrictions on the unknown parameters.

In Eq. (7.9) μ is a mean, α_i is the effect due to the ith level of the α factor, β_j is the effect of the jth level of the β factor, and θ_{ij} is the interaction effect due to the ith level of the α factor and the jth level of the β factor.

With a balanced data set, every one of the ab cells in the incidence matrix $\mathbf{N}$ would have m observations and there would be ab levels of the interaction factor θ_{ij} in the data. However, when there are missing observations (some cells are missing), there are only as many θ levels in the data as there are occupied cells.

Let $\mathbf{Y}$ denote the $N\times 1$ vector consisting of the Y_{ijq}'s ordered lexicographically and $\mathbf{1}, \mathbf{A}, \mathbf{B}, \mathbf{T}$, and $\mathbf{X}=(\mathbf{1}, \mathbf{A}, \mathbf{B}, \mathbf{T})$ denote matrices defined so that

$$E(\mathbf{Y})=\mathbf{1}\mu+\mathbf{A}\boldsymbol{\alpha}+\mathbf{B}\boldsymbol{\beta}+\mathbf{T}\boldsymbol{\theta}=\mathbf{X}\boldsymbol{\psi} \tag{7.10}$$

where $\boldsymbol{\alpha}=(\alpha_1,\ldots,\alpha_a)'$, $\boldsymbol{\beta}=(\beta_1,\ldots,\beta_b)'$, $\boldsymbol{\theta}=(\theta_{11}, \theta_{12},\ldots,\theta_{ab})'$, and $\boldsymbol{\psi}=(\mu, \boldsymbol{\alpha}', \boldsymbol{\beta}', \boldsymbol{\theta}')'$.

In a two-way classification model with interaction, it is known that when there are no restrictions on the parameters, there are no estimable α contrasts or β contrasts. So we concentrate on the estimable θ contrasts. Let $\bar{\mathscr{G}}$ denote the vector space of estimable θ contrasts.

Observe that the linear model (7.9) is in fact (7.8) if we let $\mathbf{k}_{ij}\cdot\boldsymbol{\eta}=\theta_{ij}$ and $K=\{\mathbf{k}_{ij}|n_{ij}\neq 0\}$. We note that the hypotheses of Lemma 7.3 are satisfied and hence $\bar{\mathscr{G}}$ is spanned by D^* together with the θ contrasts obtained from the loops in $\bar{\mathbf{M}}$.

The column indexed by $\mathbf{k}_{ij}\in K$ in the matrix $\mathbf{W}$ (to which the R process is applied to obtain D^*) has a nonzero entry only in the row indexed by the pair (i, j). Therefore, for each $\mathbf{k}=\mathbf{k}_{ij}\in K$, $\mathbf{N}(\mathbf{k})$ has a 1 in cell (i, j) and zero, elsewhere. Thus, there are no direct differences, so that $S=K$, $\mathbf{M}(\mathbf{k})=\mathbf{N}(\mathbf{k})$ for all $\mathbf{k} \in K$, $\bar{\mathscr{D}}^*=\{0\}$ and $\bar{\mathscr{G}}=\bar{\mathscr{W}}$.

In forming the matrix $\bar{\mathbf{M}}$ we can choose θ_s to be the numeral 1 for all $\mathbf{s} \in S$ because the position of a 1 in $\bar{\mathbf{M}}$ tells which $\mathbf{s} \in S$ is associated with it. Hence we can obtain $\bar{\mathbf{M}}$ from the incidence matrix $\mathbf{N}=(n_{ij})$ by changing all nonzero entries to 1.

Suppose a loop in $\bar{\mathbf{M}}$ is selected by applying the Q process. Then each θ contrast, denoted by $\theta(L)$, derived from a loop L by applying the Q process contains a term θ_{ij} which does not occur in any contrasts derived from loops selected later. Therefore, these θ contrasts not only form a spanning set for $\bar{\mathscr{G}}$ but in fact form a basis.

It is easily seen that the degrees of freedom, r, for the regression sum of squares $SS(\mu, \alpha, \beta, \theta)$ is equal to the number of nonzero entries in the incidence matrix. Let $r_\theta=\dim \bar{\mathscr{G}}$, then $r=r(\mathbf{X})=r(\mathbf{A}, \mathbf{B})+r_\theta$. Thus, an alternative method for calculating the rank of the design matrix for an additive two-way classification model with incidence matrix $\mathbf{N}$ is to subtract the number of loops obtained by applying the Q process in $\mathbf{N}$ from the number of nonzero entries in $\mathbf{N}$. That is, $r(\mathbf{A}, \mathbf{B})=r-r_\theta$.

EXAMPLE 7.14 Consider a two-way classification model with interaction having the expectation of the form

$$E(Y_{ijq})=\mu+\alpha_i+\beta_j+\theta_{ij}, \quad i=1,\ldots,4, \quad j=1,\ldots,5, \quad q=1,\ldots,n_{ij}.$$

Let the associated incidence matrix $\mathbf{N}=(n_{ij})$ for the above model be as:

$$\mathbf{N}=\begin{bmatrix} 3 & 1 & 1 & 0 & 0 \\ 0 & 1 & 2 & 1 & 0 \\ 1 & 0 & 1 & 0 & 0 \\ 1 & 3 & 0 & 1 & 2 \end{bmatrix}.$$

In matrix notation we can write our model as

$$E(\mathbf{Y})=\mathbf{X}\psi$$

where $\mathbf{X}$ is a 18×22 matrix, and ψ is a 22×1 vector of parameters. If there were no missing observations, ψ would have been a 30×1 vector of parameters. The model equation $E(\mathbf{Y})=\mathbf{X}\psi$ is

$$
\begin{bmatrix} Y_{111} \\ Y_{112} \\ Y_{113} \\ Y_{121} \\ Y_{131} \\ Y_{221} \\ Y_{231} \\ Y_{232} \\ Y_{241} \\ Y_{311} \\ Y_{331} \\ Y_{411} \\ Y_{421} \\ Y_{422} \\ Y_{423} \\ Y_{441} \\ Y_{451} \\ Y_{452} \end{bmatrix} =
$$

$$
\begin{array}{cccccccccccccccccccccc}
\mu & \alpha_1 & \alpha_2 & \alpha_3 & \alpha_4 & \beta_1 & \beta_2 & \beta_3 & \beta_4 & \beta_5 & \theta_{11} & \theta_{12} & \theta_{13} & \theta_{22} & \theta_{23} & \theta_{24} & \theta_{31} & \theta_{33} & \theta_{41} & \theta_{42} & \theta_{44} & \theta_{45}
\end{array}
$$

$$
\begin{bmatrix}
1 & 1 & 0 & 0 & 0 & 1 & 0 & 0 & 0 & 0 & 1 & 0 & 0 & 0 & 0 & 0 & 0 & 0 & 0 & 0 & 0 & 0 \\
1 & 1 & 0 & 0 & 0 & 1 & 0 & 0 & 0 & 0 & 1 & 0 & 0 & 0 & 0 & 0 & 0 & 0 & 0 & 0 & 0 & 0 \\
1 & 1 & 0 & 0 & 0 & 1 & 0 & 0 & 0 & 0 & 1 & 0 & 0 & 0 & 0 & 0 & 0 & 0 & 0 & 0 & 0 & 0 \\
1 & 1 & 0 & 0 & 0 & 0 & 1 & 0 & 0 & 0 & 0 & 1 & 0 & 0 & 0 & 0 & 0 & 0 & 0 & 0 & 0 & 0 \\
1 & 1 & 0 & 0 & 0 & 0 & 0 & 1 & 0 & 0 & 0 & 0 & 1 & 0 & 0 & 0 & 0 & 0 & 0 & 0 & 0 & 0 \\
1 & 0 & 1 & 0 & 0 & 0 & 1 & 0 & 0 & 0 & 0 & 0 & 0 & 1 & 0 & 0 & 0 & 0 & 0 & 0 & 0 & 0 \\
1 & 0 & 1 & 0 & 0 & 0 & 0 & 1 & 0 & 0 & 0 & 0 & 0 & 0 & 1 & 0 & 0 & 0 & 0 & 0 & 0 & 0 \\
1 & 0 & 1 & 0 & 0 & 0 & 0 & 1 & 0 & 0 & 0 & 0 & 0 & 0 & 1 & 0 & 0 & 0 & 0 & 0 & 0 & 0 \\
1 & 0 & 1 & 0 & 0 & 0 & 0 & 0 & 1 & 0 & 0 & 0 & 0 & 0 & 0 & 1 & 0 & 0 & 0 & 0 & 0 & 0 \\
1 & 0 & 0 & 1 & 0 & 1 & 0 & 0 & 0 & 0 & 0 & 0 & 0 & 0 & 0 & 0 & 1 & 0 & 0 & 0 & 0 & 0 \\
1 & 0 & 0 & 1 & 0 & 0 & 0 & 1 & 0 & 0 & 0 & 0 & 0 & 0 & 0 & 0 & 0 & 1 & 0 & 0 & 0 & 0 \\
1 & 0 & 0 & 0 & 1 & 1 & 0 & 0 & 0 & 0 & 0 & 0 & 0 & 0 & 0 & 0 & 0 & 0 & 1 & 0 & 0 & 0 \\
1 & 0 & 0 & 0 & 1 & 0 & 1 & 0 & 0 & 0 & 0 & 0 & 0 & 0 & 0 & 0 & 0 & 0 & 0 & 1 & 0 & 0 \\
1 & 0 & 0 & 0 & 1 & 0 & 1 & 0 & 0 & 0 & 0 & 0 & 0 & 0 & 0 & 0 & 0 & 0 & 0 & 1 & 0 & 0 \\
1 & 0 & 0 & 0 & 1 & 0 & 1 & 0 & 0 & 0 & 0 & 0 & 0 & 0 & 0 & 0 & 0 & 0 & 0 & 1 & 0 & 0 \\
1 & 0 & 0 & 0 & 1 & 0 & 0 & 0 & 1 & 0 & 0 & 0 & 0 & 0 & 0 & 0 & 0 & 0 & 0 & 0 & 1 & 0 \\
1 & 0 & 0 & 0 & 1 & 0 & 0 & 0 & 0 & 1 & 0 & 0 & 0 & 0 & 0 & 0 & 0 & 0 & 0 & 0 & 0 & 1 \\
1 & 0 & 0 & 0 & 1 & 0 & 0 & 0 & 0 & 1 & 0 & 0 & 0 & 0 & 0 & 0 & 0 & 0 & 0 & 0 & 0 & 1
\end{bmatrix}
\begin{bmatrix} \mu \\ \alpha_1 \\ \alpha_2 \\ \alpha_3 \\ \alpha_4 \\ \beta_1 \\ \beta_2 \\ \beta_3 \\ \beta_4 \\ \beta_5 \\ \theta_{11} \\ \theta_{12} \\ \theta_{13} \\ \theta_{22} \\ \theta_{23} \\ \theta_{24} \\ \theta_{31} \\ \theta_{33} \\ \theta_{41} \\ \theta_{42} \\ \theta_{44} \\ \theta_{45} \end{bmatrix}
$$

The rank of the design matrix $r(\mathbf{X})$ is 12, the number of nonzero entries of the incidence matrix $\mathbf{N}$. The matrix $\bar{\mathbf{M}}$ is obtained from the incidence matrix $\mathbf{N}$ by changing the nonzero entries of $\mathbf{N}$ to 1. So we have

$$\bar{\mathbf{M}}=\begin{bmatrix} 1 & 1 & 1 & 0 & 0 \\ 0 & 1 & 1 & 1 & 0 \\ 1 & 0 & 1 & 0 & 0 \\ 1 & 1 & 0 & 1 & 1 \end{bmatrix}.$$

To obtain $r_\theta=\dim \bar{\mathscr{G}}$, we apply the Q process to $\bar{\mathbf{M}}$. From cells (1, 2)(1, 1) (4, 1)(4, 2) in $\bar{\mathbf{M}}_1$ (below) (index of $\bar{\mathbf{M}}$ denotes only the steps), we find the corresponding θ contrast to be

$$\theta_{12}-\theta_{11}+\theta_{41}-\theta_{42}.$$

$$\bar{\mathbf{M}}_1=\begin{array}{|c|c|c|c|c|} \hline 1 & 1 & 1 & 0 & 0 \\ \hline 0 & 1 & 1 & 1 & 0 \\ \hline 1 & 0 & 1 & 0 & 0 \\ \hline 1 & 1 & 0 & 1 & 1 \\ \hline \end{array}.$$

Change the cell (4, 2) to a zero. So we obtain

$$\bar{\mathbf{M}}_2=\begin{array}{|c|c|c|c|c|} \hline 1 & 1 & 1 & 0 & 0 \\ \hline 0 & 1 & 1 & 1 & 0 \\ \hline 1 & 0 & 1 & 0 & 0 \\ \hline 1 & 0 & 0 & 1 & 1 \\ \hline \end{array}.$$

From cells (1, 3)(1, 1)(3, 1)(3, 3) in $\bar{\mathbf{M}}_2$ we find the corresponding θ contrast to be

$$\theta_{13}-\theta_{11}+\theta_{31}-\theta_{33}.$$

In $\bar{\mathbf{M}}_2$ change the cell (3, 3) to a 0. So we have

$$\bar{\mathbf{M}}_3 = \begin{array}{|c|c|c|c|c|}\hline 1 & 1 & 1 & 0 & 0 \\ \hline 0 & 1 & 1 & 1 & 0 \\ \hline 1 & 0 & 0 & 0 & 0 \\ \hline 1 & 0 & 0 & 1 & 1 \\ \hline \end{array}.$$

From (1, 3)(1, 2)(2, 2)(2, 3) in $\bar{\mathbf{M}}_3$ we find the corresponding θ contrast to be

$$\theta_{13}-\theta_{12}+\theta_{22}-\theta_{23}.$$

Change the cell (2, 3) to a zero, and form $\bar{\mathbf{M}}_4$.

$$\bar{\mathbf{M}}_4 = \begin{array}{|c|c|c|c|c|}\hline 1 & 1 & 1 & 0 & 0 \\ \hline 0 & 1 & 0 & 1 & 0 \\ \hline 1 & 0 & 0 & 0 & 0 \\ \hline 1 & 0 & 0 & 1 & 1 \\ \hline \end{array}.$$

In $\bar{\mathbf{M}}_4$ we see that there is a loop (2, 4)(2, 2)(1, 2)(1, 1)(4, 1)(4, 4) which corresponds to

$$\theta_{24}-\theta_{22}+\theta_{12}-\theta_{11}+\theta_{41}-\theta_{44}.$$

If we replace the entry of the cell (4, 4) by a zero, we are left with no more loops. Thus, the Q process applied to $\bar{\mathbf{M}}$ yields four loops with corresponding contrasts

$$\theta_{12}-\theta_{11}+\theta_{41}-\theta_{42},$$

$$\theta_{13}-\theta_{11}+\theta_{31}-\theta_{33},$$

$$\theta_{13}-\theta_{12}+\theta_{22}-\theta_{23},$$

and

$$\theta_{24}-\theta_{22}+\theta_{12}-\theta_{11}+\theta_{41}-\theta_{44}.$$

Hence $r_\theta = 4$. Note that, as mentioned above, each θ contrast $\theta(L)$ derived from a loop L contains a term θ_{ij} which does not occur in any contrasts derived from loops selected later. Therefore, these contrasts form a basis for $\bar{\mathscr{G}}$.

Alternatively, we can find r_θ by first finding $r(\mathbf{A}, \mathbf{B})$ using the methods developed in Chapter 5 and then obtain r_θ by $r_\theta = r - r(\mathbf{A}, \mathbf{B})$. To see this, apply the R process to the matrix $\mathbf{N} = (n_{ij})$ to obtain the final matrix $\bar{\mathbf{M}}$ associated with the two-way additive model. We see that in the final matrix $\bar{\mathbf{M}}$ all entries will be filled, so $r(\mathbf{A}, \mathbf{B}) = a + \bar{f}_\beta = 4 + 4 = 8$, and hence $r_\theta = 12 - 8 = 4$.

So far we have assumed that there are no constraints on the parameters. Now suppose $\mathbf{Y}$ is an $N \times 1$ random vector whose covariance matrix is $\sigma^2\mathbf{I}$ and whose expectation may be written as

$$E(\mathbf{Y}) = \mathbf{H}\boldsymbol{\xi} + \mathbf{T}\boldsymbol{\theta}, \qquad \boldsymbol{\Lambda}'\boldsymbol{\xi} = \mathbf{0}, \qquad \boldsymbol{\Gamma}'\boldsymbol{\theta} = \mathbf{0} \tag{7.11}$$

where $\boldsymbol{\xi} = (\mu, \alpha_1, \ldots, \alpha_a, \beta_1, \ldots, \beta_b)'$, and that $\mathbf{H}$, $\mathbf{T}$, and $\boldsymbol{\theta}$ are defined in the obvious way, for example, $\mathbf{X} = (\mathbf{H}, \mathbf{T})$.

For notational purposes it will sometimes be convenient to disregard the partitioned form of $E(\mathbf{Y})$ and think of $E(\mathbf{Y})$ as

$$E(\mathbf{Y}) = \mathbf{X}\boldsymbol{\psi}, \qquad \boldsymbol{\Delta}'\boldsymbol{\psi} = \mathbf{0}. \tag{7.12}$$

For this model a linear parametric function $\mathbf{u}'\boldsymbol{\psi} = \mathbf{g}'\boldsymbol{\xi} + \mathbf{a}'\boldsymbol{\theta}$ is a linear functional on the vector space $N(\boldsymbol{\Delta}')$ of possible parameter vectors. And a linear parametric function $\mathbf{u}'\boldsymbol{\psi}$ is said to be estimable provided there exists a linear unbiased estimator for $\mathbf{u}'\boldsymbol{\psi}$, or equivalently

$$\begin{aligned}&\mathbf{u}'\boldsymbol{\psi} \text{ is estimable if and only if there is some } \mathbf{v} \in R(\mathbf{X}')\\ &\text{such that } \mathbf{u}'\boldsymbol{\psi} = \mathbf{v}'\boldsymbol{\psi} \text{ for all } \boldsymbol{\psi} \in N(\boldsymbol{\Delta}').\end{aligned} \tag{7.13}$$

Let $\bar{\Theta}$ and $\bar{\mathscr{G}}$ denote the vector spaces of estimable linear parametric functions of the form $\mathbf{u}'\boldsymbol{\psi}$ and $\mathbf{a}'\boldsymbol{\theta} = \mathbf{0}'\boldsymbol{\xi} + \mathbf{a}'\boldsymbol{\theta}$, respectively, and let $r = \dim \bar{\Theta}$ (=degrees of freedom for regression), $r_\theta = \dim \bar{\mathscr{G}}$ (=degrees of freedom for the θ effects adjusted for the ξ effects), and let $\bar{f}_\xi$ denote the degrees of freedom for regression for the submodel $E(\mathbf{Y}) = \mathbf{H}\boldsymbol{\xi}, \boldsymbol{\Lambda}'\boldsymbol{\xi} = \mathbf{0}$. For determining when our results are applicable to model (7.11) (when we have any of the "usual" constraints) we introduce the conditions

$$R(\mathbf{X}') \cap R(\boldsymbol{\Delta}) = \{\mathbf{0}\}, \tag{7.14}$$

and

$$R(\mathbf{H}') \cap R(\boldsymbol{\Lambda}) = \{\mathbf{0}\}. \tag{7.15}$$

Condition (7.14) says that the parametric vector $\boldsymbol{\Delta}'\boldsymbol{\psi}$ is nonestimable in the unconstrained model $E(\mathbf{Y}) = \mathbf{X}\boldsymbol{\psi}$, whereas condition (7.15) says that the parametric vector $\boldsymbol{\Lambda}'\boldsymbol{\xi}$ is nonestimable in the unconstrained model $E(\mathbf{Y}) =$

$\mathbf{H}\boldsymbol{\xi}$. The condition $\boldsymbol{\Delta}=\mathbf{0}$ is equivalent to saying that the parameters in model (7.11) are unrestricted. Observe that $\boldsymbol{\Delta}=\mathbf{0}$ satisfies (7.14) and (7.15).

A classification model, with any of the "usual constraints" imposed as restrictions on the parameters will satisfy (7.14), but (7.15) will be satisfied only for particular partitions of the parameters. We can use (7.13) to establish some facts concerning linear parametric functions of the form $\mathbf{a}'\boldsymbol{\theta}$: $\mathbf{a}'\boldsymbol{\theta}$ is estimable if and only if there exists $\mathbf{b} \in \{\mathbf{T}'\mathbf{z}|\mathbf{H}'\mathbf{z} \in R(\boldsymbol{\Lambda})\}$ such that $\mathbf{a}'\boldsymbol{\theta}=\mathbf{b}'\boldsymbol{\theta}$ for all $\boldsymbol{\theta} \in N(\boldsymbol{\Gamma}')$. Since (7.15) implies $\{\mathbf{T}'\mathbf{z}|\mathbf{H}'\mathbf{z} \in R(\boldsymbol{\Lambda})\}=\{\mathbf{T}'\mathbf{z}|\mathbf{H}'\mathbf{z}=\mathbf{0}\}$, we get the following: if $\mathbf{a}_1'\boldsymbol{\theta}, \ldots, \mathbf{a}_m'\boldsymbol{\theta}$ would constitute a spanning set for $\bar{\mathscr{G}}$ under the assumption $\boldsymbol{\Delta}=\mathbf{0}$, then $\mathbf{a}_1'\boldsymbol{\theta}, \ldots, \mathbf{a}_m'\boldsymbol{\theta}$ is a spanning set for $\bar{\mathscr{G}}$ whenever (7.15) is true. It can be established that $r=\bar{f}_\xi+r_\theta$.

Moreover, suppose (7.14) and (7.15) are satisfied. Then:

1. If $\mathbf{a}_1'\boldsymbol{\theta}, \ldots, \mathbf{a}_m'\boldsymbol{\theta}$ would constitute a basis for $\bar{\mathscr{G}}$ under the assumption $\boldsymbol{\Delta}=\mathbf{0}$, then $\mathbf{a}_1'\boldsymbol{\theta}, \ldots, \mathbf{a}_m'\boldsymbol{\theta}$ is a basis for $\bar{\mathscr{G}}$.
2. If we set $\mathbf{A}=(\mathbf{a}_1, \ldots, \mathbf{a}_m)$, where the $\mathbf{a}_i$'s are as in part (1), and set $\mathbf{W}=\mathbf{TA}$, then the regression space $\{\mathbf{X}\boldsymbol{\psi}: \boldsymbol{\Delta}'\boldsymbol{\psi}=\mathbf{0}\}$ may be expressed as a direct sum $R(\mathbf{H})\oplus R(\mathbf{W})$ and $\mathbf{W}$ has full column rank.

In a two-way classification model with interaction the conditions (7.14) and (7.15) are true so that a basis for $\bar{\mathscr{G}}$ can be obtained from the θ contrasts derived from loops by applying the Q process. However, unlike additive classification models (see Chapters 5 and 6), the statements (1) and (2) above will not apply here if we interchange the roles of $\boldsymbol{\theta}$ and $\boldsymbol{\xi}$, because $R(\mathbf{T}')\cap R(\boldsymbol{\Gamma})\neq\{\mathbf{0}\}$.

The analysis of variance for the two-way classification model with interaction is similar to that of the additive two-way classification model discussed in Chapter 5 except for the inclusion of an interaction term $SS(\theta|\mu, \alpha, \beta)$ in the A.O.V. table.

The regression sum of squares $SS(\mu, \alpha, \beta, \theta)$ can be partitioned as

$$SS(\mu, \alpha, \beta, \theta)=SS(\mu, \alpha, \beta)+SS(\theta|\mu, \alpha, \beta).$$

The term $SS(\mu, \alpha, \beta)=SS(\mu)+SS(\alpha|\mu)+SS(\beta|\mu, \alpha, \beta)$ can be obtained by fitting the additive two-way classification model

$$E(Y_{ijq})=\mu+\alpha_i+\beta_j.$$

Calculation of $SS(\mu)$, $SS(\alpha|\mu)$, and $SS(\beta|\mu, \alpha)$ is the same as for that of the additive two-way classification model.

An easy way to obtain the regression sum of squares is to set the "overparameterized" model

$$E(Y_{ijq})=\mu+\alpha_i+\beta_j+\theta_{ij}$$

in the form

$$E(Y_{ijq})=\mu_{ij} \tag{7.16}$$

or in matrix notation as

$$E(\mathbf{Y})=\mathbf{C}\boldsymbol{\omega}$$

where the matrix **C** is $N\times r$ (r=number of occupied cells in the incidence matrix **N**). The model (7.16) is always of full rank, and hence the normal equations solution is simply obtained by

$$\hat{\boldsymbol{\omega}}=(\mathbf{C}'\mathbf{C})^{-1}\mathbf{C}'\mathbf{Y}$$

and thus the regression sum of squares is

$$\text{SS}(\mu, \alpha, \beta, \theta)=\text{SS}(\omega)=\hat{\boldsymbol{\omega}}'\mathbf{C}'\mathbf{Y}.$$

Observe that the matrix $\mathbf{C}'\mathbf{C}$ is an $r\times r$ diagonal matrix and that its diagonal elements are equal to the lexicographically ordered numbers of observations in the cells (i, j) for which $n_{ij}\neq 0$. That is,

$$\mathbf{C}'\mathbf{C}=\text{diag}(n_{11}, n_{12}, \ldots, n_{1b}, n_{21}, \ldots, n_{2b}, \ldots, n_{a1}, \ldots, n_{ab}),$$

only including those n_{ij} which are nonzero. Also, the vector $\mathbf{C}'\mathbf{Y}$ is

$$\mathbf{C}'\mathbf{Y}=(Y_{11.}, \ldots, Y_{1b.}, Y_{21.}, \ldots, Y_{2b.}, \ldots, Y_{a1.}, \ldots, Y_{ab.})',$$

only including those $Y_{ij.}$ for which $n_{ij}\neq 0$, and hence

$$\hat{\boldsymbol{\omega}}=(\bar{Y}_{11.}, \ldots, \bar{Y}_{ab.})'.$$

Once we have $\text{SS}(\mu, \alpha, \beta, \theta)$ and $\text{SS}(\mu, \alpha, \beta)$ (by fitting an additive two-way classification model), we can obtain $\text{SS}(\theta|\mu, \alpha, \beta)$ by subtraction, that is,

$$\text{SS}(\theta|\mu, \alpha, \beta)=\text{SS}(\mu, \alpha, \beta, \theta)-\text{SS}(\mu, \alpha, \beta).$$

The expression $\text{SS}(\theta|\mu, \alpha, \beta)$ has $r_\theta=\dim \bar{\mathscr{G}}$ degrees of freedom.

EXAMPLE 7.15 In order to compare the results of this section to that of Chapter 5, let us consider Example 5.7 once again in which we had an 8×6

factorial experiment. We had

$$\mathbf{N}=(n_{ij})=\begin{array}{c} \\ \alpha_1 \\ \alpha_2 \\ \alpha_3 \\ \alpha_4 \\ \alpha_5 \\ \alpha_6 \\ \alpha_7 \\ \alpha_8 \end{array}\begin{array}{c} \begin{array}{cccccc} \beta_1 & \beta_2 & \beta_3 & \beta_4 & \beta_5 & \beta_6 \end{array} \\ \begin{bmatrix} 0 & 0 & 2 & 0 & 1 & 0 \\ 2 & 4 & 0 & 0 & 0 & 0 \\ 0 & 0 & 3 & 0 & 0 & 0 \\ 0 & 3 & 0 & 1 & 0 & 2 \\ 0 & 0 & 0 & 0 & 2 & 0 \\ 0 & 0 & 0 & 0 & 4 & 0 \\ 0 & 0 & 0 & 5 & 0 & 1 \\ 0 & 0 & 1 & 0 & 2 & 0 \end{bmatrix} \end{array}$$

and the observations on the cells were as

(1, 3): 2.3, 4.5
(1, 5): 1.6
(2, 1): 4.2, 7.1
(2, 2): 3.6, 2.4, 1.9, 4.6
(3, 3): 3.9, 7.9, 8.5
(4, 2): 5.6, 6.1, 2.1
(4, 4): 3.4
(4, 6): 5.2, 6.9
(5, 5): 9.1, 11.9
(6, 5): 3.5, 4.2, 6.8, 6.7
(7, 4): 8.7, 12.9, 6.4, 8.6, 2.3
(7, 6): 1.4
(8, 3): 2.4
(8, 5): 4.2, 1.5.

Estimability and Analysis

The basic estimable functions for the two-way classification model with interaction are $E(Y_{ijq})=\mu+\alpha_i+\beta_j+\theta_{ij}$, $n_{ij}\neq 0$. Set

$$E(Y_{ijq})=\mu+\alpha_i+\beta_j+\theta_{ij}=\mu_{ij},$$

or in matrix notation

$$E(\mathbf{Y})=\mathbf{C}\boldsymbol{\omega}.$$

There are 14 nonzero entries in the incidence matrix **N**, so $r = r(\mathbf{X}) = r(\mathbf{C}) = 14$. To obtain BLUEs for μ_{ij} for $n_{ij} \neq 0$ we form the normal equations by

$$\mathbf{C}'\mathbf{C}\hat{\omega} = \mathbf{C}'\mathbf{Y}$$

where

$$\mathbf{C}'\mathbf{C} = \begin{bmatrix} 2 & 0 & 0 & 0 & 0 & 0 & 0 & 0 & 0 & 0 & 0 & 0 & 0 & 0 \\ 0 & 1 & 0 & 0 & 0 & 0 & 0 & 0 & 0 & 0 & 0 & 0 & 0 & 0 \\ 0 & 0 & 2 & 0 & 0 & 0 & 0 & 0 & 0 & 0 & 0 & 0 & 0 & 0 \\ 0 & 0 & 0 & 4 & 0 & 0 & 0 & 0 & 0 & 0 & 0 & 0 & 0 & 0 \\ 0 & 0 & 0 & 0 & 3 & 0 & 0 & 0 & 0 & 0 & 0 & 0 & 0 & 0 \\ 0 & 0 & 0 & 0 & 0 & 3 & 0 & 0 & 0 & 0 & 0 & 0 & 0 & 0 \\ 0 & 0 & 0 & 0 & 0 & 0 & 1 & 0 & 0 & 0 & 0 & 0 & 0 & 0 \\ 0 & 0 & 0 & 0 & 0 & 0 & 0 & 2 & 0 & 0 & 0 & 0 & 0 & 0 \\ 0 & 0 & 0 & 0 & 0 & 0 & 0 & 0 & 2 & 0 & 0 & 0 & 0 & 0 \\ 0 & 0 & 0 & 0 & 0 & 0 & 0 & 0 & 0 & 4 & 0 & 0 & 0 & 0 \\ 0 & 0 & 0 & 0 & 0 & 0 & 0 & 0 & 0 & 0 & 5 & 0 & 0 & 0 \\ 0 & 0 & 0 & 0 & 0 & 0 & 0 & 0 & 0 & 0 & 0 & 1 & 0 & 0 \\ 0 & 0 & 0 & 0 & 0 & 0 & 0 & 0 & 0 & 0 & 0 & 0 & 1 & 0 \\ 0 & 0 & 0 & 0 & 0 & 0 & 0 & 0 & 0 & 0 & 0 & 0 & 0 & 2 \end{bmatrix}$$

and

$$\mathbf{C}'\mathbf{Y} = \begin{bmatrix} 2.3+4.5 \\ 1.6 \\ 4.2+7.1 \\ 3.6+2.4+1.9+4.6 \\ 3.9+7.9+8.5 \\ 5.6+6.1+2.1 \\ 3.4 \\ 5.2+6.9 \\ 9.1+11.9 \\ 3.5+4.2+6.8+6.7 \\ 8.7+12.9+6.4+8.6+2.3 \\ 1.4 \\ 2.4 \\ 4.2+1.5 \end{bmatrix} = \begin{bmatrix} 6.8 \\ 1.6 \\ 11.3 \\ 12.5 \\ 20.5 \\ 13.8 \\ 3.4 \\ 12.1 \\ 21.0 \\ 21.2 \\ 38.9 \\ 1.4 \\ 2.4 \\ 5.7 \end{bmatrix}.$$

So we obtain

$$\hat{\omega}=(\mathbf{C}'\mathbf{C})^{-1}\mathbf{C}'\mathbf{Y}=(\bar{Y}_{13.}, \bar{Y}_{15.}, \ldots, \bar{Y}_{85.})'=\begin{bmatrix} 3.4000 \\ 1.6000 \\ 5.6500 \\ 3.1250 \\ 6.7667 \\ 4.6000 \\ 3.4000 \\ 6.0500 \\ 10.5000 \\ 5.3000 \\ 7.7800 \\ 1.4000 \\ 2.4000 \\ 2.8500 \end{bmatrix}$$

or simply

$$\hat{\mu}_{ij}=\bar{Y}_{ij.}\text{(cell mean)}.$$

The regression sum of squares is then obtained by

$$\text{SS}(\mu, \alpha, \beta, \theta)=\text{SS}(\omega)=\hat{\omega}'\mathbf{C}'\mathbf{Y}=(3.40, 1.60, \ldots, 2.85)\begin{bmatrix} 6.8 \\ 1.6 \\ \vdots \\ 5.7 \end{bmatrix}=1073.6620.$$

With no restrictions on the parameters in the original model, it is not possible to obtain an estimable function which is a function of α's or β's. The θ's are always present. However, it is possible to obtain an estimable function that is only a function of the θ's. To do this we form the matrix $\bar{\mathbf{M}}$ from $\mathbf{N}=(n_{ij})$ by changing all nonzero entries to 1:

$$\bar{\mathbf{M}}=\begin{bmatrix} 0 & 0 & 1 & 0 & 1 & 0 \\ 1 & 1 & 0 & 0 & 0 & 0 \\ 0 & 0 & 1 & 0 & 0 & 0 \\ 0 & 1 & 0 & 1 & 0 & 1 \\ 0 & 0 & 0 & 0 & 1 & 0 \\ 0 & 0 & 0 & 0 & 1 & 0 \\ 0 & 0 & 0 & 1 & 0 & 1 \\ 0 & 0 & 1 & 0 & 1 & 0 \end{bmatrix}.$$

We now apply the Q process to $\bar{\mathbf{M}}$. From the loop (1,3)(1,5)(8,5)(8,3) in $\bar{\mathbf{M}}$ we find the corresponding θ contrast to be

$$\theta_{13}-\theta_{15}+\theta_{85}-\theta_{83}.$$

In $\bar{\mathbf{M}}$ change the cell (8,3) to a zero. From the loop (4,4)(4,6)(7,6)(7,4) in $\bar{\mathbf{M}}$ we find the corresponding θ contrast to be

$$\theta_{44}-\theta_{46}+\theta_{76}-\theta_{74}.$$

Now change the cell (7,4) to a zero. Thus we have

$$\bar{\mathbf{M}}=\begin{bmatrix} 0 & 0 & 1 & 0 & 1 & 0 \\ 1 & 1 & 0 & 0 & 0 & 0 \\ 0 & 0 & 1 & 0 & 0 & 0 \\ 0 & 1 & 0 & 1 & 0 & 1 \\ 0 & 0 & 0 & 0 & 1 & 0 \\ 0 & 0 & 0 & 0 & 1 & 0 \\ 0 & 0 & 0 & 0 & 0 & 1 \\ 0 & 0 & 0 & 0 & 1 & 0 \end{bmatrix}.$$

We see that no more loops can be found in $\bar{\mathbf{M}}$. So the Q process applied to $\bar{\mathbf{M}}$ yields two loops with corresponding contrasts

$$\theta_{13}-\theta_{15}+\theta_{85}-\theta_{83},$$

$$\theta_{44}-\theta_{46}+\theta_{76}-\theta_{74}.$$

Hence $r_\theta=2$. This is the degrees of freedom associated with $SS(\theta|\mu, \alpha, \beta)$. Observe that these contrasts form a basis for $\bar{\mathscr{G}}$ (always true).

Alternatively, we can find r_θ by first finding the regression degrees of freedom for the additive two-way classification model

$$E(Y_{ijq})=\mu+\alpha_i+\beta_j,$$

that is, $r(\mathbf{A}, \mathbf{B})=a+\bar{f}_\beta$ obtain $r_\theta=r-r(\mathbf{A}, \mathbf{B})$.

To obtain $\bar{f}_\beta$, the degrees of freedom associated with $SS(\beta|\mu, \alpha)$ we use the methods developed in Chapter 5:

1 Apply the R process to the incidence matrix $\mathbf{N}$ to obtain the final matrix $\mathbf{M}$:

$$\mathbf{M}=\begin{bmatrix} 0 & 0 & 1 & 0 & 1 & 0 \\ 1 & 1 & 0 & 1 & 0 & 1 \\ 0 & 0 & 1 & 0 & 1 & 0 \\ 1 & 1 & 0 & 1 & 0 & 1 \\ 0 & 0 & 1 & 0 & 1 & 0 \\ 0 & 0 & 1 & 0 & 1 & 0 \\ 1 & 1 & 0 & 1 & 0 & 1 \\ 0 & 0 & 1 & 0 & 1 & 0 \end{bmatrix}.$$

2 Form the counter triangle $\mathbf{C}_{\bar{\beta}}$ which is the lower triangular portion of $\mathbf{M}'\mathbf{M}$ with all nonzero entries set equal to 1:

$$\mathbf{C}_{\bar{\beta}} = \begin{array}{c|cccccc} & 1 & 2 & 3 & 4 & 5 & 6 \\ \hline 2 & 1 & & & & & \\ 3 & 0 & 0 & & & & \\ 4 & 1 & 1 & 0 & & & \\ 5 & 0 & 0 & 1 & 0 & & \\ 6 & 1 & 1 & 0 & 1 & 0 & \end{array} .$$

REMARK Observe that, contrary to the additive two-way classification model, where we could find estimable functions involving only β effects from $\mathbf{C}_{\bar{\beta}}$, we are *not* able to do so when interactions are present. Forming $\mathbf{C}_{\bar{\beta}}$ at this stage is only a means for finding degrees of freedom $\bar{f}_\beta$ associated with $SS(\beta|\mu, \alpha)$. Because of the presence of θ_{ij}'s in μ_{ij}, differences between columns (β effects) are not estimable. For example, from the incidence matrix $\mathbf{N}$, we see that

$$\mu + \alpha_2 + \beta_1 + \theta_{21} = \mu_{21},$$

and

$$\mu + \alpha_2 + \beta_2 + \theta_{22} = \mu_{22}$$

are estimable. Their difference

$$\mu_{21} - \mu_{22} = \beta_1 - \beta_2 + \theta_{21} - \theta_{22}$$

is also estimable. But $\beta_1 - \beta_2$ is not. This fact is also true for the rows (α effects).

From $\mathbf{C}_{\bar{\beta}}$ we find $\bar{f}_\beta = 4$. So $r(\mathbf{A}, \mathbf{B}) = 8 + 4 = 12$, and hence $r_\theta = 14 - 12 = 2$. From the additive two-way classification model we also obtain $SS(\mu)$, $SS(\alpha|\mu)$, and $SS(\beta|\mu, \alpha)$ exactly in the same way as described in Chapter 5 (see Example 5.7 for details):

$$SS(\mu) = N\bar{Y}_{\cdots}^2 = 900.66,$$

$$SS(\alpha|\mu) = 1023.78 - 900.66 = 123.12,$$

$$SS(\beta|\mu, \alpha) = 17.99,$$

and

$$\begin{aligned} SS(\mu, \alpha, \beta) &= SS(\mu) + SS(\alpha|\mu) + SS(\beta|\mu, \alpha) \\ &= 900.66 + 123.12 + 17.99 \\ &= 1041.77. \end{aligned}$$

So we obtain $SS(\theta|\mu, \alpha, \beta)$ by subtraction:

$$\begin{aligned} SS(\theta|\mu, \alpha, \beta) &= SS(\mu, \alpha, \beta, \theta) - SS(\mu, \alpha, \beta) \\ &= 1073.66 - 1041.77 \\ &= 31.89. \end{aligned}$$

The total sum of squares SS_{tot} remains the same as in Example 5.7, and

$$\begin{aligned} \hat{R} &= SS_{tot} - SS(\mu, \alpha, \beta, \theta) \\ &= 1184.06 - 1073.66 \\ &= 110.40. \end{aligned}$$

These results are shown in Table 7.1.

Note that in Chapter 5, we have already shown how to obtain the sum of squares and the associated degrees of freedom for the *interaction* term under the name *lack of fit* (see computer output for the same example at the end of Chapter 5). Here, we have introduced the interaction term as a third factor. In this context, one has to be careful about the interpretation of the sums of squares introduced in this section, and their meaning in terms of testing hypotheses.

For the two-way classification model with interaction, contrary to the additive two-way classification model, it is not so evident as to what we are testing when we use $SS(\alpha|\mu)$ or $SS(\beta|\mu)$ as the numerator sum of squares for the F ratio.

The F ratio with $SS(\alpha|\mu)$ in the numerator is testing a hypothesis involving the α's in the presence of weighted means of the β effects and the θ effects, that is, their interpretation depends on the n_{ij} values along with the fact that other effects are always present. It appears that the only meaningful testable

Table 7.1 Analysis of Variance for Two-Way Classification with Interaction for Example 7.15

Source	Degrees of Freedom		Sums of Squares	
Regression	14		1073.66	
Fitting overall mean		1		900.66
Factor A adjusted for the overall mean		7		123.12
Factor B adjusted for A		4		17.99
Interaction		2		31.89
Residual	19		110.40	
Total	33		1184.06	

hypotheses so far as interpretation is concerned are those involving the θ effects alone. The problem becomes more complex for the term SS($\beta|\alpha, \mu$). This is why we used the symbol $\bar{f}_\beta$ for the degrees of freedom associated with the β effects when the interaction terms are present. (For the additive two-way model we used the symbol f_β.)

Using the linear model (7.16), we have seen that it is possible to obtain from the μ_{ij}'s an estimable function which is only a function of θ's by applying the Q process. For example, we have obtained

$$\theta(L_1)=\theta_{13}-\theta_{15}+\theta_{85}-\theta_{83}$$

which is an estimable θ contrast, with its BLUE

$$\hat{\theta}(L_1)=\bar{Y}_{13.}-\bar{Y}_{15.}+\bar{Y}_{85.}-\bar{Y}_{83.}$$

and

$$\operatorname{Var}(\hat{\theta}(L_1))=\left(\frac{1}{n_{13}}+\frac{1}{n_{15}}+\frac{1}{n_{85}}+\frac{1}{n_{83}}\right)\sigma^2.$$

Suppose it is desired to test the hypothesis of the form

$$H_0\colon \mathbf{H}'\boldsymbol{\theta}=\mathbf{0},$$

where

$$\mathbf{H}'=\begin{bmatrix} 1 & -1 & 0 & 0 & 0 & 0 & 0 & 0 & 0 & 0 & 0 & 0 & -1 & 1 \\ 0 & 0 & 0 & 0 & 0 & 0 & 1 & -1 & 0 & 0 & -1 & 1 & 0 & 0 \end{bmatrix}.$$

Observe that the components of $\mathbf{H}'\boldsymbol{\theta}$ are estimable parametric functions. For making such a hypothesis, we know that if $\hat{\boldsymbol{\theta}}$ is any solution to the normal equations and if $\mathbf{V_H}$ is such that $\sigma^2\mathbf{V_H}=\operatorname{Cov}(\mathbf{H}'\hat{\boldsymbol{\theta}})$, then the appropriate numerator SS for this hypothesis is

$$SS_{H_0}=(\mathbf{H}'\hat{\boldsymbol{\theta}})'\mathbf{V}_{\mathbf{H}}^{-1}(\mathbf{H}'\boldsymbol{\theta}).$$

The degrees of freedom associated with this hypothesis is $r(\mathbf{V_H})$ which is in this case $r_\theta=\dim\bar{\mathscr{G}}=2$. Furthermore, if $\hat{\boldsymbol{\theta}}$ is a solution for the normal equations and if $\mathbf{V}$ is such that $\operatorname{Cov}(\hat{\boldsymbol{\theta}})=\sigma^2\mathbf{V}$, then $\mathbf{V_H}=\mathbf{H}'\mathbf{V}\mathbf{H}$. Note that for the model (7.16)

$$\operatorname{Cov}(\hat{\boldsymbol{\theta}})=\operatorname{diag}\left(\frac{1}{n_{ij}}\right)\sigma^2,$$

where the diagonal matrix only includes those n_{ij} which are nonzero. Therefore

$$\mathbf{H}'\hat{\boldsymbol{\theta}} = \begin{bmatrix} 1 & -1 & 0 & 0 & 0 & 0 & 0 & 0 & 0 & 0 & 0 & 0 & -1 & 1 \\ 0 & 0 & 0 & 0 & 0 & 0 & 1 & -1 & 0 & 0 & -1 & 1 & 0 & 0 \end{bmatrix} \times \begin{bmatrix} 3.4000 \\ 1.6000 \\ 5.6500 \\ 3.1250 \\ 6.7667 \\ 4.6000 \\ 3.4000 \\ 6.0500 \\ 10.5000 \\ 5.3000 \\ 7.7800 \\ 1.4000 \\ 2.4000 \\ 2.8500 \end{bmatrix} = \begin{bmatrix} 2.25 \\ -9.03 \end{bmatrix}.$$

Also

$$\mathbf{V} = (\mathbf{C}'\mathbf{C})^{-1} = \text{diag}[\tfrac{1}{2}, 1, \tfrac{1}{2}, \tfrac{1}{4}, \tfrac{1}{3}, \tfrac{1}{3}, 1, \tfrac{1}{2}, \tfrac{1}{2}, \tfrac{1}{4}, \tfrac{1}{5}, 1, 1, \tfrac{1}{2}]$$

so that

$$\mathbf{V_H} = \mathbf{H}'\mathbf{V}\mathbf{H} = \begin{bmatrix} 3.0 & 0.0 \\ 0.0 & 2.7 \end{bmatrix}$$

and

$$\mathbf{V_H^{-1}} = \begin{bmatrix} 0.3333 & 0.0000 \\ 0.0000 & 0.3704 \end{bmatrix}.$$

Therefore for testing the hypothesis

$$H_0: \begin{cases} \theta_{13} - \theta_{15} + \theta_{85} - \theta_{83} = 0 \\ \theta_{44} - \theta_{46} + \theta_{76} - \theta_{74} = 0 \end{cases} \tag{7.17}$$

we obtain

$$\begin{aligned} SS_{H_0} &= [2.25 \quad -9.03] \begin{bmatrix} 0.3333 & 0.0000 \\ 0.0000 & 0.3704 \end{bmatrix} \begin{bmatrix} 2.25 \\ -9.03 \end{bmatrix} \\ &= 31.89 \\ &= SS(\theta \mid \mu, \alpha, \beta) \end{aligned}$$

of Table 7.1. This sum of squares has 2 degrees of freedom.

See computer output at the end of this chapter for the same example. Also compare these results with those obtained at the end of Chapter 5.

PROGRAM INTER

General Description

This interactive program performs the calculations required for analyzing a set of data following a two-way classification model with interaction.

The only restriction on the number of observations per cell is that each row, as well as each column, must have at least one observation.

Currently, the program is limited by a $a, b < 50$ incidence matrix, although this could easily be changed.

1 The input for the program includes:

   (a) Dimension of the incidence matrix.
   (b) The incidence matrix.
   (c) The observations.

2 The output of this program includes:

   (a) A table of sums of squares and degrees of freedom.
   (b) The contrasts found for calculating the degrees of freedom of interaction.

Input Order and Format Specification

The input is in a file which name is chosen by the user, the file extension has to be ".dat", and the file name is asked by the program. All data are read in following the Fortran-free formats.

1 The dimensions of the incidence matrix, that is, a, b, the number of rows and columns, respectively.
2 The rows of the incidence matrix.
3 The next lines contain the observations, each line contains one observation, and the data are read in considering the incidence matrix column by column and row-wise.

Output Description

The output is in a file which name is the same as the input file's, but the extension is ".imp".

Following is a description of the output from the program:

1 The table of analysis has the following structure:

Source of Variation	Degrees of Freedom	Sums of Squares
Regression	r	$SS(\mu, \alpha, \beta)$
Fitting overall mean	1	$SS(\mu)$
Factor A adjusted for mean	$a-1$	$SS(\alpha\|\mu)$
Factor B adjusted for A	$\bar{f}_\beta$	$SS(\beta\|\alpha, \mu)$
Interaction	r_θ	$SS(\theta\|\beta, \alpha, \mu)$
Residual	$N-r$	Subtraction
Total	N	$\mathbf{Y}'\mathbf{Y}$

2 The contrasts found for calculating the degrees of freedom of interaction.

Each nonzero cell of the incidence matrix gets a number from 1 (for the first nonzero element) to k for the last element, considering the incidence matrix column by column and row-wise.

One line contains one contrast. For example we might have:

Contrasts found

$$\begin{matrix} 1 & 2 & 3 & 4 & 5 & 6 & 7 & 8 & 9 & 10 & 11 & 12 & 13 & 14 \end{matrix}$$
$$\begin{bmatrix} 1 & -1 & 0 & 0 & 0 & 0 & 0 & 0 & 0 & 0 & 0 & 0 & -1 & 1 \\ 0 & 0 & 0 & 0 & 0 & 0 & 1 & -1 & 0 & 0 & -1 & 1 & 0 & 0 \end{bmatrix}.$$

BIBLIOGRAPHICAL NOTES

7.2–7.7 The material in these sections is from Dodge (1974).

7.7 Theorems 7.2, 7.3, and Corollaries 7.5, 7.6, and 7.7, along with examples on Latin and Graeco-Latin squares of side 3 and 4 are from Dodge and Shah (1977), reprinted from Dodge, Y. and Shah, K. R. (1977), *Communications in Statistics, Theory and Methods*, 1465–1472, by courtesy of Marcel Dekker, Inc.

7.8 Main results of this section are from Birkes, Dodge, and Seely (1976).

REFERENCES

Birkes, D., Dodge, Y. and Seely, J. (1976). Spanning sets for estimable contrasts in classification models. *Ann. Statistis.* **4**, 86–107.

Dodge, Y. (1974). Estimability considerations of *N*-way classification experimental arrangements with missing observations. Ph.D. thesis, Department of Statistics, Oregon State University.

Dodge, Y. and Shah, K. R. (1977). Estimation of parameters in Latin squares and Greco-Latin squares with missing observations. *Commun. Statist., Theory and Methods.* **A6**(15), 1465–1472.

PROGRAM INTER

```
C
C       THIS INTERACTIVE PROGRAM PERFORMS THE CALCULATIONS REQUIRED
C       FOR ANALYSING A SET OF DATA FOLLOWING A TWO-WAY CLASSIFICATION
C       MODEL WITH INTERACTION AND WITH MISSING DATA, USING THE
C       INCIDENCE MATRIX N AND THE VECTOR OF OBSERVATIONS.
C       RESTRICTIONS:
C       GIVEN A TWO-WAY INCIDENCE MATRIX N, EACH ROW AS WELL AS EACH
C       COLUMN MUST HAVE AT LEAST ONE OBSERVATION.
C
C       INPUT:
C       THE FIRST INPUT LINE CONSISTS OF DIMENSION OF THE INCIDENCE
C       MATRIX N (NAxNB). THIS FOLLOWS BY THE ACTUAL OBSERVATIONS.
C       FOR EACH NON EMPTY CELL, THE ACTUAL OBSERVATIONS ARE READ IN
C       SO THAT EACH DATA TAKES ONE LINE. THE NON EMPTY CELL ARE READ
C       IN FROM LEFT TO RIGHT, ROW WISE.
C
C       STRUCTURE
C       VARIABLES
C       IDFA    OUTPUT: DEGREES OF FREEDOM OF FACTOR A
C       IDFB    OUTPUT: DEGREES OF FREEDOM OF FACTOR B
C       IDFG    OUTPUT: DEGREES OF FREEDOM OF INTERACTION
C       IDFREG  OUTPUT: DEGREES OF FREEDOM OF REGRESSION
C       IFAUX   INPUT:  ERROR TEST
C       NA      INPUT:  NUMBER OF ENTRIES OF FACTOR A
C       NB      INPUT:  NUMBER OF ENTRIES OF FACTOR B
C       NTOT    INPUT:  TOTAL DATA NUMBER
C       SSALP   OUTPUT: SUM OF SQUARES OF FACTOR A
C       SSBET   OUTPUT: SUM OF SQUARES OF FACTOR B
C       SSGAM   OUTPUT: SIM OF SQUARES FOR INTERACTION
C       SSREG   OUTPUT: SUM OF SQUARES OF REGRESSION
C       SSTOT   OUTPUT: TOTAL SUM OF SQUARES
C       WEB     INPUT:  NAME OF INPUT FILE
C
C       ARRAYS
C       N       INPUT:  INCIDENCE MATRIX
C       Y       INPUT:  DATAS
C       YSUM    INPUT:  SUM OF EACH CELL
C
C       SUBROUTINES
C       COMPUTE    COMPUTES EVERY SUMS OF SQUARES
C       DEGRE      COMPUTES EVERY DEGREES OF FREEDOM
C       OUTPUT     DRAWS TABLE
C
        PROGRAM INTERACTION
        PARAMETER IA=50,IB=IA**2
        DIMENSION N(IA,IA),YSUM(IA,IA)
        CHARACTER WEB*7
        COMMON /BLOCK1/SSALP,SSBET,SSGAM
C
C       A NAME OF INPUT FILE IS ASKED (WITHOUT FILE EXTENSION
C       WHICH MUST BE ".DAT")
C
        TYPE 1
1       FORMAT(/,1X,'NAME OF INPUT FILE ? ',$)
        ACCEPT 2,WEB
2       FORMAT(A)
        OPEN(UNIT=1,FILE=WEB//'.DAT',STATUS='OLD')
```

```
      OPEN(UNIT=2,FILE=WEB//'.IMP',STATUS='NEW')
C
C     READ IN DIMENSION OF THE INCIDENCE MATRIX
C
      READ(1,*) NA,NB
      IF(NA.GT.IA.OR.NB.GT.IA) THEN
      TYPE 8
8     FORMAT(//,1X,'THE INCIDENCE MATRIX IS TOO BIG')
      STOP
      END IF
C
C     READ IN INCIDENCE MATRIX
C
      DO K=1,NA
            READ(1,*) (N(K,L),L=1,NB)
      END DO
      SSTOT=0.
      NTOT=0
      IDFREG=0
      FOM=0.
C
C     READ IN OBSERVATIONS
C
      DO K=1,NA
            DO 3 L=1,NB
                  YSUM(K,L)=0.
                  NTOT=NTOT+N(K,L)
                  IF(N(K,L).EQ.0) GOTO 3
                  IDFREG=IDFREG+1
                  DO I2=1,N(K,L)
                        READ(1,*) Y
                        SSTOT=SSTOT+Y**2
                        YSUM(K,L)=YSUM(K,L)+Y
                  END DO
                  FOM=FOM+YSUM(K,L)
3           CONTINUE
      END DO
      CLOSE(UNIT=1)
      FOM=FOM**2/FLOAT(NTOT)
C
C     SUBROUTINE DEGRE COMPUTES DEGREES OF FREEDOM OF BOTH FACTOR
C
      CALL DEGRE(N,NA,NB,IDFA,IDFB)
C
C     DEGREE OF FREEDOM OF INTERACTION IS CALCULATED BY SUBTRACTION
C
      IDFG=IDFREG-NA-IDFB
C
C     SUBROUTINE COMPUTE CALCULATES SUMS OF SQUARES
C
      CALL COMPUTE(N,YSUM,NA,NB,IDFREG,IDFB,IDFG,IFAUX)
      SSREG=SSBET+SSGAM+SSALP
C
C     ERROR TEST
C
```

```
      IF(IFAUX.EQ.1) THEN
            WRITE(2,6)
6     FORMAT('1',///,1X,'ERROR IN COMPUTING TABLE')
            GOTO 7
      END IF
C
C     SUBROUTINE OUTPUT DRAWS THE TABLEAU
C
      CALL OUTPUT(IDFREG,SSREG,NTOT,SSTOT,NA,IDFB,IDFG,FOM)
C
C     SUBROUTINE FIND FINDS THE CONTRASTS
C
      CALL FIND(N,NA,NB,IDFREG)
7     CLOSE(UNIT=2)
      STOP
      END
C
C
C
      SUBROUTINE DEGRE(N,NA,NB,IDFA,IDFB)
C     ARRAYS
C     M        INPUT: SUM OF ALL INCIDENCE MATRIX
C
C     SUBROUTINES
C     BETAT    COMPUTES DEGREES OF FREEDOM OF FACTOR A AND B
C
      PARAMETER IA=50
      DIMENSION N(IA,IA),M(IA,IA)
C
C     CONSTRUCT M
C
      DO K=1,NB
            DO L=1,NA
                  IF(N(L,K).GT.0) THEN
                        M(L,K)=1
            ELSE
                        M(L,K)=0
                  END IF
            END DO
      END DO
      CALL BETAT(M,NA,NB,IDFB,IDFA)
      RETURN
      END
C
C
C
      SUBROUTINE COMPUTE(N,Y,NA,NB,NC,IDFB,IDFG,IFAUX)
C
C     SUBROUTINES
C     ARRAY    CHANGES VECTORS TO MATRIX AND MATRIX TO VECTORS
C     BASIS    COMPUTES BASIS OF FACTOR B AND C
C     MADD     ADDS TO MATRIX
C     MINV     COMPUTES INVERSE MATRIX
C     MPRO     MULTIPLICATES TWO MATRIX
C     MSUB     SUBTRACTS TWO MATRIX
C     MTRA     TRANPOSES A MATRIX
C
```

```
      PARAMETER IA=50,IB=IA**2
      DIMENSION APA(IA),APAI(IA),BPB(IA),Y(IA,IA),BPY(IB),
     1 APY(IB),P1(IB),P2(IB),CPY(IB),CPC(IB),P3(IB),ZN(IA,IA),
     2 P4(IB),ATA(IA,IA),N(IA,IA)
      COMMON /BLOCK1/SSALP,SSBET,SSGAM
      DO I=1,NA
              DO J=1,NB
                      ZN(I,J)=FLOAT(N(I,J))
              END DO
      END DO
C
C     COMPUTE A'A, A'Y AND INVERSE
C
      DO I=1,NA
              APA(I)=0.
              APY(I)=0.
              DO J=1,NB
                      APY(I)=APY(I)+Y(I,J)
                      APA(I)=APA(I)+ZN(I,J)
              END DO
              APAI(I)=1./APA(I)
      END DO
C
C     COMPUTE B'B, B'Y AND INVERSE
C
      DO I=1,NB
              BPB(I)=0.
              DO J=1,NA
                      BPY(I)=BPY(I)+Y(J,I)
                      BPB(I)=BPB(I)+ZN(J,I)
              END DO
      END DO
C
C     COMPUTE SUM OF SQUARES OF ALPHA FACTOR
C
      CALL MPRO(APAI,APY,P1,NA,NA,1,3,IA,IB,IB)
      CALL MPRO(P1,APY,P2,1,NA,1,1,IB,IB,IB)
      SSALP=P2(1)
C
C     COMPUTE SUM OF SQUARES OF BETA FACTOR
C
      CALL ARRAY(ZN,P1,NA,NB,1,IA)
      CALL MPRO(APAI,P1,P2,NA,NA,NB,3,IA,IB,IB)
      CALL MTRA(P1,P3,NA,NB,IB,IB)
      CALL MPRO(P3,P2,P1,NB,NA,NB,1,IB,IB,IB)
      CALL MSUB(BPB,P1,P2,NB,NB,2,IA,IB,IB)
      CALL BASIS(P1,IDFB,NB)
      CALL MPRO(P1,P2,P3,IDFB,NB,NB,1,IB,IB,IB)
      CALL MTRA(P1,P2,IDFB,NB,IB,IB)
      CALL MPRO(P3,P2,P1,IDFB,NB,IDFB,1,IB,IB,IB)
      CALL ARRAY(ATA,P1,IDFB,IDFB,2,IA)
      CALL MINV(ATA,IDFB,IFAUX)
      IF(IFAUX.EQ.1) THEN
              WRITE(2,*) 'ERROR IN COMPUTING INVERSE'
              RETURN
      END IF
```

```
      CALL ARRAY(ATA,P1,IDFB,IDFB,1,IA)
      CALL ARRAY(ZN,P2,NA,NB,1,IA)
      CALL MTRA(P2,P3,NA,NB,IB,IB)
      CALL MPRO(P3,APAI,P2,NB,NA,NA,2,IB,IA,IB)
      CALL MPRO(P2,APY,P3,NB,NA,1,1,IB,IB,IB)
      CALL MSUB(BPY,P3,P2,NB,1,1,IB,IB,IB)
      CALL BASIS(P3,IDFB,NB)
      CALL MPRO(P3,P2,P4,IDFB,NB,1,1,IB,IB,IB)
      CALL MPRO(P1,P4,P2,IDFB,IDFB,1,1,IB,IB,IB)
      CALL MPRO(P2,P4,P3,1,IDFB,1,1,IB,IB,IB)
      SSBET=P3(1)
      IO=0
C
C     COMPUTE SUM OF SQUARES OF INTERACTION (GAMMA FACTOR)
C
      DO I=1,IA
            DO K=1,IA
                  ATA(I,K)=0.
            END DO
      END DO
      DO I=1,NA
            DO 1 K=1,NB
                  IF(N(I,K).LE.0) GOTO 1
                  IO=IO+1
                  ATA(IO,IO)=ZN(I,K)
                  CPY(IO)=Y(I,K)
1           CONTINUE
      END DO
      CALL MINV(ATA,IO,IFAUX)
      IF(IFAUX.EQ.1) THEN
            WRITE(2,*) 'ERROR IN COMPUTING INVERSE'
            RETURN
      END IF
      CALL ARRAY(ATA,P1,IO,IO,1,IA)
      CALL MPRO(P1,CPY,P2,IO,IO,1,1,IB,IB,IB)
      CALL MPRO(P2,CPY,P3,1,IO,1,1,IB,IB,IB)
      SSGAM=P3(1)-SSALP-SSBET
      RETURN
      END
C
C
C
      SUBROUTINE BASIS(BASE,NA,NB)
      PARAMETER IA=50,IB=IA**2
C
C     SUBROUTINE BASIS COMPUTES BASIS MATRICES IN ORDER TO
C     CALCULATE SUMS OF SQUARES
C
      DIMENSION BASE(IB),Z(IA,IA)
      ID=1
      DO I1=1,NA
            DO I2=1,NB
                  Z(I1,I2)=0.
            END DO
            Z(I1,1)=1.
            ID=ID+1
            Z(I1,ID)=-1.
```

```
          END DO
          CALL ARRAY(Z,BASE,NA,NB,1,IA)
          RETURN
          END
C
C
C
          SUBROUTINE BETAT(M,NA,NB,IDFB,IDFA)
C         SUBROUTINE
C         MPM     MULTIPLICATES TRANSPOSED M MATRIX BY M MATRIX
C
          PARAMETER IA=50
          DIMENSION M(IA,IA),M2(IA,IA),MT(IA,IA)
C
C         INITIALIZE MATRIX M2
C
          DO I=1,NA
                  DO J=1,NB
                          M2(I,J)=0.
                          IF(M(I,J).GT.0) M2(I,J)=1
                  END DO
          END DO
C
C         COMPUTE R-PROCESS
C
1         DO I=1,NA
                  DO J=1,NB
                  IL=0
                  DO K=1,NA
                          DO L=1,NB
                                  IL=IL+M2(I,L)*M2(K,L)*M2(K,J)
                          END DO
                  END DO
                  IF(IL.GT.0) MT(I,J)=1
                  END DO
          END DO
          I2=0
          IT=0
          DO I=1,NA
                  DO J=1,NB
                          I2=I2+M2(I,J)
                          IT=IT+MT(I,J)
                  END DO
          END DO
          IF(I2.EQ.IT) GOTO 2
          DO I=1,NA
                  DO J=1,NB
                          M2(I,J)=MT(I,J)
                  END DO
          END DO
          GOTO 1
2         CALL MPM(MT,NA,NB,IDFB,IDFA)
          RETURN
          END
```

```
C
C
C
      SUBROUTINE MPM(MT,NA,NB,IDFB,IDFA)
      PARAMETER IA=50
      DIMENSION M2(IA,IA),MT(IA,IA)
      DO I=1,NB
              DO J=1,NB
                      M2(I,J)=0
                      DO K=1,NA
                              M2(I,J)=M2(I,J)+MT(K,I)*MT(K,J)
                      END DO
              END DO
      END DO
      IDFB=0
      DO I=2,NB
              IL=0
              DO J=1,I-1
                      IF(M2(I,J).GT.0) IL=IL+1
              END DO
              IF(IL.GT.0) IDFB=IDFB+1
      END DO
      DO I=1,NA
              DO J=1,NA
                      M2(I,J)=0
                      DO K=1,NB
                              M2(I,J)=M2(I,J)+MT(I,K)*MT(J,K)
                      END DO
              END DO
      END DO
      IDFA=0
      DO I=2,NA
              IL=0
              DO J=1,I-1
                      IF(M2(I,J).GT.0) IL=IL+1
              END DO
              IF(IL.GT.0) IDFA=IDFA+1
      END DO
      RETURN
      END
C
C
C
      SUBROUTINE OUTPUT(IDFREG,SSREG,NTOT,SSTOT,NA,IDFB,IDFG,FOM)
      COMMON /BLOCK1/SSALP,SSBET,SSGAM
      IDFRES=NTOT-IDFREG
      SSRES=SSTOT-SSREG
      WRITE(2,2) IDFREG,SSREG
2     FORMAT('1',//,1X,72('*'),/,1X,35X,'*',12X,'*',/,3X,'SOURCE OF ',
      1 'VARIATION',14X,'*',4X,'D.F.',4X,'*',3X,'SUMS OF SQUARES',/,
      2 36X,'*',12X,'*',/,1X,72('*'),/,36X,'*',12X,'*',/,2X,
      3 'REGRESSION',24X,'*',2X,I3,7X,'*',2X,E15.7,/,36X,'*',12X,'*',
      4 /,36X,'*',12X,'*')
```

```
      WRITE(2,1) FOM,NA-1,SSALP-FOM,IDFB,SSBET,IDFG,SSGAM
1     FORMAT(6X,'FITTING OVER ALL MEAN',9X,'*',8X,'1',3X,'*',6X,E15.7,/,
     1 36X,'*',12X,'*',/,6X,'FACTOR A UNADJUSTED',11X,'*',6X,I3,3X,'*'
     2 ,6X,E15.7,/,36X,'*',12X,'*',/,6X,'FACTOR B ADJUSTED FOR A',7X,
     3 '*',6X,I3,3X,'*',6X,E15.7,/,36X,'*',12X,'*',/,6X,'INTERACTION'
     4 ,19X,'*',6X,I3,3X,'*',6X,E15.7,/,36X,'*',12X,'*')
      WRITE(2,3) IDFRES,SSRES,NTOT,SSTOT
3     FORMAT(2X,'RESIDUAL',26X,'*',2X,I3,7X,'*',2X,E15.7,
     1 /,36X,'*',12X,'*',/,1X,72('*'),/,36X,'*',12X,'*',/,2X,
     2 'TOTAL',29X,'*',2X,I3,7X,'*',2X,E15.7)
      RETURN
      END
C
C
C
      SUBROUTINE FIND(M,NA,NB,NC)
C     SUBROUTINES
C     LOOP     FINDS THE LOOPS (CONTRASTS)
C
      PARAMETER IA=50
      DIMENSION M(IA,IA),BASE(IA,IA),ZIND(IA)
C
C     INITIALIZE MATRIX M
C
      IT=0
      I1=0
      DO I=1,NA
              DO K=1,NB
                      IF(M(I,K).GT.0) THEN
                      IT=IT+1
                      M(I,K)=IT
                      END IF
              END DO
      END DO
C
C     FIND SIMPLE LOOPS
C
      DO I=1,NB-1
      DO 3 IL=1,I
              K=0
2             K=K+1
              IF(K.GE.NA) GOTO 3
              L=K
              IF(M(K,IL).EQ.0.OR.M(K,I+1).EQ.0) GOTO 2
4             L=L+1
              IF(L.GT.NA) GOTO 3
              IF(M(L,IL).EQ.0.OR.M(L,I+1).EQ.0) GOTO 4
              I1=I1+1
              I2=0
              J1=K
5             CONTINUE
              I2=I2+1
              BASE(I1,I2)=FLOAT(M(J1,IL))
              I2=I2+1
              BASE(I1,I2)=FLOAT(M(J1,I+1))
```

```
          IF(J1.EQ.K) THEN
                  J1=L
                  GOTO 5
          END IF
          M(K,IL)=0
          GOTO 2
3     CONTINUE
      END DO
C
C     SORT MATRIX BASIS AND FORM THE BASE
C
      DO I=1,I1
          DO K=1,IA
                  ZIND(K)=0
          END DO
          DO K=1,NC
                  MN=0
                  DO L=1,4
                          MN=MN+1
                          IF(L.GT.2) THEN
                          IO=1
                          ELSE
                          IO=0
                          END IF
                          IF(BASE(I,L).EQ.K) ZIND(K)=ZIND(K)+
      1 (-1.)**(MN+1+IO)
                  END DO
          END DO
          DO K=1,NC
                  BASE(I,K)=ZIND(K)
          END DO
      END DO
C
C     COMPUTE X-PROCESS
C
      CALL XPROCESS(M,NA,NB,IA)
C
C     SUBROUTINE LOOP FINDS COMPLEX LOOPS
C
      CALL LOOP(M,BASE,NA,NB,I1,NC)
      IF(I1.LE.0) GOTO 6
      WRITE(2,1)
1     FORMAT(///,1X,'CONTRASTS FOUND',//)
      DO I=1,IT
          ZIND(I)=FLOAT(I)
      END DO
      WRITE(2,8) (INT(ZIND(I)),I=1,IT)
8     FORMAT(1X,<IT>(2X,I2),/)
      DO I=1,I1
          WRITE(2,7) (INT(BASE(I,K)),K=1,IT)
7     FORMAT(/,1X,<IT>(2X,I2))
      END DO
6     RETURN
      END
C
```

```
C
C
      SUBROUTINE LOOP(M,BASE,NA,NB,I1,NC)
      PARAMETER IB=50
      DIMENSION M(IB,IB),IOU(IB),BASE(IB,IB)
      L1=0
C
C     SEARCH FIRST NON-ZERO ELEMENT
C
1     L1=L1+1
      L2=0
      IF(L1.GT.NA) GOTO 2
3     L2=L2+1
      IF(L2.GT.NB) GOTO 1
      IF(M(L1,L2).EQ.0) GOTO 3
C
C     TEST IN ORDER TO DETECT A ROW WITH ONLY ONE NON-ZERO ELEMENT
C
      ITEST=0
      DO I=1,NB
              IF(M(L1,I).GT.0) ITEST=ITEST+1
      END DO
      IF(ITEST.LE.1) GOTO 1
      ID=0
      IOU(1)=0
      GOTO 4
5     IF(IOU(2).EQ.L1.AND.IOU(3).EQ.L2.AND.IOU(1).GE.3.AND.
      1 ID.NE.MEM.AND.ID.NE.MEM+2) THEN
C
C     TEST TO DETECT A LOOP
C
      ISI=1
      I1=I1+1
      DO J2=1,NC
              BASE(I1,J2)=0.
      END DO
C
C     RECORD THE LOOP
C
      DO J1=1,IOU(1)
              N=2*J1
              L=N+1
              ISI=ISI+1
              IMS=ABS(M(IOU(N),IOU(L)))
              BASE(I1,IMS)=BASE(I1,IMS)+(-1.)**ISI
      END DO
      M(IOU(2),IOU(3))=0
C
C     COMPUTE X-PROCESS
C
      CALL XPROCESS(M,NA,NB,IB)
6     L1=IOU(2)
      L2=IOU(3)
      GOTO 3
      END IF
```

```
C
C       TEST TO DETECT IF THE ALGORITHM ALREADY PASSED THIS POINT
C
        DO K=1,IOU(1)
                J=2*K
                L=J+1
                IF(IOU(J).EQ.L1.AND.IOU(L).EQ.L2) THEN
                IF(ID.EQ.1.OR.ID.EQ.2.OR.ID.EQ.3) GOTO 7
8               IOU(1)=IOU(1)-1
                IF(IOU(1).EQ.0) GOTO 6
                J=2*IOU(1)
                L=J+1
                L1=IOU(J)
                L2=IOU(L)
                J=2*(IOU(1)+1)
                L=J+1
                ID=0
                IF(IOU(1)-1.GT.0) THEN
                IJ=2*(IOU(1)-1)
                IL=IJ+1
                IF(L1.LT.IOU(IJ)) IK=1
                IF(L1.GT.IOU(IJ)) IK=3
                IF(L2.GT.IOU(IL)) IK=2
                IF(L2.LT.IOU(IL)) IK=4
                END IF
                IF(L1.LT.IOU(J)) GOTO 12
                IF(L1.GT.IOU(J)) GOTO 10
                IF(L2.GT.IOU(L)) GOTO 8
                IF(L2.LT.IOU(L)) GOTO 11
                END IF
        END DO
4       IOU(1)=IOU(1)+1
C
C       RECORD THE POINT
C
        J=2*IOU(1)
        L=J+1
        IOU(J)=L1
        IOU(L)=L2
        IF(IOU(1).LE.1) THEN
        IK=0
        GOTO 9
        END IF
        J=2*(IOU(1)-1)
        L=J+1
C
C       TEST IN ORDER TO CHOOSE THE RIGHT DIRECTION
C
        IF(IOU(J).LT.L1) IK=3
        IF(IOU(J).GT.L1) IK=1
        IF(IOU(L).LT.L2) IK=2
        IF(IOU(L).GT.L2) IK=4
        IF(IOU(1).EQ.2) THEN
        IF(IK.EQ.2.OR.IK.EQ.4) MEM=2
        IF(IK.EQ.1.OR.IK.EQ.3) MEM=1
        END IF
C
```

```
C        DIRECTION 1: UP
C
9        IF(ID.EQ.3.OR.IK.EQ.1.OR.IK.EQ.3) GOTO 10
         IF(L1-1.LE.0.AND.M(L1,L2).EQ.0) GOTO 7
         IF(L1-1.LE.0) GOTO 10
         L1=L1-1
         ID=1
         IF(M(L1,L2).GT.0) GOTO 5
         GOTO 9
C
C        DIRECTION 2: RIGHT
C
10       IF(ID.EQ.4.OR.IK.EQ.2.OR.IK.EQ.4) GOTO 11
         IF(L2+1.GT.NB.AND.M(L1,L2).EQ.0) GOTO 7
         IF(L2+1.GT.NB) GOTO 11
         L2=L2+1
         ID=2
         IF(M(L1,L2).GT.0) GOTO 5
         GOTO 10
C
C        DIRECTION 3: DOWN
C
11       IF(ID.EQ.1.OR.IK.EQ.1.OR.IK.EQ.3) GOTO 12
         IF(L1+1.GT.NA.AND.M(L1,L2).EQ.0) GOTO 7
         IF(L1+1.GT.NA) GOTO 12
         L1=L1+1
         ID=3
         IF(M(L1,L2).GT.0) GOTO 5
         GOTO 11
C
C        DIRECTION 4: LEFT
C
12       IF(ID.EQ.2.OR.IK.EQ.2.OR.IK.EQ.4) GOTO 8
         IF(L2-1.LE.0.AND.M(L1,L2).EQ.0) GOTO 7
         IF(L2-1.LE.0) GOTO 8
         L2=L2-1
         ID=4
         IF(M(L1,L2).GT.0) GOTO 5
         GOTO 12
7        CONTINUE
         L1B=L1
         L2B=L2
         J=2*IOU(1)
         L=J+1
         L1=IOU(J)
         L2=IOU(L)
         J=2*(IOU(1)+1)
         L=J+1
         ID=0
         IF(L1.LT.L1B) GOTO 12
         IF(L1.GT.L1B) GOTO 10
         IF(L2.GT.L2B) GOTO 8
         IF(L2.LT.L2B) GOTO 11
2        RETURN
         END
C
C
```

```
C
        SUBROUTINE ARRAY(A,B,N,M,ICODE,IA)
C
C       SUBROUTINE ARRAY TRANSFORMS A MATRIX INTO A VECTOR BY
C       USING ICODE 1 AND TRANSFORMS A VECTOR INTO A MATRIX BY
C       USING ICODE 2
C
        DIMENSION A(IA,IA),B(IA**2)
        IF(ICODE.EQ.2) GOTO 1
        K=1
        DO I=1,N
        DO L=1,M
                B(K)=A(I,L)
                K=K+1
        END DO
        END DO
        GOTO 2
1       I=1
        L=1
        DO K=1,N*M
                A(I,L)=B(K)
                IF(L.LT.M) GOTO 3
                L=0
                I=I+1
3               L=L+1
        END DO
2       RETURN
        END
C
C
C
        SUBROUTINE MADD(A,B,C,N,IA)
C
C       SUBROUTINE MADD ADDS A DIAGONAL MATRIX A TO A NORMAL MATRIX B
C
        DIMENSION A(IA),B(IA**2),C(IA**2)
        K=0
        DO 2 I=1,N**2
                IF(I.EQ.1+K*(N+1)) GOTO 1
                C(I)=B(I)
                GOTO 2
1               C(I)=A(K+1)+B(I)
                K=K+1
2       CONTINUE
        RETURN
        END
C
C
C
        SUBROUTINE MINV(A,M,IFAUX)
C
C       SUBROUTINE MINV INVERSES A MATRIX A
C       FOR HIGHER PRECISION USE SYSTEM SUBROUTINES
C       ERROR TESTS ARE PROVIDED
C
```

```
          PARAMETER IA=50
          DIMENSION A(IA,IA),B(IA,IA),C(IA,IA)
          IFAUX=0
          IF(M.EQ.0.OR.(M.EQ.1.AND.A(1,1).EQ.0.)) THEN
                  IFAUX=1
                  WRITE(2,*)'ERROR IN COMPUTING INVERSE'
                  RETURN
          END IF
          IF(M.EQ.1) THEN
                  A(1,1)=1./A(1,1)
                  RETURN
          END IF
          DO I=1,M
          DO K=1,M
                  C(I,K)=A(I,K)
          END DO
          END DO
          CALL DET(C,M,DETER,IFAUX)
          IF(DETER.EQ.0.OR.IFAUX.EQ.1) THEN
                  IFAUX=1
                  WRITE(2,*)'ERROR IN COMPUTING INVERSE'
                  RETURN
          END IF
          IF(M.EQ.2) GOTO 6
          DO I=1,M
          DO 5 K=1,M
                  J1=1
                  J2=0
                  DO 1 J=1,M
                  DO 2 L=1,M
                         IF(J.EQ.I) GOTO 1
                         IF(L.EQ.K) GOTO 2
                         J2=J2+1
                         IF(J2.LE.M-1) GOTO 3
                         J2=1
                         J1=J1+1
3                        C(J1,J2)=A(J,L)
2                 CONTINUE
1                 CONTINUE
                  IF(M.NE.2) GOTO 7
                  DE=C(1,1)
                  GOTO 8
7                 CALL DET(C,M-1,DE,IFAUX)
                  IF(IFAUX.EQ.1) RETURN
8                 CONTINUE
                  B(K,I)=DE/DETER*(-1.)**(I+K)
5         CONTINUE
          END DO
          DO I=1,M
          DO J=1,M
                  A(I,J)=B(I,J)
          END DO
          END DO
          GOTO 9
6         B1=A(2,2)/DETER
          B2=-A(1,2)/DETER
          B3=-A(2,1)/DETER
```

```
      B4=A(1,1)/DETER
      A(1,1)=B1
      A(1,2)=B2
      A(2,1)=B3
      A(2,2)=B4
9     RETURN
      END
C
C
C
      SUBROUTINE DET(C,N,TOTAL,IFAUX)
C
C     SUBROUTINE DET COMPUTES DETERMINANT OF MATRIX C
C     IT COMPUTES IN DOUBLE PRECISION
C     ERROR TESTS ARE PROVIDED
C
      PARAMETER IB=50
      DIMENSION C(IB,IB)
      DOUBLE PRECISION A(IB,IB),RAP
      IFAUX=0
      DO I=1,N
              DO K=1,N
                      A(I,K)=C(I,K)
              END DO
      END DO
      RAP=0.
      SIGNE=1.
      TOTAL=1.
      DO I=N,2,-1
      DO 1 L=I-1,1,-1
              IF(A(L,I).EQ.0.) GOTO 1
              IF(A(I,I).NE.0.) GOTO 4
              DO I1=L,1,-1
                      IND=I1
                      IF(A(I1,I).NE.0.) GOTO 3
              END DO
              IFAUX=1
              WRITE(2,*)'ERROR IN COMPUTING INVERSE'
              RETURN
3             DO I1=1,N
                      B=A(IND,I1)
                      A(IND,I1)=A(I,I1)
                      A(I,I1)=B
              END DO
              SIGNE=(-1.)**(IND+I)*SIGNE
4             RAP=A(L,I)/A(I,I)
              DO K=I-1,1,-1
                      A(L,K)=A(L,K)-RAP*A(I,K)
              END DO
1     CONTINUE
      END DO
      DO I=1,N
              TOTAL=TOTAL*A(I,I)
      END DO
      TOTAL=TOTAL*SIGNE
      RETURN
      END
```

```
C
C
C
      SUBROUTINE MPRO(A,B,C,N,II,M,ICODE,ICA,ICB,ICC)
      DIMENSION A(ICA),B(ICB),C(ICC)
C
C     SUBROUTINE MPRO MULTIPLICATES A MATRIX A BY A MATRIX B.
C     FOUR CASES MAY APPEAR:
C     ICODE 1: TWO NORMAL MATRICES ARE INVOLVED
C     ICODE 2: A NORMAL AND A DIAGONAL MATRIX ARE INVOLVED
C     ICODE 3: A DIAGONAL AND A NORMAL MATRIX ARE INVOLVED
C     ICODE 4: TWO VECTORS ARE MULTIPLICATED IN ORDER TO FORM
C              A MATRIX
C
      IF(ICODE.NE.1) GOTO 1
      IPAS=1
      DO K=1,N
      DO J=1,M
              L=0
              C(IPAS)=0.
              DO I=1+II*(K-1),II*K
                      C(IPAS)=A(I)*B(L*M+J)+C(IPAS)
                      L=L+1
              END DO
              IPAS=IPAS+1
      END DO
      END DO
      GOTO 2
1     IF(ICODE.NE.2) GOTO 3
      K=1
      DO I=1,N*II
              C(I)=A(I)*B(K)
              K=K+1
              IF(K.GT.II) K=1
      END DO
      GOTO 2
3     IF(ICODE.NE.3) GOTO 5
      K=1
      J=0
      DO 4 I=1,N*M
              C(I)=A(K)*B(I)
              J=J+1
              IF(J.LT.M) GOTO 4
              K=K+1
              J=0
4     CONTINUE
      GOTO 2
5     IPAS=1
      DO I=1,M
      DO K=1,N
              C(IPAS)=A(K)*B(I)
              IPAS=IPAS+1
      END DO
      END DO
2     RETURN
      END
C
```

```
C
C
      SUBROUTINE MSUB(A,B,C,N,M,ICODE,ICA,IA,IB)
C
C     SUBROUTINE MSUB SUBTRACTS A MATRIX B FROM A MATRIX A
C     ICODE 1: TWO NORMAL MATRICES ARE INVOLVED
C     ICODE 2: A DIAGONAL AND A NORMAL MATRIX ARE INVOLVED
C
      DIMENSION A(ICA),B(IA),C(IB)
      K=0
      IF(ICODE.NE.1) GOTO 3
      DO I=1,N*M
              C(I)=A(I)-B(I)
      END DO
      GOTO 4
3     DO 2 I=1,N*M
              IF(I.EQ.1+K*(M+1)) GOTO 1
              C(I)=-B(I)
              GOTO 2
1             C(I)=A(K+1)-B(I)
              K=K+1
2     CONTINUE
4     RETURN
      END
C
C
C
      SUBROUTINE MTRA(A,B,N,M,IC,ID)
C
C     SUBROUTINE MTRA TRANSPOSES A MATRIX A INTO A MATRIX B
C
      DIMENSION A(IC),B(ID)
      K=0
      IPAS=1
      DO 1 I=1,M*N
              B(I)=A(IPAS+K*M)
              K=K+1
              IF(K.LT.N) GOTO 1
              IPAS=IPAS+1
              K=0
1     CONTINUE
      RETURN
      END
C
C
C
      SUBROUTINE XPROCESS(M,NA,NB,IA)
      DIMENSION M(IA,IA)
      DO I=1,NA
              ISOM=0
              DO K=1,NB
                      IF(M(I,K).EQ.0) ISOM=ISOM+1
              END DO
              IF(ISOM.EQ.NB-1) THEN
                      DO K=1,NB
                              M(I,K)=0
                      END DO
```

```
              END IF
        END DO
        DO I=1,NB
              ISOM=0
              DO K=1,NA
                    IF(M(K,I).EQ.0) ISOM=ISOM+1
              END DO
              IF(ISOM.EQ.NA-1) THEN
                    DO K=1,NA
                          M(K,I)=0
                    END DO
              END IF
        END DO
        RETURN
        END
C
C
C
```

INPUT DATA

```
8 6
0 0 2 0 1 0
2 4 0 0 0 0
0 0 3 0 0 0
0 3 0 1 0 2
0 0 0 0 2 0
0 0 0 0 4 0
0 0 0 5 0 1
0 0 1 0 2 0
2.3
4.5
1.6
4.2
7.1
3.6
2.4
1.9
4.6
3.9
7.9
8.5
5.6
6.1
2.1
3.4
5.2
6.9
9.1
11.9
3.5
4.2
6.8
6.7
8.7
12.9
6.4
8.6
2.3
1.4
2.4
4.2
1.5
```

SOURCE OF VARIATION	D.F.	SUMS OF SQUARES
REGRESSION	14	0.1073663E+04
FITTING OVER ALL MEAN	1	0.9006593E+03
FACTOR A UNADJUSTED	7	0.1231241E+03
FACTOR B ADJUSTED FOR A	4	0.1799169E+02
INTERACTION	2	0.3188783E+02
RESIDUAL	19	0.1103970E+03
TOTAL	33	0.1184060E+04

CONTRASTS FOUND

1	2	3	4	5	6	7	8	9	10	11	12	13	14
1	-1	0	0	0	0	0	0	0	0	0	0	-1	1
0	0	0	0	0	0	1	-1	0	0	-1	1	0	0

CHAPTER 8

Generalized Inverses for Classification Models

Ecstasy is not reached by just anyone who starts to dance.
Dancing results from the soul's inner state; the inner
state of the soul does not result from dancing.

SOHRAVARDI

8.1 INTRODUCTION

In factorial experiments when there are no missing observations, the data can be analyzed in a straightforward manner by the usual analysis of variance technique. These techniques were introduced in Chapters 2, and 3. However, if there are missing observations, all the contrasts within each factor might still be estimable, in which case the design matrix is said to be of maximal rank. In the event that enough observations are missing so that the design matrix is not of maximal rank, some parametric functions of interest could become nonestimable.

In Chapter 5 we presented complete results for estimability in two-way additive classification model following two factor experiments with completely arbitrary pattern. We introduced the R process that determines what cell expectations are estimable and also an algorithm for finding a basis for each effect.

In Chapter 6 we introduced the $R3$ process, which when applied to the incidence matrix $\mathbf{N}$ of a factorial experiment with an additive three-way classification model with missing observations produces a final matrix $\mathbf{M}$, which provides sufficient condition for a cell expectation to be estimable. As we have seen in Chapter 7, such models may become so complicated in structure that it could be difficult even to identify estimable functions and the confounded effects. In such a case, if one has access to a least squares generalized inverse of the design matrix, then one can obtain BLUEs for the estimable functions and unbiased estimates of their variances. Additionally, this inverse could be used to identify the estimable functions since $\mathbf{p}'\boldsymbol{\beta}$ is estimable if and only if $\mathbf{p}'\mathbf{X}^-\mathbf{X}=\mathbf{p}'$, and one may use the $\mathbf{X}_l^-$ for $\mathbf{X}^-$.

The main purpose of this chapter is to present a method of analysis for a data set following an n-way classification model with completely arbitrary pattern via an algorithm which finds a least square g inverse for the design matrix using the incidence and the design matrices *simultaneously*.

Computational aspects of generalized inverses have not been dealt with very extensively in literature and in any event it seems that such computations are subject to rounding off errors, which might even lead to erroneous conclusions. In this chapter we shall give an algorithm for finding a least square generalized inverse for the design matrix of a data set following an n-way classification model with completely arbitrary pattern.

The algorithm will make use of the R process and some theorems on generalized inverses of partitioned matrices. It gives a g inverse without any rounding off errors. This algorithm has been programmed in standard Fortran language and has been tested over a variety of design matrices. It provides a least square g inverse for the design matrix with or without interactions for two-way classification models. The extension of the program to cover higher dimensions is straightforward.

8.2 GENERALIZED INVERSE OF A MATRIX

Consider a square matrix $\mathbf{X}$ of order $p\times p$ of rank p. Then there exists a unique matrix $\mathbf{X}^{-1}$, called the inverse of $\mathbf{X}$ with the property that

$$\mathbf{X}\mathbf{X}^{-1}=\mathbf{X}^{-1}\mathbf{X}=\mathbf{I}_p,$$

where $\mathbf{I}_p$ is the identity matrix of order p. In this case the solution of the linear equation $\mathbf{X}\mathbf{b}=\mathbf{Y}$, where $\mathbf{Y}$ is a $p\times 1$ column vector, is given by $\mathbf{b}=\mathbf{X}^{-1}\mathbf{Y}$. The question is whether a similar representation of the solution of the form $\mathbf{b}=\mathbf{X}^-\mathbf{Y}$ is possible, when $\mathbf{X}$ is a singular square or a rectangular matrix. If there exists a matrix $\mathbf{X}^-$ such that $\mathbf{b}=\mathbf{X}^-\mathbf{Y}$ is a solution of $\mathbf{X}\mathbf{b}=\mathbf{Y}$ for any $\mathbf{Y}$ such that $\mathbf{X}\mathbf{b}=\mathbf{Y}$ is consistent equation, then $\mathbf{X}^-$ behaves as the inverse of $\mathbf{X}$, hence it may be called a generalized inverse (g inverse) of $\mathbf{X}$.

Let $\mathbf{X}$ be an $n\times p$ matrix of arbitrary rank. A generalized inverse of $\mathbf{X}$ is a $p\times n$ matrix $\mathbf{X}^-$ such that $\mathbf{b}=\mathbf{X}^-\mathbf{Y}$ is a solution of $\mathbf{X}\mathbf{b}=\mathbf{Y}$ for any $\mathbf{Y}$ which makes the equation consistent.

LEMMA 8.1 $\mathbf{X}^-$ exists if and only if $\mathbf{X}\mathbf{X}^-\mathbf{X}=\mathbf{X}$.

Proof The equation $\mathbf{X}\mathbf{b}=\mathbf{Y}$ is consistent if $\mathbf{Y}=\mathbf{X}\mathbf{Z}$ where $\mathbf{Z}$ is an arbitrary $p\times 1$ vector. Therefore the existence of $\mathbf{X}^-$ implies $\mathbf{X}(\mathbf{X}^-\mathbf{X}\mathbf{Z})=\mathbf{X}\mathbf{Z}$ for all $\mathbf{Z}$ which gives $\mathbf{X}\mathbf{X}^-\mathbf{X}=\mathbf{X}$.

Conversely, assume that the equation $\mathbf{Xb}=\mathbf{Y}$ is consistent. This implies the existence of a vector $\mathbf{U}$ such that $\mathbf{XU}=\mathbf{Y}$. Since $\mathbf{XX^-X}=\mathbf{X}$, $\mathbf{XX^-XU}=\mathbf{XU}$ which implies $\mathbf{XX^-Y}=\mathbf{Y}$; that is, $\mathbf{X^-Y}$ is a solution of the equation $\mathbf{Xb}=\mathbf{Y}$. We now have an equivalent definition of a g inverse. □

If $\mathbf{X}$ is an $n\times p$ matrix, then a $p\times n$ matrix $\mathbf{X}^-$ is its generalized inverse if

$$\mathbf{XX^-X}=\mathbf{X} \tag{8.1}$$

A g inverse so defined is not unique in general. Given $\mathbf{X}^-$ since $\mathbf{XX^-X}=\mathbf{X}$, then $\mathbf{X^-XX^-X}=\mathbf{X^-X}$ or $\mathbf{X^-X}$ is idempotent. Let $r(\mathbf{X})$ be the rank of $\mathbf{X}$. Note that $r(\mathbf{X^-X})=r(\mathbf{X})$ as $r(\mathbf{X})\geqslant r(\mathbf{X^-X})\geqslant r(\mathbf{XX^-X})=r(\mathbf{X})$. So, a g inverse of $\mathbf{X}$ of order $n\times p$ is a matrix $\mathbf{X}^-$ of order $p\times n$ such that $\mathbf{X^-X}$ is idempotent and $r(\mathbf{X^-X})=r(\mathbf{X})$.

LEMMA 8.2 A necessary and sufficient condition that

$$\mathbf{BX^-X}=\mathbf{B}$$

is that $R(\mathbf{B}')\subset R(\mathbf{X}')$, that is, there exists a matrix $\mathbf{D}$ such that $\mathbf{B}=\mathbf{DX}$. Similarly, if $\mathbf{B}=\mathbf{XX^-B}$, then it is necessary and sufficient that $\mathbf{B}=\mathbf{XD}$ for some $\mathbf{D}$.

Now consider the solution of the homogeneous equation $\mathbf{Xb}=\mathbf{0}$ in terms of a g inverse. We have:

- **(a)** A general solution of the homogeneous equation $\mathbf{Xb}=\mathbf{0}$ is $(\mathbf{I}-\mathbf{X^-X})\mathbf{Z}$ where $\mathbf{Z}$ is an arbitrary vector.
- **(b)** A general solution of a consistent nonhomogeneous equation $\mathbf{Xb}=\mathbf{Y}$ is $\mathbf{X^-Y}+(\mathbf{I}-\mathbf{X^-X})\mathbf{Z}$ where $\mathbf{Z}$ is an arbitrary vector.

Part (a) follows from the fact that $\mathbf{X}(\mathbf{I}-\mathbf{X^-X})=\mathbf{0}$ and $r(\mathbf{I}-\mathbf{X^-X})=p-r(\mathbf{X})$, and (b) follows from (a) and the fact that a general solution of $\mathbf{Xb}=\mathbf{Y}$ is the sum of a particular solution of $\mathbf{Xb}=\mathbf{Y}$ and a general solution of $\mathbf{Xb}=\mathbf{0}$.

Let us consider an inconsistent equation $\mathbf{Xb}=\mathbf{Y}$. We say that $\hat{\mathbf{b}}$ is a *least squares* solution if

$$\|\mathbf{X}\hat{\mathbf{b}}-\mathbf{Y}\|=\inf_{\mathbf{b}}\|\mathbf{Xb}-\mathbf{Y}\|.$$

THEOREM 8.1 Let $\mathbf{X}^-$ be a matrix (not necessarily a g inverse) such that $\mathbf{X^-Y}$ is a least squares of $\mathbf{Xb}=\mathbf{Y}$ for $\mathbf{Y}\in R^n$. Then it is necessary and sufficient that

$$\mathbf{XX^-X}=\mathbf{X}, \qquad (\mathbf{XX^-})'=\mathbf{XX^-}. \tag{8.2}$$

Proof By hypothesis

$$\|\mathbf{X}\mathbf{X}^{-}\mathbf{Y}-\mathbf{Y}\| \leqslant \|\mathbf{X}\mathbf{b}-\mathbf{Y}\| \qquad \text{for all } \mathbf{b}, \text{ and } \mathbf{Y}$$

$$\leqslant \|\mathbf{X}\mathbf{X}^{-}\mathbf{Y}-\mathbf{Y}+\mathbf{X}\mathbf{W}\| \qquad \text{for all } \mathbf{Y}, \mathbf{W}=\mathbf{b}-\mathbf{X}^{-}\mathbf{Y}$$

if and only if

$$[\mathbf{X}\mathbf{W}, (\mathbf{X}\mathbf{X}^{-}-\mathbf{I})\mathbf{Y}]=\mathbf{0} \qquad \text{for all } \mathbf{Y}, \mathbf{W}$$

if and only if

$$\mathbf{X}'\mathbf{X}\mathbf{X}^{-}=\mathbf{X}',$$

which is equivalent to the two conditions in (8.2). Note that a least squares solution may not be unique, but min $\|\mathbf{X}\mathbf{b}-\mathbf{Y}\|$ is unique. □

If $\mathbf{X}$ is an $n \times p$ matrix, then a $p \times n$ matrix $\mathbf{X}_l^{-}$ is its *least squares g* inverse if in addition to (8.1) $\mathbf{X}_l^{-}$ satisfies

$$(\mathbf{X}\mathbf{X}_l^{-})'=\mathbf{X}\mathbf{X}_l^{-}.$$

Furthermore, if $\mathbf{X}$ is an $n \times p$ matrix, then a $p \times n$ matrix $\mathbf{X}_r^{-}$ is its *reflexive g* inverse if in addition to (8.1), $\mathbf{X}_r^{-}$ satisfies

$$\mathbf{X}_r^{-}\mathbf{X}\mathbf{X}_r^{-}=\mathbf{X}_r^{-}.$$

Given an $n \times p$ matrix $\mathbf{X}$ and a vector $\mathbf{a}$, in the following theorems we provide formula for computing several types of g inverses of $(\mathbf{X}: \mathbf{a})$ from the corresponding g inverses of $\mathbf{X}$. Similar formula are obtained for computing g inverse of $\mathbf{X}$ from those of $(\mathbf{X}: \mathbf{a})$. These formulae are useful for revising least squares g inverses when deletion or addition of an observation is required.

THEOREM 8.2 Let $(\mathbf{X}^{-}: \mathbf{b})$ be the least squares g inverse of $\begin{pmatrix}\mathbf{X}\\ \mathbf{a}'\end{pmatrix}$, where $\mathbf{a}$, $\mathbf{b}$ are column vectors and $\mathbf{X}$, $\mathbf{X}^{-}$ are matrices of appropriate dimensions. Further, let $\mathbf{a} \in R(\mathbf{X}')$ and $\mathbf{b}'\mathbf{a} \neq 1$. Then

$$\mathbf{Y}=\left(\mathbf{I}+\frac{\mathbf{b}\mathbf{a}'}{1-\mathbf{b}'\mathbf{a}}\right)\mathbf{X}^{-}$$

is a least squares g inverse of $\mathbf{X}$.

Proof Given

$$\begin{pmatrix}\mathbf{X}\\ \mathbf{a}'\end{pmatrix}(\mathbf{X}^{-}: \mathbf{b})\begin{pmatrix}\mathbf{X}\\ \mathbf{a}'\end{pmatrix}=\begin{pmatrix}\mathbf{X}\\ \mathbf{a}'\end{pmatrix}.$$

That is,

$$\begin{cases} \mathbf{X}(\mathbf{X}^-\mathbf{X}+\mathbf{ba}')=\mathbf{X} & (8.3)\\ \mathbf{a}'(\mathbf{X}^-\mathbf{X}+\mathbf{ba}')=\mathbf{a}' & (8.4)\end{cases}$$

and

$$\begin{pmatrix} \mathbf{XX}^- & \mathbf{Xb} \\ \mathbf{a}'\mathbf{X}^- & \mathbf{a}'\mathbf{b} \end{pmatrix} \text{ is symmetric.} \qquad (8.5)$$

Now

$$\begin{aligned}
\mathbf{XYX} &= \mathbf{X}\left(\mathbf{I}+\frac{\mathbf{ba}'}{1-\mathbf{b}'\mathbf{a}}\right)\mathbf{X}^-\mathbf{X}\\
&= \mathbf{XX}^-\mathbf{X}+\frac{1}{1-\mathbf{b}'\mathbf{a}}[\mathbf{Xba}'\mathbf{X}^-\mathbf{X}]\\
&= \mathbf{XX}^-\mathbf{X}+\frac{1}{1-\mathbf{b}'\mathbf{a}}[\mathbf{Xb}(\mathbf{a}'-\mathbf{a}'\mathbf{ba}')] \text{ by (8.4)}\\
&= \mathbf{XX}^-\mathbf{X}+\mathbf{Xba}'\\
&= \mathbf{X} \text{ by (8.3)}
\end{aligned}$$

and

$$\mathbf{XY}=\mathbf{XX}^-+\left(\frac{1}{1-\mathbf{b}'\mathbf{a}}\right)\mathbf{Xba}'\mathbf{X}^-.$$

$\mathbf{XX}^-$ is symmetric from (8.5). Also from (8.5), $(\mathbf{Xb})'=\mathbf{a}'\mathbf{X}^-$. Hence $\mathbf{Xba}'\mathbf{X}^-$ is symmetric, so $\mathbf{XY}$ is symmetric. □

THEOREM 8.3 Let $\mathbf{X}^-$ be a least squares g inverse of $\mathbf{X}$, and let $\mathbf{a}$ be a column vector such that $\mathbf{a}\in R(\mathbf{X}')$. Let

$$\mathbf{d}=\mathbf{X}^{-\prime}\mathbf{a} \quad \text{and} \quad \mathbf{b}=\frac{\mathbf{X}^-\mathbf{X}^{-\prime}\mathbf{a}}{1+\mathbf{a}'\mathbf{X}^-\mathbf{X}^{-\prime}\mathbf{a}}.$$

Then

$$\mathbf{G}=(\mathbf{X}^- - \mathbf{bd}' : \mathbf{b})$$

is a least squares g inverse of $\begin{pmatrix}\mathbf{X}\\ \mathbf{a}'\end{pmatrix}$.

Proof Given

$$\mathbf{XX}^-\mathbf{X}=\mathbf{X} \qquad (8.6)$$

$$\mathbf{XX}^- \text{ is symmetric} \qquad (8.7)$$

$$\mathbf{a}\in R(\mathbf{X}'),$$

and $\mathbf{d}=\mathbf{X}^{-\prime}\mathbf{a}$ and $\mathbf{b}=(1+\mathbf{a}'\mathbf{X}^{-}\mathbf{X}^{-\prime}\mathbf{a})^{-1}\mathbf{X}^{-}\mathbf{X}^{-\prime}\mathbf{a}$.

Now

$$\begin{pmatrix}\mathbf{X}\\ \mathbf{a}'\end{pmatrix}(\mathbf{X}^{-}-\mathbf{bd}':\mathbf{b})=\begin{pmatrix}\mathbf{XX}^{-}-\mathbf{Xbd}' & \mathbf{Xb}\\ \mathbf{a}'\mathbf{X}^{-}-\mathbf{a}'\mathbf{bd}' & \mathbf{a}'\mathbf{b}\end{pmatrix},$$

$$\begin{aligned}(\mathbf{a}'\mathbf{X}^{-}-\mathbf{a}'\mathbf{bd}')'&=\mathbf{X}^{-\prime}\mathbf{a}-(\mathbf{a}'\mathbf{b})\mathbf{d}=(1-\mathbf{a}'\mathbf{b})\mathbf{d}\\
&=\left(1-\frac{\mathbf{a}'\mathbf{X}^{-}\mathbf{X}^{-\prime}\mathbf{a}}{1+\mathbf{a}'\mathbf{X}^{-}\mathbf{X}^{-\prime}\mathbf{a}}\right)\mathbf{d}\\
&=\frac{1}{1+\mathbf{a}'\mathbf{X}^{-}\mathbf{X}^{-\prime}\mathbf{a}}\mathbf{d},\\
\mathbf{Xb}&=\left(\frac{1}{1+\mathbf{a}'\mathbf{X}^{-}\mathbf{X}^{-\prime}\mathbf{a}}\right)\mathbf{XX}^{-}\mathbf{X}^{-\prime}\mathbf{a}\\
&=\left(\frac{1}{1+\mathbf{a}'\mathbf{X}^{-}\mathbf{X}^{-\prime}\mathbf{a}}\right)\mathbf{X}^{-\prime}\mathbf{X}'\mathbf{X}^{-\prime}\mathbf{a}\quad\text{by (8.6)}\\
&=\left(\frac{1}{1+\mathbf{a}'\mathbf{X}^{-}\mathbf{X}^{-\prime}\mathbf{a}}\right)\mathbf{X}^{-\prime}\mathbf{a}\quad\text{by (8.7) and Lemma 8.2,}\\
&=(\mathbf{a}'\mathbf{X}^{-}-\mathbf{a}'\mathbf{bd}')',\end{aligned}$$

and

$$\mathbf{XX}^{-}-\mathbf{Xbd}'=\mathbf{XX}^{-}-\frac{1}{1+\mathbf{a}'\mathbf{X}^{-}\mathbf{X}^{-\prime}\mathbf{a}}\mathbf{dd}'$$

is symmetric by (8.6), which implies

$$\begin{pmatrix}\mathbf{X}\\ \mathbf{a}'\end{pmatrix}(\mathbf{X}^{-}-\mathbf{bd}':\mathbf{b})\text{ is symmetric.}$$

Now,

$$\begin{pmatrix}\mathbf{X}\\ \mathbf{a}'\end{pmatrix}(\mathbf{X}^{-}-\mathbf{bd}':\mathbf{b})\begin{pmatrix}\mathbf{X}\\ \mathbf{a}'\end{pmatrix}=\begin{pmatrix}\mathbf{XX}^{-}\mathbf{X}-\mathbf{Xbd}'\mathbf{X}+\mathbf{Xba}'\\ \mathbf{a}'\mathbf{X}^{-}\mathbf{X}-\mathbf{a}'\mathbf{bd}'\mathbf{X}+\mathbf{a}'\mathbf{ba}'\end{pmatrix}.$$

But

$$\begin{aligned}\mathbf{Xba}'-\mathbf{Xbd}'\mathbf{a}&=\mathbf{Xba}'-\mathbf{Xba}'\mathbf{X}^{-}\mathbf{X}\\
&=\mathbf{Xba}'-\mathbf{Xba}'\quad\text{by (8.7) and Lemma 8.2}\\
&=\mathbf{0},\end{aligned}$$

and

$$\mathbf{XX}^{-}\mathbf{X}=\mathbf{X}\quad\text{by (8.1)}$$

and

$$\begin{aligned}\mathbf{a}'\mathbf{X}^-\mathbf{X} - \mathbf{a}'\mathbf{b}\mathbf{d}'\mathbf{X} + \mathbf{a}'\mathbf{b}\mathbf{a}' &= \mathbf{d}'\mathbf{X} - \mathbf{a}'\mathbf{b}\mathbf{d}'\mathbf{X} + \mathbf{a}'\mathbf{b}\mathbf{a}' \\ &= \mathbf{a}' - \mathbf{a}'\mathbf{b}\mathbf{a}' + \mathbf{a}'\mathbf{b}\mathbf{a}' = \mathbf{a}' \quad \text{since} \quad \mathbf{d}'\mathbf{X} = \mathbf{a}'\mathbf{X}^-\mathbf{X} = \mathbf{a}' \quad \text{by (8.7).}\end{aligned}$$

Hence

$$\begin{pmatrix}\mathbf{X}\\ \mathbf{a}'\end{pmatrix}(\mathbf{X}^- - \mathbf{b}\mathbf{d}' : \mathbf{b})\begin{pmatrix}\mathbf{X}\\ \mathbf{a}'\end{pmatrix} = \begin{pmatrix}\mathbf{X}\\ \mathbf{a}'\end{pmatrix}. \quad \square$$

THEOREM 8.4 Let $\mathbf{X}^-$ be a least squares g inverse of $\mathbf{X}$, and let $\mathbf{a}$ be a column vector such that $\mathbf{a} \in R(\mathbf{X})$. Let $\mathbf{d} = \mathbf{X}^-\mathbf{a}$. Then

$$\mathbf{G} = \begin{pmatrix}\mathbf{X}^- - \mathbf{d}\mathbf{b}' \\ \mathbf{b}'\end{pmatrix},$$

with $\mathbf{b}$ arbitrary is a least squares g inverse of $(\mathbf{X} : \mathbf{a})$.

Proof Given

$$\mathbf{X}\mathbf{X}^-\mathbf{X} = \mathbf{X}$$

$$\mathbf{X}\mathbf{X}^- \text{ is symmetric}$$

$$\mathbf{a} \in R(\mathbf{X})$$

also $\mathbf{d} = \mathbf{X}^-\mathbf{a}$, $\mathbf{b}$ arbitrary. Now

$$\begin{aligned}(\mathbf{X} : \mathbf{a})\begin{pmatrix}\mathbf{X}^- - \mathbf{d}\mathbf{b}' \\ \mathbf{b}'\end{pmatrix} &= \mathbf{X}\mathbf{X}^- - \mathbf{X}\mathbf{d}\mathbf{b}' + \mathbf{a}\mathbf{b}' \\ &= \mathbf{X}\mathbf{X}^- - \mathbf{X}\mathbf{X}^-\mathbf{a}\mathbf{b}' + \mathbf{a}\mathbf{b}' \\ &= \mathbf{X}\mathbf{X}^- - \mathbf{a}\mathbf{b}' + \mathbf{a}\mathbf{b}' \quad \text{since } \mathbf{a} \in R(\mathbf{X}) \text{ and Lemma 8.2} \\ &= \mathbf{X}\mathbf{X}^-.\end{aligned}$$

$\mathbf{X}\mathbf{X}^-$ is symmetric by hypothesis. And

$$\begin{aligned}(\mathbf{X} : \mathbf{a})\begin{pmatrix}\mathbf{X}^- - \mathbf{d}\mathbf{b}' \\ \mathbf{b}'\end{pmatrix}(\mathbf{X} : \mathbf{a}) &= (\mathbf{X}\mathbf{X}^-\mathbf{X} : \mathbf{X}\mathbf{X}^-\mathbf{a}) \\ &= (\mathbf{X} : \mathbf{a}),\end{aligned}$$

since $\mathbf{X}\mathbf{X}^-\mathbf{X} = \mathbf{X}$, $\mathbf{a} \in R(\mathbf{X})$ and by Lemma 8.2. Hence the result. $\square$

THEOREM 8.5 Let $\mathbf{X}^-$ be a least squares g inverse of $\mathbf{X}$ and let $\mathbf{a}$ be a column vector such that $\mathbf{a} \notin R(\mathbf{X})$. Let $\mathbf{d} = \mathbf{X}^-\mathbf{a}$, $\mathbf{c} = (\mathbf{I} - \mathbf{X}\mathbf{X}^-)\mathbf{a}$, and $\mathbf{b} = \mathbf{c}/\mathbf{c}'\mathbf{a}$. Then

$$\mathbf{G}=\begin{pmatrix}\mathbf{X}^- - \mathbf{d}\mathbf{b}' \\ \mathbf{b}'\end{pmatrix}$$

is a least squares g inverse of $(\mathbf{X}\colon \mathbf{a})$.

Proof Given

$$\mathbf{X}\mathbf{X}^-\mathbf{X}=\mathbf{X}$$

$$\mathbf{X}\mathbf{X}^- \text{ is symmetric}$$

$$\mathbf{a}\notin R(\mathbf{X})$$

also $\mathbf{d}=\mathbf{X}^-\mathbf{a}$, $\mathbf{c}=(\mathbf{I}-\mathbf{X}\mathbf{X}^-)\mathbf{a}$, and $\mathbf{b}=\mathbf{c}/\mathbf{c}'\mathbf{a}$. Note that $\mathbf{c}'\mathbf{a}=\mathbf{a}'(\mathbf{I}-\mathbf{X}\mathbf{X}^-)\mathbf{a}\neq 0$ because $\mathbf{a}\notin R(\mathbf{X})$. Now

$$\begin{aligned}(\mathbf{X}\colon \mathbf{a})\begin{pmatrix}\mathbf{X}^- - \mathbf{d}\mathbf{b}' \\ \mathbf{b}'\end{pmatrix}&=\mathbf{X}\mathbf{X}^- - \mathbf{X}\mathbf{d}\mathbf{b}'+\mathbf{a}\mathbf{b}'\\ &=\mathbf{X}\mathbf{X}^- - \mathbf{X}\mathbf{X}^-\mathbf{a}\mathbf{b}'+\mathbf{a}\mathbf{b}'\\ &=\mathbf{X}\mathbf{X}^- +(\mathbf{I}-\mathbf{X}\mathbf{X}^-)\mathbf{a}\mathbf{b}'\\ &=\mathbf{X}\mathbf{X}^- +\mathbf{c}\mathbf{b}'\\ &=\mathbf{X}\mathbf{X}^- +\mathbf{c}\mathbf{c}'/\mathbf{c}'\mathbf{a},\end{aligned}$$

which is symmetric by hypothesis. Also

$$(\mathbf{X}\colon \mathbf{a})\begin{pmatrix}\mathbf{X}^- - \mathbf{d}\mathbf{b}' \\ \mathbf{b}'\end{pmatrix}(\mathbf{X}\colon \mathbf{a})=\left(\mathbf{X}\mathbf{X}^-\mathbf{X}+\left(\frac{1}{\mathbf{c}'\mathbf{a}}\right)\mathbf{c}\mathbf{c}'\mathbf{X}\colon \mathbf{X}\mathbf{X}^-\mathbf{a}+\mathbf{c}\right),$$

but

$$\mathbf{X}\mathbf{X}^-\mathbf{a}+\mathbf{c}=\mathbf{X}\mathbf{X}^-\mathbf{X}+(\mathbf{I}-\mathbf{X}\mathbf{X}^-)\mathbf{a}=\mathbf{a},$$

and

$$\mathbf{X}\mathbf{X}^-\mathbf{X}+\left(\frac{1}{\mathbf{c}'\mathbf{a}}\right)\mathbf{c}\mathbf{c}'\mathbf{X}=\mathbf{X}$$

since

$$\begin{aligned}\mathbf{c}'\mathbf{X}&=\mathbf{a}'[\mathbf{I}-(\mathbf{X}\mathbf{X}^-)']\mathbf{X}\\ &=\mathbf{a}'(\mathbf{I}-\mathbf{X}\mathbf{X}^-)\mathbf{X}, \quad \text{since } \mathbf{X}\mathbf{X}^- \text{ is symmetric}\\ &=\mathbf{0} \quad \text{because} \quad \mathbf{a}\notin R(\mathbf{X}).\end{aligned}$$

Hence

$$(\mathbf{X}\colon \mathbf{a})\begin{pmatrix}\mathbf{X}^- - \mathbf{d}\mathbf{b}' \\ \mathbf{b}'\end{pmatrix}(\mathbf{X}\colon \mathbf{a})=(\mathbf{X}\colon \mathbf{a}). \quad \square$$

THEOREM 8.6 Let $\mathbf{X}^-$ be a least squares g inverse of $\mathbf{X}$ and let $\mathbf{a}$ be a column vector such that $\mathbf{a} \notin R(\mathbf{X}')$. Let $\mathbf{d} = \mathbf{X}^{-\prime}\mathbf{a}$, $\mathbf{c} = (\mathbf{I} - \mathbf{X}^-\mathbf{X})(\mathbf{I} - \mathbf{X}^-\mathbf{X})'\mathbf{a}$ and $\mathbf{b} = \mathbf{c}/\mathbf{c}'\mathbf{a}$. Then

$$\mathbf{G} = (\mathbf{X}^- - \mathbf{b}\mathbf{d}' : \mathbf{b})$$

is a least squares g inverse of $\begin{pmatrix} \mathbf{X} \\ \mathbf{a}' \end{pmatrix}$.

Proof Given

$$\mathbf{X}\mathbf{X}^-\mathbf{X} = \mathbf{X}$$

$$\mathbf{X}\mathbf{X}^- \text{ is symmetric, and}$$

$$\mathbf{a} \notin R(\mathbf{X}'),$$

also $\mathbf{d} = \mathbf{X}^{-\prime}\mathbf{a}$, $\mathbf{c} = (\mathbf{I} - \mathbf{X}^-\mathbf{X})(\mathbf{I} - \mathbf{X}^-\mathbf{X})'\mathbf{a}$, $\mathbf{b} = \mathbf{c}/\mathbf{c}'\mathbf{a}$. Now

$$\begin{pmatrix} \mathbf{X} \\ \mathbf{a}' \end{pmatrix} (\mathbf{X}^- - \mathbf{b}\mathbf{d}' : \mathbf{b}) = \begin{pmatrix} \mathbf{X}\mathbf{X}^- - \mathbf{X}\mathbf{b}\mathbf{d}' & \mathbf{X}\mathbf{b} \\ \mathbf{a}'\mathbf{X}^- - \mathbf{a}'\mathbf{b}\mathbf{d}' & \mathbf{a}'\mathbf{b} \end{pmatrix}.$$

Now,

$$\mathbf{X}\mathbf{X}^- - \mathbf{X}\mathbf{b}\mathbf{d}' = \mathbf{X}\mathbf{X}^- - \left(\frac{1}{\mathbf{c}'\mathbf{a}}\right) \mathbf{X}\mathbf{c}\mathbf{a}'\mathbf{X}^- = \mathbf{X}\mathbf{X}^-$$

which is symmetric by the hypotheses, and since

$$\mathbf{X}\mathbf{c} = (\mathbf{I} - \mathbf{X}^-\mathbf{X})(\mathbf{I} - \mathbf{X}^-\mathbf{X})'\mathbf{a} = \mathbf{0},$$

and

$$\begin{aligned} (\mathbf{a}'\mathbf{X}^- - \mathbf{a}'\mathbf{b}\mathbf{d}')' &= \mathbf{X}^{-\prime}\mathbf{a} - (\mathbf{a}'\mathbf{b})\mathbf{d} \\ &= (1 - \mathbf{a}'\mathbf{b})\mathbf{d} = \mathbf{0} \end{aligned}$$

since $\mathbf{a}'\mathbf{b} = \mathbf{a}'\mathbf{c}/\mathbf{c}'\mathbf{a} = 1$, and $\mathbf{X}\mathbf{b} = \mathbf{X}(\mathbf{I} - \mathbf{X}^-\mathbf{X})(\mathbf{I} - \mathbf{X}^-\mathbf{X})'\mathbf{a}/\mathbf{c}'\mathbf{a} = \mathbf{0}$. So

$$\begin{pmatrix} \mathbf{X} \\ \mathbf{a}' \end{pmatrix} (\mathbf{X}^- - \mathbf{b}\mathbf{d}' : \mathbf{b}) = \begin{pmatrix} \mathbf{X}\mathbf{X}^- & \mathbf{0} \\ \mathbf{0} & 1 \end{pmatrix} \text{ is symmetric,}$$

and

$$\begin{pmatrix} \mathbf{X} \\ \mathbf{a}' \end{pmatrix} (\mathbf{X}^- - \mathbf{b}\mathbf{d}' : \mathbf{b}) \begin{pmatrix} \mathbf{X} \\ \mathbf{a}' \end{pmatrix} = \begin{pmatrix} \mathbf{X}\mathbf{X}^-\mathbf{X} \\ \mathbf{a}' \end{pmatrix} = \begin{pmatrix} \mathbf{X} \\ \mathbf{a}' \end{pmatrix}.$$

Hence the result. □

We now proceed to find least squares g inverses for classification models following a factorial experiment with arbitrary pattern.

8.3 LEAST SQUARES GENERALIZED g INVERSES FOR FACTORIAL EXPERIMENTS

Let $Y_{ijk...wh}$ be a collection of independent random variables with a common unknown variance σ^2 and each having expectation of the form

$$E(Y_{ijk...wh})=\alpha_i+\beta_j+\delta_k+\cdots+\eta_w+(\alpha\beta)_{ij}+\cdots+(\alpha\beta\ldots\eta)_{ij...w} \quad (8.8)$$

where $i=1, 2,\ldots, a$; $j=1, 2,\ldots, b$; $\ldots$; $w=1, 2,\ldots, p$. The index h ranges from 1 to $n_{ijk...w}$. If $n_{ijk...w}=0$, then no random variable with subscripts $ijk\ldots w$ occurs in the collection. Thus we are working with a fixed effect n-way classification model with arbitrary pattern.

Let $\mathbf{X}$ denote the design matrix associated with the above model. Thus $\mathbf{X}=(\mathbf{A}\colon \mathbf{B}\colon\ldots\colon \mathbf{P}\colon \mathbf{X}_1)$, where $\mathbf{A}, \mathbf{B}, \mathbf{C},\ldots, \mathbf{P}$ are the submatrices associated with the α, β, $\delta,\ldots,\eta$ effects respectively, and $\mathbf{X}_1$ corresponds to the interactions.

In matrix notation, we can write (8.8) as

$$E(\mathbf{Y})=\mathbf{X}\boldsymbol{\theta}$$

where $\mathbf{Y}$ is an $N\times 1$ vector of observations and

$$\boldsymbol{\theta}'=[\alpha_1,\ldots,\alpha_a,\beta_1,\ldots,\beta_b,\delta_1,\ldots,\delta_c,\ldots,\eta_1,\ldots,\eta_p,(\alpha\beta)_{11},\ldots,(\alpha\beta\ldots\eta)_{ab...p}].$$

The problem is to find least squares solutions for the parameters in models with any arbitrary pattern. That is, we have to find an $\mathbf{X}_l^-$ for the design matrix $\mathbf{X}$. Due to the specific structure of two-way designs first we consider such designs in the following section, and then we expand the problem to the general case of n-way.

Least Squares Generalized Inverses for Additive Two-Way Models

Here the model would be

$$E(Y_{ijk})=\alpha_i+\beta_j,$$

the design matrix is $\mathbf{X}=(\mathbf{A}\colon \mathbf{B})$, and the incidence matrix $\mathbf{N}=(n_{ij})=\mathbf{A}'\mathbf{B}$. We assume that

$$n_{i.}=\sum_j n_{ij}\neq 0 \quad \text{for } i=1, 2,\ldots, a$$

$$n_{.j}=\sum_i n_{ij}\neq 0 \quad \text{for } j=1, 2,\ldots, b.$$

that is, there is at least one observation in each row and column.

The Algorithm

In this section we introduce an algorithm for finding the least square g inverse of the design matrix which employs the R process which was established in Chapter 5, and Theorems 8.2–8.6. The algorithm consists of the following steps.

Step 1 Apply the R process to the incidence matrix $\mathbf{N}$ in order to obtain a final matrix $\mathbf{M}$.

Step 2 Construct the design matrix $\mathbf{X}^*$ corresponding to $\mathbf{M}$.

Note that $\mathbf{X}^*$ differs from the original design matrix $\mathbf{X}$ in having some extra rows corresponding to those cells which were filled by the R process and with no row being repeated. We can partition the final matrix $\mathbf{M}$ into s sets of connected portions as defined in Chapter 5 such that the final matrix will be of the following form:

$$\mathbf{M}=\begin{bmatrix}\mathbf{M}_1 & \mathbf{0} & \cdots & \mathbf{0}\\ \mathbf{0} & \mathbf{M}_2 & \cdots & \mathbf{0}\\ \mathbf{0} & \mathbf{0} & \cdots & \mathbf{M}_s\end{bmatrix}.$$

This induces a corresponding partition on $\mathbf{X}^*$.

$$\mathbf{X}^*=\begin{bmatrix}\mathbf{X}_1^* & \mathbf{0} & \cdots & \mathbf{0}\\ \mathbf{0} & \mathbf{X}_2^* & \cdots & \mathbf{0}\\ \mathbf{0} & \mathbf{0} & \cdots & \mathbf{X}_s^*\end{bmatrix}$$

where $\mathbf{X}_i^*=(\mathbf{A}_i^*:\mathbf{B}_i^*)$ is the design matrix associated with $\mathbf{M}_i=(m_{jh}^{(i)})(i=1,2,\ldots,s)$. Let $\mathbf{M}_i$ use rows $j_1,\ldots,j_s$ and columns $k_1,\ldots,k_t$ of $\mathbf{N}$. Then define $\mathbf{N}_i=(n_{jh}^{(i)})$ as the submatrix using these rows and columns of $\mathbf{N}$. Also define

$$T_i=\{(j,h)|n_{jh}^{(i)}=k_p\geqslant 2\},$$

$$S_i=\{(j,h)|m_{jh}^{(i)}=1,\ n_{jh}^{(i)}=0\},\qquad i=1,2,\ldots,s.$$

If there are d_i elements in T_i, then $p=1,2,\ldots,d_i$. For each connected portion (i.e., for each i, $i=1,2,\ldots,s$), execute steps 3 through 8.

Step 3 In $\mathbf{X}_i^{*\prime}$ replace zeros by -1 and ones by $2a_i-1$ for $\mathbf{A}_i^{*\prime}$ and $2b_i-1$ for $\mathbf{B}_i^{*\prime}$ where a_i is the number of levels of the α effects in $\mathbf{M}_i$ and b_i the number of levels of β effects in $\mathbf{M}_i$. Multiply the resultant matrix by the scalar $(1/2a_ib_i)$ to obtain $\mathbf{Z}_{i1}$. A justification for this step is given in Theorem 8.7.

In the cells for which $n_{jh}=k\geqslant 2$ in the original incidence matrix, they correspond to k repetitions of the same row in the design matrix.

Step 4 Choose (j_1, h_1) from T_i. Let $\mathbf{a}_1' = \mathbf{f}_1'\mathbf{X}_i^*$ be the row of $\mathbf{X}_i^*$ corresponding to (j_1, h_1) where $\mathbf{f}_1$ is a vector of zeros except for one 1 suitably placed. Set $\mathbf{f} = \mathbf{f}_1$, $k = k_1 - 1$, $\mathbf{a} = \mathbf{a}_1$, $\mathbf{U} = \mathbf{Z}_{i1}$, and compute

$$\mathbf{g} = \mathbf{Uf}, \qquad \mathbf{d} = \mathbf{U}'\mathbf{a}, \qquad c_k = \frac{1}{1 + k\mathbf{g}'\mathbf{a}}.$$

Step 5 Compute the least squares g inverse

$$\mathbf{V} = (\mathbf{U} - kc_k\mathbf{gd}' : c_k\mathbf{g} : \ldots : c_k\mathbf{g})$$

where the column $c_k\mathbf{g}$ is repeated k times. (In fact one column of $\mathbf{U} - kc_k\mathbf{gd}'$ will also be $c_k\mathbf{g}$.) Change $\mathbf{X}_i^*$ to $\mathbf{E}_i$ which is now augmented by k extra rows corresponding to k repetitions. This should be done any time step 5 is applied.

Step 6 Choose (j_2, h_2) from T_i. Set $k = k_2 - 1$, $\mathbf{a} = \mathbf{a}_2 (= \mathbf{f}_2'\mathbf{E}_i)$, $\mathbf{f} = \mathbf{f}_2$, $\mathbf{U} = \mathbf{V}$ and compute steps 4 and 5. Continue until T_i is exhausted. Call final $\mathbf{V}$ as $\mathbf{Z}_{i2}$.

At the end of step 6, $\mathbf{X}_i^*$ is augmented to include all repeated rows. Call this matrix $\mathbf{X}_i^{**}$.

REMARK 8.1 The above steps 4, 5, and 6 have been obtained by applying Theorem 8.3 repeatedly. To arrive at this form one has to use the fact that the $\mathbf{U}$'s will be reflexive g inverses. This follows from the property that $\mathbf{Z}_{i1}$ is reflexive, a fact that is proved in Corollary 8.1.

Now suppose there are l elements in S_i. Let the rows of $\mathbf{X}_i^*$ corresponding to these be $\mathbf{a}_1', \mathbf{a}_2', \ldots, \mathbf{a}_l'$ and the columns of $\mathbf{Z}_{i2}$ be $\mathbf{g}_1, \mathbf{g}_2, \ldots, \mathbf{g}_l$: Let

$$\mathbf{X}_i^{**\prime} = [\mathbf{A}_i' : \mathbf{a}_1, \ldots, \mathbf{a}_l]$$

and

$$\mathbf{Z}_{i2} = [\mathbf{Q} : \mathbf{g}_1, \ldots, \mathbf{g}_l].$$

Note that $\mathbf{a}_j \in R(\mathbf{A}_i')$ for each j, $j = 1, 2, \ldots, l$. By repeated application of Theorem 8.2 we get the least squares generalized inverse of $\mathbf{A}_i$ in the following two steps.

Step 7 Set $\mathbf{C}_1 = \mathbf{I}$, $\mathbf{b}_1 = \mathbf{C}_1\mathbf{g}_1$. Compute

$$\mathbf{C}_i = \left(\mathbf{I} + \frac{\mathbf{b}_{i-l}\mathbf{a}_{i-l}'}{1 - \mathbf{b}_{i-l}'\mathbf{a}_{i-l}}\right)\mathbf{C}_{i-l}, \qquad \mathbf{b}_i = \mathbf{C}_i\mathbf{g}_i, \qquad i = 2, 3, \ldots$$

Continue until $\mathbf{C}_{l+1}$ is obtained.

Step 8 Compute $\mathbf{Y} = \mathbf{C}_{l+1}\mathbf{Q}$.

REMARK 8.2 In case one is willing to sacrifice some accuracy of this procedure one can substitute steps 7 and 8 by the following:

Let

$$\mathbf{X}_i^* = \begin{pmatrix} \mathbf{A}_1 \\ \mathbf{A}_2 \end{pmatrix},$$

where $\mathbf{A}_2$ corresponds to the missing cells of $\mathbf{N}_i$. Corresponding to this let $\mathbf{Z}_{i2} = (\mathbf{Q} : \mathbf{B})$. Then

$$\mathbf{Y} = \mathbf{Q} + \mathbf{B}(\mathbf{I} - \mathbf{A}_2\mathbf{B})^{-1}\mathbf{A}_2\mathbf{Q}$$

is a least squares generalized inverse of $\mathbf{A}_1$ (see Theorem 8.8 for justification). This involves the inversion of a matrix whose dimension is the number of missing observations in $\mathbf{N}_i$.

Step 9 If the resultant matrices at step 8 were $\mathbf{G}_i (i = 1, 2, \ldots, s)$ then the least squares g inverse of $\mathbf{X}^*$ takes the form

$$\mathbf{G} = \begin{bmatrix} \mathbf{G}_1 & \mathbf{0} & \cdots & \mathbf{0} \\ \mathbf{0} & \mathbf{G}_2 & \cdots & \mathbf{0} \\ \vdots & \vdots & & \vdots \\ \mathbf{0} & \mathbf{0} & & \mathbf{G}_s \end{bmatrix}.$$

The least squares g inverse $\mathbf{X}_l^-$ of $\mathbf{X}$ is obtained from $\mathbf{G}$ by suitably permuting rows and columns.

REMARK 8.3 In many situations especially if the number of missing cells in any connected portion is greater than the number of occupied cells it might be easier to adopt a different strategy.

Consider an 8×8 design with 48 missing cells having the following pattern:

	β_1	β_2	β_3	β_4	β_5	β_6	β_7	β_8
α_1	1							
α_2		1	1					
α_3			1	1			1	
α_4				1	1			
α_5					1	1		
α_6						1	1	
α_7							1	1
α_8	1							1

First find the least squares g inverses for the design matrices corresponding to boxes 1, 2, 3, and 4, where box 1 consists of cells {(2, 2), (2, 3), (3, 2), (3, 3)}, box 2 of {(4, 4), (4, 5), (5, 4), (5, 5)}, box 3 of {(6, 6), (6, 7), (7, 6), (7, 7)}, and finally box 4 of {(1, 1), (1, 8), (8, 1), (8, 8)}, by steps 1 through 8, for each box separately. Connect these g inverses by step 9. Add the rows corresponding to cells (3, 4), (5, 6), (7, 8) to the design matrix for the boxes and find the least squares g inverse by Theorem 8.6, since these additional rows do not belong to the row span of the existing design matrix. However, the row corresponding to cell (3, 7) now belongs to the row span. So, add this row and find the g inverse by Theorem 8.3. At any stage if the R process applied to the incidence matrix of the existing design matrix can fill a cell then the corresponding row is in the range space of the existing design matrix, otherwise not.

THEOREM 8.7 Let $\mathbf{N}$ denote the incidence matrix of an $a \times b$ additive two-way classification model, such that $n_{ij}=1$ for all i and j. Let $\mathbf{X}=(\mathbf{A}:\mathbf{B})$ denote the corresponding design matrix. Then a least squares generalized inverse $\mathbf{X}_l^-$ of $\mathbf{X}$ is obtained from $(1/2ab)\mathbf{X}'$ by replacing all zeros by -1 and ones by $(2a-1)$ in $\mathbf{A}'$ and $2b-1$ in $\mathbf{B}'$.

Proof Let $\mathbf{X}=[\mathbf{x}_1, \mathbf{x}_2, \ldots, \mathbf{x}_a, \mathbf{x}_{a+1}, \ldots, \mathbf{x}_{a+b}]$ where $\mathbf{x}_i$ is an $ab \times 1$ column vector. Let

$$T_i=\mathbf{x}_i'\mathbf{Y}, \qquad i=1, 2, \ldots, a$$

$$B_j=\mathbf{x}_{a+j}'\mathbf{Y}, \qquad j=1, 2, \ldots, b$$

$$G=\sum_{i=1}^{a} T_i=\sum_{j=1}^{b} B_j.$$

Then the normal equations would be

$$b\alpha_i+\beta_1+\beta_2+\cdots+\beta_b=T_i, \qquad i=1, 2, \ldots, a$$

$$\alpha_1+\alpha_2+\cdots+\alpha_a+a\beta_j=B_j, \qquad j=1, \ldots, b.$$

Solving these under the restriction $\sum_{j=1}^{b} \beta_j=G/2a$, we obtain

$$\hat{\alpha}_i=\frac{1}{2ab}[(2a-1)T_i-(G-T_i)], \qquad i=1, \ldots, a$$

$$\hat{\beta}_j=\frac{1}{2ab}[(2b-1)B_j-(G-B_j)], \qquad j=1, \ldots, b.$$

Since these are the least squares solutions the theorem follows. □

Now we exhibit a property of the g inverse obtained in Theorem 8.7.

COROLLARY 8.1 The generalized inverse obtained in Theorem 8.7 is reflexive.

Proof Let $\mathbf{X}_l^{-1}=(\mathbf{G}_\alpha: \mathbf{G}_\beta)$, where $\mathbf{G}_\alpha$ and $\mathbf{G}_\beta$ correspond to α and β effects, respectively. First note that $b\mathbf{G}_\alpha\mathbf{e}=a\mathbf{G}_\beta\mathbf{e}$ where $\mathbf{e}$ is a vector of ones. Hence $r(\mathbf{X}_l^-)\leqslant a+b-1$. Since $\mathbf{X}_l^-$ is a g inverse of $\mathbf{X}$, $r(\mathbf{X}_l^-)\geqslant r(\mathbf{X})=a+b-1$. Thus $r(\mathbf{X}_l^-)=r(\mathbf{X})$. The theorem is now proved by appealing to Lemma 2.5.1 of Rao and Mitra (1971). □

The following theorem is a generalization of Theorem 8.2.

THEOREM 8.8 Let

$$\mathbf{A}=\begin{pmatrix}\mathbf{A}_1\\ \mathbf{A}_2\end{pmatrix}$$

be such that $R(\mathbf{A}_2')\subset R(\mathbf{A}_1')$. Further suppose $(\mathbf{G}: \mathbf{B})$ is a least squares g inverse of $\mathbf{A}$. Then $\mathbf{Y}=\mathbf{G}+\mathbf{B}(\mathbf{I}-\mathbf{A}_2\mathbf{B})^{-1}\mathbf{A}_2\mathbf{G}$ is a least squares g inverse of $\mathbf{A}_1$.

Proof We have

$$\mathbf{A}_1\mathbf{G}\mathbf{A}_1+\mathbf{A}_1\mathbf{B}\mathbf{A}_2=\mathbf{A}_1 \tag{8.9}$$

$$\mathbf{A}_2\mathbf{G}\mathbf{A}_1+\mathbf{A}_2\mathbf{B}\mathbf{A}_2=\mathbf{A}_2 \tag{8.10}$$

and

$$\begin{pmatrix}\mathbf{A}_1\mathbf{G} & \mathbf{A}_1\mathbf{B}\\ \mathbf{A}_2\mathbf{G} & \mathbf{A}_2\mathbf{B}\end{pmatrix} \tag{8.11}$$

is symmetric.

From (8.10)

$$(\mathbf{I}-\mathbf{A}_2\mathbf{B})\mathbf{A}_2=\mathbf{A}_2\mathbf{G}\mathbf{A}_1,$$

that is,

$$\mathbf{A}_2=(\mathbf{I}-\mathbf{A}_2\mathbf{B})^{-1}\mathbf{A}_2\mathbf{G}\mathbf{A}_1$$

provided $(\mathbf{I}-\mathbf{A}_2\mathbf{B})^{-1}$ exists.

Thus from (8.9) we have

$$\mathbf{A}_1\mathbf{G}\mathbf{A}_1+\mathbf{A}_1\mathbf{B}(\mathbf{I}-\mathbf{A}_2\mathbf{B})^{-1}\mathbf{A}_2\mathbf{G}\mathbf{A}_1=\mathbf{A}_1.$$

So $\mathbf{Y}=\mathbf{G}+\mathbf{B}(\mathbf{I}-\mathbf{A}_2\mathbf{B})^{-1}\mathbf{A}_2\mathbf{G}$ is a g inverse of $\mathbf{A}_1$. Also using (8.11) $\mathbf{A}_1\mathbf{Y}=\mathbf{A}_1\mathbf{G}+\mathbf{A}_1\mathbf{B}(\mathbf{I}-\mathbf{A}_2\mathbf{B})^{-1}\mathbf{A}_2\mathbf{G}$ is seen to be symmetric. Hence $\mathbf{Y}$ is a least squares generalized inverse of $\mathbf{A}_1$. Now it only remains to be shown that $(\mathbf{I}-\mathbf{A}_2\mathbf{B})$ is nonsingular.

We shall show that $(\mathbf{I}-\mathbf{A}_2\mathbf{B})\mathbf{x}=\mathbf{0}=>\mathbf{x}=\mathbf{0}$. Now $\mathbf{x}=\mathbf{A}_2\mathbf{B}\mathbf{x}=\mathbf{A}_2\mathbf{B}\mathbf{A}_2\mathbf{B}\mathbf{x}=(\mathbf{A}_2-\mathbf{A}_2\mathbf{G}\mathbf{A}_1)\mathbf{B}\mathbf{x}$ from (8.10). Thus,

$$\mathbf{A}_2\mathbf{G}\mathbf{A}_1\mathbf{B}\mathbf{x}=\mathbf{0}$$

that is,

$$(\mathbf{A}_1\mathbf{B})'(\mathbf{A}_1\mathbf{B})\mathbf{x}=\mathbf{0} \quad \text{from (8.11)},$$

that is,

$$\mathbf{A}_1\mathbf{B}\mathbf{x}=\mathbf{0}.$$

Since $R(\mathbf{A}_2')\subset R(\mathbf{A}_1')$, we have

$$\mathbf{x}=\mathbf{A}_2\mathbf{B}\mathbf{x}=\mathbf{0}. \quad \square$$

In step 7 of the algorithm, where Theorem 8.2 is repeatedly applied, we have to check at each stage that $b_i'a_i\neq 1$. This is equivalent to saying that $|\mathbf{I}-\mathbf{A}_2\mathbf{B}|\neq 0$, a proof of which has just been given.

Now we give an example to demonstrate the algorithm.

EXAMPLE 8.1 We consider an example in which some $n_{ij}\geqslant 2$ and there are many missing observations. Here we also demonstrate the advantage of finding least squares g inverses for different sets of connected portions, and then connecting these disconnected parts. Suppose we have an additive 4×5 classification model with the following pattern:

$$\mathbf{N}=\begin{array}{c} \\ \alpha_1 \\ \alpha_2 \\ \alpha_3 \\ \alpha_4 \end{array}\begin{array}{c} \begin{array}{ccccc} \beta_1 & \beta_2 & \beta_3 & \beta_4 & \beta_5 \end{array} \\ \begin{bmatrix} 1 & 1 & 0 & 0 & 0 \\ 0 & 0 & 0 & 1 & 2 \\ 1 & 0 & 1 & 0 & 0 \\ 0 & 1 & 0 & 0 & 0 \end{bmatrix}\end{array}.$$

The corresponding design matrix is

$$\mathbf{X}=\begin{bmatrix} 1 & 0 & 0 & 0 & 1 & 0 & 0 & 0 & 0 \\ 1 & 0 & 0 & 0 & 0 & 1 & 0 & 0 & 0 \\ 0 & 1 & 0 & 0 & 0 & 0 & 0 & 1 & 0 \\ 0 & 1 & 0 & 0 & 0 & 0 & 0 & 0 & 1 \\ 0 & 1 & 0 & 0 & 0 & 0 & 0 & 0 & 1 \\ 0 & 0 & 1 & 0 & 1 & 0 & 0 & 0 & 0 \\ 0 & 0 & 1 & 0 & 0 & 0 & 1 & 0 & 0 \\ 0 & 0 & 0 & 1 & 0 & 1 & 0 & 0 & 0 \end{bmatrix}.$$

Then we briefly sketch the working of the algorithm as follows:

Steps 1 and 2 Applying the R process to $\mathbf{N}$, the final matrix $\mathbf{M}$ (after a permutation of rows) comes out as

$$\mathbf{M}=\begin{array}{c} \\ \alpha_1 \\ \alpha_3 \\ \alpha_4 \\ \alpha_2 \end{array}\begin{array}{c}\begin{array}{ccccc}\beta_1 & \beta_2 & \beta_3 & \beta_4 & \beta_5\end{array}\\ \begin{bmatrix} 1 & 1 & 1 & 0 & 0 \\ 1 & 1 & 1 & 0 & 0 \\ 1 & 1 & 1 & 0 & 0 \\ 0 & 0 & 0 & 1 & 1 \end{bmatrix}\end{array}=\begin{bmatrix} \mathbf{M}_1 & \mathbf{0} \\ \mathbf{0} & \mathbf{M}_2 \end{bmatrix}.$$

Thus there are two connected portions. Here

$$T_1=\emptyset, \qquad T_2=\{(1,2)\}$$
$$S_1=\{(1,3),(2,2),(3,1),(3,3)\}, \qquad S_2=\emptyset.$$

The corresponding partition to the final matrix $\mathbf{M}$ is

$$\mathbf{X}^*=\begin{bmatrix} 1 & 0 & 0 & 1 & 0 & 0 & 0 & 0 & 0 \\ 1 & 0 & 0 & 0 & 1 & 0 & 0 & 0 & 0 \\ 1 & 0 & 0 & 0 & 0 & 1 & 0 & 0 & 0 \\ 0 & 1 & 0 & 1 & 0 & 0 & 0 & 0 & 0 \\ 0 & 1 & 0 & 0 & 1 & 0 & 0 & 0 & 0 \\ 0 & 1 & 0 & 0 & 0 & 1 & 0 & 0 & 0 \\ 0 & 0 & 1 & 1 & 0 & 0 & 0 & 0 & 0 \\ 0 & 0 & 1 & 0 & 1 & 0 & 0 & 0 & 0 \\ 0 & 0 & 1 & 0 & 0 & 1 & 0 & 0 & 0 \\ 0 & 0 & 0 & 0 & 0 & 0 & 1 & 1 & 0 \\ 0 & 0 & 0 & 0 & 0 & 0 & 1 & 0 & 1 \end{bmatrix}=\begin{bmatrix} \mathbf{X}_1^* & \mathbf{0} \\ \mathbf{0} & \mathbf{X}_2^* \end{bmatrix}.$$

Step 3 In $\mathbf{X}_1^{*\prime}$ replace zeros by -1 and ones by 5 for $\mathbf{A}_1^{*\prime}$ and 5 for $\mathbf{B}_1^{*\prime}$. And in $\mathbf{X}_2^{*\prime}$ replace zeros by -1 and ones by 1 in $\mathbf{A}_2^{*\prime}$ and 3 in $\mathbf{B}_2^{*\prime}$. ($a_1=3$, $b_1=3$, $a_2=1$, $b_2=2$.) So we obtain

$$\mathbf{Z}_{11}=\frac{1}{18}\begin{bmatrix} 5 & 5 & 5 & -1 & -1 & -1 & -1 & -1 & -1 \\ -1 & -1 & -1 & 5 & 5 & 5 & -1 & -1 & -1 \\ -1 & -1 & -1 & -1 & -1 & -1 & 5 & 5 & 5 \\ 5 & -1 & -1 & 5 & -1 & -1 & 5 & -1 & -1 \\ -1 & 5 & -1 & -1 & 5 & -1 & -1 & 5 & -1 \\ -1 & -1 & 5 & -1 & -1 & 5 & -1 & -1 & 5 \end{bmatrix},$$

$$\mathbf{Z}_{21}=\frac{1}{4}\begin{bmatrix} 1 & 1 \\ 3 & -1 \\ -1 & 3 \end{bmatrix}.$$

Step 4 Here $\mathbf{Z}_{11}$ remains unaltered since $T_1=\emptyset$. There is only one element in $T_2=\{(1,2)\}$ corresponding to (α_2, β_5), or last row in $\mathbf{X}_2^*$. We have

$$\mathbf{a}_1'=(1 \quad 0 \quad 1)=\mathbf{f}_1'\begin{bmatrix}1 & 1 & 0\\ 1 & 0 & 1\end{bmatrix}$$

which gives $\mathbf{f}_1'=(0 \quad 1)$. Set $\mathbf{f}=\mathbf{f}_1$, $k=2-1=1$, $\mathbf{a}=\mathbf{a}_1=(1 \quad 0 \quad 1)$, $\mathbf{U}=\mathbf{Z}_{21}$, and compute

$$\mathbf{g}=\mathbf{U}\mathbf{f}=\frac{1}{4}\begin{bmatrix}1 & 1\\ 3 & -1\\ -1 & 3\end{bmatrix}\begin{bmatrix}0\\ 1\end{bmatrix}=\frac{1}{4}\begin{bmatrix}1\\ -1\\ 3\end{bmatrix},$$

$$\mathbf{d}=\mathbf{U}'\mathbf{a}=\frac{1}{4}\begin{bmatrix}1 & 3 & -1\\ 1 & -1 & 3\end{bmatrix}\begin{bmatrix}1\\ 0\\ 1\end{bmatrix}=\frac{1}{4}\begin{bmatrix}0\\ 4\end{bmatrix}=\begin{bmatrix}0\\ 1\end{bmatrix},$$

$$c_1=\frac{1}{1+\frac{1}{4}[1 \quad -1 \quad 3]\begin{bmatrix}1\\ 0\\ 1\end{bmatrix}}=\frac{1}{2}.$$

Step 5 Compute

$$\mathbf{Z}_{22}=(\mathbf{U}-kc_k\mathbf{g}\mathbf{d}' : c_k\mathbf{g})=\frac{1}{8}\begin{bmatrix}2 & 1 & 1\\ 6 & -1 & -1\\ -2 & 3 & 3\end{bmatrix}.$$

Step 6 This is not required here since T_2 has only one element.

There are four elements in $S_1=\{(1,3), (2,2), (3,1), (3,3)\}$. The rows of $\mathbf{X}_1^*$ corresponding to these four elements in S_1 are

$$\mathbf{a}_1'=(1 \quad 0 \quad 0 \quad 0 \quad 0 \quad 1),$$

$$\mathbf{a}_2'=(0 \quad 1 \quad 0 \quad 0 \quad 1 \quad 0),$$

$$\mathbf{a}_3'=(0 \quad 0 \quad 1 \quad 1 \quad 0 \quad 0),$$

and

$$\mathbf{a}_4'=(0 \quad 0 \quad 1 \quad 0 \quad 0 \quad 1).$$

Let

$$\mathbf{X}_1^{**\prime}=\begin{bmatrix}1&1&0&0&0&0&0&0&1\\0&0&1&1&0&0&0&1&0\\0&0&0&0&1&1&1&0&0\\1&0&1&0&0&0&1&0&0\\0&1&0&0&1&0&0&1&0\\0&0&0&1&0&1&0&0&1\end{bmatrix}=[\mathbf{A}_1':\mathbf{a}_4,\,\mathbf{a}_3,\,\mathbf{a}_2,\,\mathbf{a}_1],$$

and

$$\mathbf{Z}_{12}=\frac{1}{18}\begin{bmatrix}5&5&-1&-1&-1&-1&-1&-1&5\\-1&-1&5&5&-1&-1&-1&5&-1\\-1&-1&-1&-1&5&5&5&-1&-1\\5&-1&5&-1&-1&-1&5&-1&-1\\-1&5&-1&-1&5&-1&-1&5&-1\\-1&-1&-1&5&-1&5&-1&-1&5\end{bmatrix}=[\mathbf{Q}:\mathbf{g}_4,\,\mathbf{g}_3,\,\mathbf{g}_2,\,\mathbf{g}_1].$$

Step 7 Set $\mathbf{C}_1=\mathbf{I}_6$,

$$\mathbf{b}_1=\mathbf{I}_6,\qquad \mathbf{g}_1=\frac{1}{18}\begin{bmatrix}5\\-1\\-1\\-1\\-1\\5\end{bmatrix}.$$

Compute

$$\mathbf{C}_i=\left(\mathbf{I}_6+\frac{\mathbf{b}_{i-1}\mathbf{a}_{i-1}'}{1-\mathbf{b}_{i-1}'\mathbf{a}_{i-1}}\right)\mathbf{C}_{i-1},\qquad \mathbf{b}_i=\mathbf{C}_i\mathbf{g}_i\qquad \text{for } i=2,3,4,5.$$

For example,

$$\mathbf{C}_2=\frac{1}{8}\begin{bmatrix}13&0&0&0&0&5\\-1&8&0&0&0&-1\\-1&0&8&0&0&-1\\-1&0&0&8&0&-1\\-1&0&0&0&8&-1\\5&0&0&0&0&13\end{bmatrix}$$

and

$$\mathbf{C}_3 = \frac{1}{20}\begin{bmatrix} 39 & -6 & 0 & 0 & -6 & 12 \\ -6 & 39 & 0 & 0 & 12 & -6 \\ -3 & 2 & 20 & 0 & 2 & -3 \\ -3 & 2 & 0 & 20 & 2 & -3 \\ -6 & 12 & 0 & 0 & 39 & -6 \\ -12 & -6 & 0 & 0 & -6 & 39 \end{bmatrix}$$

proceed until $\mathbf{C}_5$ is obtained.

Step 8 For the first connected portion $\mathbf{Z}_{11}$ becomes

$$\mathbf{Y} = \mathbf{C}_5\mathbf{Q} = \frac{1}{6}\begin{bmatrix} 3 & 2 & -2 & 1 & -1 \\ -3 & 2 & 4 & -1 & -1 \\ 3 & -4 & -2 & 1 & 5 \\ 3 & -2 & 2 & -1 & 1 \\ -3 & 4 & 2 & -1 & 1 \\ 3 & -2 & -4 & 5 & 1 \end{bmatrix}.$$

It can be verified that the same result is obtained by doing steps 7 and 8 by the methods proposed in Remark 8.2 or Remark 8.3. In Remark 8.3,

$\mathbf{N}_1 =$

	1	2	3
1	1	1	0
2	1	0	1
3	0	1	0

Take box $1 = \{(1, 1), (1, 2), (3, 1), (3, 2)\}$ and box $2 = \{(2, 3)\}$.

After computing the least squares g inverses for the two boxes and connecting them one has to add the row corresponding to the cell (2, 1) of $\mathbf{N}_1$ by using Theorem 8.6. Note that while finding the g inverse for box 1 one has to apply Theorem 8.2 only once. This method is thus very quick. $\mathbf{Z}_{22}$ remains unaltered here since $S_2 = \emptyset$.

Step 9 The matrix $\mathbf{G}$ is

$$\mathbf{G}=\begin{bmatrix}\mathbf{Y} & \mathbf{0}\\ \mathbf{0} & \mathbf{Z}_{22}\end{bmatrix}=\frac{1}{24}\begin{bmatrix}12 & 8 & -8 & 4 & -4 & 0 & 0 & 0\\ -12 & 8 & 16 & 4 & -4 & 0 & 0 & 0\\ 12 & -16 & -8 & 4 & 20 & 0 & 0 & 0\\ 12 & -8 & 8 & -4 & 4 & 0 & 0 & 0\\ -12 & 16 & 8 & -4 & 4 & 0 & 0 & 0\\ 12 & -8 & -16 & 20 & 4 & 0 & 0 & 0\\ 0 & 0 & 0 & 0 & 0 & 6 & 3 & 3\\ 0 & 0 & 0 & 0 & 0 & 18 & -3 & -3\\ 0 & 0 & 0 & 0 & 0 & -6 & 9 & 9\end{bmatrix}.$$

The least squares g inverse of $\mathbf{X}$ is obtained from $\mathbf{G}$ by appropriate permutations of rows and columns. Thus,

$$\mathbf{X}_l^- = \frac{1}{24}\begin{bmatrix}12 & 8 & 0 & 0 & 0 & -8 & 4 & -4\\ 0 & 0 & 6 & 3 & 3 & 0 & 0 & 0\\ -12 & 8 & 0 & 0 & 0 & 16 & 4 & -4\\ 12 & -16 & 0 & 0 & 0 & -8 & 4 & 20\\ 12 & -8 & 0 & 0 & 0 & 8 & -4 & 4\\ -12 & 16 & 0 & 0 & 0 & 8 & -4 & 4\\ 12 & -8 & 0 & 0 & 0 & -16 & 20 & 4\\ 0 & 0 & 18 & -3 & -3 & 0 & 0 & 0\\ 0 & 0 & -6 & 9 & 9 & 0 & 0 & 0\end{bmatrix}.$$

A least squares g inverse such as the one found above could be used to identify the estimable parametric functions using the fact that $\boldsymbol{\lambda}'\boldsymbol{\theta}$ is estimable if and only if $\boldsymbol{\lambda}'\mathbf{X}_l^-\mathbf{X}=\boldsymbol{\lambda}'$.

For an additive two-way classification model

$$E(Y_{ijk})=\alpha_i+\beta_j$$

$$i=1,\ldots,a, \qquad j=1,\ldots,b, \qquad k=1,\ldots,n_{ij},$$

let $\mathbf{P}'\boldsymbol{\theta}$ denote an $[a(a-1)/2+b(b-1)/2]\times 1$ vector of estimable parametric functions which constitute all α and β contrasts of the form $\alpha_i-\alpha_j$, $i<j$, $(1\leqslant i, j\leqslant a)$ and $\beta_{i'}-\beta_{j'}$, $i'<j'$, $(1\leqslant i', j'\leqslant b)$, and $\boldsymbol{\theta}=(\alpha_1,\ldots,\alpha_a,\beta_1,\ldots,\beta_b)'$, with $r(\mathbf{P})=a+b$ (which is number of columns of the matrix $\mathbf{P}'$). Then premultiplying the matrix $\mathbf{P}'$ by $\mathbf{X}_l^-\mathbf{X}$ will provide us with estimable parametric functions (if any exists) of the form $\alpha_i-\alpha_j$ and $\beta_{i'}-\beta_{j'}$. Observe that these estimable functions are not in general linearly independent.

For Example 8.5 above we have $a=4$ and $b=5$. Thus the matrix $\mathbf{P}'$ is a 16×9 matrix as follows:

$$
\mathbf{P}' = \left[\begin{array}{cccc:ccccc}
1 & -1 & 0 & 0 & & & & & \\
1 & 0 & -1 & 0 & & & & & \\
1 & 0 & 0 & -1 & & & \mathbf{O} & & \\
0 & 1 & -1 & 0 & & & & & \\
0 & 1 & 0 & -1 & & & & & \\
0 & 0 & 1 & -1 & & & & & \\
\hdashline
 & & & & 1 & -1 & 0 & 0 & 0 \\
 & & & & 1 & 0 & -1 & 0 & 0 \\
 & & & & 1 & 0 & 0 & -1 & 0 \\
 & & & & 1 & 0 & 0 & 0 & -1 \\
 & \mathbf{O} & & & 0 & 1 & -1 & 0 & 0 \\
 & & & & 0 & 1 & 0 & -1 & 0 \\
 & & & & 0 & 1 & 0 & 0 & -1 \\
 & & & & 0 & 0 & 1 & -1 & 0 \\
 & & & & 0 & 0 & 1 & 0 & -1 \\
 & & & & 0 & 0 & 0 & 1 & -1
\end{array}\right]
$$

and hence

$$
\mathbf{P}'\mathbf{X}_l^- \mathbf{X} = \frac{1}{24}\begin{bmatrix}
20 & -12 & -4 & -4 & 4 & 4 & 4 & -6 & -6 \\
24 & 0 & -24 & 0 & 0 & 0 & 0 & 0 & 0 \\
24 & 0 & 0 & -24 & 0 & 0 & 0 & 0 & 0 \\
4 & 12 & -20 & 4 & -4 & -4 & -4 & 6 & 6 \\
4 & 12 & 4 & -20 & -4 & -4 & -4 & 6 & 6 \\
0 & 0 & 24 & -24 & 0 & 0 & 0 & 0 & 0 \\
0 & 0 & 0 & 0 & 24 & -24 & 0 & 0 & 0 \\
0 & 0 & 0 & 0 & 24 & 0 & -24 & 0 & 0 \\
4 & -12 & 4 & 4 & 20 & -4 & -4 & -18 & 6 \\
4 & -12 & 4 & 4 & 20 & -4 & -4 & 6 & -18 \\
0 & 0 & 0 & 0 & 0 & 24 & -24 & 0 & 0 \\
4 & -12 & 4 & 4 & -4 & 20 & -4 & -18 & 6 \\
4 & -12 & 4 & 4 & -4 & 20 & -4 & 6 & -18 \\
4 & -12 & 4 & 4 & -4 & -4 & 20 & -18 & 6 \\
4 & -12 & 4 & 4 & -4 & -4 & 20 & 6 & -18 \\
0 & 0 & 0 & 0 & 0 & 0 & 0 & 24 & -24
\end{bmatrix}.
$$

From the above matrix it is easily seen (rows 2, 3 and 6) that $\alpha_1-\alpha_3$, $\alpha_1-\alpha_4$ and $\alpha_3-\alpha_4$ and from rows 7, 8, 11 and 16 that $\beta_1-\beta_2$, $\beta_1-\beta_3$, $\beta_2-\beta_3$ and $\beta_4-\beta_5$ are estimable.

Least Squares Generalized Inverses for Additive *n*-Way Classification Models

In this section we consider the additive *n*-way classification model of the form (8.8) without interactions. Without loss of generality we assume that

$$
\begin{aligned}
n_{i\ldots} &= \sum_{jk\ldots w} n_{ijk\ldots w} \neq 0, & i &= 1, 2, \ldots, a \\
n_{.j..} &= \sum_{ik\ldots w} n_{ijk\ldots w} \neq 0, & j &= 1, 2, \ldots, b \\
&\vdots \qquad\qquad \vdots & &\vdots \\
n_{\ldots w} &= \sum_{ijk\ldots} n_{ijk\ldots w} \neq 0, & w &= 1, 2, \ldots, p.
\end{aligned}
$$

When the design is such that all contrasts within each factor are estimable (though the pattern may be arbitrary otherwise), we find the least squares generalized inverse for the design matrix by using Theorem 8.9. The situation when the design matrix is not of maximal rank will be treated as a separate case.

Case I The design matrix is of maximal rank. Here the rows corresponding to the missing observations are in the range space of the rows corresponding to the available observations.

THEOREM 8.9 Let $\mathbf{X}$ be the design matrix for an additive *n*-way classification model with no missing observation. Then a least squares generalized inverse of $\mathbf{X}$ is obtained from $(1/nab\ldots p)\mathbf{X}'$ by replacing zeros by $-(n-1)$ and by $na-(n-1)$ for $\mathbf{A}'$, $nb-(n-1)$ for $\mathbf{B}', \ldots,$ and $np-(n-1)$ for $\mathbf{P}'$.

Proof follows exactly on the same lines as in Theorem 8.7.

After obtaining the least squares g inverse for the complete design matrix we drop the rows corresponding to the missing observations and find the g inverse by applying Theorem 8.2.

If any row is repeated in $\mathbf{X}$ then we apply Theorem 8.3 to find the least squares generalized inverse.

COROLLARY 8.2 For a complete fractional design, say $1/q$ of a q^n design we find the least squares generalized inverse of $\mathbf{X}$ from $q/nq^n\mathbf{X}'$ by replacing zeros by $-(n-1)$ and the ones by $nq-(n-1)$.

This corollary can be very useful for Latin square designs with no missing observations.

EXAMPLE 8.2 Consider the following 3×3 Latin square design:

	c_1	c_2	c_3
r_1	1	2	3
r_2	3	1	2
r_3	2	3	1

This is a 1/3 replicate of a 3^3 factorial design with the design matrix $\mathbf{X}$ as

$$\mathbf{X} = \begin{bmatrix} 1 & 0 & 0 & 1 & 0 & 0 & 1 & 0 & 0 \\ 1 & 0 & 0 & 0 & 1 & 0 & 0 & 1 & 0 \\ 1 & 0 & 0 & 0 & 0 & 1 & 0 & 0 & 1 \\ 0 & 1 & 0 & 1 & 0 & 0 & 0 & 0 & 1 \\ 0 & 1 & 0 & 0 & 1 & 0 & 1 & 0 & 0 \\ 0 & 1 & 0 & 0 & 0 & 1 & 0 & 1 & 0 \\ 0 & 0 & 1 & 1 & 0 & 0 & 0 & 1 & 0 \\ 0 & 0 & 1 & 0 & 1 & 0 & 0 & 0 & 1 \\ 0 & 0 & 1 & 0 & 0 & 1 & 1 & 0 & 0 \end{bmatrix}.$$

Then $\mathbf{X}_l^-$ is obtained from $\frac{1}{27}\mathbf{X}'$ by replacing zeros by -2 and ones by 7 as

$$\mathbf{X}_l^- = \frac{1}{27}\begin{bmatrix} 7 & -2 & -2 & 7 & -2 & -2 & 7 & -2 & -2 \\ 7 & -2 & -2 & -2 & 7 & -2 & -2 & 7 & -2 \\ 7 & -2 & -2 & -2 & -2 & 7 & -2 & -2 & 7 \\ -2 & 7 & -2 & 7 & -2 & -2 & -2 & -2 & 7 \\ -2 & 7 & -2 & -2 & 7 & -2 & 7 & -2 & -2 \\ -2 & 7 & -2 & -2 & -2 & 7 & -2 & 7 & -2 \\ -2 & -2 & 7 & 7 & -2 & -2 & -2 & 7 & -2 \\ -2 & -2 & 7 & -2 & 7 & -2 & -2 & -2 & 7 \\ -2 & -2 & 7 & -2 & -2 & 7 & 7 & -2 & -2 \end{bmatrix}.$$

Case II The design matrix is not of maximal rank. In this case some effects of one factor are confounded with the effects of some other factor(s).

In such a situation when there are confounded effects the method of Case I cannot be applied directly. However, the least squares generalized inverse can be obtained by modifying our techniques. Consider a three-way classification model with design matrix $\mathbf{X}=(\mathbf{A}:\mathbf{B}:\mathbf{C})$. Suppose some of the α effects are confounded with some β and δ effects. Then we first find the least squares generalized inverse $\mathbf{G}_1$ for $\mathbf{X}_0=(\mathbf{B}:\mathbf{C})$ by considering it as a two-way classification model. If this is a case of total confounding, that is, all α effects are confounded with β and δ effects, then $R(\mathbf{A})\subset R(\mathbf{X}_0)$. Here the least squares generalized inverse of $\mathbf{X}$ is obtained by adding a null rows to $\mathbf{G}_1$ (see Theorem 8.4). However, if the α effects are only partially confounded with the β and δ effects, then some columns of $\mathbf{A}$ will not be in $R(\mathbf{X}_0)$. Here also we first compute the least squares g inverse of $\mathbf{X}_0$. When we add a column of $\mathbf{A}$ to $\mathbf{X}_0$, we find the corresponding g inverse by the formula in Theorem 8.5. If this column is not in the range then this formula will go through. Otherwise $\mathbf{c}=\mathbf{0}$. In this case put $\mathbf{b}=\mathbf{0}$. This conforms with Theorem 8.4. In this way we add the columns of $\mathbf{A}$ to $\mathbf{X}_0$ successively and find the least squares g inverse of $\mathbf{X}$.

Since in some designs with arbitrary patterns it is difficult to find its nature (e.g., the confounded effects), the algorithm is such that one can avoid such complications. Thus, we first compute the least squares g inverse for $\mathbf{X}_0$ by the two-way techniques and then add the columns of $\mathbf{A}$ one by one and find the corresponding least squares g inverses by Theorem 8.5, putting $\mathbf{b}=\mathbf{0}$ whenever $\mathbf{c}=\mathbf{0}$.

The n-way classification model can be treated in a similar manner.

EXAMPLE 8.3 Consider the following $2\times2\times2$ design where the α effects are totally confounded with the β effects.

	δ_1		δ_2	
	β_1	β_2	β_1	β_2
α_1	1	0	1	0
α_2	0	1	0	1

Writing the design matrix as $\mathbf{X}=(\mathbf{A}:\ \mathbf{B}:\ \mathbf{C})$, we notice that $\mathbf{B}=\mathbf{A}$. So we find the least squares g inverse for $\mathbf{X}_0=(\mathbf{A}:\ \mathbf{C})$ by the usual techniques for two-way and add two null rows as proposed in Theorem 8.4. Thus we obtain

$$\mathbf{X}_l^- = \frac{1}{8}\begin{bmatrix} 3 & -1 & 3 & -1 \\ -1 & 3 & -1 & 3 \\ 0 & 0 & 0 & 0 \\ 0 & 0 & 0 & 0 \\ 3 & 3 & -1 & -1 \\ -1 & -1 & 3 & 3 \end{bmatrix}.$$

EXAMPLE 8.4 Consider the following Latin square design with two missing observations:

	c_1	c_2	c_3
r_1	—	2	3
r_2	2	3	1
r_3	3	1	—

The corresponding design matrix $\mathbf{X}$ is

$$\mathbf{X} = \begin{bmatrix} 1 & 0 & 0 & 0 & 1 & 0 & 0 & 1 & 0 \\ 1 & 0 & 0 & 0 & 0 & 1 & 0 & 0 & 1 \\ 0 & 1 & 0 & 1 & 0 & 0 & 0 & 1 & 0 \\ 0 & 1 & 0 & 0 & 1 & 0 & 0 & 0 & 1 \\ 0 & 1 & 0 & 0 & 0 & 1 & 1 & 0 & 0 \\ 0 & 0 & 1 & 1 & 0 & 0 & 0 & 0 & 1 \\ 0 & 0 & 1 & 0 & 1 & 0 & 1 & 0 & 0 \end{bmatrix} = (\mathbf{A} : \mathbf{B} : \mathbf{C}),$$

where $\mathbf{A}$ corresponds to rows, $\mathbf{B}$ to columns, and $\mathbf{C}$ to treatment effects. The least squares g inverse for $\mathbf{X}_0 = (\mathbf{A} : \mathbf{B})$ is

$$(\mathbf{X}_0)_l^- = \frac{1}{30}\begin{bmatrix} 14 & 11 & 4 & -3 & -6 & 1 & -6 \\ -2 & -3 & 8 & 9 & 8 & -3 & -2 \\ -6 & 1 & -6 & -3 & 4 & 11 & 14 \\ 4 & 1 & 14 & -3 & -6 & 11 & -6 \\ 8 & -3 & -2 & 9 & -2 & -3 & 8 \\ 6 & 11 & -6 & -3 & 14 & 1 & 4 \end{bmatrix}.$$

Then we add the columns of $\mathbf{C}$ one by one to $\mathbf{X}_0$ and find the least squares g inverses by Theorem 8.5.

For the first column $\mathbf{c} \neq \mathbf{0}$ [hence it is not in $R(\mathbf{X}_0)$]. But for the next two, $\mathbf{c} = \mathbf{0}$ and thus these columns are in the range of the existing design matrices. For both of these we take $\mathbf{b} = \mathbf{0}$. It is also shown that there are two confounded effects. The final least squares g inverse comes out as

$$\mathbf{X}_l^- = \frac{1}{6}\begin{bmatrix} 4 & 1 & 2 & -3 & 0 & -1 & 0 \\ -1 & 0 & 1 & 3 & 1 & 0 & -1 \\ -3 & 2 & -3 & 3 & -1 & 4 & 1 \\ 2 & -1 & 4 & -3 & 0 & 1 & 0 \\ 1 & 0 & -1 & 3 & -1 & 0 & 1 \\ -3 & 4 & -3 & 3 & 1 & 2 & -1 \\ 3 & -3 & 3 & -6 & 3 & -3 & 3 \\ 0 & 0 & 0 & 0 & 0 & 0 & 0 \\ 0 & 0 & 0 & 0 & 0 & 0 & 0 \end{bmatrix}.$$

Least Squares g Inverses for the Most General Classification Models

We now consider the completely general model allowing the presence of interaction effects. As stated in Section 8.3 the model is

$$E(\mathbf{Y}) = \mathbf{X}\boldsymbol{\theta} \quad \text{with} \quad \mathbf{X} = (\mathbf{X}_0 \colon \mathbf{X}_1),$$

where $\mathbf{X}_0$ is the design matrix under the assumption of additivity and $\mathbf{X}_1$ corresponds to the interactions we believe might be present.

To get $\mathbf{X}_l^-$, we first find $(\mathbf{X}_0)_l^-$ using the methods of the previous subsections. Then we append the columns of $\mathbf{X}_1$, one by one, to $\mathbf{X}_0$ and find the least squares g inverse by Theorem 8.5, putting $\mathbf{b} = \mathbf{0}$ whenever $\mathbf{c} = \mathbf{0}$ (this conforms with Theorem 8.4).

If N is the total number of observations, then $\mathbf{c}$ can be non-null at most $[N - r(\mathbf{X}_0)]$ times. When $\mathbf{c} = \mathbf{0}$ we only add a null row to the existing g inverse and the $\mathbf{XX}^-$ of Theorem 8.5 remains unchanged. Thus computationally this process is expected to be very fast.

EXAMPLE 8.5 Suppose we have a 3×4 factorial experiment with the following incidence matrix:

$$\mathbf{N} = \begin{array}{c} \\ \alpha_1 \\ \alpha_2 \\ \alpha_3 \end{array} \begin{array}{c} \begin{array}{cccc} \beta_1 & \beta_2 & \beta_3 & \beta_4 \end{array} \\ \begin{bmatrix} 1 & 0 & 1 & 0 \\ 1 & 3 & 0 & 0 \\ 0 & 2 & 1 & 2 \end{bmatrix} \end{array}.$$

Let the model associated with this factorial experiment be

$$E(Y_{ij})=\alpha_i+\beta_j+(\alpha\beta)_{ij} \qquad \begin{aligned} i&=1,\ldots,3\\ j&=1,\ldots,4. \end{aligned}$$

The corresponding design matrix $\mathbf{X}$ associated to this experiment and the model has the form

$$\mathbf{X}=\begin{bmatrix} 1&0&0&1&0&0&0&1&0&0&0&0&0&0\\ 1&0&0&0&0&1&0&0&1&0&0&0&0&0\\ 0&1&0&1&0&0&0&0&0&1&0&0&0&0\\ 0&1&0&0&1&0&0&0&0&0&1&0&0&0\\ 0&1&0&0&1&0&0&0&0&0&1&0&0&0\\ 0&1&0&0&1&0&0&0&0&0&1&0&0&0\\ 0&0&1&0&1&0&0&0&0&0&0&1&0&0\\ 0&0&1&0&1&0&0&0&0&0&0&1&0&0\\ 0&0&1&0&0&1&0&0&0&0&0&0&1&0\\ 0&0&1&0&0&0&1&0&0&0&0&0&0&1\\ 0&0&1&0&0&0&1&0&0&0&0&0&0&1 \end{bmatrix}=(\mathbf{X}_0\colon\,\mathbf{X}_1).$$

We first find a least squares g inverse for $\mathbf{X}_0$ using the algorithm for two-way additive model. This gives

$$(\mathbf{X}_0)_l^-=\frac{1}{1392}$$

$$\times\begin{bmatrix} 568&568&-394&54&54&54&6&6&-418&87&87\\ -248&-248&422&246&246&246&-282&-282&158&87&87\\ -8&-8&182&-138&-138&-138&294&294&398&87&87\\ 536&536&682&-150&-150&-150&138&138&130&-87&-87\\ 152&152&-326&186&186&186&330&330&-254&-87&-87\\ -280&-280&106&42&42&42&-150&-150&706&-87&-87\\ 8&8&-182&138&138&138&-294&-294&-398&609&609 \end{bmatrix}$$

We now add the columns of $\mathbf{X}_1$ one by one to $\mathbf{X}_0$ and find the least squares g inverse by applying Theorem 8.5, putting $\mathbf{b}=\mathbf{0}$ whenever $\mathbf{c}=\mathbf{0}$. Now the first column is

$$\mathbf{a}'=(1\quad 0\quad 0\quad 0\quad 0\quad 0\quad 0\quad 0\quad 0\quad 0\quad 0).$$

$$\mathbf{d}=(\mathbf{X}_0)_l^-\mathbf{a}=\frac{1}{1392}\begin{bmatrix}568\\-248\\-8\\536\\152\\-280\\8\end{bmatrix}.$$

$$\mathbf{c}=(\mathbf{I}-\mathbf{X}\mathbf{X}_l^-)\mathbf{a}=\frac{1}{1392}\begin{bmatrix}288\\-288\\-288\\96\\96\\-144\\-144\\288\\0\\0\end{bmatrix}.$$

$$\mathbf{c}'\mathbf{a}=\frac{288}{1392}.$$

$$\mathbf{b}'=\frac{1}{288}(288 \quad -288 \quad -288 \quad 96 \quad 96 \quad -144 \quad -144 \quad 288 \quad 0 \quad 0).$$

Now using Theorem 8.5 and forming

$$\mathbf{G}=\begin{bmatrix}\mathbf{X}^- - \mathbf{d}\mathbf{b}'\\ \mathbf{b}'\end{bmatrix}$$

we find

$$\mathbf{G}=\frac{1}{1392}\begin{bmatrix}0 & 1160 & 174 & -135 & -135 & -135 & 290 & 290 & -986 & 87 & 87\\ 0 & -232 & 174 & 329 & 329 & 329 & -406 & -406 & 406 & 87 & 87\\ 0 & -232 & 174 & -135 & -135 & -135 & 290 & 290 & 406 & 87 & 87\\ 0 & 232 & 1218 & -329 & -329 & -329 & 406 & 406 & -406 & -87 & -87\\ 0 & 232 & -174 & 135 & 135 & 135 & 406 & 406 & -406 & -87 & -87\\ 0 & 232 & -174 & 135 & 135 & 135 & -290 & -290 & 986 & -87 & -87\\ 0 & 232 & -174 & 135 & 135 & 135 & -290 & -290 & -406 & 609 & 609\\ 1392 & -1392 & -1392 & 464 & 464 & 464 & -696 & -696 & 1392 & 0 & 0\end{bmatrix}$$

to be the g inverse of

$$\mathbf{X}=\begin{bmatrix} 1 & 0 & 0 & 1 & 0 & 0 & 0 & 1 \\ 1 & 0 & 0 & 0 & 0 & 1 & 0 & 0 \\ 0 & 1 & 0 & 1 & 0 & 0 & 0 & 0 \\ 0 & 1 & 0 & 0 & 1 & 0 & 0 & 0 \\ 0 & 1 & 0 & 0 & 1 & 0 & 0 & 0 \\ 0 & 1 & 0 & 0 & 1 & 0 & 0 & 0 \\ 0 & 0 & 1 & 0 & 1 & 0 & 0 & 0 \\ 0 & 0 & 1 & 0 & 1 & 0 & 0 & 0 \\ 0 & 0 & 1 & 0 & 0 & 1 & 0 & 0 \\ 0 & 0 & 1 & 0 & 0 & 0 & 1 & 0 \\ 0 & 0 & 1 & 0 & 0 & 0 & 1 & 0 \end{bmatrix}=(\mathbf{X}:\mathbf{a}).$$

We continue this process by adding other columns of $\mathbf{X}_1$ one by one and applying Theorem 8.5. In this way we find the final g inverse of $\mathbf{X}$ to be

$$\mathbf{G}=\frac{1}{144}\begin{bmatrix} 0 & 120 & 18 & -14 & -14 & -14 & 30 & 30 & -120 & 9 & 9 \\ 0 & -24 & 18 & 34 & 34 & 34 & -42 & -42 & 42 & 9 & 9 \\ 0 & -24 & 18 & -14 & -14 & -14 & 30 & 30 & 42 & 9 & 9 \\ 0 & 24 & 126 & -34 & -34 & -34 & 42 & 42 & -42 & -9 & -9 \\ 0 & 24 & -18 & 14 & 14 & 14 & 42 & 42 & -42 & -9 & -9 \\ 0 & 24 & -18 & 14 & 14 & 14 & -30 & -30 & 102 & -9 & -9 \\ 0 & 24 & -18 & 14 & 14 & 14 & -30 & -30 & -42 & 63 & 63 \\ 144 & -144 & -144 & 48 & 48 & 48 & -72 & -72 & 144 & 0 & 0 \\ 0 & 0 & 0 & 0 & 0 & 0 & 0 & 0 & 0 & 0 & 0 \\ 0 & 0 & 0 & 0 & 0 & 0 & 0 & 0 & 0 & 0 & 0 \\ 0 & 0 & 0 & 0 & 0 & 0 & 0 & 0 & 0 & 0 & 0 \\ 0 & 0 & 0 & 0 & 0 & 0 & 0 & 0 & 0 & 0 & 0 \\ 0 & 0 & 0 & 0 & 0 & 0 & 0 & 0 & 0 & 0 & 0 \\ 0 & 0 & 0 & 0 & 0 & 0 & 0 & 0 & 0 & 0 & 0 \end{bmatrix}.$$

When multiplying this least squares g inverse by the vector of observations we obtain a least squares solution for the α, β, and the $(\alpha\beta)$ effects. However, as stated in Chapter 7, because of the presence of the interaction terms in the model, we cannot find the estimable functions involving only α, and the β effects. The first seven rows of this least squares g inverse are for estimating the α and β effects and the eight row is estimating the interaction effect. Now multiplying the eight row by the vector of observations we obtain

$$\frac{1}{144}[144\ -144\ -144\ 48\ 48\ 48\ -72\ -72\ 144\ 0\ 0]\begin{bmatrix}\mu+\alpha_1+\beta_1+(\alpha\beta)_{11}\\ \mu+\alpha_1+\beta_3+(\alpha\beta)_{13}\\ \mu+\alpha_2+\beta_1+(\alpha\beta)_{21}\\ \mu+\alpha_2+\beta_2+(\alpha\beta)_{22}\\ \mu+\alpha_2+\beta_2+(\alpha\beta)_{22}\\ \mu+\alpha_2+\beta_2+(\alpha\beta)_{22}\\ \mu+\alpha_3+\beta_2+(\alpha\beta)_{32}\\ \mu+\alpha_3+\beta_2+(\alpha\beta)_{32}\\ \mu+\alpha_3+\beta_3+(\alpha\beta)_{33}\\ \mu+\alpha_3+\beta_4+(\alpha\beta)_{34}\\ \mu+\alpha_3+\beta_4+(\alpha\beta)_{34}\end{bmatrix}$$

$$=(\alpha\beta)_{11}-(\alpha\beta)_{13}-(\alpha\beta)_{21}+(\alpha\beta)_{22}-(\alpha\beta)_{32}+(\alpha\beta)_{33}$$

which is an estimable function involving only the interaction terms, and corresponds to a loop in $\bar{\mathbf{M}}$ of the method used in Section 7.8. We consider this example with hypothetical data with and without interaction effects for the computer program at the end of this chapter.

8.4 OTHER COMPUTATIONAL METHODS

In this section we provide some formulae for computing g inverses (not necessary a least squares g inverse). In some problems of estimation from simple linear models by the method of least squares, the rank of the normal equations may be known, and it is also possible to identify independent rows and columns. In such situations the computation of g inverse is easy, and it involves only the inversion of suitably chosen nonsingular matrices.

1 Let $\mathbf{X}$ be an $n\times p$ matrix of rank r. Then there exists a nonsingular submatrix $\mathbf{A}$ of order r. If $\mathbf{X}$ can be partitioned as

$$\mathbf{X}=\begin{bmatrix}\mathbf{A} & \mathbf{C}\\ \mathbf{F} & \mathbf{E}\end{bmatrix},\quad \text{then}\quad \mathbf{X}^-=\begin{bmatrix}\mathbf{A}^{-1} & \mathbf{0}\\ \mathbf{0} & \mathbf{0}\end{bmatrix},\quad \text{or}\quad \begin{bmatrix}\mathbf{A}^{-1} & -\mathbf{A}^{-1}\mathbf{C}\\ \mathbf{0} & \mathbf{I}\end{bmatrix}.$$

Here $\mathbf{X}^-\mathbf{X}\mathbf{X}^-=\mathbf{X}^-$ when we choose

$$\mathbf{X}^-=\begin{bmatrix}\mathbf{A}^{-1} & \mathbf{0}\\ \mathbf{0} & \mathbf{0}\end{bmatrix},$$

that is, $\mathbf{X}^-$ is a reflexive g inverse of $\mathbf{X}$. When

$$\mathbf{X}^-=\begin{bmatrix}\mathbf{A}^{-1} & -\mathbf{A}^{-1}\mathbf{C}\\ \mathbf{0} & \mathbf{I}\end{bmatrix},$$

then rank $\mathbf{X}^- = \min(n, p)$ and $\mathbf{X}^-$ is a g inverse with maximum rank.

2 Let $\mathbf{X}$ be an $n \times p$ matrix of rank r, and let it be possible to partition $\mathbf{X}$ as

$$\mathbf{X} = (\mathbf{A}_1 : \mathbf{A}_2)$$

by a permutation of columns, if necessary, such that $\mathbf{A}_1$ is a $n \times r$ matrix of rank r and $\mathbf{A}_2$ is a $n \times (p-r)$ matrix. Then one choice of g inverse of $\mathbf{X}$ is

$$\mathbf{X}^- = \begin{bmatrix} (\mathbf{A}_1'\mathbf{A}_1)^{-1} & \mathbf{A}_1' \\ & \mathbf{0} \end{bmatrix}.$$

This inverse is reflexive and provide a *basic solution* to $\mathbf{Xb} = \mathbf{Y}$. [A solution $\mathbf{b}$ is said to be a basic solution of the equation $\mathbf{Xb} = \mathbf{Y}$ if (1) $\mathbf{Xb} = \mathbf{Y}$, and (2) $\mathbf{b}$ has utmost r nonzero components, where $r = r(\mathbf{X})$.]

Least Squares Solution

We now briefly discuss three alternatives for computing a least squares solution of $\mathbf{Y} = \mathbf{X}\boldsymbol{\theta}$.

1 Normal Equation Approach First obtain the normal equation $\mathbf{X}'\mathbf{X}\hat{\boldsymbol{\theta}} = \mathbf{X}'\mathbf{Y}$. Append the unit matrix $\mathbf{I}$, and reduce the $(\mathbf{X}'\mathbf{X} : \mathbf{X}'\mathbf{Y} : \mathbf{I})$ by Gauss–Doolittle or square root methods. By doing this, we obtain a solution to $\mathbf{X}'\mathbf{X}\hat{\boldsymbol{\theta}} = \mathbf{X}'\mathbf{Y}$ and also a g inverse of $\mathbf{X}'\mathbf{X}$ which is needed for further analysis.

2 Reduction by householder transformation In this method we consider the $(\mathbf{X} : \mathbf{Y})$ matrix and reduce it by a series of householder transformations. Computationally this method has some advantages over the first method, but in problems where the normal equation is well conditioned the normal equation method is preferable.

Consider the first column of $\mathbf{X}$. If it has a nonzero element, apply the Householder transformation to have a nonzero value in the first position and zero elsewhere in the first column. If all the elements of the first column are zero, move to the nearest column which has a nonzero value, say the ith, and apply the householder transformation to have a nonzero value in the first position and zero elsewhere in the jth column. Now omit the first row and repeat the process on the reduced matrix, and so on, until all the columns of $\mathbf{X}$ are covered. After rearrangement of columns, the reduced matrix is of the form

$$\begin{bmatrix} \mathbf{T} & \mathbf{U} & \mathbf{Q}_1 \\ \mathbf{0} & \mathbf{0} & \mathbf{Q}_2 \end{bmatrix}$$

where $\mathbf{T}$ is an upper triangular and nonsingular matrix of rank r, $\mathbf{Q}_1$ is r vector, $\mathbf{Q}_2$ is $(p-r)$ vector, and $\mathbf{U}$ is a $r\times(p-r)$ matrix. A least squares solution is

$$(\hat{\theta}_1, \hat{\theta}_2, \ldots, \hat{\theta}_r)' = \mathbf{T}^{-1}\mathbf{Q}_1$$
$$(\hat{\theta}_{r+1}, \ldots, \hat{\theta}_n) = \mathbf{0}.$$

A g inverse of $\mathbf{X}'\mathbf{X}$ is

$$\begin{bmatrix} \mathbf{T}^{-1}(\mathbf{T}^{-1})' & \mathbf{0} \\ \mathbf{0} & \mathbf{0} \end{bmatrix}.$$

3 Mathematical programming approach With this approach we can also obtain a solution to normal equations and also a g inverse of $\mathbf{X}'\mathbf{X}$. Given the linear model $\mathbf{Y}=\mathbf{X}\boldsymbol{\theta}$, the normal equation is given by $\mathbf{X}'\mathbf{X}\hat{\boldsymbol{\theta}}=\mathbf{X}'\mathbf{Y}$. Let $\boldsymbol{\theta}_a$ denote the vector of artificial variables. Add one artificial variable to each row of $\mathbf{X}'\mathbf{X}$. We obtain

$$[\mathbf{X}'\mathbf{X}:\mathbf{I}]\begin{bmatrix} \hat{\boldsymbol{\theta}} \\ \boldsymbol{\theta}_a \end{bmatrix} = \mathbf{X}'\mathbf{Y}.$$

A readily available basis for this system is $\mathbf{I}$, corresponding to the artificial variables. Now to find a basis containing a maximum number of columns from $\mathbf{X}'\mathbf{X}$, we proceed as follows:

$$\text{Maximize } -\boldsymbol{\theta}_a'\mathbf{e}$$

$$\text{subject to } [\mathbf{X}'\mathbf{X}:\mathbf{I}]\begin{bmatrix} \hat{\boldsymbol{\theta}} \\ \boldsymbol{\theta}_a \end{bmatrix} = \mathbf{X}'\mathbf{Y}$$

$$\boldsymbol{\theta}_a \geqslant \mathbf{0}, \qquad \hat{\boldsymbol{\theta}} \text{ unrestricted in sign.}$$

This is a linear programming problem, and the objective function has a maximum equal to zero, corresponding to $\boldsymbol{\theta}_a=\mathbf{0}$. If we have to use the ordinary simplex method $\hat{\boldsymbol{\theta}}$ must be expressed as the difference of two non-negative vectors.

Computationally, method 2 has some advantages, but in problems where the normal equation is well conditioned, method 1 is more simple. In method 3 if we express $\hat{\boldsymbol{\theta}}$ as the difference of two non-negative vectors the size of the problem will increase. This increase may be avoided by the modified simplex method for unrestricted variables.

8.5 LEAST SQUARES SOLUTION AND SUM OF SQUARES

Suppose we want to estimate the unknown parameters from a factorial experiment, assuming the linear model

$$\mathbf{Y} = \mathbf{X}\boldsymbol{\theta} + \mathbf{e}$$

where $\mathbf{Y}$ is an $N \times 1$ vector of observations Y_i, $\boldsymbol{\theta}$ is a $p \times 1$ vector of parameters, $\mathbf{X}$ is an $N \times p$ design matrix consisting of zeros and ones, and $\mathbf{e}$ is an $N \times 1$ vector of random error terms. Also assume $\text{Cov}(\mathbf{Y}) = \sigma^2 \mathbf{I}$ with normality being introduced subsequently, when needed for hypothesis testing.

Now having obtained a least squares g inverse $\mathbf{X}_l^-$ of $\mathbf{X}$ by the algorithm described in Section 8.3, we can immediately obtain a solution to the normal equations by

$$\hat{\boldsymbol{\theta}} = \mathbf{X}_l^- \mathbf{Y}.$$

Note that this solution provides the BLUEs for estimable parametric functions.

The sum of squares for total, residual, and the model (regression) are obtained as before, that is, the total sum of squares is

$$\text{SS}_{\text{tot}} = \mathbf{Y}'\mathbf{Y},$$

and the sum of squares for residual is

$$\hat{R} = \mathbf{Y}'\mathbf{Y} - \hat{\boldsymbol{\theta}}'\mathbf{X}'\mathbf{Y},$$

and the difference

$$\text{SS}(\theta) = \text{SS}_{\text{tot}} - \hat{R} = \hat{\boldsymbol{\theta}}'\mathbf{X}'\mathbf{Y}$$

provides the sum of squares due to the model. These results are being summarized in Table 8.1.

Table 8.1 Partitioning Total Sum of Squares for the Model $\mathbf{Y} = \mathbf{X}\boldsymbol{\theta} + \mathbf{e}$

Source	Degrees of Freedom	Sum of Squares
Model	$r = r(\mathbf{X})$	$\text{SS}(\theta) = \hat{\boldsymbol{\theta}}'\mathbf{X}'\mathbf{Y}$
Residual	$N - r(\mathbf{X})$	$\hat{R} = \mathbf{Y}'\mathbf{Y} - \hat{\boldsymbol{\theta}}'\mathbf{X}'\mathbf{Y}$
Total	N	$\mathbf{Y}'\mathbf{Y}$

EXAMPLE 8.6 Consider the following 4×5 factorial experiments having the incidence matrix of the form

$$\mathbf{N}=\begin{matrix} \\ \alpha_1 \\ \alpha_2 \\ \alpha_3 \\ \alpha_4 \end{matrix}\begin{matrix} \beta_1 & \beta_2 & \beta_3 & \beta_4 & \beta_5 \\ \end{matrix}\begin{bmatrix} 1 & 1 & 0 & 0 & 0 \\ 0 & 0 & 0 & 1 & 2 \\ 1 & 0 & 1 & 0 & 0 \\ 0 & 1 & 0 & 0 & 0 \end{bmatrix}$$

and assuming the additive model $E(Y_{ijk})=\alpha_i+\beta_j$ with the corresponding observations

$$\mathbf{Y}'=(Y_{111}\quad Y_{121}\quad Y_{241}\quad Y_{251}\quad Y_{252}\quad Y_{311}\quad Y_{331}\quad Y_{421})$$
$$=(2\ \ 3\ \ 1\ \ 4\ \ 5\ \ 3\ \ 2\ \ 4).$$

Then the solution of normal equations is

$$\hat{\boldsymbol{\theta}}=\mathbf{X}_l^{-}\mathbf{Y}=\frac{1}{24}\begin{bmatrix} 12 & 8 & 0 & 0 & 0 & -8 & 4 & -4 \\ 0 & 0 & 6 & 3 & 3 & 0 & 0 & 0 \\ -12 & 8 & 0 & 0 & 0 & 16 & 4 & -4 \\ 12 & -16 & 0 & 0 & 0 & -8 & 4 & 20 \\ 12 & -8 & 0 & 0 & 0 & 8 & -4 & 4 \\ -12 & 16 & 0 & 0 & 0 & 8 & -4 & 4 \\ 12 & -8 & 0 & 0 & 0 & -16 & 20 & 4 \\ 0 & 0 & 18 & -3 & -3 & 0 & 0 & 0 \\ 0 & 0 & -6 & 9 & 9 & 0 & 0 & 0 \end{bmatrix}\begin{bmatrix} 2 \\ 3 \\ 1 \\ 4 \\ 5 \\ 3 \\ 2 \\ 4 \end{bmatrix}$$

which gives

$$\hat{\alpha}_1=0.667,\qquad \hat{\alpha}_2=1.375,\qquad \hat{\alpha}_3=1.667,\qquad \hat{\alpha}_4=1.667,$$

$$\hat{\beta}_1=1.333,\qquad \hat{\beta}_2=2.333,\qquad \hat{\beta}_3=0.333,\qquad \hat{\beta}_4=-0.375,\ \text{and}\ \hat{\beta}_5=3.125.$$

The BLUEs for the estimable functions are

$$\hat{\alpha}_3-\hat{\alpha}_1=1,\qquad \hat{\alpha}_4-\hat{\alpha}_1=1,$$
$$\hat{\beta}_2-\hat{\beta}_1=1,\qquad \hat{\beta}_3-\hat{\beta}_1=-1,\qquad \hat{\beta}_5-\hat{\beta}_4=3.5.$$

It follows that $r(\mathbf{X})=a+f_\beta=4+3=7$.

The reduction sum of squares due to fitting the model (regression) is

$$SS(\theta)=\hat{\boldsymbol{\theta}}'\mathbf{X}'\mathbf{Y}=83.5$$

and the residual sum of squares is

$$\hat{R}=\mathbf{Y}'\mathbf{Y}-\hat{\boldsymbol{\theta}}'\mathbf{X}'\mathbf{Y}=84.0-83.5=0.5.$$

Table 8.2a Analysis of Variance for the Model $\mathbf{Y}=\mathbf{X}\boldsymbol{\theta}+\mathbf{e}$

Source of Variation	Degrees of Freedom	Sum of Squares
Regression (the model)	7	83.5
Residual	1	0.5
Total	8	84.0

These results are summarized into the analysis of variance of Table 8.2a.

To test the linear hypothesis of $\boldsymbol{\Lambda}'\boldsymbol{\alpha}=\mathbf{0}$ or $\boldsymbol{\Gamma}'\boldsymbol{\beta}=\mathbf{0}$, we require $\mathrm{SS}(\alpha|\beta)$ and $\mathrm{SS}(\beta|\alpha)$, respectively. These two sums of squares can be obtained by first partitioning our original model into

$$E(\mathbf{Y})=\mathbf{X}_1\boldsymbol{\alpha}+\mathbf{X}_2\boldsymbol{\beta}=\mathbf{X}\boldsymbol{\theta}$$

where $\mathbf{X}_1$ and $\mathbf{X}_2$ are design matrices corresponding to α and β effects, respectively, and then fitting the models in terms of α and β effects separately.

First, let us consider the α effect in terms of fitting the model

$$E(\mathbf{Y})=\mathbf{X}_1\boldsymbol{\alpha},$$

where

$$\mathbf{X}_1=\begin{bmatrix}1&0&0&0\\1&0&0&0\\0&1&0&0\\0&1&0&0\\0&1&0&0\\0&0&1&0\\0&0&1&0\\0&0&0&1\end{bmatrix}.$$

Since this is just a model for a one-way classification, the sum of squares for fitting it is

$$\mathrm{SS}(\alpha)=\hat{\boldsymbol{\alpha}}'\mathbf{X}_1'\mathbf{Y}=74.33.$$

This leads us to

$$\mathrm{SS}(\beta|\alpha)=83.5-74.33=9.17.$$

This is the sum of squares attributable to fitting the β's after α (blocks adjusted).

Now to obtain $\mathrm{SS}(\alpha|\beta)$, we fit the model

$$E(\mathbf{Y})=\mathbf{X}_2\boldsymbol{\beta},$$

where

$$\mathbf{X}_2=\begin{bmatrix} 1 & 0 & 0 & 0 & 0 \\ 0 & 1 & 0 & 0 & 0 \\ 0 & 0 & 0 & 1 & 0 \\ 0 & 0 & 0 & 0 & 1 \\ 0 & 0 & 0 & 0 & 1 \\ 1 & 0 & 0 & 0 & 0 \\ 0 & 0 & 1 & 0 & 0 \\ 0 & 1 & 0 & 0 & 0 \end{bmatrix}$$

so

$$SS(\beta)=\hat{\boldsymbol{\beta}}'\mathbf{X}_2'\mathbf{Y}=82.5,$$

and

$$SS(\alpha|\beta)=83.5-82.5=1.0,$$

which is sum of squares attributable to α's (treatments) after fitting β. These calculations are summarized into the analysis of variance of Table 8.2b.

Other forms of analysis are also possible. For example, if it is required to see the adequacy of the full model fitted, assume that the model had no α or β effects in it but simply had been $E(\mathbf{Y})=\mathbf{I}_N\mu$. The sum of squares due to this model is simply corrected for the mean and it is

$$SS(\mu)=N\bar{Y}_{..}^2=72.0$$

and sum of squares under deviation from this model, the regression sum of squares corrected for the mean, becomes

$$SS(\alpha,\beta|\mu)=83.5-72.0=11.5$$

with 6 degrees of freedom $[r(\mathbf{X})-1]$.

Table 8.2b Analysis of Variance for the Model $\mathbf{Y}=\mathbf{X}\boldsymbol{\theta}+\mathbf{e}$

Source of Variation	Degrees of Freedom		Sum of Squares	
Regression	7		83.5	
Hypothesis $\boldsymbol{\Gamma}'\boldsymbol{\beta}=\mathbf{0}$		3		9.17
Treatments unadjusted		4		74.33
Hypothesis $\boldsymbol{\Lambda}'\boldsymbol{\alpha}=\mathbf{0}$		2		1.00
Blocks unadjusted		5		82.50
Residual	1		0.5	
Total	8		84.0	

When interactions are present similar analysis can be done using the least squares g inverse. For example, consider Example 7.15 again. We found the estimates for the α, β, and $(\alpha\beta)$ effects to be

$$\hat{\boldsymbol{\theta}} = \mathbf{X}^{-}\mathbf{Y} = \begin{bmatrix} -1.259 \\ 3.321 \\ 4.358 \\ 4.796 \\ 7.641 \\ 2.441 \\ 0.146 \\ -0.009 \\ 2.329 \\ -0.196 \\ 2.409 \\ 7.634 \\ 2.859 \\ 1.254 \\ 2.250 \\ 0.000 \\ 0.000 \\ 0.000 \\ 0.000 \\ 0.000 \\ -9.030 \\ 0.000 \\ 0.000 \\ 0.000 \\ 0.000 \\ 0.000 \\ 0.000 \\ 0.000 \end{bmatrix}.$$

Now multiplying the transpose of this vector by $\mathbf{X}'\mathbf{Y}$ we obtain

$$\hat{\boldsymbol{\theta}}'\mathbf{X}'\mathbf{Y} = 1073.7$$

which is the SS due to regression, and is the same as what we have obtained for Example 7.15 using different method. Observe that the estimates obtained for the interaction terms are also the same as in Example 7.15. As noted in Example 8.5 these estimates are in fact BLUEs for the estimable parametric functions. Moreover, to test the hypotheses of the form $\mathbf{H}'\boldsymbol{\gamma}=\mathbf{0}$, where $\boldsymbol{\gamma}$ denote the $(\alpha\beta)$ terms, one can use the 15th and 21th rows of the least squares g inverse for rows of $\mathbf{H}'$, which are the same as those found before. The rest of the analysis is similar to Example 7.15.

We have seen that if one has access to a least squares generalized inverse of the design matrix $\mathbf{X}$, then one can obtain BLUEs for the estimable parametric functions, unbiased estimates of their variances, and can identify the estimable parametric functions.

PROGRAM INVERSE

General Description

This interactive program performs the calculations required to compute a least squares generalized inverse for two-way classification models.

Currently, the program is limited in memory space as follows:

The incidence matrix is limited by a 25×25 matrix.

The design matrix of the incidence matrix is limited by a 250×250 matrix.

The generalized inverse matrix is limited by a 500×250 matrix.

The memory space could not be exactly defined because of the algorithm's specifications.

The program stops if memory space lacks.

Each row as well as each column of the incidence matrix must have at least one observation.

1 The input for this program includes:

(a) Dimensions of the incidence matrix and the test for computing a least squares solution.
(b) The incidence matrix.
(c) The vectors of observations.

2 The output of this program includes:

(a) The initial matrix.
(b) The design matrix of the incidence matrix.
(c) A least squares generalized inverse.
(d) A least squares solution, that is $\hat{\boldsymbol{\theta}}=\mathbf{X}_l^-\mathbf{Y}$.

Input Order and Format Specification

The input is in a file which name is chosen by the user; the file extension has to be ".dat". The file name is asked by the program.

It also asks if the user wants the interaction or not and the number of decimals wanted in the output.

All data are read in following the Fortran-free formats.

1 The dimension of the incidence matrix (i.e., number of rows and number of columns) and the test for computing the estimators. If the test number is equal to zero, there is no computing; for any other value, the program performs the calculations.
2 The incidence matrix.
3 The vector of observations, one per line, and ordered in a row-wise fashion [i.e., the observations for cell (1, 1) are in row 1, those for cell (1, 2) are in row 2, etc.].

Output Description

The output is in a file which name is the same as the input file's, but the extension is ".imp".

The output matrices are written on 100-character rows, therefore a matrix row can take more than one output line.

Following is a description of the output from the program:

1 The incidence matrix.
2 The design matrix of the incidence matrix.
3 A least squares generalized inverse of **X**.
4 A least squares solution.

BIBLIOGRAPHICAL NOTES

8.1 Moore (1920) extended the notion of inverse of a nonsingular matrix to singular matrices and discussed this concept in detail in (1935). Moore's definition of an inverse of **X** is equivalent to the existence of a matrix $\mathbf{X}^-$ such that

$$\mathbf{X}\mathbf{X}^- = \mathbf{P}_\mathbf{X}, \qquad \mathbf{X}^-\mathbf{X} = \mathbf{P}_{\mathbf{X}^-}$$

where $\mathbf{P}_\mathbf{X}$ stands for the projection operator onto $R(\mathbf{X})$. Penrose (1955) independently defined an inverse $\mathbf{X}^-$ of $\mathbf{X}$ satisfying

$$\mathbf{X}\mathbf{X}^-\mathbf{X}=\mathbf{X}, \qquad (\mathbf{X}\mathbf{X}^-)'=\mathbf{X}\mathbf{X}^-$$

$$\mathbf{X}^-\mathbf{X}\mathbf{X}^-=\mathbf{X}^-, \qquad (\mathbf{X}^-\mathbf{X})'=\mathbf{X}\mathbf{X}^-$$

which are equivalent to Moore's conditions.

Tseng (1949a, 1949b, and 1956) considered the problem of defining inverses of singular operators which are more general than matrices. Rao (1955) constructed an inverse of a singular matrix that occurs in normal equations in the least squares theory, and showed that it serves the same purpose as the regular inverse of a nonsingular matrix in solving normal equations and also in computing standard errors of least squares estimators. The only property required by his definition was that $\mathbf{X}^-$ be such that $\mathbf{b}=\mathbf{X}^-\mathbf{Y}$ provide a solution to $\mathbf{X}\mathbf{b}=\mathbf{Y}$ for any $\mathbf{Y}$, such that $\mathbf{X}\mathbf{b}=\mathbf{Y}$ is consistent, that is, $\mathbf{X}^-$ be such that $\mathbf{X}\mathbf{X}^-\mathbf{X}=\mathbf{X}$. Rao (1962) called a matrix $\mathbf{X}^-$ satisfying $\mathbf{X}\mathbf{X}^-\mathbf{X}=\mathbf{X}$ generalized inverse of $\mathbf{X}$ and studied its properties. Some principal contributors to the subject are Greville (1957), Bjerhammer (1957 and 1958), Ben-Israel and Charnes (1963), Chipman (1964), Scroggs and Odell (1966), and Chernoff (1953). Mitra (1968a, 1968b) introduced new classes of g inverses. For books on generalized inverses see Rao and Mitra (1971), Boullion and Odell (1971), Bjerhammer (1973), and Ben-Israel and Greville (1974). For a complete list of references on generalized inverses which are excellently documented see Nashed and Rall (1976). There are 1776 references in the list. See also Chapter 1 of Searle (1971).

8.2 Theorems 8.2 through 8.6 are derived from Mitra and Bhrimasankaram (1971). Lemma 8.2 is from Rao and Mitra (1971).

8.3 Complete section is derived from Dodge and Majumdar (1979).

8.4 Mathematical programming approach to computation of g inverses and solution to normal equations was first introduced by Arthanari and Dodge (1981). The householder transformation approach was advocated by Golub (1965), Businger and Golub (1965), and Björck and Golub (1968). For the singular value decomposition approach see Golub and Reinsch (1969). See also Chapter 11 of Rao and Mitra (1971) for other computational methods on g inverses. Mazumdar, Li, and Bryce (1980) consider the problem of deriving a generalized inverse of $\mathbf{X}'\mathbf{X}$, denoted by say $\mathbf{G}_r$ for a linear model $\mathbf{Y}=\mathbf{X}\boldsymbol{\beta}+\mathbf{e}$ with restriction $\mathbf{P}'\boldsymbol{\beta}=\mathbf{0}$, such that a solution to the least squares equations for the restricted model is $\hat{\boldsymbol{\beta}}=\mathbf{G}_r\mathbf{X}'\mathbf{y}$. They found this by using full rank submatrices of $\mathbf{P}'$ and $\mathbf{X}'\mathbf{X}$. Searle (1984) suggested an alternative procedure that is based only on generalized inverses and that requires no partitioning of matrices.

REFERENCES

Arthanari, T. S. and Dodge, Y. (1981). *Mathematical Programming in Statistics*. New York: Wiley.

Ben-Israel, A. and Charnes, A. (1963). Contributions to the theory of generalized inverse. *SIAM J. Appl. Math.* **11**, 667–699.

Ben-Israel, A. and Greville, T. N. E. (1974). *Generalized Inverses: Theory and Applications*. New York: Wiley.

Bjerhammer, A. (1957). Application of calculus of matrices to method of least squares with special reference to geodetic calculations. *Kungl. Tekn. Högsk. Handl. Stockholm* **49**, 1–86.

Bjerhammer, A. (1958). A generalized matrix algebra. *Kungl. Tekn. Högsk. Handl. Stockholm*. **124**, 1–32.

Bjerhammer, A. (1973). *Theory of Errors and Generalized Matrix Inverses*. New York: Elsevier.

Björck, A. and Golub, G. (1968). Iterative refinements of linear least squares solution by householder transformations. Technical Report No. CS 83, Stanford University.

Boullion, T. L. and Odell, P. L. (1971). *Generalized Inverse Matrices*. New York: Wiley.

Businger, P. and Golub, G. H. (1965). Linear least squares solutions by householder transformations. *Numer. Math.* **7**, 269–276.

Chernoff, H. (1953). Locally optimal designs for estimating parameters. *Ann. Math. Statist.* 24, 586–602.

Chipman, J. S. (1964). On least-squares with insufficient observations. *J. Am. Statist. Assoc.* **59**, 1078–1111.

Dodge, Y. and Majumdar, D. (1979). An algorithm for finding least squares generalized inverses for classification models with arbitrary Patterns. *J. Stat. Comput. Simul.* **9**, 1.

Golub, G. (1965). Numerical methods for solving linear least squares problems. *Numer. Math.* **7**, 206–216.

Golub, G. Reinsch, C. (1969). Singular value decomposition and least squares solutions. Technical Report, CS133, Computer Science Department, Stanford University.

Greville, T. N. E. (1957). The pseudoinverse of a rectangular or singular matrix and its application to the solution of systems of linear equations. *SIAM NEWS LETT.* **5**, 3–6.

Mazumdar, S., Li, C. C., and Bryce, G. R. (1980). Correspondence between a linear restriction and a generalized inverse in linear model analysis. *Am. Statist.* **34**, 103–105.

Mitra, S. K. (1968a). On a generalized inverse of a matrix and applications. *Sankhyā Ser. A* **30**, 107–114.

Mitra, S. K. (1968b). A new class of g-inverse of square matrices. *Sankhyā Ser. A* **30**, 323–330.

Mitra, S. K. and Bhimasankaram, P. (1971). Generalized inverse of partitioned matrices and recalculation of least squares estimates for data or model changes. *Sankhyā Ser. A* **33**, 395–410.

Moore, E. H. (1920). On the reciprocal of the general algebraic matrix (abstract). *Bull. Am. Math. Soc.* **26**, 394–395.

Moore, E. H. (1935). *General Analysis*. Philadelphia: American Philosophical Society.

Nashed, M. Z. and Rall, L. B. (1976). Annotated bibliography on generalized inverses and applications. In *Proceedings of an Advanced Seminar*, M. Z. Nashed, Ed., p. 771. New York: Academic Press.

Penrose, R. (1955). A generalized inverse for matrices. *Proc. Cambridge Philos. Soc.* **51**, 406–413.

Rao, C. R. (1955). Analysis of dispersion for multiply classified data with unequal numbers in cells. *Sankhyā* **15**, 253–280.

Rao, C. R. (1962). A note on a generalized inverse of a matrix with applications to problems in mathematical statistics. *J. Roy. Statist. Soc. Ser. B* **24**, 152–158.

Rao, C. R. and Mitra, S. K. (1971). *Generalized Inverse of Matrices and Applications*. New York: Wiley.

Scroggs, J. E. and Odell, P. L. (1966). An alternative definition of the pseudoinverse of a matrix. *SIAM J. Appl. Math.* **14**, 796–810.

Searle, R. L. (1971). *Linear Models*. New York: Wiley.

Searle, S. R. (1984). Restrictions and generalized inverses in linear models. *Am. Statist.* **38**, 53–54.

Tseng, Y. (1949a). Generalized inverses of unbounded operators between two unitary spaces. *Dokl. Akad. Nauk, USSR (N.S.)* **67**, 431–434.

Tseng, Y. (1949b). Properties and classifications of generalized inverses of closed operators. *Dokl. Akad. Nauk, USSR* **67**, 607–610.

Tseng, Y. (1956). Virtual solutions and generalized inversions. *Usp. Math. Nauk (N.S.)* **11**, 213–215.

PROGRAM INVERSE

```
C
C        THIS INTERACTIVE PROGRAM PERFORMS THE CALCULATIONS REQUIRED
C        FOR FINDING A LEAST SQUARES GENERALIZED INVERSE OF THE
C        DESIGN MATRIX CORRESPONDING TO A TWO-WAY CLASSIFICATION
C        MODEL WITH OR WITHOUT INTERACTION, USING THE INCIDENCE
C        MATRIX N WITH ARBITRARY PATTERN AND THE VECTOR OF
C        OBSERVATIONS. GIVEN THE VECTOR OF OBSERVATIONS, IT ALSO
C        PROVIDES A LEAST SQUARES SOLUTION FOR THE UNKNOWN PARAMETER.
C        RESTRICTIONS:
C        GIVEN A TWO-WAY INCIDENCE MATRIX N, EACH ROW AS WELL AS
C        EACH COLUMN MUST HAVE AT LEAST ONE OBSERVATION.
C
C        INPUT:
C        THE FISRT INPUT LINE CONSISTS OF DIMENSION OF THE INCIDENCE
C        MATRIX N (NBLxNBC) AND THE TEST FOR COMPUTING A LEAST SQUARES
C        SOLUTION. IF THE TEST IS EQUAL TO ZERO, IT WILL NOT PROVIDE
C        A SOLUTION FOR THE UNKNOWN PARAMETER; FOR ANY OTHER VALUE,
C        THEY ARE CALCULATED.
C        NEXT INPUT LINES ARE THE ENTRIES OF THE INCIDENCE MATRIX
C        ROW BY ROW. THIS FOLLOWS BY THE ACTUAL OBSERVATIONS IF THEY
C        ARE NEEDED. FOR EACH NON EMPTY CELL, THE ACTUAL OBSERVATIONS
C        ARE READ IN IN ONE LINE WITH A BLANK SPACE BETWEEN TWO
C        OBSERVATIONS. THE NON EMPTY CELLS ARE READ IN FROM LEFT TO
C        RIGHT, ROW WISE.
C
C
C        STRUCTURE
C
C        VARIABLES
C        ESTIM    INPUT:   TEST FOR DOING OR NOT DOING SUBROUTINE ESTIMATE
C        INTER    INPUT:   ANSWER FOR INTERACTION
C        IPLUS    INPUT:   COLUMNS TO ADD FOR INTERACTION
C        IPOS     INPUT:   NUMBER OF PARTITIONS
C        ISUMM    INPUT:   SUM OF THE INCIDENCE MATRIX ELEMENTS
C        ISUMN    INPUT:   SUM OF THE PARTITIONED MATRIX ELEMENTS
C        ITOUS    INPUT:   NUMBER OF COLUMNS OF PARTITIONED MATRIX
C        MA       OUTPUT:   NUMBER OF LINES OF GENERALIZED INVERSE
C        MB       OUTPUT:   NUMBER OF COLUMNS OF GENERALIZED INVERSE
C        NBC      INPUT:   NUMBER OF COLUMNS OF INCIDENCE MATRIX
C        NBL      INPUT:   NUMBER OF LINES OF INCIDENCE MATRIX
C        NUA      INPUT:   NUMBER OF LINES OF ZI1 MATRIX
C        NUB      INPUT:   NUMBER OF COLUMNS OF ZI1 MATRIX
C        NUMBER   INPUT:   NUMBER OF DECIMALS IN RESULTS
C
C        ARRAYS
C        ES       INPUT:   DATA
C        IAWAY    INPUT:   INDEX OF IDENTICAL LINES
C        INDI     INPUT:   INDEX OF LINES BEFORE PARTITIONING
C        INDJ     INPUT:   INDEX OF COLUMNS BEFORE PARTITIONING
C        ITC      INPUT:   INDEX OF COLUMNS FOR TI DATA AFTER
C                             PARTITIONING
C        ITL      INPUT:   INDEX OF LINES FOR TI DATA AFTER
C                             PARTITIONING
C        IX       INPUT:   DESIGN MATRIX FOR INCIDENCE MATRIX
C        IXS      INPUT:   PARTITIONED MATRIX
C        M        INPUT:   FINAL MATRIX
C        N        INPUT:   INCIDENCE MATRIX
```

```
C         X         INPUT:  ZI1 MATRIX
C         Y         INPUT:  ZI2 MATRIX
C         Z         INPUT:  GENERALIZED INVERSE BEFORE COLUMN
C                              AND ROW PERMUTATIONS
C
C         CONSTANTS
C         IA        INPUT:  STORAGE PARAMETER
C
C         SUBROUTINES
C         DO_ZI1            CALCULATES MATRIX ZI1
C         ESTIMATE          CALCULATES THE ESTIMATES OF PARAMETERS
C         MPART             FINDS PARTITIONED MATRIX
C         PERCO             MAKES FIRST APPROPRIATE COLUMN PERMUTATIONS
C         PERMUCO           MAKES FINAL APPROPRIATE COLUMN PERMUTATIONS
C         PERMULI           MAKES APPROPRIATE LINE PERMUTATIONS
C         RPROCESS          COMPUTES R-PROCESS
C         SI_TREAT          TREATS SI DATA CASE
C         TI_SI             FINDS SI AND TI DATA
C         TI_TREAT          TREATS TI DATA CASE
C         XMATRIX           FINDS DESIGN MATRIX OF INCIDENCE MATRIX
C         XSMATRIX          FINDS DESIGN MATRIX OF PARTITIONED MATRIX
C
          PROGRAM INVERSE
          PARAMETER IA=250
          CHARACTER INTER,WEB*7
          DIMENSION IAWAY(IA),N(IA/10,IA/10),M(IA/10,IA/10),INDI(IA/10,IA),
          1 INDJ(IA/10,IA),ITL(IA),ITC(IA),IX(IA,IA*2)
          2 ,IXS(IA,IA),X(IA,IA),Z(IA,IA),Y(IA,IA),ES(IA)
C
C         A INPUT FILE IS ASKED (WITHOUT FILE EXTENSION WHICH
C         MUST BE ".DAT")
C
          TYPE 18
18        FORMAT(//,1X,'NAME OF INPUT FILE ? ',$)
          ACCEPT 19,WEB
19        FORMAT(A)
          OPEN(UNIT=1,FILE=WEB//'.DAT',STATUS='OLD')
          OPEN(UNIT=2,FILE=WEB//'.IMP',STATUS='NEW')
C
C         IT IS ASKED IF USER WANTS INTERACTION OR NOT
C
          TYPE 6
6         FORMAT(//,1X,'WITH INTERACTION ? (Y/N) ',$)
          ACCEPT 7,INTER
C
C         NUMBER OF DECIMAL IN OUTPUT IS ASKED
C
          TYPE 14
14        FORMAT(//,1X,'NUMBER OF DECIMALS ? ',$)
          ACCEPT *,NUMBER
7         FORMAT(A1)
C
C         READ IN DIMENSION OF INCIDENCE MATRIX AND A NON-ZERO
C         ESTIM VALUE MEANS THAT USER WANTS TO COMPUTE THE ESTIMATORS
C
          READ(1,*) NBL,NBC,ESTIM
```

```
      IF(NBL.GT.IA.OR.NBC.GT.IA) THEN
      TYPE 2
2     FORMAT(//,1X,'THE INCIDENCE MATRIX IS TOO BIG')
      STOP
      END IF
      WRITE(2,8)
8     FORMAT(///,<NBC+5>X,'INCIDENCE MATRIX',//)
      K0=0
      L0=25
      ITEST=0
      IF(NBC.LE.25) GOTO 10
      ZA=FLOAT(NBC)
      ITEST=INT(ZA/25.)
      IF(ITEST.EQ.NBC/25) GOTO 11
      ITEST=ITEST-1
11    CONTINUE
      K0=1
10    CONTINUE
C
C     READ IN INCIDENCE MATRIX N
C
      DO 17 I=1,NBL
              READ(1,*) (N(I,K),K=1,NBC)
              K1=K0
              L1=L0
C
C     AN OUTPUT OF THE INCIDENCE MATRIX IS PROVIDED
C
              IF(NBC.LE.25) GOTO 12
              DO ID=1,ITEST
                      WRITE(2,13) (N(I,IE),IE=K1,L1)
13                    FORMAT(1X,25(1X,I3))
                      K1=K1+25
                      L1=L1+25
              END DO
              K1=K1-1
12            IDIF=NBC-ITEST*25
              IF(IDIF.LE.0) GOTO 17
              WRITE(2,9) (N(I,IE),IE=K1+1,NBC)
9             FORMAT(1X,<IDIF>(1X,I3),/)
17    CONTINUE
      J=1
      IF(ESTIM.EQ.0.) GOTO 16
C
C     READ IN OBSERVATIONS TO COMPUTE THE ESTIMATORS
C
      DO I=1,NBL
              DO 15 K=1,NBC
                      IF(N(I,K).EQ.0) GOTO 15
C
C     ONE LINE CONTAINS THE OBSERVATIONS OF ONE CELL
C
                      READ(1,*) (ES(L),L=J,J+N(I,K)-1)
                      J=J+N(I,K)
15            CONTINUE
      END DO
16    CONTINUE
```

```
C
C         SUBROUTINE RPROCESS COMPUTE THE R-PROCESS
C
          CALL RPROCESS(N,M,NBL,NBC)
C
C         COUNT ELEMENTS IN M AND INITIALIZE IAWAY AT 0
C
          ISUMM=0
          ISUMN=0
          DO I=1,NBL
                  IAWAY(I)=0
                  DO K=1,NBC
                          ISUMM=ISUMM+M(I,K)
                          ISUMN=ISUMN+N(I,K)
                  END DO
          END DO
C
C         SUBROUTINE MPART PARTITIONS THE INCIDENCE MATRIX N
C
          CALL MPART(IPOS,IAWAY,M,INDI,INDJ,NBL,NBC,IA)
C
C         SUBROUTINE XMATRIX CONSTRUCTS THE X MATRIX (IX)
C
          CALL XMATRIX(N,IX,INDI,INDJ,NBL,NBC,ISUMN,IPOS,ITOUS,IA,INTER,
          1 IPLUS)
C
C         SUBROUTINE XSMATRIX CONSTRUCTS THE X STAR MATRIX (IXS)
C
          CALL XSMATRIX(IXS,INDI,INDJ,IPOS,NBL,NBC,ISUMM,ITOUS,IA)
C
C         TRANSPOSE IXS INTO X
C
          DO I=1,ISUMM
                  DO K=1,ITOUS
                          X(K,I)=FLOAT(IXS(I,K))
                  END DO
          END DO
C
C         STEP 4 TO 8 IN DODGE'S ALGORITHM FOR EACH PARTITION FOUND
C
          ISTARTL=1
          IENDC=0
          MA=0
          MB=0
          ISTARTC=1
          DO 1 I=1,IPOS
C
C         SUBROUTINE TI_SI FINDS ELEMENTS OF THE INCIDENCE MATRIX
C         WICH ARE GREATER THAN 1
C
                  CALL TI_SI(N,I,INDI,INDJ,ITL,ITC,ISL,IPOS,IA)
C
C         SUBROUTINE DO_ZI1 CONSTRUCTS MATRIX ZI1
C
                  CALL DO_ZI1(X,INDI,INDJ,ISTARTL,ISTARTC,I,IA,IENDC)
                  ISTOPC=ISTARTC-1+INDI(I,1)*INDJ(I,1)
                  ISTOPL=ISTARTL-1+INDI(I,1)+INDJ(I,1)
```

```
            IF(ITL(1).EQ.0) THEN
C
C       IN THIS CASE THERE ARE NO TI DATA
C
            NUA=0
            DO K=ISTARTL,ISTOPL
                  NUA=NUA+1
                  NUB=0
                  DO L=ISTARTC,ISTOPC
                        NUB=NUB+1
                        IF(NUA.GT.IA.OR.NUB.GT.IA) THEN
                        TYPE 20
20      FORMAT(//,1X,'GENERALIZED INVERSE MATRIX IS TOO BIG')
                        STOP
                        END IF
                        Z(NUA,NUB)=X(K,L)
                  END DO
            END DO
            ELSE
C
C       SUBROUTINE TI_TREAT COMPUTES ALGORITHM CONCERNING TI DATA
C
            CALL TI_TREAT(N,IXS,X,Z,INDI,INDJ,ITL,ITC,ISTARTL
     1 ,ISTARTC,ISTOPL,ISTOPC,NUA,NUB,I)
            END IF
C
C       TREATMENT OF SI DATA
C
            IF(ISL.NE.0) GOTO 4
C
C       SUBROUTINE MAKES APPROPRIATE COLUMN PERMUTATIONS
C       IF THERE ARE ANY TI_DATA
C
            IF(ITL(1).GT.0) CALL PERCO(N,Z,INDI,INDJ,NUA,
     1 NUB,I)
C
C       IN THIS CASE MATRIX ZI2 CAN BE EASILY FOUND
C       Y = Zi2
C
            MA2=MA+1
            DO I1=1,NUA
                  MA=MA+1
                  NO=0
                  DO I2=1,NUB
                        NO=NO+1
                        IF(MA.GT.IA.OR.(NO+MB).GT.IA) THEN
                        TYPE 20
                        STOP
                        END IF
                        Y(MA,NO+MB)=Z(I1,I2)
                  END DO
            END DO
            GOTO 5
```

```
C
C       SUBROUTINE SI_TREAT COMPUTES THE ALGORITHM CONCERNING SI DATA
C
4               CALL SI_TREAT(N,M,IXS,Z,Y,INDI,INDJ,ITL,ISL,NUA,NUB,
        1 MA,MB,ISTARTL,MA2,I,NO)
C
C       SUBROUTINE PERMULI MAKES APPROPRIATE ROW PERMUTATIONS
C
5               CALL PERMULI(Y,INDI,INDJ,MA,MB,MA2,ISTARTL,ISTARTC,I,
        1 NO,NBL)
                ISTARTL=ISTARTL+INDI(I,1)+INDJ(I,1)
                ISTARTC=ISTARTC+INDI(I,1)*INDJ(I,1)
1       CONTINUE
C
C       SUBROUTINE PERMUCO MAKES APPROPRIATE COLUMN PERMUTATIONS
C
        CALL PERMUCO(N,Y,IX,INDI,INDJ,IPOS,MA,MB,ITOUS,IPLUS,INTER,NUMBER,
        1 Z)
C
C       IF ESTIME IS A NON ZERO SCALAR, SUBROUTINE ESTIMATE
C       CALCULATES THE ESTIMATORS
C
        IF(ESTIM.NE.0) CALL ESTIMATE(Z,ES,NBL,NBC,MA,MB,INTER)
        CLOSE(UNIT=1)
        CLOSE(UNIT=2)
        STOP
        END
C
C
C
        SUBROUTINE RPROCESS(N,M,NBL,NBC)
C
C       VARIABLES
C       IXZ     OUTPUT:  TEST FOR ENDIND
C       MZ      OUTPUT:  TEST FOR ENDING
C
C       ARRAYS
C       IX      OUTPUT:  INTERMEDIATE MATRIX
C       M       OUTPUT:  FINAL MATRIX
C
        PARAMETER IB=250
        DIMENSION N(IB/10,IB/10),M(IB/10,IB/10),IX(IB/10,IB/10)
C
C       INITIALIZE MATRIX IX
C
        DO I=1,NBL
                DO J=1,NBC
                        IX(I,J)=N(I,J)
                        IF(IX(I,J).GT.0) IX(I,J)=1
                END DO
        END DO
C
C       COMPUTE R-PROCESS
C
1       DO I=1,NBL
                DO L=1,NBC
                        IL=0
```

```
                        DO K=1,NBL
                                DO J=1,NBC
                                        IL=IL+IX(I,J)*IX(K,J)*IX(K,L)

                                END DO
                        END DO
                        IF(IL.GT.0) M(I,L)=1
                END DO
        END DO
        MZ=0
        IXZ=0
        DO I=1,NBL
                DO K=1,NBC
                        IXZ=IXZ+IX(I,K)
                        MZ=MZ+M(I,K)
                END DO
        END DO
        IF(IXZ.EQ.MZ) GOTO 2
        DO I=1,NBL
                DO K=1,NBC
                        IX(I,K)=M(I,K)
                END DO
        END DO
        GOTO 1
2       RETURN
        END
C
C
        SUBROUTINE MPART(IPOS,IAWAY,M,INDI,INDJ,NBL,NBC,IA)
C
C       ARRAYS
C       IAWAY   INPUT:  INDEX OF IDENTICAL ROWS
C       M       OUPUT:  PARTITIONED MATRIX
C
        DIMENSION IAWAY(IA),M(IA/10,IA/10),INDI(IA/10,IA),INDJ(IA/10,IA)
C
C       CONSTRUCTION OF PARTITIONED MATRIX
C
C
C       RECOGNIZE IDENTICAL LINES
C
        IPOS=0
        K=1
        IAWAY(K)=K
1       CONTINUE
        DO 2 I=1,NBL
                IF(IAWAY(I).EQ.K) GOTO 2
                DO L=1,NBC
                        IF(M(I,L).NE.M(K,L)) GOTO 2
                END DO
                IAWAY(I)=K
2       CONTINUE
C
C       SEARCH OF THE POSITION OF THE FIRST NUMBER 1
C
        DO I=1,NBL
                IND=I
                IF(IAWAY(I).EQ.K) GOTO 3
```

```
      END DO
3     CONTINUE
C
C     RECORD COLUMN POSITION
C
      IPOS=IPOS+1
      INC=1
      DO 4 I=1,NBC
            IF(M(IND,I).EQ.0) GOTO 4
            INC=INC+1
            INDJ(IPOS,INC)=I
4     CONTINUE
      INDJ(IPOS,1)=INC-1
C
C     RECORD ROW POSITION
C
      INL=1
      DO 5 I=1,NBL
            IF(IAWAY(I).NE.K) GOTO 5
            INL=INL+1
            INDI(IPOS,INL)=I
5     CONTINUE
      INDI(IPOS,1)=INL-1
      DO 6 I=1,NBL
            IF(IAWAY(I).NE.0) GOTO 6
            K=I
            GOTO 1
6     CONTINUE
C
C     INITIALIZING  M AT 0
C
      DO I=1,NBC
            DO K=1,NBL
                  M(K,I)=0
            END DO
      END DO
C
C     CONSTRUCT PARTIONNED MATRIX M
C
      ITEST=0
      ISTARTL=1
      IENDL=INDI(1,1)
      ISTARTC=1
      IENDC=INDJ(1,1)
7     CONTINUE
      DO I=ISTARTC,IENDC
            DO K=ISTARTL,IENDL
                  M(K,I)=1
            END DO
      END DO
      ITEST=ITEST+1
      IF(ITEST.GE.IPOS) GOTO 10
      ISTARTL=IENDL+1
      ISTARTC=IENDC+1
      IENDC=IENDC+INDJ(ITEST+1,1)
      IENDL=IENDL+INDI(ITEST+1,1)
```

```
      GOTO 7
10    RETURN
      END
C
C
C
      SUBROUTINE XMATRIX(N,IX,INDI,INDJ,NBL,NBC,ISUMN,IPOS,ITOUS,IA,
     1 INTER,IPLUS)
C
C     VARIABLES
C     INTER    INPUT:   ANSWER FOR INTERACTION
C     IPLUS    INPUT:   NUMBER OF COLUMNS TO ADD FOR INTERACTION
C     ISUMN    INPUT:   SUM OF THE INCIDENCE MATRIX ELEMENTS
C
C     ARRAYS
C     IX       INPUT:   DESIGN MATRIX OF INCIDENCE MATRIX
C
      CHARACTER INTER
      DIMENSION IX(IA,IA*2),INDI(IA/10,IA),INDJ(IA/10,IA),
     1 N(IA/10,IA/10)
C
C     CONSTRUCT X
C
C     INITIALIZE AT 0
C
      ITOUS=0
      IPLUS=0
      DO 7 I=1,IPOS
              ITOUS=ITOUS+INDI(I,1)+INDJ(I,1)
              IF(INTER.EQ.'n'.OR.INTER.EQ.'N') GOTO 7
C
C     THIS IS THE CASE IN WHICH USER WANTS INTERACTION
C     INTERACTION COLUMNS MUST BE ADDED
C
              DO I1=2,INDI(I,1)+1
                      DO I2=2,INDJ(I,1)+1
                              IF(N(INDI(I,I1),INDJ(I,I2)).GE.1) IPLUS
     1 =IPLUS+1
                      END DO
              END DO
7     CONTINUE
      ICALCUL=NBC+NBL
      IF(INTER.EQ.'y'.OR.INTER.EQ.'Y') ICALCUL=ICALCUL+IPLUS
C
C     INITIALIZE IX AT 0
C
      DO I=1,ISUMN
              DO K=1,ICALCUL
                      IX(I,K)=0
              END DO
      END DO
C
C     FILLING UP THE MATRIX IX
C
      IZA=ITOUS+IPLUS
      ITITLE=IZA*3/2-3
      IF(ITITLE.GE.45) ITITLE=45
```

```
      WRITE(2,1)
1     FORMAT('1',///,<ITITLE>X,'DESIGN MATRIX',//)
      ISTART=0
      K0=0
      L0=30
      ITEST=0
      IF(IZA.LE.30) GOTO 4
      ZA=FLOAT(IZA)
      ITEST=INT(ZA/30.)
      IF(ITEST.EQ.IZA/30) GOTO 5
      ITEST=ITEST-1
5     CONTINUE
      K0=1
4     CONTINUE
      IOU=0
      DO I=1,NBL
              DO 2 K=1,NBC
                      IF(N(I,K).EQ.0) GOTO 2
                      IOU=IOU+1
                      DO 9 L=1,N(I,K)
                              ISTART=ISTART+1
                              IX(ISTART,I)=1
                              IX(ISTART,NBL+K)=1
      IF(ISTART.GT.IA.OR.(NBL+K).GT.IA.OR.(INTER.EQ.'Y'.AND.
     1 ITOUS+IOU.GT.IA).OR.(INTER.EQ.'y'.AND.ITOUS+IOU.GT.IA)) THEN
      TYPE 10
10    FORMAT(//,1X,'DESIGN MATRIX OF INCIDENCE MATRIX IS TOO BIG')
      STOP
      END IF
                              IF(INTER.EQ.'y'.OR.INTER.EQ.'Y')
     1 IX(ISTART,ITOUS+IOU)=1
C
C     AN OUTPUT OF THE X MATRIX IS PROVIDED
C
                              L1=L0
                              K1=K0
                              IF(IZA.LE.30) GOTO 6
                              DO ID=1,ITEST
                                      WRITE(2,3) (IX(ISTART,IE),IE=K1,L1)

3                                     FORMAT(1X,30(2X,I1))

                                      L1=L1+30

                                      K1=K1+30


                              END DO
                              K1=K1-1
6                             CONTINUE
                              IDIF=IZA-ITEST*30
                              IF(IDIF.LE.0) GOTO 9
                              WRITE(2,8) (IX(ISTART,IE),IE=K1+1,IZA)
8                             FORMAT(1X,<IDIF>(2X,I1),/)
9                     CONTINUE
2             CONTINUE
      END DO
      RETURN
      END
C
C
C
```

```
      SUBROUTINE TI_SI(N,I,INDI,INDJ,ITL,ITC,ISL,IPOS,IA)
C
C     VARIABLES
C     ISL     OUTPUT:  NUMBER OF SI ELEMENTS FOR EACH PARTITION
C     ITLSUM  OUTPUT:  NUMBER OF TI ELEMENTS FOR EACH PARTITION
C
C     ARRAYS
C     ITC     OUTPUT:  COLUMN INDEX OF DATA(>1) FOR EACH PARTITION
C     ITL     OUTPUT:  LINE INDEX OF DATA(>1) FOR EACH PARTITION
C
      DIMENSION N(IA/10,IA/10),INDI(IA/10,IA),INDJ(IA/10,IA),ITL(IA),
      1 ITC(IA)
C
C     SUBROUTINE TI_SI FINDS TI AND SI DATA, THEIR PLACE AND
C     TOTAL OF EACH KIND
C
      IO1=1
      IPL1=0
      ITLSUM=0
      ISL=0
      DO K=2,INDI(I,1)+1
              IPL1=IPL1+1
              IPC1=0
              DO 1 L=2,INDJ(I,1)+1
                      IPC1=IPC1+1
C
C     TEST FOR TI DATA
C
                      IF(N(INDI(I,K),INDJ(I,L)).LT.2) GOTO 2
                      IO1=IO1+1
                      ITLSUM=ITLSUM+1
                      ITL(IO1)=IPL1
                      ITC(IO1)=IPC1
C
C     TEST FOR SI DATA
C
2                     IF(N(INDI(I,K),INDJ(I,L)).NE.0) GOTO 1
                      ISL=ISL+1
1             CONTINUE
      END DO
      ITL(1)=ITLSUM
      RETURN
      END
C
C
C
      SUBROUTINE XSMATRIX(IXS,INDI,INDJ,IPOS,NBL,NBC,ISUMM,ITOUS,IA)
C
C     ARRAYS
C     IXS     OUTPUT:  DESIGN MATRIX OF PARTITIONED MATRIX
C
      DIMENSION IXS(IA,IA),INDI(IA/10,IA),INDJ(IA/10,IA)
C
C     CONSTRUCT X* = IXS
C
C     INITIALIZE AT 0
C
```

```
      IF(ISUMM.GT.IA.OR.(NBL+NBC).GT.IA) THEN
      TYPE 1
1     FORMAT(//,1X,'DESIGN MATRIX OF MATRIX M IS TOO BIG')
      STOP
      END IF
      DO I=1,ISUMM
              DO K=1,NBC+NBL
                      IXS(I,K)=0
              END DO
      END DO
      ISTARTL=0
      ISTARTC=INDI(1,1)
      IO=0
      DO I=1,IPOS
              DO K=1,INDI(I,1)
                      JA=0
                      DO L=1,INDJ(I,1)
                              JA=JA+1
                              IO=IO+1
                              IXS(IO,K+ISTARTL)=1
                              IXS(IO,JA+ISTARTC)=1
                      END DO
              END DO
              ISTARTL=ISTARTL+INDI(I,1)+INDJ(I,1)
              ISTARTC=ISTARTL+INDI(I+1,1)
      END DO
      RETURN
      END
C
C
C
      SUBROUTINE DO_ZI1(X,INDI,INDJ,ISTARTL,ISTARTC,I,IA,IENDC)
C
C     ARRAYS
C     X       I-O:   PARTITIONED AND ZI1 MATRIX
C
      DIMENSION X(IA,IA),INDI(IA/10,IA),INDJ(IA/10,IA)
C
C     SUBROUTINE DO_ZI1 CONSTRUCTS ZI1
C
      IENDC=INDI(I,1)*INDJ(I,1)+IENDC
      DO K=ISTARTL,INDI(I,1)+INDJ(I,1)+ISTARTL-1
              DO 1 L=ISTARTC,IENDC
                      IF(X(K,L).EQ.0.) GOTO 2
                      IF(K.GT.ISTARTL-1+INDI(I,1)) GOTO 3
                      X(K,L)=(2.*FLOAT(INDI(I,1))-1.)/
     1 2./FLOAT(INDI(I,1))/FLOAT(INDJ(I,1))
                      GOTO 1
3                     CONTINUE
                      X(K,L)=(2.*FLOAT(INDJ(I,1))-1.)/
     1 2./FLOAT(INDI(I,1))/FLOAT(INDJ(I,1))
                      GOTO 1
2                     CONTINUE
                      X(K,L)=-1./2./FLOAT(INDI(I,1)
     1 )/FLOAT(INDJ(I,1))
1             CONTINUE
      END DO
```

```
      RETURN
      END
C
C
C
      SUBROUTINE SI_TREAT(N,M,IXS,Z,Y,INDI,INDJ,ITL,ISL,NUA,NUB,
     1 MA,MB,ISTARTL,MA2,I,NO)
C
C     VARIABLES
C     MA       INPUT:  NUMBER OF ROWS OF GENERALIZED INVERSE
C     MB       INPUT:  NUMBER OF COLUMNS OF GENERALIZED INVERSE
C     NUA      INPUT:  NUMBER OF ROWS OF ZI2 MATRIX
C     NUB      INPUT:  NUMBER OF COLUMNS OF ZI2 MATRIX
C
C     ARRAYS           B       I  VECTOR IN DODGE'S ALGORITHM
C     B        INPUT:  VECTOR ( SEE DODGE'S ALGORITHM )
C     BAP      INPUT:  B*A MATRIX
C     BAPZC    INPUT:  B*A*C MATRIX
C     IAWAY    INPUT:  INDEX FOR G VECTORS
C     IXS      INPUT:  DESIGN MATRIX OF PARTITIONED MATRIX
C     M        INPUT:  PARTITIONED MATRIX
C     Y        INPUT:  INTERMEDIATE MATRIX
C     Z        OUTPUT:  ZI2 MATRIX
C     ZC       INPUT:  MATRIX C ( SEE DODGE'S ALGORITHM )
C     ZQ       INPUT:  MATRIX Q ( SEE DODGE'S ALGORITHM )
C
C
C     ALL NOTATIONS USED IN THIS PROGRAM ARE THE SAME AS THOSE
C     USED IN DODGE'S REPORT
C
      PARAMETER IB=250
      DIMENSION IXS(IB,IB),Z(IB,IB),Y(IB,IB),INDI(IB/10,IB),
     1 INDJ(IB/10,IB),IAWAY(IB),ZQ(IB,IB),ZC(IB,IB),B(IB),
     2 BAP(IB,IB),BAPZC(IB,IB),N(IB/10,IB/10),M(IB/10,IB/10),ITL(IB)
      ISUML=0
      L2=0
      L3=0
      IANFANG1=0
      IANFANG2=0
C
C     SEARCH FOR THE CORRECT EMPLACEMENT OF DATA
C
      IF(I.EQ.1) GOTO 1
      DO I1=1,I-1
              DO I2=1,INDI(I1,1)
                      DO I3=1,INDJ(I1,1)
                              ISUML=ISUML+M(L2+I2,L3+I3)
                      END DO
              END DO
              L2=L2+INDI(I1,1)
              L3=L3+INDJ(I1,1)
              IANFANG1=IANFANG1+INDI(I1,1)
              IANFANG2=IANFANG2+INDJ(I1,1)
      END DO
1     CONTINUE
      NF=0
      L1=IANFANG1
```

```
          ITOTAL=0
          DO I1=2,INDI(I,1)+1
                  L1=L1+1
                  L2=IANFANG2
                  DO 8 I2=2,INDJ(I,1)+1
                          L2=L2+1
                          ITOTAL=ITOTAL+M(L1,L2)
                          IF(N(INDI(I,I1),INDJ(I,I2)).NE.0) GOTO 8
                          NF=NF+1
                          IAWAY(NF)=ITOTAL
 8                CONTINUE
          END DO
C
C         ISOLATE Q = ZQ
C
          NG=1
          NJ=0
          DO 2 I1=1,NUB
                  IF(IAWAY(NG).EQ.I1) GOTO 3
                  NI=0
                  NJ=NJ+1
                  DO I2=1,NUA
                          NI=NI+1
                          ZQ(NI,NJ)=Z(I2,I1)
                  END DO
                  GOTO 2
 3                NG=NG+1
 2        CONTINUE
C
C         INITIALIZE C = ZC
C
          DO I1=1,NI
                  DO 4 I2=1,NI
                          IF(I1.EQ.I2) GOTO 5
                          ZC(I1,I2)=0.
                          GOTO 4
 5                        ZC(I1,I2)=1.
 4                CONTINUE
          END DO
C
C         COMPUTE B1
C
          DO I1=1,NI
                  B(I1)=Z(I1,IAWAY(1))
          END DO
          NG=0
          DO 6 I1=2,ISL+1
                  NG=NG+1
C
C         Bi-1*A'i-1 = BAP
C
                  DO I2=1,NI
                          DO I3=1,NI
                                  BAP(I2,I3)=B(I2)*FLOAT(IXS(IAWAY
     1 (NG)+ISUML,ISTARTL-1+I3))
                          END DO
                  END DO
```

```
C
C       COMPUTE BPA
C
                SOMME=0.
                DO I2=1,NI
                        SOMME=SOMME+B(I2)*FLOAT(IXS(IAWAY(NG)+
     1 ISUML,ISTARTL-1+I2))
                END DO
                SOMME=1.-SOMME
C
C       COMPUTE BAP/SOMME+I
C
                DO I2=1,NI
                        DO 7 I3=1,NI
                                BAP(I2,I3)=BAP(I2,I3)/SOMME
                                IF(I2.NE.I3) GOTO 7
                                BAP(I2,I3)=BAP(I2,I3)+1.
7                       CONTINUE
                END DO
C
C       COMPUTE BAP*ZC
C
                DO I2=1,NI
                        DO I3=1,NI
                                BAPZC(I2,I3)=0.
                                DO I4=1,NI
                                        BAPZC(I2,I3)=BAPZC(I2,I3)+BAP(I2

     1 ,I4)*ZC(I4,I3)
                                END DO
                        END DO
                END DO
                DO K1=1,NI
                        DO K2=1,NI
                                ZC(K1,K2)=BAPZC(K1,K2)
                        END DO
                END DO
                IF(I1.EQ.ISL+1) GOTO 6
C
C       COMPUTE Bi = Ci Gi
C
                NK=0
                DO I2=1,NI
                        NK=NK+1
                        B(NK)=0.
                        DO I3=1,NI
                                B(NK)=B(NK)+ZC(I2,I3)*Z(I3
     1 ,IAWAY(NG+1))
                        END DO
                END DO
6       CONTINUE
C
C       COMPUTE STEP 8    Y = ZC * ZQ
C
        DO I2=1,NI
                DO I3=1,NJ
                        BAP(I2,I3)=0.
```

```
                    DO I4=1,NI
                         BAP(I2,I3)=BAP(I2,I3)+ZC(I2,I4)*ZQ(I4,I3)
                    END DO
               END DO
          END DO
C
C         TEST FOR INTERMEDIATE COLUMN PERMUTATIONS
C
          IF(ITL(1).GT.0) CALL PERCO(N,BAP,INDI,INDJ,NI,NJ,I)
          MA2=MA+1
          DO I2=1,NI
               MA=MA+1
               NO=0
               DO I3=1,NJ
                    NO=NO+1
                    IF(MA.GT.IB.OR.(MB+NO).GT.IB) THEN
                    TYPE 9
9                   FORMAT(//,1X,'GENERALIZED INVERSE MATRIX IS '
          1 ,'TOO BIG')
                    STOP
                    END IF
                    Y(MA,MB+NO)=BAP(I2,I3)
               END DO
          END DO
          RETURN
          END
C
C
C
          SUBROUTINE PERMULI(Y,INDI,INDJ,MA,MB,MA2,ISTARTL,ISTARTC,I,
          1 NO,NBL)
C
C         VARIABLES
C         MA       INPUT:  NUMBER OF ROWS OF GENERALIZED INVERSE
C         MB       INPUT:  NUMBER OF COLUMNS OF GENERALIZED INVERSE
C
C         ARRAYS
C         IAWAY    INPUT:  INDEX OF ROWS
C         Y        OUTPUT:  GENERALIZED INVERSE AFTER PERMUTATIONS
C
          PARAMETER IB=250
          DIMENSION INDI(IB/10,IB),INDJ(IB/10,IB),Y(IB,IB),B(IB),IAWAY(IB)
C
C         RECORD ALL ALFHAS AND BETAS OF THIS PARTITION
C
          NA=0
          DO I2=2,INDI(I,1)+1
               NA=NA+1
               B(NA)=FLOAT(INDI(I,I2))
          END DO
          DO I2=2,INDJ(I,1)+1
               NA=NA+1
               B(NA)=FLOAT(INDJ(I,I2)+NBL)
          END DO
```

```
C
C         SORT THE ELEMENTS
C
          DO I2=2,NA-1
                  DO 1 I3=NA,I2,-1
                          IF(B(I3-1).LE.B(I3)) GOTO 1
                          BB=B(I3)
                          B(I3)=B(I3-1)
                          B(I3-1)=BB
1                 CONTINUE
          END DO
C
C         COPY THE ELEMENTS WITHOUT REDONDANCY
C
          NB=1
          IAWAY(1)=INT(B(1))
          DO 2 I2=2,NA
                  IF(B(I2).EQ.B(I2-1)) GOTO 2
                  NB=NB+1
                  IAWAY(NB)=INT(B(I2))
2         CONTINUE
          DO I2=MB+1,MB+NO
                  NC=0
                  DO I3=MA2,MA
                          NC=NC+1
                          B(NC)=Y(I3,I2)
                          Y(I3,I2)=0.
                  END DO
                  DO I3=1,NB
                  Y(IAWAY(I3),I2)=B(I3)
                  END DO
          END DO
          MB=MB+NO
          RETURN
          END
C
C
C
          SUBROUTINE PERMUCO(N,Y,IX,INDI,INDJ,IPOS,MA,MB,ITOUS,IPLUS,INTER
          1 ,NUMBER,BAP)
C
C         VARIABLES
C         INTER    INPUT:  ANSWER FOR INTERACTION
C         IPLUS    INPUT:  COLUMNS TO ADD FOR INTERRACTION
C
C         ARRAYS
C         BAP      INPUT:  GENERALIZED INVERSE
C         IL1      INPUT:  SORTED COPY OF INDI
C         IL2      INPUT:  SORTED COPY OF INDJ
C         IL3      INPUT:  INDEX OF THE INCIDENCE MATRIX ELEMENTS
C         IX       INPUT:  DESIGN MATRIX OF INCIDENCE MATRIX
C
C         SUBROUTINE       ACTION      COMPUTES INTERACTION
C
          PARAMETER IB=250
          CHARACTER INTER
          DIMENSION N(IB/10,IB/10),Y(IB,IB),INDI(IB/10,IB),INDJ(IB/10,IB),
```

```
     1 BAP(IB,IB),IL1(IB),IL2(IB),IL3(IB),IX(IB,IB*2)
C
C      MAKE APROPRIATE ROW PERMUTATIONS
C
C      COPY OF INDI AND INDJ INTO IL1 AND IL2
C
       NA=0
       ISOMME=0
       DO I1=1,IPOS
               DO I2=2,INDI(I1,1)+1
                       DO 1 I3=2,INDJ(I1,1)+1
                               IF(N(INDI(I1,I2),INDJ(I1,I3)).EQ.0)
     1 GOTO 1
                               NA=NA+1
                               IL1(NA)=INDI(I1,I2)
                               IL2(NA)=INDJ(I1,I3)
                               IL3(NA)=ISOMME+1
                               ISOMME=N(INDI(I1,I2),INDJ(I1,I3))+ISOMME
1                      CONTINUE
               END DO
       END DO
C
C      SORT THE IL1 AND IL2
C
       DO I1=2,NA-1
               DO 2 I2=NA,I1,-1
                       IF(IL1(I2-1).LT.IL1(I2)) GOTO 2
                       IF(IL1(I2-1).EQ.IL1(I2).AND.IL2(I2-1).LE.IL2(I2))
     1 GOTO 2
                       IL=IL1(I2)
                       IL1(I2)=IL1(I2-1)
                       IL1(I2-1)=IL
                       IC=IL2(I2)
                       IL2(I2)=IL2(I2-1)
                       IL2(I2-1)=IC
                       IS=IL3(I2)
                       IL3(I2)=IL3(I2-1)
                       IL3(I2-1)=IS
2              CONTINUE
       END DO
C
C      REWRITE MATRIX Y INTO MATRIX BAP WHILE SORTING
C
       MB=0
       DO I1=1,NA
               DO I2=1,N(IL1(I1),IL2(I1))
                       MB=MB+1
                       DO I3=1,MA
                               BAP(I3,MB)=Y(I3,IL3(I1)-1+I2)
                       END DO
               END DO
       END DO
```

```
C
C       SUBROUTINE ACTION IS CALLED IN ORDER TO TREAT INTERACTION CASE
C
        IF(INTER.EQ.'y'.OR.INTER.EQ.'Y') CALL ACTION(IX,BAP,ITOUS,
        1 IPLUS,MA,MB)
        NUMERO=100/(4+NUMBER)
        K0=0
        L0=NUMERO
        ITEST=0
        IF(MB.LE.NUMERO) GOTO 3
        ZA=FLOAT(MB)
        ZB=FLOAT(NUMERO)
        ITEST=INT(ZA/ZB)
        IF(ITEST.EQ.MB/NUMERO) GOTO 4
        ITEST=ITEST-1
4       CONTINUE
C
C       AN OUTPUT OF THE GENERALIZED INVERSE IS PROVIDED
C
        K0=1
3       CONTINUE
        ITITLE=MB*(4+NUMBER)/2-6
        IF(ITITLE.GE.45) ITITLE=45
        WRITE(2,8)
8       FORMAT('1',///,<ITITLE>X,'GENERALIZED INVERSE',/)
        DO 9 I=1,MA
                K1=K0
                L1=L0
                IF(MB.LE.NUMERO) GOTO 5
                DO ID=1,ITEST
                        WRITE(2,6) (BAP(I,IE),IE=K1,L1)
                        K1=K1+NUMERO
                        L1=L1+NUMERO
6                       FORMAT(1X,<NUMERO>(1X,F<3+NUMBER>.<NUMBER>))
                END DO
                K1=K1-1
5               CONTINUE
                IDIF=MB-ITEST*NUMERO
                IF(IDIF.LE.0) GOTO 9
                WRITE(2,7) (BAP(I,IE),IE=K1+1,MB)
7               FORMAT(1X,<IDIF>(1X,F<3+NUMBER>.<NUMBER>),/)
9       CONTINUE
        RETURN
        END
C
C
C
        SUBROUTINE TI_TREAT(N,IXS,X,Z,INDI,INDJ,ITL,ITC,ISTARTL,ISTARTC
        1 ,ISTOPL,ISTOPC,K1,K2,I)
C
C       ARRAYS
C       A       INPUT:  VECTOR TO ADD AT EACH ITERATION
C       D       INPUT:  VECTOR G*A IN DODGE'S ALGORITHM
C       E       INPUT:  VECTOR G*D IN DODGE'S ALGORITHM
C       IXS     INPUT:  DESIGN MATRIX OF PARTITIONED MATRIX
C       X       INPUT:  MATRIX ZI1
```

```
C         Z        INPUT:   MATRIX ZI2
C
          PARAMETER IB=250
          DIMENSION N(IB/10,IB/10),IXS(IB,IB),X(IB,IB),INDI(IB/10,IB),
          1 Z(IB,IB),INDJ(IB/10,IB),ITL(IB),ITC(IB),A(IB),E(IB),D(IB)
C
C         ALL NOTATIONS USED IN THIS PROGRAM ARE THE SAME AS THOSE
C         USED IN DODGE'S REPORT
C
C         REWRITE CORRECT PART OF X INTO MATRIX Z
C
          K1=0
          DO I1=ISTARTL,ISTOPL
                 K1=K1+1
                 K2=0
                 DO I2=ISTARTC,ISTOPC
                        K2=K2+1
                        IF(K1.GT.IB.OR.K2.GT.IB) THEN
                        TYPE 2
2         FORMAT(//,1X,'GENERALIZED INVERSE MATRIX IS TOO BIG')
                        STOP
                        END IF
                        Z(K1,K2)=X(I1,I2)
                 END DO
          END DO
          DO KA=2,ITL(1)+1
C
C         FIND THE CORRECT ROW OF MATRIX X
C
                 ISOMME=0
                 IBEGIN=0
                 IF(I.LT.2) GOTO 1
                 DO K=1,I-1
                        ISOMME=ISOMME+INDI(K,1)*INDJ(K,1)
                        IBEGIN=IBEGIN+INDI(K,1)+INDJ(K,1)
                 END DO
1                ISOMME=ISOMME+ITC(KA)+INDJ(I,1)*(ITL(KA)-1)
                 IBEGIN=IBEGIN+1
                 ISTOP=IBEGIN-1+INDJ(I,1)+INDI(I,1)
                 NBA=0
                 DO K=IBEGIN,ISTOP
                        NBA=NBA+1
                        A(NBA)=FLOAT(IXS(ISOMME,K))
                 END DO
                 DO KB=1,N(INDI(I,ITL(KA)+1),INDJ(I,ITC(KA)+1))-1
C
C         COMPUTE D=GPA
C
                        NUA=0
                        DO I1=1,K2
                               NUA=NUA+1
                               D(NUA)=0.
                               NUC=0
                               DO I2=1,K1
                                      NUC=NUC+1

                                      D(NUA)=D(NUA)+Z(I2,I1)*A(NUC)

                               END DO
```

```
                     END DO
C
C       COMPUTE E=GD
C
                     NUA=0
                     DO I1=1,K1
                             NUA=NUA+1
                             E(NUA)=0
                             NUC=0
                             DO I2=1,K2
                                     NUC=NUC+1

                                     E(NUA)=E(NUA)+Z(I1,I2)*D(NUC)

                             END DO
                     END DO
C
C       COMPUTE APE=ZK
C
                     ZK=0.
                     DO I1=1,NUA
                             ZK=ZK+A(I1)*E(I1)
                     END DO
                     ZK=ZK+1
                     DO I1=1,NUA
                             E(I1)=E(I1)/ZK
                     END DO
C
C       COMPUTE Z, THE NEW INTERMEDIATE MATRIX
C
                     DO I1=1,K1
                             DO I2=1,K2
                                     Z(I1,I2)=Z(I1,I2)-E(I1)*D(I2)

                             END DO
                     END DO
                     K2=K2+1
                     IF(K2.GT.IB) THEN
                     TYPE 2
                     STOP
                     END IF
                     DO I1=1,K1
                             Z(I1,K2)=E(I1)
                     END DO
             END DO
        END DO
        RETURN
        END
C
C
        SUBROUTINE PERCO(N,A,INDI,INDJ,NUA,NUB,I)
C
C       VARIABLES
C       NUA     INPUT:  NUMBER OF ROWS OF GENERALIZED INVERSE
C       NUB     INPUT:  NUMBER OF COLUMNS OF GENERALIZED INVERSE
C
C       ARRAYS
C       A       I-O:    GENERALIZED INVERSE
C       IAWAY1  INPUT:  INDEX OF ELEMENTS
C       AWAY2   INPUT:  INDEX OF ELEMENTS
```

```
C
      PARAMETER IB=250
      DIMENSION N(IB/10,IB/10),INDI(IB/10,IB),INDJ(IB/10,IB),
     1 IAWAY1(IB),IAWAY2(IB),A(IB,IB),B(IB,IB)
C
C     COPY THE MATRIX A INTO MATRIX B
C
      DO I1=1,NUA
              DO I2=1,NUB
                      B(I1,I2)=A(I1,I2)
              END DO
      END DO
      K1=0
      K2=0
      NF=0
C
C     RECORD THE EMPLACEMENT OF THE COLUMNS WHICH HAS BEEN ADDED
C     BY R-PROCESS
C
      DO I1=2,INDI(I,1)+1
              DO 2 I2=2,INDJ(I,1)+1
                      K1=K1+1
                      IF(N(INDI(I,I1),INDJ(I,I2)).EQ.0) GOTO 2
                      NF=NF+1
                      K2=K2+1
                      IAWAY1(NF)=K1
                      IAWAY2(NF)=K2
2             CONTINUE
      END DO
C
C     RECORD THE EMPLACEMENT OF THE COLUMNS CORRESPONDING
C     TO ELEMENTS IN THE INCIDENCE MATRIX WHICH ARE LOWER THAN 2
C
      K1=0
      DO I1=2,INDI(I,1)+1
              DO 3 I2=2,INDJ(I,1)+1
                      K1=K1+1
                      IF(N(INDI(I,I1),INDJ(I,I2)).LT.2) GOTO 3
                      DO I3=1,N(INDI(I,I1),INDJ(I,I2))-1
                              K2=K2+1
                              NF=NF+1
                              IAWAY1(NF)=K1
                              IAWAY2(NF)=K2
                      END DO
3             CONTINUE
      END DO
C
C     SORT IAWAY1 AND IAWAY2 IN ORDER TO SORT THE COLUMNS
C
      DO I1=2,NUB-1
              DO 4 I2=NUB,I1,-1
                      IF(IAWAY1(I2-1).LE.IAWAY1(I2)) GOTO 4
                      K=IAWAY1(I2-1)
                      IAWAY1(I2-1)=IAWAY1(I2)
                      IAWAY1(I2)=K
                      K=IAWAY2(I2-1)
                      IAWAY2(I2-1)=IAWAY2(I2)
```

```
                IAWAY2(I2)=K
4           CONTINUE
        END DO
C
C       COPY MATRIX B INTO MATRIX A
C
        DO I1=1,NF
                DO I2=1,NUA
                        A(I2,I1)=B(I2,IAWAY2(I1))
                END DO
        END DO
        RETURN
        END
C
C
        SUBROUTINE ACTION(IX,BAP,ITOUS,IPLUS,MA,MB)
C
C       VARIABLES
C       E       INPUT:  IS EQUAL TO C'*A IN DODGE'S ALGORITHM
C       IPLUS   INPUT:  NUMBER OF COLUMNS TO ADD FOR INTERACTON
C       MA      OUTPUT:  NUMBER OF ROWS OF GENERALIZED INVERSE
C       MB      OUTPUT:  NUMBER OF COLUMNS OF GENERALIZED INVERSE
C
C       ARRAYS
C       BAO     INPUT:  GENERALIZED INVERSE BEFORE INTERACTION
C       C       INPUT:  C VECTOR IN DODGE'S ALGORITHM
C       D       INPUT:  D VECTOR IN DODGE'S ALGORITHM
C       IX      INPUT:  INCIDENCE MATRIX
C       PAREN   INPUT:  EQUALS TO (I-A*G) IN DODGE'S ALGORITHM
C
        PARAMETER IB=250
        DIMENSION IX(IB,IB*2),BAP(IB,IB),PAREN(IB,IB),D(IB),C(IB)
        DO I1=1,IPLUS
C
C       ALL NOTATIONS USED IN THIS PROGRAM ARE THE SAME AS THOSE
C       USED IN DODGE'S REPORT
C
C       COMPUTE I-A*G
C
                DO I2=1,MB
                        DO I3=1,MB
                                PAREN(I2,I3)=0.
                                DO I4=1,MA
                                        PAREN(I2,I3)=PAREN(I2,I3)+

        1 FLOAT(IX(I2,I4))*BAP(I4,I3)
                                END DO
                                PAREN(I2,I3)=-PAREN(I2,I3)
                                IF(I2.EQ.I3) PAREN(I2,I3)=1.+PAREN(I2,I3)
                        END DO
                END DO
C
C       C=PAREN * A
C
                ITEST=0
                DO I2=1,MB
                        C(I2)=0.
```

```
                    DO I3=1,MB
                            C(I2)=C(I2)+PAREN(I2,I3)*FLOAT(IX
     1 (I3,ITOUS+I1))
                    END DO
                    IF(C(I2).GT.-1.E-6.AND.C(I2).LT.1.E-6)
     1 ITEST=ITEST+1
            END DO
            IF(ITEST.EQ.MB) GOTO 1
C
C       COMPUTE D=G*A
C
            DO I2=1,MA
                    D(I2)=0.
                    DO I3=1,MB
                            D(I2)=D(I2)+BAP(I2,I3)*
     1 FLOAT(IX(I3,ITOUS+I1))
                    END DO
            END DO
C
C       COMPUTE E=C*A
C
            E=0.
            DO I2=1,MB
                    E=E+C(I2)*FLOAT(IX(I2,ITOUS+I1))
            END DO
C
C       COMPUTE G=G-D*B
C
            DO I2=1,MA
                    DO I3=1,MB
                            BAP(I2,I3)=BAP(I2,I3)-D(I2)*C(I3)/E
                    END DO
            END DO
1           MA=MA+1
            DO I2=1,MB
                    BAP(MA,I2)=0.
                    IF(ITEST.NE.MB) BAP(MA,I2)=C(I2)/E
            END DO
        END DO
        RETURN
        END
C
C
        SUBROUTINE ESTIMATE(Z,Y,NBL,NBC,MA,MB,INTER)
C
C       ARRAYS
C       F         OUTPUT:  ESTIMATES OF PARAMETER
C       Y         INPUT:  DATA
C       Z         INPUT:  GENERALIZED INVERSE
C
        PARAMETER IA=250
        DIMENSION Z(IA,IA),Y(IA),F(IA)
        CHARACTER INTER
C
C       COMPUTE F=Z*Y
C
```

```
      DO I=1,MA
              F(I)=0.
              DO K=1,MB
                      F(I)=F(I)+Z(I,K)*Y(K)
              END DO
      END DO
      WRITE(2,1)
1     FORMAT('1',//,10X,'ESTIMATE OF PARAMETERS',//,15X,'TREATMENTS',/)
      DO I=1,NBL
              WRITE(2,2) F(I)
2             FORMAT(15X,F9.3)
      END DO
      WRITE(2,3)
3     FORMAT(//,15X,'BLOCKS',/)
      DO I=NBL+1,NBL+NBC
              WRITE(2,2) F(I)
      END DO
      IF(INTER.EQ.'n'.OR.INTER.EQ.'N') GOTO 4
      WRITE(2,5)
5     FORMAT(//,15X,'INTERACTION',/)
      DO I=NBC+NBL+1,MA
              WRITE(2,2) F(I)
      END DO
4     RETURN
      END
```

INPUT DATA

```
3 4 1
1 0 1 0
1 3 0 0
0 2 1 2
2.0
4.0
3.0
5.0 6.0 2.0
3.0 6.0
9.0
5.0 4.0
```

INCIDENCE MATRIX

```
1   0   1   0

1   3   0   0

0   2   1   2
```

DESIGN MATRIX

1	0	0	1	0	0	0
1	0	0	0	0	1	0
0	1	0	1	0	0	0
0	1	0	0	1	0	0
0	1	0	0	1	0	0
0	1	0	0	1	0	0
0	0	1	0	1	0	0
0	0	1	0	1	0	0
0	0	1	0	0	1	0
0	0	1	0	0	0	1
0	0	1	0	0	0	1

GENERALIZED INVERSE

0.41	0.43	-0.28	0.04	0.04	0.04	0.00	0.00	-0.30	0.06	0.06
-0.18	0.01	0.30	0.18	0.18	0.18	-0.20	-0.20	0.11	0.06	0.06
-0.01	-0.16	0.13	-0.10	-0.10	-0.10	0.21	0.21	0.29	0.06	0.06
0.39	-0.22	0.49	-0.11	-0.11	-0.11	0.10	0.10	0.09	-0.06	-0.06
0.11	0.06	-0.23	0.13	0.13	0.13	0.24	0.24	-0.18	-0.06	-0.06
-0.20	0.37	0.08	0.03	0.03	0.03	-0.11	-0.11	0.51	-0.06	-0.06
0.01	0.16	-0.13	0.10	0.10	0.10	-0.21	-0.21	-0.29	0.44	0.44

ESTIMATE OF PARAMETERS

TREATMENTS

0.071
2.657
3.485

BLOCKS

1.136
1.412
4.722
1.015

INPUT DATA

```
3 4 1
1 0 1 0
1 3 0 0
0 2 1 2
2.0
4.0
3.0
5.0 6.0 2.0
3.0 6.0
9.0
5.0 4.0
```

INCIDENCE MATRIX

1	0	1	0
1	3	0	0
0	2	1	2

DESIGN MATRIX

1	0	0	1	0	0	0	1	0	0	0	0	0	0
1	0	0	0	0	1	0	0	1	0	0	0	0	0
0	1	0	1	0	0	0	0	0	1	0	0	0	0
0	1	0	0	1	0	0	0	0	0	1	0	0	0
0	1	0	0	1	0	0	0	0	0	1	0	0	0
0	1	0	0	1	0	0	0	0	0	1	0	0	0
0	0	1	0	1	0	0	0	0	0	0	1	0	0
0	0	1	0	1	0	0	0	0	0	0	1	0	0
0	0	1	0	0	1	0	0	0	0	0	0	1	0
0	0	1	0	0	0	1	0	0	0	0	0	0	1
0	0	1	0	0	0	1	0	0	0	0	0	0	1

GENERALIZED INVERSE

0.00	0.83	0.13	-0.10	-0.10	-0.10	0.21	0.21	-0.71	0.06	0.06
0.00	-0.17	0.12	0.24	0.24	0.24	-0.29	-0.29	0.29	0.06	0.06
0.00	-0.17	0.13	-0.10	-0.10	-0.10	0.21	0.21	0.29	0.06	0.06
0.00	0.17	0.88	-0.24	-0.24	-0.24	0.29	0.29	-0.29	-0.06	-0.06
0.00	0.17	-0.12	0.10	0.10	0.10	0.29	0.29	-0.29	-0.06	-0.06
0.00	0.17	-0.13	0.10	0.10	0.10	-0.21	-0.21	0.71	-0.06	-0.06
0.00	0.17	-0.13	0.10	0.10	0.10	-0.21	-0.21	-0.29	0.44	0.44
1.00	-1.00	-1.00	0.33	0.33	0.33	-0.50	-0.50	1.00	0.00	0.00
0.00	0.00	0.00	0.00	0.00	0.00	0.00	0.00	0.00	0.00	0.00
0.00	0.00	0.00	0.00	0.00	0.00	0.00	0.00	0.00	0.00	0.00
0.00	0.00	0.00	0.00	0.00	0.00	0.00	0.00	0.00	0.00	0.00
0.00	0.00	0.00	0.00	0.00	0.00	0.00	0.00	0.00	0.00	0.00
0.00	0.00	0.00	0.00	0.00	0.00	0.00	0.00	0.00	0.00	0.00
0.00	0.00	0.00	0.00	0.00	0.00	0.00	0.00	0.00	0.00	0.00

ESTIMATE OF PARAMETERS

TREATMENTS

-1.493
3.340
3.507

BLOCKS

-0.340
0.993
5.493
0.993

INTERACTION

3.833
0.000
0.000
0.000
0.000
0.000
0.000

CHAPTER 9

Minimally Connected Factorial Experiments and the Problem of Selecting a Factorial Experiment

Alas, that Spring should vanish with the Rose
That Youth's sweet-scented Manuscript should close
The Nightingale that in the Branches sang,
Ah, whence, and whither flown again, who knows.

KHAYYAM NAISHAPURI-RUBAIYAT

9.1 INTRODUCTION

In 2^n factorial experiments, as the number of factors increases, the number of treatment combinations required in a full experiment becomes so large that the experiment may not be feasible. Designs such as regular 2^n fractional factorials and their derivatives are already available for running experiments with only a fraction of the full number of treatment combinations. In this chapter we present some irregular designs with even smaller numbers of points. Irregular designs can also be beneficial in situations where certain expensive treatment combinations should be avoided.

When an experiment is designed, the experimenter must decide what parameters to include in the model and which combinations of levels of the factors to observe. The choice of the parameters depends on what the experimenter is willing to assume about the interactions among the factors, and the choice of the interactions to be included in the model often influences which combinations are to be observed.

Suppose the experimenter's interest lies in obtaining information on a given set of parameters using a factorial design D. In practice, the number of observations in D is subject to the number of parameters to be estimated and economical and physical constraints. For a given set of parameters, there may be many choices for the treatment combinations in D which will allow the desired parameters to be estimated. To be more precise, for a given linear model on the factor space, there is a class of d competing designs $D_1, D_2, \ldots, D_d$ which lead to the same information on the estimability of the given set of parameters.

Our primary concern in this chapter is to exhibit the class of all possible choices of distinct designs which lead to the estimability of the main effects in a 2^n or an $a \times b$ factorial experiment with the minimal number of observations when none of the interactions are present in the model. In order to investigate the estimability of main effects in 2^n factorial experiments, we provide a simplified version of the algorithms presented in Chapter 7. Using this algorithm, we are able to provide the class of all *minimally connected* designs. A catalogue of such designs is furnished at the end of this chapter. We also give a computer program for generating minimally connected $a \times b$ designs, based on the R process presented in Chapter 5.

We begin this chapter by considering the estimability of the main effects in 2^n factorial experiments. Following an algorithm which finds the estimable contrasts involving a single effect, via an example, we show how the estimability of the main effects can be reduced to designs with fewer factors. Later we consider the effect of changing the position of a missing cell on the estimability of the main effects. The class of minimally connected designs will be considered next. Finally, we give a selection criterion which compares the efficiency of two designs with equal numbers of observations. This measure is based on estimable functions of interest and takes into account the precision with which these functions have been estimated.

9.2 ESTIMABILITY

Let $Y_{ijk\ldots w}$ be a collection of N independently distributed random variables each having a common variance σ^2 and each having an expectation of the form

$$E(Y_{ijk\ldots we}) = \mu + \alpha_i + \beta_j + \gamma_k + \cdots + \eta_w \tag{9.1}$$

where $i, j, \ldots, w$ range from 1 to 2 and for a given $i, j, k, \ldots, w$ the index e ranges from 1 to $n_{ijk\ldots w}$. If $n_{ij\ldots w} = 0$, no random variables with the first n subscripts $i, j, k, \ldots, w$ occur in the collection. We assume a fixed n-way classification model with each factor at two levels, that is, 2^n factorial experiment without interaction.

Given the above assumptions, the first problem is to find the rank of the design matrix and consequently the degree of freedom for each effect when the experimenter is faced with a completely arbitrary missing experiment.

Let $\mathbf{N}=(n_{ijk\ldots w})$ be a $2\times 2\times \cdots \times 2$ incidence matrix consisting of 2^n treatment combinations of (9.1), that is, given n factors each at two levels coded by 1 and 2. Then, of course, if all treatment combinations are available (all entries of the incidence matrix are filled) then there is 1 degree of freedom for each effect. But when there are missing points the problem is to find whether the design matrix is of maximal rank. In the event that the design is not of maximal rank then not all parametric functions are estimable. It is then important to find which parametric functions are estimable.

We first consider a 2^2 factorial experiment with one missing point and later we generalize the problem to the case of 2^n factorial experiment with completely arbitrary pattern.

2^2 Factorial Experiment with One Missing Point Cell

For an additive 2^2 factorial arrangement with arbitrary incidence matrix $\mathbf{N}$ (see Table 9.1), it is assumed that Y_{ijk} is a collection of independent and normally distributed random variables each having a common unknown variance σ^2 and each having the expectation of the form

$$E(Y_{ijk})=\mu+\alpha_i+\beta_j \tag{9.2}$$

where $i=1, 2, j=1, 2$ and $k=1,\ldots, n_{ij}$ with the usual interpretation that when $n_{ij}=0$ no random variables with the first two subscripts i, j occur in the collection. Note that no restrictions are imposed upon the unknown parameters occurring in (9.2). It is further assumed, and without loss in generality, that

$$n_{i.}=\sum_j n_{ij}\neq 0 \quad \text{for } i=1, 2,$$

$$n_{.j}=\sum_i n_{ij}\neq 0 \quad \text{for } j=1, 2.$$

Table 9.1 Typical Pattern of 2^2 Factorial Experiment with One Missing Cell

		Level of β factor 1	Level of β factor 2
Level of α factor	1	n_{11}	n_{12}
	2	n_{21}	—

Thus we are assuming a fixed effects two-way classification model without interaction, and each factor has two levels; and when the data pattern is viewed in the form of a two-way table we suppose there are n_{ij} observations in cell (i, j) and each row as well as each column, has at least one observation.

THEOREM 9.1 If 2^n designs has $2^{n-1}+1$ cells filled then all cells are estimable.

Proof For $n=2$: $2^{2-1}+1=3$, then we have

	β_1	β_2
α_1	1	1
α_2	1	

assumed true for $n-1$ proceed by induction. Consider the two levels of the last factor. They are both 2^{n-1} design. Then one must have at least $2^{n-2}+1$ cells filled (if both have 2^{n-2} then the whole design would have 2^{n-1}).

By induction this level has all cells estimable. Then the other level has at least one filled cell. So all cells are estimable. □

Given the above theorem we see that in the model given above with three cells filled and one missing we have no problem of finding an estimate for the missing ones. If it is given that we have the following cells $(\mu+\alpha_1+\beta_1)$, $(\mu+\alpha_1+\beta_2)$, and $(\mu+\alpha_2+\beta_1)$ we can obtain an estimate for $(\mu+\alpha_2+\beta_2)$ by

$$-(\mu+\alpha_1+\beta_1)+(\mu+\alpha_2+\beta_1)+(\mu+\alpha_1+\beta_2)=\mu+\alpha_2+\beta_2.$$

The position of the missing cell does not affect the result. As a result, we have 1 degree of freedom for each effect.

In the event that two cells are missing, two situations could happen:

1 When the occupied cells are from the same factor.

2 When the occupied cells are from both factors.

In case 1, we have only 1 degree of freedom for that particular factor, and in case 2 nothing is estimable.

In case 1 we may have the following situations:

$$\{11, 21\} \quad \text{or} \quad \{11, 12\} \quad \text{or} \quad \{12, 22\} \quad \text{and finally } \{21, 22\}.$$

In all the above situations we actually have a 2^1 factorial arrangements and the problem of 2^2 has been reduced to a 2^1 factorial. For example, 11, 21 is in fact {1, 2} and the other factor can be assumed not existant. Therefore we have

$$(\mu+\alpha_1+\beta_1)-(\mu+\alpha_2+\beta_1)=\alpha_1-\alpha_2 \tag{9.3}$$

which is the same as

$$(\mu+\alpha_1)-(\mu+\alpha_2)=\alpha_1-\alpha_2.$$

So we have 1 degree of freedom for factor A. This matter is true for other situations.

In case 2 we have {12, 21} or {11, 22}. The design matrix corresponding to {11, 22} is as follows:

$$\begin{array}{c} \\ \mathbf{X}= \end{array}\begin{array}{c} \begin{array}{ccccc} \mu & \alpha_1 & \alpha_2 & \beta_1 & \beta_2 \end{array} \\ \begin{bmatrix} 1 & 1 & 0 & 1 & 0 \\ 1 & 0 & 1 & 0 & 1 \end{bmatrix} \end{array}.$$

After a row-reduction we will have

$$\mathbf{X}=\begin{bmatrix} 1 & 1 & 0 & 1 & 1 \\ 0 & 1 & -1 & 1 & -1 \end{bmatrix}.$$

The rank of $\mathbf{X}$ is 2 and we have the following table of degrees of freedom:

Source of Variation	Degrees of Freedom
Mean	1
A	0
B	0
Residual	$N-r(\mathbf{X})=2-2=0$
Total	$N=2$

As we see from the above table, we are missing 1 degree of freedom. The missing degree of freedom has the form

$$\alpha_1-\alpha_2+\beta_1-\beta_2$$

or inseparable contrast. This contrast is the estimate of both A and B effect but none separately. Thus the correct table of degrees of freedom is as follows:

Source of Variation	Degrees of Freedom
Mean	1
A	0
B	0
Inseparable	1
Residual	0
Total	2

2^n Factorial Experiment with Arbitrary Patterns

Algorithm We now give an algorithm for finding degrees of freedom for each effect in a 2^n factorial experiment with an arbitrary pattern. We employ the incidence matrix $\mathbf{N}=(n_{ijk\ldots w})$ which is an n-dimensional matrix. The algorithm consists of the following steps:

Step 1 Begin to form the matrix $\mathbf{M}$ by changing every nonzero entry of $\mathbf{N}$ to 1.

Step 2 For $t \neq h$ if $m_{ijk\ldots zh}=1$ and $m_{ijk\ldots zt}=1$ for some $i, j, k, \ldots, z$ then $\eta_1-\eta_2$ is estimable. That is, if for any two occupied cells all but one factor is at the same level, then that particular factor is estimable, and there is 1 degree of freedom for that effect. For example, in a 2^4 factorial if $\mu+\alpha_2+\beta_2+\gamma_2+\delta_1$ and $\mu+\alpha_2+\beta_2+\gamma_2+\delta_2$ are both estimable, their difference $\delta_1-\delta_2$ is estimable. We refer to these as *directly* estimable factors.

Step 3 For t and h as in step 2 define a 2^{n-1} submatrix $\mathbf{M}_t$ of $\mathbf{M}$ by

$$m_{ijk\ldots zt}=\begin{cases}1 & \text{if } m_{ijk\ldots zh}=1 \quad \text{or} \quad m_{ijk\ldots zt}=1\\ 0 & \text{otherwise.}\end{cases}$$

Eliminate the submatrix $\mathbf{M}_h$.

REMARK At the above step if $\eta_1-\eta_2$ is estimable, then the design will be reduced to a design with one less factor. Thus we can eliminate one submatrix and keep the other one. Since there are only two levels for each factor the last index can also be eliminated.

Step 4 For the new incidence matrix, repeat steps 2 and 3 until no more changes can be made in the matrix $\mathbf{M}$. Collect all direct differences that can be obtained from steps 2 and 4. Thus these constitute D, the set of direct estimable factors. If D contains s elements the problem has been reduced to a 2^{n-s} factorial.

Step 5 Let $\theta_{(k,r,\ldots,t)}=\gamma_k+\delta_r+\cdots+\eta_t$. For $(k, r, \ldots, t)\neq(u, v, \ldots, h)$, if $m_{fgk\ldots t}=1$ and $m_{fguv\ldots h}=1$ for some f, g, then $\theta_{(k,r,\ldots,t)}-\theta_{(u,v,\ldots,h)}$ is estimable. At this step we are actually eliminating μ, α, and β effects. This, however, could be done for any two effects; for simplicity we took the first two. We refer to these as "direct θ differences."

Step 6 For $(k, r, \ldots, t)$ and $(u, v, \ldots, h)$ as in step 5, redefine the submatrix $\mathbf{M}_{(k,r,\ldots,t)}$ of $\mathbf{M}$ by

$$m_{ijk\ldots t}=\begin{cases}1 & \text{if } m_{ijk\ldots t}=1 \quad \text{or} \quad m_{ijuv\ldots h}=1\\ 0 & \text{otherwise.}\end{cases}$$

Eliminate the submatrix $\mathbf{M}_{(u,v,\ldots,h)}$.

Step 7 Repeat steps 5 and 6 until no more changes can be made in the matrix $\mathbf{M}$. Collect all direct θ differences that can be obtained by steps 5–7. These contrasts form a set E.

Note that at the end of step 7 we are left with at most four 2×2 submatrices.

Step 8 Form a matrix $\bar{\mathbf{M}}$ such that

$$\bar{m}_{ij}=\begin{cases}1 & \text{if } m_{kr\ldots t}=1 \quad \text{or} \quad m_{ijk\ldots t}=1\\ 0 & \text{otherwise.}\end{cases}$$

The above matrix is a 2×2 matrix.

Step 9 If all cells in $\bar{\mathbf{M}}$ are occupied cells then there is a loop (1, 1) (1, 2) (2, 2) (2, 1). Corresponding to this loop we have a contrast of the form

$$\begin{aligned}(\mu+\alpha_1+\beta_1+\gamma_{k_1}+\cdots+\gamma_{t_1})-(\mu+\alpha_1+\beta_2+\gamma_{k_2}+\cdots+\eta_{t_2})\\ +(\mu+\alpha_2+\beta_2+\gamma_{k_3}+\cdots+\eta_{t_3})-(\mu+\alpha_2+\beta_2+\gamma_{k_4}+\cdots+\eta_{t_4})\\ =\gamma_{k_1}+\cdots+\eta_{t_1}-\gamma_{k_2}-\cdots-\eta_{t_3}+\gamma_{k_3}+\cdots+\eta_{t_3}-\gamma_{k_4}-\cdots-\eta_{t_4}.\end{aligned}$$

We shall call this contrast G. If $\bar{\mathbf{M}}$ has any empty cell no such contrast is available. In this case G is empty.

THEOREM 9.2 Let $\bar{\mathscr{G}}$ be the vector space of estimable θ contrasts, and let $\bar{\mathscr{D}}$, $\bar{\mathscr{E}}$, and $\bar{\mathscr{W}}$ be the vector spaces spanned by the sets, D, E, and G respectively, then $\bar{\mathscr{G}}=\bar{\mathscr{D}}\oplus\{\bar{\mathscr{E}}+\bar{\mathscr{W}}\}$.

Proof Evidently every effect in $\bar{\mathscr{D}}$, $\bar{\mathscr{E}}$, or $\bar{\mathscr{W}}$ is estimable; since indices for the factors corresponding to the elements of D have been suppressed it follows that $\bar{\mathscr{D}} \cap \{\bar{\mathscr{E}} + \bar{\mathscr{W}}\} = \varnothing$. If $m_{ijk\ldots t_1} = 1$, and $m_{ijk\ldots t_2} = 1$, then $\eta_1 - \eta_2$ is directly estimable. Once we have $\eta_1 - \eta_2$ we reduce the estimability problem to a model with one less factor by keeping the submatrix for η effect at level one, and replacing every zero entry by 1 if the corresponding entry in the submatrix for η effect at level two is 1. Here we do not lose any cell expectations, since $m_{ijk\ldots t_1} = 1$ and we have $\eta_1 - \eta_2$, so

$$\mu + \alpha_i + \beta_j + \gamma_k + \cdots + \eta_t + \eta_2 = \mu + \alpha_i + \beta_j + \gamma_k + \cdots + \eta_t + \eta_1 - (\eta_1 - \eta_2). \tag{9.4}$$

This argument can be repeated to cover all elements of D.

At the next stage of our algorithm if $m_{ijk\ldots p} = 1$ and $m_{iju,\ldots,h} = 1$, we discard α, β submatrix for $(u, \ldots, h)$ and transfer any nonzero entries to the corresponding position in the α, β submatrix for $(k, \ldots, p)$. Clearly $\theta_{(k,\ldots,p)} - \theta_{(u,\ldots,h)}$ which is estimable has been transferred to the set E, and hence it is not lost. Any difference between the cell expectations for the nonzero entries in the α, β submatrix for $(u, \ldots, h)$ will be the same as the difference between the cell expectations of the corresponding entries in the new α, β submatrix for $(k, \ldots, p)$, and hence will continue to be available. Thus we have not altered the space of estimable parametric functions at any stage of the algorithm. Since at the end of the algorithm the estimation consists of $\bar{\mathscr{D}} \oplus \{\bar{\mathscr{E}} + \bar{\mathscr{W}}\}$ the proof follows. □

We now give two examples in order to show how the algorithm works. We will see how the estimability problem can be reduced to models with fewer factors. Also we provide some miscellaneous shortcuts that we find useful in many cases. These shortcuts bypass some of the steps of the algorithm.

EXAMPLE 9.1 Consider a 2^5 factorial design with the following incidence matrix:

$\mathbf{N}_1 =$

	η_1							
	δ_1				δ_2			
	γ_1		γ_2		γ_1		γ_2	
	β_1	β_2						
α_1	1							1
α_2			1		1			

,

$N_2=$

	η_2							
	δ_1				δ_2			
	γ_1		γ_2		γ_1		γ_2	
	β_1	β_2						
α_1								1
α_2		1						

Step 1 Since all the nonzero entries of **N** are equal to 1, thus $\mathbf{M}=\mathbf{N}$.

Step 2 Because $m_{12221}=1$ and $m_{12222}=1$, then $\eta_1-\eta_2$ is estimable.

Step 3 Form the new matrix **M** by eliminating the second level of η and replacing the nonzero entries of the second level of η into the appropriate cell in η_1. The set $D=\{\eta_1-\eta_2\}$, and we have the following pattern:

	δ_1				δ_2			
	γ_1		γ_2		γ_1		γ_2	
	β_1	β_2						
α_1	1							1
α_2		1	1		1			

Step 4 Since $m_{2121}=1$, and $m_{2112}=1$, then their differences $\mu+\alpha_2+\beta_1+\gamma_2+\delta_1-(\mu+\alpha_2+\beta_1+\gamma_1+\delta_2)=\gamma_2+\delta_1-\gamma_1-\delta_2$ are directly estimable. We see that there are no more direct differences of this form, and therefore the set $E=\{\gamma_2+\delta_1-\gamma_1-\delta_2\}$.

Step 5 Eliminate the submatrix $\mathbf{M}_{21}$, and keep $\mathbf{M}_{12}$. Now we are left with the following submatrices of **M**:

$\mathbf{M}_{11}=$

1	
	1

, $\mathbf{M}_{12}=$

1	

, $\mathbf{M}_{22}=$

	1

.

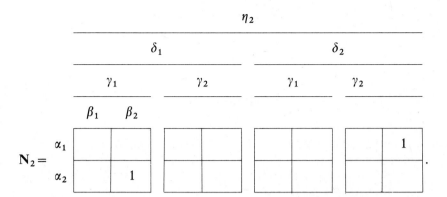

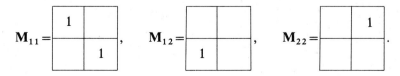

Step 6 Form the matrix $\bar{\mathbf{M}}$ as defined in algorithm,

(1, 1)	(2, 2)
(1, 2)	(1, 1)

Since all cells in $\bar{\mathbf{M}}$ are filled they form a loop from which we obtain

$$(\gamma_1+\delta_1)-(\gamma_2+\delta_2)+(\gamma_1+\delta_1)-(\gamma_1+\delta_2)=\gamma_1-\gamma_2+2\delta_1-2\delta_2.$$

Note that at this step we are eliminating η, α, and β effects. The set $G=\{\gamma_1-\gamma_2+2\delta_1-2\delta_2\}$. Thus we have

$$D=\{\eta_1-\eta_2\},$$

$$E=\{\gamma_2-\gamma_1+\delta_1-\delta_2\},$$

$$G=\{\gamma_1-\gamma_2+2\delta_1-2\delta_2\}.$$

By Theorem 9.2, $\bar{\mathscr{G}}$ is spanned by D, E, and G. After reduction we see that $\bar{\mathscr{G}}$ is spanned by $\{\gamma_1-\gamma_2, \delta_1-\delta_2, \eta_1-\eta_2\}$. Thus dim $\bar{\mathscr{G}}=3$, and $r(\mathbf{X})=r(\mathbf{A}, \mathbf{B})+\dim \bar{\mathscr{G}}=3+3=6$, which implies $\mathbf{X}$ is of maximal rank. The rank of $(\mathbf{A}, \mathbf{B})$ is obtained by regarding it as the design matrix of an additive two-way classification model in which $\mathbf{A}$ and $\mathbf{B}$ are the submatrices associated with α and β effects, respectively.

EXAMPLE 9.2 In order to show how a problem can be reduced to a smaller one, let us consider a 2^4 factorial experiment with the following pattern:

$\mathbf{N}_1 =$

	δ_1			
	γ_1		γ_2	
	β_1	β_2	β_1	β_2
α_1	1			1
α_2		1		

$\mathbf{N}_2 =$

	δ_2			
	γ_1		γ_2	
	β_1	β_1	β_1	β_2
α_1	1			
α_2			1	

We see that from cells (1, 1, 1, 1) and (1, 1, 1, 2), $\delta_1 - \delta_2$ is directly estimable. Consider any occupied cells in $\mathbf{N}_2$, such as (2, 1, 2, 2). We see that

$$\mu + \alpha_2 + \beta_1 + \gamma_2 + \delta_1 = \mu + \alpha_2 + \beta_1 + \gamma_2 + \delta_2 + (\delta_1 - \delta_2)$$

is estimable; we can indicate this by placing a 1 in cell (2, 1, 2, 1).

Now we can place a zero in cell (2, 1, 2, 2) without losing any information about estimability, because

$$\mu + \alpha_2 + \beta_1 + \gamma_2 + \delta_1 = \mu + \alpha_2 + \beta_1 + \gamma_2 + \delta_1 - (\delta_1 - \delta_2).$$

In general, knowing that $\delta_1 - \delta_2$ is estimable, we can change the entries in all occupied cells on $\mathbf{N}_2$ to zero if we place a 1 in the corresponding cells in $\mathbf{N}_1$. Thus we can eliminate $\mathbf{N}_2$ and reduce the problem to a new model with only three effects. The new incidence matrix $\mathbf{N}$ obtained from $\mathbf{N}_1$ as just described, is shown below:

$\mathbf{N} =$

	γ_1		γ_2	
	β_1	β_2	β_1	β_2
α_1	1			1
α_2		1	1	

Now $\mathbf{N}$ is a 2×2 matrix and it follows immediately that $\gamma_1 - \gamma_2$ is estimable. We see this by forming the matrix $\bar{\mathbf{M}}$.

$\bar{\mathbf{M}} =$

1	2
2	1

Now we see that there is a loop, namely (1, 1)(1, 2)(2, 2)(2, 1) on $\bar{\mathbf{M}}$. We get $2(\gamma_1 - \gamma_2)$ to be estimable. Therefore we can now work with the following pattern by noting that γ levels have been dropped from the model:

	β_1	β_2
α_1	1	1
α_2	1	1

Again the problem can be considered as a two-way model, from which it immediately follows that all estimable differences of α's and β's exist.

It is important to notice that each $\mathbf{N}_t$ can be considered as the design matrix of a three-way classification model, so it is sometimes wise and efficient to see if we can find some estimable contrasts from each submatrices.

9.3 THE EFFECT OF CHANGING THE POSITION OF AN OCCUPIED CELL

In a 2^n factorial experiment with n factors each at two levels we need $n+1$ cells filled for the design matrix to be of maximal rank *given that* these filled cells have some certain positions. We consider a 2^6 factorial experiment with 57 missing cells in order to demonstrate the following facts, keeping in mind that we are only concerned about the *main effects*. These facts are as follows:

1 The effect of changing the position of a occupied cell on the rank of the design matrix, and consequently the estimability of the main effects, may be dramatic.

2 After finding some estimable contrasts, the problem of estimability may be reduced to considering a design with fewer factors.

3 The determination of $r(\mathbf{X})$ in a 2^n factorial experiment is relatively simple compared to other factorial designs that have been discussed before.

We consider the linear model

$$E(Y_{ijktuve}) = \mu + \alpha_i + \beta_j + \gamma_k + \delta_t + \tau_u + \xi_v$$
$$i, j, k, t, u, v = 1, 2$$

where μ is a mean, and $\alpha_i, \beta_j, \ldots, \xi_v$ are the main effects. We assume therefore none of the interactions are present. The incidence matrix $\mathbf{N}$ is presented in pattern 9.1.

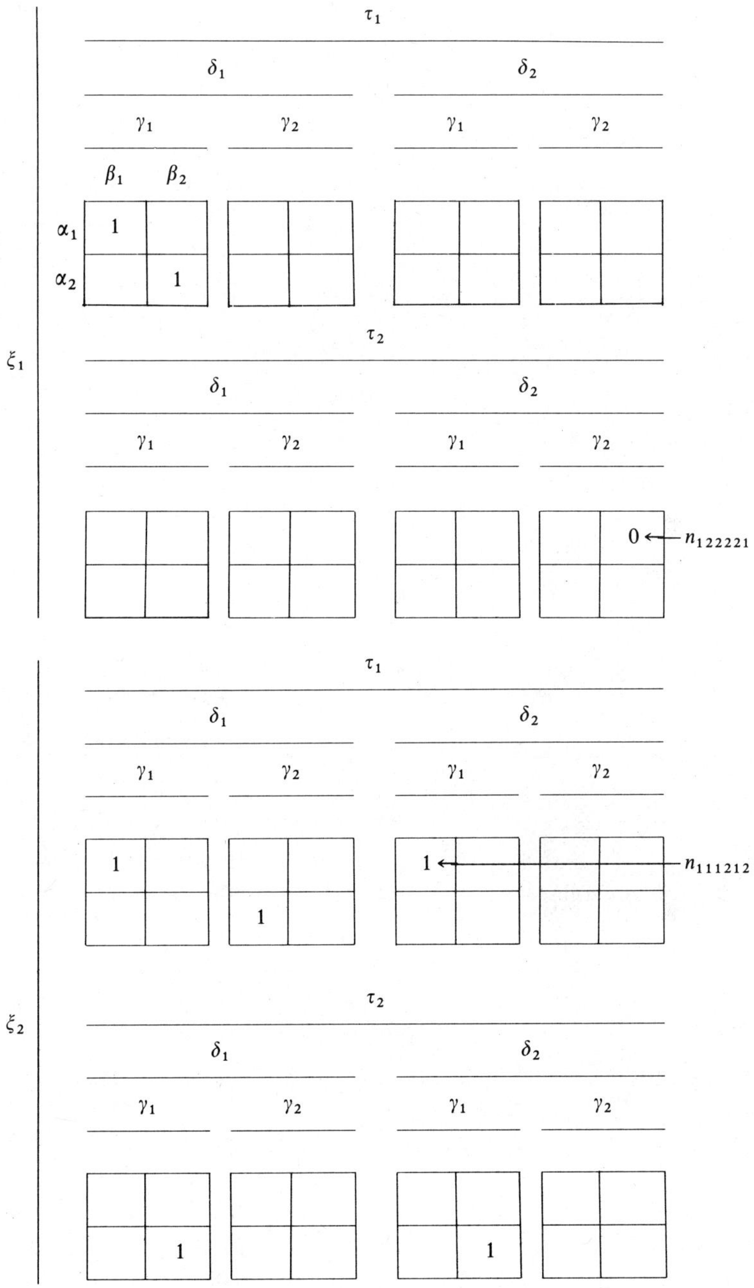
ξ_1
ξ_2
τ_1
τ_2
δ_1
δ_2
γ_1
γ_2
β_1
β_2
α_1
α_2
n_{122221}
n_{111212}

Pattern 9.1

Since $n_{111111} \neq 0$ and $n_{111112} \neq 0$, $\xi_1 - \xi_2$ is directly estimable. Now the problem collapses to a 2^5 factorial having the following pattern:

τ_1

	δ_1				δ_2			
	γ_1		γ_2		γ_1		γ_2	
	β_1	β_2						
α_1	1				1			
α_2		1	1					

τ_2

	δ_1				δ_2			
	γ_1		γ_2		γ_1		γ_2	
	β_1	β_2						
α_1								
α_2		1				1		

Since $n_{22111} \neq 0$ and $n_{22112} \neq 0$, $\tau_1 - \tau_2$ is directly seen to be estimable. Now the problem collapses to a 2^4 factorial having the following pattern:

	δ_1				δ_2			
	γ_1		γ_2		γ_1		γ_2	
	β_1	β_2						
α_1	1				1			
α_2		1	1			1		

We see that $n_{1111} \neq 0$ and $n_{1112} \neq 0$ which implies that $\delta_1 - \delta_2$ is directly estimable and therefore the problem can be considered as a 2^3 factorial having the following pattern:

	γ_1: β_1	γ_1: β_2	γ_2: β_1	γ_2: β_2
α_1	1			
α_2		1	1	

We are left with no more direct differences. Now we form the matrix $\bar{\mathbf{M}}$.

$\bar{\mathbf{M}}=$

1	
2	1

Applying the Q process to the above matrix $\bar{\mathbf{M}}$, we see that there is no loop, which implies there is no estimable γ contrast.

In order to find degrees of freedom for α we switch the role of γ with α. In doing this we have the following pattern:

	α_1: β_1	α_1: β_2	α_2: β_1	α_2: β_2
γ_1	1			1
γ_2			1	

Because there is no direct difference we form the matrix $\bar{\mathbf{M}}$.

$\bar{\mathbf{M}}=$

1	2
2	

Applying the Q process to the above matrix $\bar{\mathbf{M}}$, we see that there is no loop, which implies there is no estimable α contrast. By the same procedure, we we find no β contrast. Therefore the $r(\mathbf{X})=r(\mathbf{A}, \mathbf{B})+\dim \bar{\mathscr{G}}=3+3=6$, which is not of maximal rank, and the table of degrees of freedom is as follows:

Source of Variation	Degrees of Freedom
Mean	1
α	0
β	0
γ	0
δ	1
τ	1
ξ	1
Confounded	2
Residual	1
Total	7

Now let us change the position of the cell (1, 1, 1, 2, 1, 2) to (1, 2, 2, 2, 2, 1). By this we mean having *no* observation in the position (1, 1, 1, 2, 1, 2) and one or more observations in the position (1, 2, 2, 2, 2, 1). Therefore $n_{111212}=0$ and $n_{122221}\neq 0$. See Pattern 9.2.

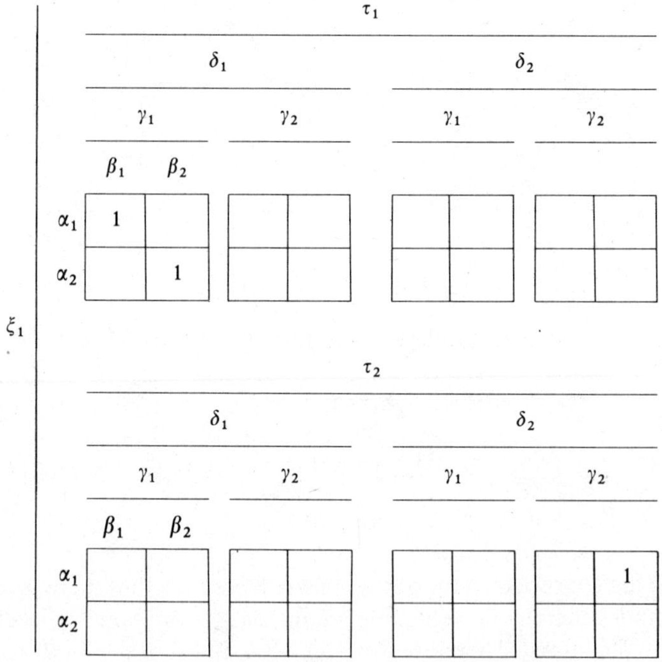

ξ_2, τ_1:

	δ_1				δ_2			
	γ_1		γ_2		γ_1		γ_2	
	β_1	β_2						
α_1	1				0			
α_2				1				

ξ_2, τ_2:

	δ_1				δ_2			
	γ_1		γ_2		γ_1		γ_2	
	β_1	β_2						
α_1								
α_2		1				1		

Pattern 9.2

Because $n_{111111} \neq 0$ and $n_{111112} \neq 0$, then $\xi_1 - \xi_2$ is directly estimable and the problem collapses to a 2^5 factorial having the following pattern:

τ_1:

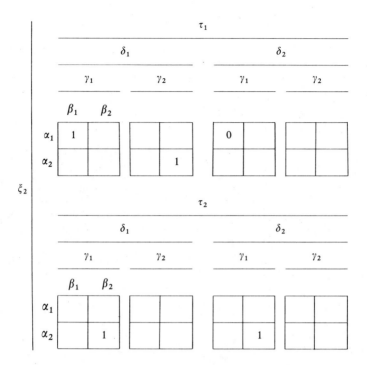

	δ_1				δ_2			
	γ_1		γ_2		γ_1		γ_2	
	β_1	β_2						
α_1	1							
α_2		1	1					

	τ_2							
	δ_1				δ_2			
	γ_1		γ_2		γ_1		γ_2	
	β_1	β_2						
α_1								1
α_2		1				1		

Since $n_{22111} \neq 0$ and $n_{22112} \neq 0$, $\tau_1 - \tau_2$ is directly estimable and the problem collapses to a 2^4 factorial having the following pattern:

	δ_1				δ_2			
	γ_1		γ_2		γ_1		γ_2	
	β_1	β_2						
α_1	1							1
α_2		1	1			1		

We see that $n_{2211} \neq 0$ and $n_{2212} \neq 0$, and therefore $\delta_1 - \delta_2$ is seen to be directly estimable. Now the problem collapses to a 2^3 factorial with the following pattern:

	γ_1		γ_2	
	β_1	β_2		
α_1	1			1
α_2		1	1	

We are left with no direct differences. We now form the matrix $\bar{\mathbf{M}}$.

$\bar{\mathbf{M}} =$

1	2
2	1

By applying the Q process we find that $2\gamma_1 - 2\gamma_2$ is estimable and thus the problem collapses to a 2^2 factorial with the following pattern:

	β_1	β_2
α_1	1	1
α_2	1	1

We see that $\alpha_1 - \alpha_2$ and $\beta_1 - \beta_2$ are directly estimable. The design matrix $\mathbf{X}$ is of maximal rank 7 and the table of degrees of freedom is as follows:

Source of Variation	Degrees of Freedom
Mean	1
α	1
β	1
γ	1
δ	1
τ	1
ξ	1
Residual	0
Total	7

Again note how the *position* of a filled cell changes the estimability of each effect and consequently the rank of the design matrix. Moreover, only with 7 filled cells and 57 missing cells, all factors are estimable, and this is the minimum number of filled cells needed for the maximality of $r(\mathbf{X})$. In Section 9.4 we consider the construction of such designs.

9.4 MINIMALLY CONNECTED DESIGNS

Consider a 2^n factorial experiment with the additive model

$$E(Y_{ij\ldots w}) = \mu + \alpha_i + \beta_j + \cdots + \eta_w,$$
$$0 = \alpha_{.} = \beta_{.} = \cdots = \eta_{.}, \tag{9.5}$$

where μ is a mean and $\alpha_i, \beta_j, \ldots,$ and η_w are the main effects, $i, j, \ldots, w = 1, 2$.

In matrix notation, we can write (9.5) as

$$E(\mathbf{Y}) = \mathbf{X}\boldsymbol{\theta}, \qquad \boldsymbol{\Delta}'\boldsymbol{\theta} = \mathbf{0} \tag{9.6}$$

where **X** is the $N \times (2n+1)$ design matrix, Δ' is the $n \times (2n+1)$ constraint matrix, and $\boldsymbol{\theta}$ is the $(2n+1) \times 1$ vector of unknown parameters.

Let $\mathscr{D}(2^n, m)$ be the class of all 2^n factorial designs such that the number of distinct treatment combinations is a fixed number m.

Since the domain consists of 2^n treatment combinations, it follows that there are $C(2^n, m)$ distinct designs in $\mathscr{D}(2^n, m)$.

EXAMPLE 9.3 Consider a 2^3 factorial. Let the coded levels for the ith factor be indicated by $\{1, 2\}$ for $i = 1, 2, 3$. Assume the linear model associated with this 2^3 factorial is

$$E(Y_{ijk}) = \mu + \alpha_i + \beta_j + \gamma_k$$

$$i, j, k = 1, 2.$$

The complete minimal factorial is

$$\{111, 211, 121, 112, 221, 212, 122, 222\}.$$

Now consider the class $\mathscr{D}(2^3, 4)$ of designs such that each design consists of $m = 4$ distinct treatment combinations. Since the factorial domain consists of eight treatments, it follows that the cardinality of $\mathscr{D}(2^3, 4)$ is equal to $C(8, 4) = 70$ and hence there are 70 competing designs. Each design in $\mathscr{D}(2^3, 4)$ is capable of providing some information relative to $\boldsymbol{\theta}$.

Suppose now that the problem consists of selecting a fractional factorial design consisting of m distinct treatment combinations such that all main effects are estimable.

Let $\mathscr{D}^*(2^n, m)$ be the class of all designs in $\mathscr{D}(2^n, m)$ in which all main effects are estimable.

A 2^n fractional factorial experiment is said to be *minimally connected* if it has $n+1$ observations and all the main effects are estimable.

The class of all minimally connected 2^n designs in $\mathscr{D}_n^* = \mathscr{D}^*(2^n, n+1)$. These designs are in fact the smallest designs that permit the estimation of all independent linear parametric functions in $\boldsymbol{\theta}$ (all main effects and the mean).

Our concern now is to find all such designs in the class $\mathscr{D}_n^*$ from $\mathscr{D}(2^n, n+1)$. Given all 2^n fractional factorial designs in $\mathscr{D}(2^n, n+1)$ we can apply Algorithm 9.1 to each of them to find whether the design is connected or not.

EXAMPLE 9.4 Consider Example 9.3. We have seen in Example 9.3 that there are exactly $\binom{8}{4} = 70$ possible designs in $\mathscr{D}(2^3, 4)$, out of which 58 are of full rank and 12 of less than full rank. Hence $\mathscr{D}_3^*$ consists of 58 designs that are minimally connected 2^3 designs.

A complete list of 58 minimally connected 2^3 is presented in Table 9.4. Notice that the classical $\frac{1}{2}$ fraction of 2^3 is one such design. We have also provided some of the minimally connected designs for $n \leqslant 9$ at the end of this chapter.

Table 9.2 Number of Connected Designs in $\mathscr{D}_n^*$

n	2	3	4	5	6	7
Number of 2^n designs	4	70	4368	906192	621216192	1429702652400
Number of connected designs	4	58	3008	556192	366179200	858240222176

The number of minimal 2^n factorial designs and minimal connected 2^n factorial designs are given in Table 9.2.

As can be seen from Table 9.2, when n increases the number of designs to be generated increases rapidly, and that is why we need some rules and methods in order to reduce the number of designs in the catalogue. The following theorem greatly reduces the number of designs that need to be listed.

THEOREM 9.3 A minimally connected 2^n factorial design remains connected if the levels of any factor are interchanged.

Proof is straightforward and therefore it is omitted.

Another method of generating connected factorial designs with minimal points can be given as follows:

In 2^1, we need two treatments; therefore using the codes 1 and 2 for the first and second levels of each factor we have

$$\begin{matrix} 1 \\ 2 \end{matrix}$$

as the only connected 2^1 design. To obtain a minimally connected 2^2 design we put the pattern of the 2^1 design in the first level of the second factor, thus obtaining the points 11 and 21, and then we add the point 22:

$$\begin{matrix} 11 \\ 21 \\ 22 \end{matrix}$$

Pictorially, we have

		B	
		β_1	β_2
A	α_1	1	
	α_2	1	1

.

Table 9.3 A set of 2^n Minimally Connected Designs

2^1	2^2	2^3	2^4	2^5	2^6	2^7
1	11	111	1111	11111	111111	1111111
2	21	211	2111	21111	211111	2111111
	22	221	2211	22111	221111	2211111
		222	2221	22211	222111	2221111
			2222	22221	222211	2222111
				22222	222221	2222211
					222222	2222221
						2222222

We see that the fourth cell can be estimated by using the three available cells, that is,

$$(\mu+\alpha_1+\beta_2)=(\mu+\alpha_1+\beta_1)-(\mu+\alpha_2+\beta_1)+(\mu+\alpha_2+\beta_2).$$

For a minimally connected 2^3 design, we put the pattern of the preceding 2^2 design in the first level of the third factor and then add the point 222 to obtain

111
211
221
222

This process can be continued to as many factors as we wish, in constructing connected 2^n factorial designs with $n+1$ observations. Thus we have Table 9.3.

Note that these minimally connected designs are only representative of the rest. One can make use of Theorem 9.3 to obtain more minimally connected 2^n factorial designs. (See Table 9.4).

Minimally Connected Block Designs

In block designs the problem of determining minimally connected designs is much simpler than that of 2^n factorial designs. Recall from Chapter 5 that for an additive two-way classification model with arbitrary $a \times b$ incidence matrix $\mathbf{N}$, the design matrix associated with the incidence matrix is of maximal rank, that is, $r(\mathbf{X})=a+b-1$, if and only if the design is connected. Therefore the minimum number of observations that are required for an $a \times b$ block design to be *connected is* $a+b-1$. *Given an* $a \times b$ *block design with* $a+b-1$ *observations, it is connected if and only if after the applying R process to* $\mathbf{N}$ *the*

Table 9.4 All Possible Minimally Connected 2^3 Designs

1	1	1	1	1	1	1	1	1	1	1	1	1	1	1	1	1	1
1	2	1	1	2	1	1	2	1	1	2	1	1	2	1	1	2	1
2	1	1	2	1	1	2	1	1	2	1	1	2	1	2	2	1	2
2	2	2	1	2	2	1	1	2	2	1	2	1	1	2	1	2	2
1	1	1	1	1	1	1	1	1	1	1	1	1	1	1	1	1	1
1	2	1	1	2	1	1	2	1	1	2	1	1	2	1	1	2	1
2	1	2	2	2	2	2	2	2	2	2	2	2	2	1	2	2	1
2	2	1	1	2	2	1	1	2	2	2	1	1	1	2	1	2	2
1	1	1	1	1	1	1	1	1	1	1	1	1	1	1	1	1	1
1	2	2	1	2	2	1	2	2	1	2	2	1	2	2	1	2	2
1	1	2	1	1	2	1	1	2	1	1	2	2	2	1	2	2	1
2	2	2	2	1	2	2	1	1	2	2	1	2	1	1	2	1	2
1	1	1	1	1	1	1	1	1	1	1	1	1	1	1	1	1	1
1	2	2	1	1	2	1	2	2	1	2	2	1	1	2	1	1	2
2	2	1	2	1	1	2	1	2	2	2	2	2	1	2	2	1	1
2	2	2	2	2	1	2	1	1	2	1	2	2	2	2	2	2	2
1	1	1	1	1	1	1	1	1	1	1	1	1	1	2	1	1	1
1	1	2	2	1	1	2	1	1	2	1	1	2	2	1	2	1	2
2	1	2	2	2	2	2	2	1	2	2	1	1	2	2	2	2	2
2	2	1	2	1	2	2	2	2	2	1	2	2	2	2	2	2	1
1	1	2	1	1	2	1	1	2	1	1	2	1	1	2	1	1	2
2	2	1	2	2	1	2	2	1	2	2	1	2	2	1	2	2	1
1	2	2	1	2	1	1	2	1	1	2	2	2	2	2	2	2	2
2	1	2	2	2	2	2	1	1	1	2	1	2	1	2	2	1	1
1	1	2	1	1	2	1	1	2	1	1	2	1	1	2	1	1	2
2	2	1	2	2	2	2	2	2	2	2	2	2	2	2	2	2	2
2	1	2	2	1	1	2	1	1	2	1	1	1	2	1	2	1	2
2	1	1	2	1	2	1	2	1	1	2	2	1	2	2	1	2	1
1	1	2	1	1	2	1	1	2	1	1	2	1	2	2	1	2	2
2	1	2	2	1	1	2	1	2	2	1	2	2	2	2	2	2	2
1	2	2	1	2	1	1	2	1	1	2	2	2	1	1	2	1	1
1	2	1	1	2	2	2	1	1	2	1	1	2	1	2	1	2	1
1	2	2	1	2	2	1	2	2	1	2	2	1	2	2	1	2	2
2	2	2	2	2	1	2	2	1	2	2	1	2	2	1	2	2	1
2	1	2	2	2	2	1	2	1	1	2	1	2	1	2	2	2	2
1	2	1	2	1	2	2	1	2	2	1	1	2	1	1	2	1	1
1	2	1	1	2	1	1	2	1	1	2	1						
2	2	1	2	2	2	2	2	1	2	2	1						
2	1	1	2	1	2	2	1	2	2	1	1						
2	2	2	2	1	1	2	2	2	2	1	2						

final matrix **M** *has no nonzero entries.* Any connected block design having the total number of observations $N=a+b-1$ will be referred to as a minimally connected block design.

Since in practical situations it is useless to have a level of factor which is never used in any treatment combination, we will assume, for an $a\times b$ binary incidence matrix **N**, that each row, as well as each column, has at least one observation.

Although the R process provides a necessary and sufficient condition for connectedness of a block design, we may equivalently state the following theorem.

THEOREM 9.4 The design D is connected if and only if its incidence matrix **N** cannot be partitioned as follows (after permutation of rows and columns):

$$\mathbf{N}=\begin{bmatrix} \mathbf{N}_1 & & & \mathbf{0} \\ & \mathbf{N}_2 & & \\ & & \ddots & \\ \mathbf{0} & & & \mathbf{N}_s \end{bmatrix}, \qquad 1<s\leqslant a,$$

where the $\mathbf{N}_i$'s are the connected portions.

To make this design connected, we would require $s-1$ additional observations in appropriate cells.

Let k_j ($j=1,\ldots,b$) be the number of treatments assigned to blocks j and let r_i ($i=1,\ldots,a$) be the number of times treatment i is replicated.

The following theorem can be useful for construction of minimally connected block designs.

THEOREM 9.5 If $b\geqslant a$, then there does not exist a minimally connected block design with $k_j\geqslant 2$ for all $j=1,\ldots,b$.

Proof To be minimal the design should have $N=a+b-1$. However, $N=\sum_{j=1}^{b} k_j \geqslant 2b\geqslant a+b>a+b-1$. Hence the proof. □

When $k_1=\cdots=k_b=2$, then it is also required to have $a=b+1$ for the design to be connected, because if $a\geqslant b+2$, then $N=2b\leqslant a+2-2<a+b-1$.

From Theorem 9.5 we see that the number of blocks must be less than the number of treatments under the condition $k_j\geqslant 2$ in order to have a minimally connected block design.

Next consider minimally connected designs in the following situations:

1 If $k_j = k$ for all j, then $bk = a + b - 1$ which yields $k = 1 + (a-1)/b$. Hence necessary conditions for the existence of a minimally connected block design with parameters (a, b, r_i, k) are:

(a) $k = 1 + (a-1)/b$.
(b) $k - 1$ divides $a - 1$; b divides $a - 1$.
(c) $b < a$ unless $a = 1$.

2 If $r_i = r$ for all i, then $ar = a + b - 1$, which yields $r = 1 + (b-1)/a$. Hence necessary conditions for the existence of a minimally connected block design with parameters (a, b, r, k_j) are:

(a) $r = 1 + (b-1)/a$.
(b) $r - 1$ divides $b - 1$; a divides $b - 1$.
(c) $a < b$ unless $b = 1$.

From statements (c) of 1 and 2 above it follows immediately that a minimally connected block design with $r_i = r$ and $k_j = k$ does not exist except for the trivial case when $a = 1$ or $b = 1$. This means BIBs are not minimally connected.

Based on Theorem 9.5 and the discussion that followed we now present some minimally connected block designs using the incidence matrix $\mathbf{N}$. Let $\mathbf{e}_k$ denote a k-dimensional vector with 1 everywhere.

1 For $k = 2$ in $b - 1$ blocks and $k = a - b + 1$ in the last block (here $a \geqslant b$), we have

$$\mathbf{N} = \begin{bmatrix} 1 & & & & \mathbf{0} \\ 1 & 1 & & & \\ & 1 & & & \\ & & \ddots & & \\ \mathbf{0} & & & 1 & \\ & & & 1 & 1 \\ & & & & 1 \\ & & & & \vdots \\ \mathbf{0} & & & & 1 \end{bmatrix} \begin{matrix} \\ \\ \\ \\ \\ \left.\begin{matrix} \\ \\ \\ \\ \end{matrix}\right\} a-b+1 \end{matrix}.$$

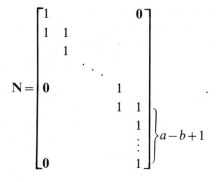

2 For $k=1+(a-1)/b$, and $r_i=1$ or 2,

$$\mathbf{N}=\begin{bmatrix} \mathbf{e}_{k-1} & & & & & \mathbf{0} \\ 1 & 1 & & & & \\ & \mathbf{e}_{k-2} & & & & \\ & 1 & 1 & & & \\ & & \mathbf{e}_{k-2} & & & \\ & & 1 & & & \\ & & & \ddots & & 1 \\ \mathbf{0} & & & & & \mathbf{e}_{k-1} \end{bmatrix}$$

3 For $r=1+(b-1)/a$, and $k_j=1$ or 2

$$\mathbf{N}=\begin{bmatrix} \mathbf{e}'_{r-1} & 1 & & & & & \\ & 1 & \mathbf{e}'_{r-2} & 1 & & & \\ & & & 1 & \mathbf{e}'_{r-2} & 1 & \\ & & & & \ddots & & \ddots \\ & & & & & 1 & \mathbf{e}'_{r-1} \end{bmatrix}.$$

Minimally connected block designs with different r_i and k_j can be constructed similarly.

4 Finally, if in terms of the incidence matrix, we have the following partitioned form (Theorem 9.4)

$$\mathbf{N}=\begin{bmatrix} \mathbf{N}_1 & & & \mathbf{0} \\ & \mathbf{N}_2 & & \\ & & \ddots & \\ \mathbf{0} & & & \mathbf{N}_s \end{bmatrix}$$

where $\mathbf{N}_i$'s are connected portions for all $i=1,\ldots,s$ $(s\geqslant 2)$, and in each $\mathbf{N}_i$ we have $n_i=a_i+b_i-1$ observations, and we want to make this design minimally connected, then we need exactly $s-1$ observations such that any two treatments belonging to two different connected portions are connected. As a special case, when $s=2$, we have

$$\mathbf{N}=\begin{bmatrix} \mathbf{N}_1 & \\ & \mathbf{N}_2 \end{bmatrix}.$$

Then

$$\mathbf{N}=\begin{bmatrix} \mathbf{N}_1 & \mathbf{E} \\ \mathbf{0} & \mathbf{N}_2 \end{bmatrix} \quad \text{or} \quad \mathbf{N}=\begin{bmatrix} \mathbf{N}_1 & \mathbf{0} \\ \mathbf{E} & \mathbf{N}_2 \end{bmatrix}$$

are minimally connected block designs, where $\mathbf{E}$ is a matrix of an appropriate dimension in which only one element is 1 and zeros elsewhere.

In Table 9.5 a complete list of minimally connected 2×3, 2×4 and 3×3 designs are provided. The total number of minimally connected for 2×3, 2×4 and 3×3 are 12, 32 and 81, respectively. For each 5×5, 4×5, 3×3 and 2×3 block design, a list of 10 minimally connected designs is provided as a sample at the end of this chapter. The statement "NUMBER OF MINIMALLY CONNECTED DESIGNS FOUND 10" at the end of each listing signifies that we requested only 10 of such designs.

Table 9.5 All Possible Minimally Connected 2×3, 2×4 and 3×3 Designs

All Possible Minimally Connected 2×3 Designs

13	12	12	12	11	11	11	11	11	11	11	11
21	21	13	13	21	13	13	12	12	12	12	12
22	22	21	21	22	22	21	22	21	13	13	13
23	23	23	22	23	23	22	23	23	23	22	21

All Possible Minimally Connected 2×4 Designs

14	13	13	13	12	12	12	12	12	12	12	12	11	11
21	21	14	14	21	14	14	13	13	13	13	13	21	14
22	22	21	21	22	21	21	21	21	14	14	14	22	22
23	23	22	22	23	23	22	23	22	21	21	21	23	23
24	24	24	23	24	24	23	24	24	24	23	22	24	24

11	11	11	11	11	11	11	11	11	11	11	11	11	11
14	13	13	13	13	13	12	12	12	12	12	12	12	12
21	22	21	14	14	14	22	21	14	14	14	13	13	13
22	23	22	22	22	21	23	23	23	22	21	23	22	21
23	24	24	24	23	22	24	24	24	23	23	24	24	24

11	11	11	11
12	12	12	12
13	13	13	13
14	14	14	14
24	23	22	21

All Possible Minimally Connected 3×3 Designs

13	13	13	13	13	13	13	13	13	13	13	13	12	12
23	22	22	22	21	21	21	21	21	21	21	21	23	22
31	31	23	23	31	23	23	22	22	22	22	22	31	31
32	32	31	31	32	32	31	32	31	23	23	23	32	32
33	33	33	32	33	33	32	33	33	33	32	31	33	33

Table 9.5 (*Continued*)

12	12	12	12	12	12	12	12	12	12	12	12	12	12
22	22	21	21	21	21	21	21	21	21	13	13	13	13
23	23	31	23	23	22	22	22	22	22	23	23	22	22
31	31	32	32	31	32	31	23	23	23	31	31	31	31
33	32	33	33	32	33	33	33	32	31	33	32	33	32
12	12	12	12	12	12	12	12	11	11	11	11	11	11
13	13	13	13	13	13	13	13	23	22	22	22	21	21
21	21	21	21	21	21	21	21	31	31	23	23	31	23
31	31	23	23	23	22	22	22	32	32	31	31	32	32
33	32	33	32	31	33	32	31	33	33	33	32	33	33
11	11	11	11	11	11	11	11	11	11	11	11	11	11
21	21	21	21	21	21	13	13	13	13	13	13	13	13
23	22	22	22	22	22	23	23	22	22	22	22	22	21
31	32	31	23	23	23	32	31	32	31	23	23	23	32
32	33	33	33	32	31	33	32	33	32	33	32	31	33
11	11	11	11	11	11	11	11	11	11	11	11	11	11
13	13	13	13	12	12	12	12	12	12	12	12	12	12
21	21	21	21	23	23	22	22	22	22	22	21	21	21
31	22	22	22	32	31	32	31	23	23	23	32	31	23
32	33	32	31	33	33	33	33	33	32	31	33	33	33
11	11	11	11	11	11	11	11	11	11	11			
12	12	12	12	12	12	12	12	12	12	12			
21	21	13	13	13	13	13	13	13	13	13			
23	23	23	23	23	22	22	22	21	21	21			
32	31	33	32	31	33	32	31	33	32	31			

9.5 THE PROBLEM OF SELECTING A FACTORIAL EXPERIMENT

In factorial experiments one often has a situation where it is not possible to accommodate all treatment combinations in the experiment. The device of fractional replications has proved very useful in such situations. In such experiments the set of treatment effects can typically be divided into three categories. First, there is a subset of treatment effects which are of interest to the experimenter. Next, there are effects which are not of interest but which cannot be assumed to be negligible. Finally, there are effects which can be assumed to be negligible.

In evaluating the performance of a given design in such situations one may adopt different approaches. For example, one may use a measure that is based on the aliasing of the estimates of the effects of the first set due to the existence of the effects of the second set. As an alternative approach one may use a measure which takes into account only the effects of the first set which are estimable, that is, their estimates are free from the effects of the second set and of each other. This measure of information provided by the design is based on the number of such effects and the precision with which they are estimated.

Consider a $s_1 \times s_2 \times \cdots \times s_k$ factorial experiment. We have $2^k - 1$ sets of main effects and interactions and the total number of possible treatment combinations is $\prod_{i=1}^{k} s_i$. In practice, some of these effects, especially the interactions of higher orders, can be assumed to be negligible. Further, the experimenter is often interested in only a subset of the remaining effects.

Let $\boldsymbol{\theta}_{(1)}, \boldsymbol{\theta}_{(2)}, \ldots, \boldsymbol{\theta}_{(p)}$ denote the vectors of parameters corresponding to the p effects of interest. Similarly, let $\boldsymbol{\xi}_{(1)}, \ldots, \boldsymbol{\xi}_{(q)}$ be the vectors of parameters corresponding to the q effects which are not of interest but are not negligible. Let $\boldsymbol{\theta}' = (\boldsymbol{\theta}'_{(1)}, \ldots, \boldsymbol{\theta}'_{(p)})$ and let $\boldsymbol{\xi}' = (\boldsymbol{\xi}'_{(1)}, \ldots, \boldsymbol{\xi}'_{(q)})$. The model may be written as

$$E(Y_{i_1,\ldots,i_k}) = \mathbf{x}'(i_1, \ldots, i_k)\boldsymbol{\theta} + \mathbf{z}'(i_1, \ldots, i_k)\boldsymbol{\xi} \tag{9.7}$$

where $Y_{i_1,\ldots,i_k}$ refers to an observation on the treatment combination $(i_1, \ldots, i_k)$ and $\mathbf{x}(i_1, \ldots, i_k)$, $\mathbf{z}(i_1, \ldots, i_k)$ are the vectors of coefficients of $\boldsymbol{\theta}$ and $\boldsymbol{\xi}$, respectively, in the above expectation.

One would like to estimate all components of $\boldsymbol{\theta}$. The estimability of these parameters would depend on the choice of the design, that is, the set of treatment combinations included in the experiment. Some $\boldsymbol{\theta}_{(i)}$'s in $\boldsymbol{\theta}' = (\boldsymbol{\theta}'_{(1)}, \ldots, \boldsymbol{\theta}'_{(p)})$ may not be estimable. However, it may be possible to estimate $\boldsymbol{\eta}_{(i)} = \mathbf{L}_i \boldsymbol{\theta}_{(i)}$ where $\mathbf{L}_i$ is the coefficient matrix giving estimable parametric functions of $\boldsymbol{\theta}_{(i)}$. Let $r_i = \text{rank } \mathbf{L}_i$ and let $r = \sum_{i=1}^{p} r_i$. If $r_i = 0$, no component of $\boldsymbol{\theta}_{(i)}$ is estimable. On the other hand if $r_i = n_i$, where n_i is the number of elements in $\boldsymbol{\theta}_{(i)}$, all components of $\boldsymbol{\theta}_{(i)}$ are estimable. Hedayat, Raktoe, and Federer (1974) considered designs for which $\boldsymbol{\theta}'_{(i)}$'s or specified linear functions of $\boldsymbol{\theta}_{(i)}$'s are estimable under the model $E(Y_{i_1,\ldots,i_k}) = \mathbf{x}'(i_1, \ldots, i_k)\boldsymbol{\theta}$ and used the bias of the estimates obtained under this model when the true model is (9.7). They introduced the alias matrix which is the coefficient matrix of $\boldsymbol{\xi}$ in the least squares estimates obtained under the model (9.7) with $\boldsymbol{\xi} = \mathbf{0}$. The norm of this alias matrix has been proposed by them as a measure which would guide the experimenter in the selection of a design. It is pointed out by them that this measure does not take into account the precision of the estimates.

We present an alternative approach to this problem. We use model (9.7) to obtain the least squares estimates of $\boldsymbol{\eta}_{(i)}$'s. The remaining linear functions of $\boldsymbol{\theta}_{(i)}$'s are not estimable, that is, they are confounded with $\boldsymbol{\xi}$ or with each other. No useful information can be extracted from the data on these functions without making further assumptions concerning $\boldsymbol{\xi}$. Consequently, our measure of usefulness of the design is based only on the $\boldsymbol{\eta}_{(i)}$'s. Let $\hat{\boldsymbol{\eta}}_{(i)}$ be the least squares estimate of $\boldsymbol{\eta}_{(i)}$ based on model (9.7). Let $\hat{\boldsymbol{\eta}}=(\hat{\boldsymbol{\eta}}_{(1)},\ldots,\hat{\boldsymbol{\eta}}_{(p)})$. Let $\sigma^2\mathbf{V}$ be the dispersion matrix of $\boldsymbol{\eta}$. We propose to use trace $\mathbf{V}^{-1}$ as a measure uf usefulness of the design. This measure is based on estimable functions of interest and takes into account the precision with which these functions have been estimated. We shall show that this measure is invariant under different choices of $\boldsymbol{\theta}_{(i)}$, provided that they are normalized.

We shall now formally introduce a measure of information on the parametric functions of interest provided by a design. Given the estimation problem as specified by model (9.7), an experimenter would like to compare various designs each having the same number of observations. It may happen that with a given design not all parameters of interest can be estimated. The measure of information given below takes into account this confounding of the remaining effects.

For a given design the observational equations may be written as

$$E(\mathbf{Y})=\mathbf{X}\boldsymbol{\theta}+\mathbf{Z}\boldsymbol{\xi}$$

where $\boldsymbol{\theta}'=(\boldsymbol{\theta}'_{(1)},\ldots,\boldsymbol{\theta}'_{(p)})$ are the parameters of interest, and $\boldsymbol{\xi}'=(\boldsymbol{\xi}'_{(1)},\ldots,\boldsymbol{\xi}'_{(q)})$ are the remaining parameters that cannot be assumed to be negligible. It should be noted that the parameters relating to the same effect are grouped together and hence we are *not* interested in estimating parametric functions involving more than one $\boldsymbol{\theta}_{(i)}$. Suppose the design permits estimation of r_i linearly independent parametric functions of $\boldsymbol{\theta}_{(i)}$. These may be written as $\boldsymbol{\eta}_{(i)}=\mathbf{L}_i\boldsymbol{\theta}_{(i)}$. Without less of generality we can assume that $\mathbf{L}_i'\mathbf{L}_i=\mathbf{I}_{r_i}$. Let $r=\sum r_i$. Thus our design permits estimation of r linearly independent parametric functions of interest. Note that $r\leqslant$ rank $\mathbf{X}$. Let $\hat{\boldsymbol{\eta}}_{(i)}$ be the least squares estimate of $\boldsymbol{\eta}_{(i)}$. Let $\hat{\boldsymbol{\eta}}'=(\hat{\boldsymbol{\eta}}'_{(1)},\ldots,\hat{\boldsymbol{\eta}}'_{(p)})$. A measure of information contained in the design should come from the dispersion matrix of $\hat{\boldsymbol{\eta}}$. Let $\sigma^2\mathbf{V}$ denote this dispersion matrix. We shall use $U_d=trace\ \mathbf{V}^{-1}$ as a measure of usefulness of the design. In other words, one should like to choose a design that maximizes this trace. In factorial experiments there is no natural coordinate system for the $\boldsymbol{\theta}'_{(i)}$'s. The following theorem shows that this measure is invariant with respect to the choice of the coordinate system.

THEOREM 9.6 If $\boldsymbol{\delta}_{(i)}=\mathbf{P}_i\boldsymbol{\theta}_{(i)}$ where $\mathbf{P}_i$ is an orthogonal matrix, $i=1, 2, \ldots, p$, the information measure $U_d(\boldsymbol{\delta})$ with $\boldsymbol{\delta}_{(i)}$'s as parameters is the same as $U_d(\boldsymbol{\theta})$.

Proof Let $\hat{\boldsymbol{\theta}}_{(i)}$ denote the solutions to the normal equations. Consider $\hat{\boldsymbol{\eta}}_{(i)}=\mathbf{L}_i\hat{\boldsymbol{\theta}}_{(i)}$. If $\hat{\boldsymbol{\delta}}_{(i)}$ denote the solutions to the normal equations in terms of $\boldsymbol{\delta}_{(i)}$, $\hat{\boldsymbol{\eta}}_{(i)}$ can be written as $\hat{\boldsymbol{\eta}}_{(i)}=\mathbf{M}_i\hat{\boldsymbol{\delta}}_i$ where $\mathbf{M}_i=\mathbf{L}_i\mathbf{P}_i'$. Since $\mathbf{M}_i\mathbf{M}_i'=\mathbf{L}_i\mathbf{P}_i'\mathbf{P}_i\mathbf{L}_i\mathbf{L}_i'=\mathbf{L}_i\mathbf{L}_i'=\mathbf{I}_{r_i}$ the invariance follows. □

The measure $U_d(\boldsymbol{\theta})$ may be used as a measure of information on $\boldsymbol{\theta}_{(i)}$'s provided by the experiment. Thus, we would prefer d_1 to d_2 if $U_{d_1}(\boldsymbol{\theta})>U_{d_2}(\boldsymbol{\theta})$. However, if for designs d_1 and d_2 $U_{d_1}(\boldsymbol{\theta})=U_{d_2}(\boldsymbol{\theta})$, we would prefer the design that estimates more parameters.

Examples

We first consider two 4×3 factorials with no interactions. Each design has six observations. We give below the incidence matrices for the two designs.

d_1:

	β_1	β_2	β_3
α_1	1	0	0
α_2	0	1	0
α_3	1	0	1
α_4	0	1	1

d_2:

	β_1	β_2	β_3
α_1	1	0	0
α_2	0	1	0
α_3	0	1	1
α_4	0	1	1

Methods developed in Chapter 5 can be applied to show that with d_1 all the main effects are estimable. We may write the three components of $\boldsymbol{\theta}_{(1)}$ as $(\alpha_1-\alpha_2)/\sqrt{2}$, $(\alpha_1+\alpha_2-2\alpha_3)/\sqrt{6}$, and $(\alpha_1+\alpha_2+\alpha_3-3\alpha_4)/\sqrt{12}$. Similarly, the two components of $\boldsymbol{\theta}_{(2)}$ are $(\beta_1-\beta_2/\sqrt{2}$ and $(\beta_1+\beta_2-2\beta_3)/\sqrt{6}$. Here $\mathbf{L}_1=\mathbf{I}_3$ and $\mathbf{L}_2=\mathbf{I}_2$. The matrix $\mathbf{V}$ turns out to be

$$\mathbf{V}_{d_1}=\begin{bmatrix} 3 & -\frac{\sqrt{3}}{3} & \frac{\sqrt{6}}{3} & -2 & 0 \\ -\frac{\sqrt{3}}{3} & 1 & 0 & \frac{\sqrt{3}}{3} & -\frac{\sqrt{3}}{3} \\ \frac{\sqrt{6}}{3} & 0 & 1 & -\frac{\sqrt{6}}{3} & -\frac{\sqrt{2}}{6} \\ -2 & \frac{\sqrt{3}}{3} & -\frac{\sqrt{6}}{3} & 2 & 0 \\ 0 & -\frac{\sqrt{3}}{3} & -\frac{\sqrt{2}}{6} & 0 & \frac{2}{3} \end{bmatrix}.$$

The inverse of this is given by

$$\mathbf{V}_{d_1}^{-1} = \begin{bmatrix} 1 & 0 & 0 & 1 & 0 \\ 0 & \frac{14}{9} & -\frac{2\sqrt{2}}{9} & -\frac{\sqrt{3}}{3} & \frac{2}{3} \\ 0 & \frac{2\sqrt{2}}{9} & \frac{16}{9} & \frac{\sqrt{6}}{3} & \frac{\sqrt{2}}{3} \\ 1 & -\frac{\sqrt{3}}{3} & \frac{\sqrt{6}}{3} & 2 & 0 \\ 0 & \frac{2}{3} & \frac{\sqrt{2}}{3} & 0 & 2 \end{bmatrix}$$

giving $U_{d_1} = \text{trace } \mathbf{V}_{d_1}^{-1} = 8.3333$.

For d_2, not all main effects are estimable. Using the methods developed in Chapter 5 we see that two linear functions of $\boldsymbol{\theta}_{(1)}$ and one linear function of $\boldsymbol{\theta}_{(2)}$ can be estimated. We get

$$\mathbf{L}_1 = \begin{bmatrix} -\frac{1}{2} & \frac{\sqrt{3}}{2} & 0 \\ -\frac{\sqrt{3}}{6} & -\frac{1}{6} & \frac{2\sqrt{2}}{3} \end{bmatrix} \quad \text{and} \quad \mathbf{L}_2 = \begin{bmatrix} -\frac{1}{2} & \frac{\sqrt{3}}{2} \end{bmatrix}.$$

The $\mathbf{V}$ matrix turns out to be

$$\mathbf{V}_{d_2} = \frac{1}{24} \begin{bmatrix} 21 & 3\sqrt{3} & -6 \\ 3\sqrt{3} & 15 & -2\sqrt{3} \\ -6 & -2\sqrt{3} & 12 \end{bmatrix}.$$

This gives

$$\mathbf{V}_{d_2}^{-1} = \frac{1}{5} \begin{bmatrix} 7 & -\sqrt{3} & 3 \\ -\sqrt{3} & 9 & \sqrt{3} \\ 3 & \sqrt{3} & 12 \end{bmatrix},$$

giving $U_{d_2} = \text{trace } \mathbf{V}_{d_2}^{-1} = 5.6$.

Since $U_{d_2} < U_{d_1}$ we prefer design d_1 to design d_2. The relative efficiency of d_2 with respect to d_1 is 0.671995. Even though the design d_2 estimates only three parameters and the design d_1 estimates five parameters the relative efficiency of d_2 with respect to d_1 is higher than $\frac{3}{5}$ because the measure of information takes into account the precision associated with the estimable parametric functions.

As our second example we consider two 2^3 factorial each with six observations with the following patterns:

d_1:

	δ_1: β_1	δ_1: β_2		δ_2: β_1	δ_2: β_2
α_1	0	1	α_1	1	1
α_2	1	1	α_2	1	0

d_2:

	δ_1: β_1	δ_1: β_2		δ_2: β_1	δ_2: β_2
α_1	1	1	α_1	1	1
α_2	0	1	α_2	1	0

We consider the estimation of the main effects and the two-factor interactions assuming that the three-factor interactions are negligible. For both designs we set

$$\boldsymbol{\theta}' = (\boldsymbol{\theta}_{(1)}, \boldsymbol{\theta}_{(2)}, \boldsymbol{\theta}_{(3)}, \boldsymbol{\theta}_{(4)}, \boldsymbol{\theta}_{(5)}, \boldsymbol{\theta}_{(6)}) = (A, B, C, AB, AC, BC).$$

With d_1 we can estimate the three main effects but not any of the interactions. We get

$$\mathbf{V}_{d_1} = \begin{bmatrix} 4 & -2 & -2 \\ -2 & 4 & 2 \\ -2 & 2 & 4 \end{bmatrix}, \qquad \mathbf{V}_{d_1}^{-1} = \frac{1}{8}\begin{bmatrix} 3 & 1 & 1 \\ 1 & 3 & -1 \\ 1 & -1 & 3 \end{bmatrix}.$$

Thus, $U_{d_1} = 1.125$.

With the design d_2 we can estimate A and BC effects. We get

$$\mathbf{V}_{d_2} = \begin{bmatrix} 4 & -2 \\ -2 & 4 \end{bmatrix}, \qquad \mathbf{V}_{d_2}^{-1} = \frac{1}{12}\begin{bmatrix} 4 & 2 \\ 2 & 4 \end{bmatrix},$$

giving $U_{d_2} = 0.6667$. Efficiency of d_2 with respect to d_1 is 0.5926. Here, d_2 estimates two effects and d_1 estimates three effects. However, the relative efficiency of d_2 with respect to d_1 is less than $\frac{2}{3}$ which is due to higher precision for each estimated by d_1.

If one is interested in estimating only the main effects, $\boldsymbol{\theta}' = (A, B, C)$ and $\boldsymbol{\xi}' = (AB, AC, BC)$. Since d_1 estimates all the main effects U_{d_1} remains unaltered, while d_2 estimates A only and hence $U_{d_2} = \frac{1}{4}$, giving us the relative efficiency of d_2 with respect to d_1 equal to $\frac{2}{9}$.

Thirty-Two Minimally connected 2^4 Designs Selected Arbitrary

1	1	1	1	1	1	1	1	1	1	1	1	1	1	1	1
1	2	1	1	1	2	1	1	1	2	1	1	1	2	1	1
2	1	1	1	2	1	1	1	2	1	1	1	2	1	1	1
2	2	2	1	2	2	2	1	2	2	2	1	2	2	2	1
2	2	2	2	2	2	1	2	2	1	1	2	2	1	2	2
1	1	1	1	1	1	1	1	1	1	1	1	1	1	1	1
1	2	1	1	1	2	1	1	1	2	1	1	1	2	1	1
2	1	1	1	2	1	1	1	2	1	1	1	2	1	1	1
2	2	2	1	2	2	2	1	2	2	1	2	2	2	1	2
1	2	1	2	1	2	2	2	2	2	2	2	2	1	2	1
1	1	1	1	1	1	1	1	1	1	1	1	1	1	1	1
1	2	1	1	1	2	1	1	1	2	1	1	1	2	1	1
2	1	1	1	2	1	1	1	2	1	1	1	2	1	1	1
2	2	1	2	2	2	1	2	2	2	1	2	2	2	2	2
1	1	2	1	1	2	2	1	1	2	2	2	1	2	1	2
1	1	1	1	1	1	1	1	1	1	1	1	1	1	1	1
1	2	1	1	1	2	1	1	1	2	1	1	1	2	1	1
2	1	1	1	2	1	1	1	2	1	1	1	2	1	1	1
2	2	2	2	2	1	2	2	2	1	2	2	2	1	2	2
2	1	1	2	2	1	2	1	2	1	1	2	1	1	2	1
1	1	1	1	1	1	1	1	1	1	1	1	1	1	1	1
1	2	1	1	1	2	1	1	1	2	1	1	1	2	1	1
2	1	1	1	2	1	1	1	2	1	1	1	2	1	1	1
2	1	2	2	2	1	2	2	2	1	2	2	2	1	1	2
1	2	2	1	1	2	1	2	2	2	2	1	1	2	2	2
1	1	1	1	1	1	1	1	1	1	1	1	1	1	1	1
1	2	1	1	1	2	1	1	1	2	1	1	1	2	1	1
2	1	1	1	2	1	1	1	2	1	1	1	2	1	1	1
2	1	1	2	2	1	1	2	2	1	1	2	2	1	1	2
1	1	2	1	1	2	2	1	1	1	2	2	2	1	2	1
1	1	1	1	1	1	1	1	1	1	1	1	1	1	1	1
1	2	1	1	1	2	1	1	1	2	1	1	1	2	1	1
2	1	1	1	2	1	1	1	2	1	1	1	2	1	1	1
2	1	2	1	2	1	2	1	1	1	2	1	1	1	2	1
1	2	1	2	1	2	2	2	1	2	2	2	1	2	1	2
1	1	1	1	1	1	1	1	1	1	1	1	1	1	1	1
1	2	1	1	1	2	1	1	1	2	1	1	1	2	1	1
2	1	1	1	2	1	1	1	2	1	1	1	2	1	1	1
1	1	2	1	1	1	1	2	1	1	1	2	1	1	1	2
1	1	2	2	1	1	2	2	1	2	2	1	1	2	2	2

Twenty-Four Minimally Connected 2^5 Designs Selected Arbitrary

1	1	1	1	1	1	1	1	1	1	1	1	1	1	1	1	1	1	1	1
1	2	1	1	1	1	2	1	1	1	1	2	1	1	1	1	2	1	1	1
2	1	1	1	1	2	1	1	1	1	2	1	1	1	1	2	1	1	1	1
2	2	2	1	1	2	2	2	1	1	2	2	2	1	1	2	2	2	1	1
2	2	2	2	1	2	2	2	2	1	2	2	2	2	1	2	2	2	2	1
2	2	2	2	2	2	2	1	2	2	2	2	1	1	2	2	1	2	1	2

1	1	1	1	1	1	1	1	1	1	1	1	1	1	1	1	1	1	1	1
1	2	1	1	1	1	2	1	1	1	1	2	1	1	1	1	2	1	1	1
2	1	1	1	1	2	1	1	1	1	2	1	1	1	1	2	1	1	1	1
2	2	2	1	1	2	2	2	1	1	2	2	2	1	1	2	2	2	1	1
2	2	2	2	1	2	2	2	2	1	2	2	2	2	1	2	2	2	2	1
2	1	1	1	2	2	1	1	2	2	2	2	2	1	2	2	1	2	2	2

1	1	1	1	1	1	1	1	1	1	1	1	1	1	1	1	1	1	1	1
1	2	1	1	1	1	2	1	1	1	1	2	1	1	1	1	2	1	1	1
2	1	1	1	1	2	1	1	1	1	2	1	1	1	1	2	1	1	1	1
2	2	2	1	1	2	2	2	1	1	2	2	2	1	1	2	2	2	1	1
2	2	2	2	1	2	2	2	2	1	2	2	2	2	1	2	2	2	2	1
1	1	1	1	2	1	1	1	2	2	1	1	2	1	2	1	2	2	1	2

1	1	1	1	1	1	1	1	1	1	1	1	1	1	1	1	1	1	1	1
1	2	1	1	1	1	2	1	1	1	1	2	1	1	1	1	2	1	1	1
2	1	1	1	1	2	1	1	1	1	2	1	1	1	1	2	1	1	1	1
2	2	2	1	1	2	2	2	1	1	2	2	2	1	1	2	2	2	1	1
2	2	2	2	1	2	2	2	2	1	2	2	1	2	1	2	2	1	2	1
1	2	1	1	2	1	2	2	2	2	1	2	2	2	2	1	2	1	2	2

1	1	1	1	1	1	1	1	1	1	1	1	1	1	1	1	1	1	1	1
1	2	1	1	1	1	2	1	1	1	1	2	1	1	1	1	2	1	1	1
2	1	1	1	1	2	1	1	1	1	2	1	1	1	1	2	1	1	1	1
2	2	2	1	1	2	2	2	1	1	2	2	2	1	1	2	2	2	1	1
2	2	1	2	1	2	2	1	2	1	2	2	1	2	1	2	2	1	2	1
1	2	2	1	2	1	1	2	1	2	1	1	1	1	2	1	1	1	2	2

1	1	1	1	1	1	1	1	1	1	1	1	1	1	1	1	1	1	1	1
1	2	1	1	1	1	2	1	1	1	1	2	1	1	1	1	2	1	1	1
2	1	1	1	1	2	1	1	1	1	2	1	1	1	1	2	1	1	1	1
2	2	2	1	1	2	2	2	1	1	2	2	2	1	1	2	2	2	1	1
2	2	1	2	1	2	2	1	2	1	2	2	1	2	1	2	2	1	2	1
2	1	2	2	2	2	1	1	2	2	2	1	1	1	2	2	1	2	1	2

1	1	1	1	1	1		1	1	1	1	1	1		1	1	1	1	1	1
1	2	1	1	1	1		1	2	1	1	1	1		1	2	1	1	1	1
2	1	1	1	1	1		2	1	1	1	1	1		2	1	1	1	1	1
2	2	2	1	1	1		2	2	2	1	1	1		2	2	2	1	1	1
2	2	2	2	1	1		2	2	2	2	1	1		2	2	2	2	1	1
2	2	2	2	2	1		2	2	2	2	2	1		2	2	2	2	2	1
2	2	2	2	2	2		2	2	1	2	2	2		2	2	1	1	2	2

1	1	1	1	1	1		1	1	1	1	1	1		1	1	1	1	1	1
1	2	1	1	1	1		1	2	1	1	1	1		1	2	1	1	1	1
2	1	1	1	1	1		2	1	1	1	1	1		2	1	1	1	1	1
2	2	2	1	1	1		2	2	2	1	1	1		2	2	2	1	1	1
2	2	2	2	1	1		2	2	2	2	1	1		2	2	2	2	1	1
2	2	2	2	2	1		2	2	2	2	2	1		2	2	2	2	2	1
2	2	1	1	1	2		2	2	1	2	1	2		1	2	1	2	2	2

1	1	1	1	1	1		1	1	1	1	1	1		1	1	1	1	1	1
1	2	1	1	1	1		1	2	1	1	1	1		1	2	1	1	1	1
2	1	1	1	1	1		2	1	1	1	1	1		2	1	1	1	1	1
2	2	2	1	1	1		2	2	2	1	1	1		2	2	2	1	1	1
2	2	2	2	1	1		2	2	2	2	1	1		2	2	2	2	1	1
2	2	2	2	2	1		2	2	2	2	2	1		2	2	2	2	2	1
2	2	2	2	1	2		2	1	1	2	1	2		2	1	1	1	2	2

1	1	1	1	1	1		1	1	1	1	1	1		1	1	1	1	1	1
1	2	1	1	1	1		1	2	1	1	1	1		1	2	1	1	1	1
2	1	1	1	1	1		2	1	1	1	1	1		2	1	1	1	1	1
2	2	2	1	1	1		2	2	2	1	1	1		2	2	2	1	1	1
2	2	2	2	1	1		2	2	2	2	1	1		2	2	2	2	1	1
2	2	2	2	2	1		2	2	2	2	2	1		2	2	2	2	2	1
2	1	2	2	2	2		2	1	2	1	2	2		2	1	2	1	1	2

1	1	1	1	1	1		1	1	1	1	1	1		1	1	1	1	1	1
1	2	1	1	1	1		1	2	1	1	1	1		1	2	1	1	1	1
2	1	1	1	1	1		2	1	1	1	1	1		2	1	1	1	1	1
2	2	2	1	1	1		2	2	2	1	1	1		2	2	2	1	1	1
2	2	2	2	1	1		2	2	2	2	1	1		2	2	2	2	1	1
2	2	2	2	2	1		2	2	2	2	2	1		2	2	2	2	2	1
1	1	2	2	2	2		1	1	2	2	1	2		1	1	2	1	1	2

1	1	1	1	1	1		1	1	1	1	1	1		1	1	1	1	1	1
1	2	1	1	1	1		1	2	1	1	1	1		1	2	1	1	1	1
2	1	1	1	1	1		2	1	1	1	1	1		2	1	1	1	1	1
2	2	2	1	1	1		2	2	2	1	1	1		2	2	2	1	1	1
2	2	2	2	1	1		2	2	2	2	1	1		2	2	2	2	1	1
2	2	2	2	2	1		2	2	2	2	2	1		2	2	2	2	2	1
1	1	1	2	2	2		1	1	1	1	2	2		1	1	1	1	1	2

Fifteen Minimally Connected 2^7 *Designs Selected Arbitrary*

1	1	1	1	1	1	1	1	1	1	1	1	1	1	1	1	1	1	1	1	1
1	2	1	1	1	1	1	1	2	1	1	1	1	1	1	2	1	1	1	1	1
2	1	1	1	1	1	1	2	1	1	1	1	1	1	2	1	1	1	1	1	1
2	2	2	1	1	1	1	2	2	2	1	1	1	1	2	2	2	1	1	1	1
2	2	2	2	1	1	1	2	2	2	2	1	1	1	2	2	2	2	1	1	1
2	2	2	2	2	1	1	2	2	2	2	2	1	1	2	2	2	2	2	1	1
2	2	2	2	2	2	1	2	2	2	2	2	2	1	2	2	2	2	2	2	1
2	2	2	2	2	2	2	2	2	1	2	2	2	2	2	2	1	1	2	2	2

1	1	1	1	1	1	1	1	1	1	1	1	1	1	1	1	1	1	1	1	1
1	2	1	1	1	1	1	1	2	1	1	1	1	1	1	2	1	1	1	1	1
2	1	1	1	1	1	1	2	1	1	1	1	1	1	2	1	1	1	1	1	1
2	2	2	1	1	1	1	2	2	2	1	1	1	1	2	2	2	1	1	1	1
2	2	2	2	1	1	1	2	2	2	2	1	1	1	2	2	2	2	1	1	1
2	2	2	2	2	1	1	2	2	2	2	2	1	1	2	2	2	2	2	1	1
2	2	2	2	2	2	1	2	2	2	2	2	2	1	2	2	2	2	2	2	1
2	2	1	1	2	1	2	2	2	1	2	1	1	2	2	2	2	1	1	2	2

1	1	1	1	1	1	1	1	1	1	1	1	1	1	1	1	1	1	1	1	1
1	2	1	1	1	1	1	1	2	1	1	1	1	1	1	2	1	1	1	1	1
2	1	1	1	1	1	1	2	1	1	1	1	1	1	2	1	1	1	1	1	1
2	2	2	1	1	1	1	2	2	2	1	1	1	1	2	2	2	1	1	1	1
2	2	2	2	1	1	1	2	2	2	2	1	1	1	2	2	2	2	1	1	1
2	2	2	2	2	1	1	2	2	2	2	2	1	1	2	2	2	2	2	1	1
2	2	2	2	2	2	1	2	2	2	2	2	2	1	2	2	2	2	2	2	1
2	2	1	2	1	2	2	2	2	2	1	2	2	2	2	2	1	1	1	2	2

1	1	1	1	1	1	1	1	1	1	1	1	1	1	1	1	1	1	1	1	1
1	2	1	1	1	1	1	1	2	1	1	1	1	1	1	2	1	1	1	1	1
2	1	1	1	1	1	1	2	1	1	1	1	1	1	2	1	1	1	1	1	1
2	2	2	1	1	1	1	2	2	2	1	1	1	1	2	2	2	1	1	1	1
2	2	2	2	1	1	1	2	2	2	2	1	1	1	2	2	2	2	1	1	1
2	2	2	2	2	1	1	2	2	2	2	2	1	1	2	2	2	2	2	1	1
2	2	2	2	2	2	1	2	2	2	2	2	2	1	2	2	2	2	2	2	1
2	2	1	2	2	1	2	2	2	2	1	2	1	2	2	2	2	1	1	1	2

1	1	1	1	1	1	1	1	1	1	1	1	1	1	1	1	1	1	1	1	1
1	2	1	1	1	1	1	1	2	1	1	1	1	1	1	2	1	1	1	1	1
2	1	1	1	1	1	1	2	1	1	1	1	1	1	2	1	1	1	1	1	1
2	2	2	1	1	1	1	2	2	2	1	1	1	1	2	2	2	1	1	1	1
2	2	2	2	1	1	1	2	2	2	2	1	1	1	2	2	2	2	1	1	1
2	2	2	2	2	1	1	2	2	2	2	2	1	1	2	2	2	2	2	1	1
2	2	2	2	2	2	1	2	2	2	2	2	2	1	2	2	2	2	2	2	1
2	2	1	1	1	1	2	1	2	1	2	2	2	2	1	2	1	1	2	2	2

Fifteen Minimally Connected 2^8 Designs Selected Arbitrary

1	1	1	1	1	1	1	1
1	2	1	1	1	1	1	1
2	1	1	1	1	1	1	1
2	2	2	1	1	1	1	1
2	2	2	2	1	1	1	1
2	2	2	2	2	1	1	1
2	2	2	2	2	2	1	1
2	2	2	2	2	2	2	1
2	2	2	2	2	2	2	2

1	1	1	1	1	1	1	1
1	2	1	1	1	1	1	1
2	1	1	1	1	1	1	1
2	2	2	1	1	1	1	1
2	2	2	2	1	1	1	1
2	2	2	2	2	1	1	1
2	2	2	2	2	2	1	1
2	2	2	2	2	2	2	1
2	2	1	2	2	2	2	2

1	1	1	1	1	1	1	1
1	2	1	1	1	1	1	1
2	1	1	1	1	1	1	1
2	2	2	1	1	1	1	1
2	2	2	2	1	1	1	1
2	2	2	2	2	1	1	1
2	2	2	2	2	2	1	1
2	2	2	2	2	2	2	1
2	2	1	1	2	2	2	2

1	1	1	1	1	1	1	1
1	2	1	1	1	1	1	1
2	1	1	1	1	1	1	1
2	2	2	1	1	1	1	1
2	2	2	2	1	1	1	1
2	2	2	2	2	1	1	1
2	2	2	2	2	2	1	1
2	2	2	2	2	2	2	1
1	2	1	1	1	1	2	2

1	1	1	1	1	1	1	1
1	2	1	1	1	1	1	1
2	1	1	1	1	1	1	1
2	2	2	1	1	1	1	1
2	2	2	2	1	1	1	1
2	2	2	2	2	1	1	1
2	2	2	2	2	2	1	1
2	2	2	2	2	2	2	1
1	2	1	1	1	2	2	2

1	1	1	1	1	1	1	1
1	2	1	1	1	1	1	1
2	1	1	1	1	1	1	1
2	2	2	1	1	1	1	1
2	2	2	2	1	1	1	1
2	2	2	2	2	1	1	1
2	2	2	2	2	2	1	1
2	2	2	2	2	2	2	1
1	2	1	1	2	2	2	2

1	1	1	1	1	1	1	1
1	2	1	1	1	1	1	1
2	1	1	1	1	1	1	1
2	2	2	1	1	1	1	1
2	2	2	2	1	1	1	1
2	2	2	2	2	1	1	1
2	2	2	2	2	2	1	1
2	2	2	2	2	2	2	1
2	2	1	1	2	1	2	2

1	1	1	1	1	1	1	1
1	2	1	1	1	1	1	1
2	1	1	1	1	1	1	1
2	2	2	1	1	1	1	1
2	2	2	2	1	1	1	1
2	2	2	2	2	1	1	1
2	2	2	2	2	2	1	1
2	2	2	2	2	2	2	1
1	1	1	1	2	2	2	2

1	1	1	1	1	1	1	1
1	2	1	1	1	1	1	1
2	1	1	1	1	1	1	1
2	2	2	1	1	1	1	1
2	2	2	2	1	1	1	1
2	2	2	2	2	1	1	1
2	2	2	2	2	2	1	1
2	2	2	2	2	2	2	1
2	2	1	1	1	2	2	2

1	1	1	1	1	1	1	1
1	2	1	1	1	1	1	1
2	1	1	1	1	1	1	1
2	2	2	1	1	1	1	1
2	2	2	2	1	1	1	1
2	2	2	2	2	1	1	1
2	2	2	2	2	2	1	1
2	2	2	2	2	2	2	1
2	2	1	2	2	1	1	2

1	1	1	1	1	1	1	1
1	2	1	1	1	1	1	1
2	1	1	1	1	1	1	1
2	2	2	1	1	1	1	1
2	2	2	2	1	1	1	1
2	2	2	2	2	1	1	1
2	2	2	2	2	2	1	1
2	2	2	2	2	2	2	1
1	2	1	2	2	1	1	2

1	1	1	1	1	1	1	1
1	2	1	1	1	1	1	1
2	1	1	1	1	1	1	1
2	2	2	1	1	1	1	1
2	2	2	2	1	1	1	1
2	2	2	2	2	1	1	1
2	2	2	2	2	2	1	1
2	2	2	2	2	2	2	1
1	2	1	2	2	1	2	2

1	1	1	1	1	1	1	1
1	2	1	1	1	1	1	1
2	1	1	1	1	1	1	1
2	2	2	1	1	1	1	1
2	2	2	2	1	1	1	1
2	2	2	2	2	1	1	1
2	2	2	2	2	2	1	1
2	2	2	2	2	2	2	1
1	2	1	1	2	2	1	2

1	1	1	1	1	1	1	1
1	2	1	1	1	1	1	1
2	1	1	1	1	1	1	1
2	2	2	1	1	1	1	1
2	2	2	2	1	1	1	1
2	2	2	2	2	1	1	1
2	2	2	2	2	2	1	1
2	2	2	2	2	2	2	1
1	2	1	1	2	1	2	2

1	1	1	1	1	1	1	1
1	2	1	1	1	1	1	1
2	1	1	1	1	1	1	1
2	2	2	1	1	1	1	1
2	2	2	2	1	1	1	1
2	2	2	2	2	1	1	1
2	2	2	2	2	2	1	1
2	2	2	2	2	2	2	1
1	2	1	1	2	1	1	2

Twelve Minimally Connected 2^9 Designs Selected Arbitrary

1	1	1	1	1	1	1	1	1		1	1	1	1	1	1	1	1	1		1	1	1	1	1	1	1	1	1
1	2	1	1	1	1	1	1	1		1	2	1	1	1	1	1	1	1		1	2	1	1	1	1	1	1	1
2	1	1	1	1	1	1	1	1		2	1	1	1	1	1	1	1	1		2	1	1	1	1	1	1	1	1
2	2	2	1	1	1	1	1	1		2	2	2	1	1	1	1	1	1		2	2	2	1	1	1	1	1	1
2	2	2	2	1	1	1	1	1		2	2	2	2	1	1	1	1	1		2	2	2	2	1	1	1	1	1
2	2	2	2	2	1	1	1	1		2	2	2	2	2	1	1	1	1		2	2	2	2	2	1	1	1	1
2	2	2	2	2	2	1	1	1		2	2	2	2	2	2	1	1	1		2	2	2	2	2	2	1	1	1
2	2	2	2	2	2	2	1	1		2	2	2	2	2	2	2	1	1		2	2	2	2	2	2	2	1	1
2	2	2	2	2	2	2	2	1		2	2	2	2	2	2	2	2	1		2	2	2	2	2	2	2	2	1
2	2	2	2	2	2	2	2	2		2	2	1	2	2	2	2	2	2		2	2	1	1	2	2	2	2	2

1	1	1	1	1	1	1	1	1		1	1	1	1	1	1	1	1	1		1	1	1	1	1	1	1	1	1
1	2	1	1	1	1	1	1	1		1	2	1	1	1	1	1	1	1		1	2	1	1	1	1	1	1	1
2	1	1	1	1	1	1	1	1		2	1	1	1	1	1	1	1	1		2	1	1	1	1	1	1	1	1
2	2	2	1	1	1	1	1	1		2	2	2	1	1	1	1	1	1		2	2	2	1	1	1	1	1	1
2	2	2	2	1	1	1	1	1		2	2	2	2	1	1	1	1	1		2	2	2	2	1	1	1	1	1
2	2	2	2	2	1	1	1	1		2	2	2	2	2	1	1	1	1		2	2	2	2	2	1	1	1	1
2	2	2	2	2	2	1	1	1		2	2	2	2	2	2	1	1	1		2	2	2	2	2	2	1	1	1
2	2	2	2	2	2	2	1	1		2	2	2	2	2	2	2	1	1		2	2	2	2	2	2	2	1	1
2	2	2	2	2	2	2	2	1		2	2	2	2	2	2	2	2	1		2	2	2	2	2	2	2	2	1
2	2	1	1	1	1	1	1	2		1	2	1	1	1	1	1	2	2		1	2	1	1	1	1	2	2	2

1	1	1	1	1	1	1	1	1		1	1	1	1	1	1	1	1	1		1	1	1	1	1	1	1	1	1
1	2	1	1	1	1	1	1	1		1	2	1	1	1	1	1	1	1		1	2	1	1	1	1	1	1	1
2	1	1	1	1	1	1	1	1		2	1	1	1	1	1	1	1	1		2	1	1	1	1	1	1	1	1
2	2	2	1	1	1	1	1	1		2	2	2	1	1	1	1	1	1		2	2	2	1	1	1	1	1	1
2	2	2	2	1	1	1	1	1		2	2	2	2	1	1	1	1	1		2	2	2	2	1	1	1	1	1
2	2	2	2	2	1	1	1	1		2	2	2	2	2	1	1	1	1		2	2	2	2	2	1	1	1	1
2	2	2	2	2	2	1	1	1		2	2	2	2	2	2	1	1	1		2	2	2	2	2	2	1	1	1
2	2	2	2	2	2	2	1	1		2	2	2	2	2	2	2	1	1		2	2	2	2	2	2	2	1	1
2	2	2	2	2	2	2	2	1		2	2	2	2	2	2	2	2	1		2	2	2	2	2	2	2	2	1
2	2	1	1	1	2	1	1	2		2	2	1	1	1	2	1	2	2		2	2	1	1	1	2	2	2	2

1	1	1	1	1	1	1	1	1		1	1	1	1	1	1	1	1	1		1	1	1	1	1	1	1	1	1
1	2	1	1	1	1	1	1	1		1	2	1	1	1	1	1	1	1		1	2	1	1	1	1	1	1	1
2	1	1	1	1	1	1	1	1		2	1	1	1	1	1	1	1	1		2	1	1	1	1	1	1	1	1
2	2	2	1	1	1	1	1	1		2	2	2	1	1	1	1	1	1		2	2	2	1	1	1	1	1	1
2	2	2	2	1	1	1	1	1		2	2	2	2	1	1	1	1	1		2	2	2	2	1	1	1	1	1
2	2	2	2	2	1	1	1	1		2	2	2	2	2	1	1	1	1		2	2	2	2	2	1	1	1	1
2	2	2	2	2	2	1	1	1		2	2	2	2	2	2	1	1	1		2	2	2	2	2	2	1	1	1
2	2	2	2	2	2	2	1	1		2	2	2	2	2	2	2	1	1		2	2	2	2	2	2	2	1	1
2	2	2	2	2	2	2	2	1		2	2	2	2	2	2	2	2	1		2	2	2	2	2	2	2	2	1
2	2	1	1	2	2	1	2	2		2	2	1	1	2	2	1	1	2		1	2	1	1	2	2	1	1	2

PROGRAM BLOCKCON

General Description

This interactive program gives a list of $a \times b$ minimally connected designs. The program is limited at 9×9 design.

The input of this program is quite simple; it consists of three answers required by the program.

The first question is dimension of the incidence matrix or design:

$$a \times b, \qquad a < 9, b < 9.$$

The second question is the number of minimally connected designs desired. If this number is greater than the number of existing solutions, the program stops after it has given most of the existing solutions.

The last question is the output file's name which must be given by the user.

The output of this program is a list of minimally connected designs.

Output Description

The output is in a file which name is asked by the program; the file extension is always ".dat ".

The output matrices are written on 100-character rows.

At the end of the file, the number of desired or found designs is given.

BIBLIOGRAPHICAL NOTES

9.3 The material in this section is from Dodge (1976).

9.4 The material in this section is from Dodge and Afsarinejad (1984). For the problem of 2^n minimally connected designs and transportation matrices see Arthanari and Dodge (1981). A complete table of 2^3 minimally connected designs is also given in Arthanari and Dodge (1981).

9.5 The material in this section is from Dodge and Shah (1977). The pioneer article in this subject is by Hedayat, Raktoe, and Federer (1974). They proposed and developed a method for selecting a design to estimate a set of linear parametric functions in cases wherein the adequacy of the preliminary linear model is in doubt. Hedayat, Raktoe, and Federer (1974) considered designs for which $\boldsymbol{\theta}'_{(i)}$'s [see Section 9.5 for description of $\boldsymbol{\theta}_{(i)}$'s and the model (9.8)] or specified linear functions of $\boldsymbol{\theta}_{(i)}$'s are estimable under the model $E(Y_{i_1,\ldots,i_k}) = \mathbf{x}'(i_1, \ldots, i_k)\boldsymbol{\theta}$ and used the bias of the estimates obtained under this model when the true model is (9.7). They introduced the alias matrix which is the coefficient matrix of $\boldsymbol{\xi}$ in the least squares estimates obtained under the model (9.7) with $\boldsymbol{\xi}=\mathbf{0}$. The norm of this alias matrix has been proposed by them as a measure which would guide the experimenter in the selection of a design. It is pointed out by them that this measure does not take into account the precision of the estimates. See also Kuwada (1981) and Shirakura (1976, 1979a, 1979b).

REFERENCES

Arthanari, T. S. and Dodge, Y. (1981). *Mathematical Programming in Statistics.* New York: Wiley.

Dodge, Y. (1976). Estimability considerations for 2^n factorial experiments with missing observations. Technical Report 7606, Indian Statistical Institute, New Delhi.

Dodge, Y. and Afsarinejad, K. (1984). Minimal 2^n connected factorial experiments. Technical Report 24, Department of Mathematics, Linköping University, Sweden.

Dodge, Y. and Shah, K. R. (1977). On a measure of usefulness for fractionally replicated designs. Bulletin of the International Statistical Institute, 41st session, 156–159, New Delhi.

Hedayat, A., Raktoe, B. L., and Federer, W. T. (1974). On a measure of aliasing due to fitting an incomplete model. *Ann. Statist.* **2**, 650–660.

Kuwada, M. (1981). Note on the norm of some alias matrix in balanced fractional 2^m factorial designs. Technical Report 41, Hiroshima University.

Shirakura, T. (1976). A note on the norm of alias matrices in fractional replication. *Austral. J. Statist.* **18**, 158–160.

Shirakura, T. (1979a). Alias balanced and alias partially balanced fractional 2^m factorial designs of resolution $2l+1$. *Ann. Inst. Statist. Math.* **31**, 57–65.

Shirakura, T. (1979b). On the norm of alias matrices in balanced fractional 2^m factorial designs of resolution $2l+1$. *J. Statist. Planning Inf.* **3**, 337–345.

PROGRAM BLOCKCON

```
C
C       THIS INTERACTIVE PROGRAM GIVES A LIST OF AxB MINIMALLY
C       CONNECTED DESIGNS. THE SIZE (AxB) OF THE DESIGNS IS CHOOSEN
C       BY THE USER AS WELL AS THE NUMBER OF DESIGNS AND THE NAME OF
C       THE OUTPUT FILE.
C       THE PROGRAM IS LIMITED AT 9x9 DESIGNS, BUT IT COULD BE CHANGED.
C       THE FIRST DESIGN IS ALWAYS AN INCIDENCE MATRIX WHICH LAST
C       COLUMN AND LAST ROW ARE FILLED WITH NON-ZERO ELEMENTS.
C
C       STRUCTURE
C       VARIABLES
C       IFIN    INPUT:  STOP TEST
C       N       INPUT:  NUMBER OF CELLS IN THE INCIDENCE MATRIX
C       NA      INPUT:  NUMBER OF ROWS OF THE INCIDENCE MATRIX
C       NB      INPUT:  NUMBER OF COLUMNS OF THE INCIDENCE MATRIX
C       NC      INPUT:  NUMBER OF NON-ZERO CELLS
C       NOM     INPUT:  NAME OF THE OUTPUT FILE
C       ZLIMITE INPUT:  NUMBER OF DESIRED DESIGNS
C
C       ARRAYS
C       A       INPUT:  COMBINATION VECTOR
C       B       INPUT:  OUTPUT MATRIX OF COMBINATIONS
C       C       INPUT:  INCIDENCE MATRIX
C
C       SUBROUTINES
C       CALCUL  : IT PERFORMS R-PROCESS
C       COMBI   : IT FINDS EVERY COMBINATIONS OF CELLS
C       LARGEUR : IT COMPUTES THE LENGTH OF A STRING WHICH IS A INTEGER
C       MATRIX  : IT CONSTRUCTS THE INCIDENCE MATRIX
C
        PROGRAM NCOTWO
        PARAMETER  IB=81
        CHARACTER C(IB)*2,NOM*9
        DIMENSION A(IB),B(14,IB)
        TYPE 1
1       FORMAT(//,1X,'SIZE OF INCIDENCE MATRIX (axb)  ? ',$)
        ACCEPT *,NA,NB
        TYPE 6
6       FORMAT(//,1X,'NUMBER OF MINIMALLY CONNECTED DESIGNS DESIRED ? ',$)
        ACCEPT *,ZLIMITE
        TYPE 7
7       FORMAT(//,1X,'NAME OF OUTPUT FILE ? ',$)
        ACCEPT 8,NOM
8       FORMAT(A9)
        OPEN(UNIT=5,FILE=NOM//'.DAT',STATUS='NEW')
        NC=NB+NA-1
        N=NA*NB
C
C       CONSTRUCT THE INCIDENCE MATRIX
C
        CALL MATRIX(C,A,IB,NA,NB,NC)
        IFOIS=99/7
        INC=0
        ITOTAL=0.
        TEST=0.
        TOT=0.
```

```
2         CONTINUE
          TEST=TEST+1.
C
C         PERFORM R-PROCESS
C
          CALL CALCUL(C,A,IB,NA,NB,NC,IT)
          IF(TEST.GT.10.AND.IT.LT.N) THEN
C
C         SOME COMBINATIONS ARE LAID OUT
C
          TEST=0.
          DO I=1,10
                  CALL COMBI(A,NC,N,IB)
          END DO
          CALL CALCUL(C,A,IB,NA,NB,NC,IT)
          END IF
          IF(IT.GE.N) THEN
C
C         RECORD THE MATRIX WHICH IS MINIMALLY CONNECTED
C
          INC=INC+1
          TEST=0.
          ITOTAL=ITOTAL+1
          TOT=TOT+1.
          DO I=1,NC
                  B(INC,I)=A(I)
          END DO
          END IF
          IF(INC.GE.IFOIS) THEN
          DO K=1,NC
                  WRITE(5,3) (C(INT(B(I,K)))(1:2),I=1,IFOIS)
3         FORMAT(<IFOIS>(5X,A2))
          END DO
          WRITE(5,4)
4         FORMAT(1X,//)
          INC=0
          END IF
          IFIN=0
C
C         STOP TEST
C
          DO I=1,NC
                  IF(A(I).EQ.FLOAT(I)) IFIN=IFIN+1
          END DO
          IF(TOT.LT.ZLIMITE) THEN
          IF(IFIN.NE.NC) THEN
          CALL COMBI(A,NC,N,IB)
          GOTO 2
          END IF
          END IF
          IF(INC.GT.0) THEN
          DO K=1,NC
                  WRITE(5,5) (C(INT(B(I,K)))(1:2),I=1,INC)
5         FORMAT(<INC>(5X,A2))
          END DO
          END IF
```

```
      CALL LARGEUR(ITOTAL,KK)
      WRITE(5,9) ITOTAL
9     FORMAT(//,1X,'NUMBER OF MINIMALLY CONNECTED DESIGNS FOUND ',I<KK>)
      CLOSE(UNIT=5)
      STOP
      END
C
C
C
      SUBROUTINE COMBI(A,NC,N,IB)
      DIMENSION A(IB)
      DO I=NC,1,-1
            IND=I
            IF(I-1.LE.0) GOTO 1
            IF(A(I).NE.A(I-1)+1.) GOTO 1
      END DO
1     CONTINUE
      Z=FLOAT(N)
      IF(IND+1.LE.NC) THEN
      DO I=NC,IND+1,-1
            A(I)=Z
            Z=Z-1.
      END DO
      END IF
      A(IND)=A(IND)-1.
      RETURN
      END
C
C
C
      SUBROUTINE MATRIX(C,A,IB,NA,NB,NC)
      CHARACTER I1*1,I2*1,C(IB)*2,CP(81)*2
      DIMENSION A(IB)
      IO=0
      DO I=1,NA
      DO K=1,NB
            IO=IO+1
            WRITE(I1,1) I
            WRITE(I2,1) K
1           FORMAT(I1)
            C(IO)(1:2)=I1//I2
      END DO
      END DO
C
C     FIND THE FIRST COMBINATION
C
      DO I=1,IO
            CP(I)=C(I)
      END DO
      IZ=NC
      WRITE(I1,1) NA
      DO I=IO,1,-1
            IF(CP(I)(1:1).EQ.I1) THEN
            A(IZ)=FLOAT(I)
            IZ=IZ-1
            CP(I)(:)=' '
            END IF
```

```
        END DO
        WRITE(I1,1) NB
        DO I=IO,1,-1
                IF(IZ.LE.0) GOTO 2
                IF(CP(I)(2:2).EQ.I1) THEN
                A(IZ)=FLOAT(I)
                IZ=IZ-1
                CP(I)(:)=' '
                END IF
        END DO
2       RETURN
        END
C
C
C
        SUBROUTINE CALCUL(C,A,IB,NA,NB,NC,IT)
C
C       ARRAYS
C       M       INPUT : INCIDENCE MATRIX
C
        PARAMETER IA=81
        CHARACTER C(IB)*2
        DIMENSION M(IA,IA),MT(IA,IA),A(IB)
C
C       INITIALIZE M AT 0
C
        DO I=1,NA
        DO K=1,NB
                M(I,K)=0
                MT(I,K)=0.
        END DO
        END DO
C
C       CONSTRUCT M
C
        DO I=1,NC
                READ(C(INT(A(I)))(1:1),2) I1
                READ(C(INT(A(I)))(2:2),2) I2
2               FORMAT(I1)
                M(I1,I2)=1
        END DO
C
C       PERFORM R-PROCESS
C
1       CONTINUE
        DO I=1,NA
                DO J=1,NB
                IL=0
                DO K=1,NA
                        DO L=1,NB
                                IL=IL+M(I,L)*M(K,L)*M(K,J)
                        END DO
                END DO
                IF(IL.GT.0) MT(I,J)=1
                END DO
        END DO
```

```
      I2=0
      IT=0
      DO I=1,NA
              DO J=1,NB
                      I2=I2+M(I,J)
                      IT=IT+MT(I,J)
              END DO
      END DO
      IF(I2.EQ.IT) GOTO 3
      DO I=1,NA
      DO K=1,NB
              M(I,K)=MT(I,K)
      END DO
      END DO
      GOTO 1
3     RETURN
      END
C
C
C
      SUBROUTINE LARGEUR(I,K)
      CHARACTER CHAINE*12
      WRITE(CHAINE,1) I
1     FORMAT(I12)
      DO K=1,12
              IF(CHAINE(K:K).NE.' ') GOTO 2
      END DO
2     CONTINUE
      K=13-K
      RETURN
      END
C
```

15	15	15	15	15	15	15	15	15	15
25	25	25	25	25	25	25	25	25	25
35	35	35	35	35	35	35	35	35	35
45	44	44	44	43	43	43	43	43	43
51	51	45	45	51	45	45	44	44	44
52	52	51	51	52	51	51	51	51	45
53	53	52	52	53	52	52	52	52	51
54	54	53	53	54	54	53	54	53	52
55	55	55	54	55	55	54	55	55	55

NUMBER OF MINIMALLY CONNECTED DESIGNS FOUND 10

15	15	15	15	15	15	15	15	15	15
25	25	25	25	25	25	25	25	25	25
35	34	34	34	33	33	33	33	33	33
41	41	35	35	41	35	35	34	34	34
42	42	41	41	42	41	41	41	41	35
43	43	42	42	43	42	42	42	42	41
44	44	43	43	44	44	43	44	43	42
45	45	45	44	45	45	44	45	45	45

NUMBER OF MINIMALLY CONNECTED DESIGNS FOUND 10

14	14	14	14	14	14	14	14	14	14
24	24	24	24	24	24	24	24	24	24
34	33	33	33	32	32	32	32	32	32
41	41	34	34	41	34	34	33	33	33
42	42	41	41	42	41	41	41	41	34
43	43	42	42	43	43	42	43	42	41
44	44	44	43	44	44	43	44	44	44

NUMBER OF MINIMALLY CONNECTED DESIGNS FOUND 10

Index

HOLLANDER and WOLFE • Nonparametric Statistical Methods
IMAN and CONOVER • Modern Business Statistics
JAGERS • Branching Processes with Biological Applications
JESSEN • Statistical Survey Techniques
JOHNSON and KOTZ • Distributions in Statistics
Discrete Distributions
Continuous Univariate Distributions—1
Continuous Univariate Distributions—2
Continuous Multivariate Distributions
JOHNSON and KOTZ • Urn Models and Their Application: An Approach to Modern Discrete Probability Theory
JOHNSON and LEONE • Statistics and Experimental Design in Engineering and the Physical Sciences, Volumes I and II, *Second Edition*
JUDGE, HILL, GRIFFITHS, LÜTKEPOHL and LEE • Introduction to the Theory and Practice of Econometrics
JUDGE, GRIFFITHS, HILL, LÜTKEPOHL and LEE • The Theory and Practice of Econometrics, *Second Edition*
KALBFLEISCH and PRENTICE • The Statistical Analysis of Failure Time Data
KISH • Survey Sampling
KUH, NEESE, and HOLLINGER • Structural Sensitivity in Econometric Models
KEENEY and RAIFFA • Decisions with Multiple Objectives
LAWLESS • Statistical Models and Methods for Lifetime Data
LEAMER • Specification Searches: Ad Hoc Inference with Nonexperimental Data
LEBART, MORINEAU, and WARWICK • Multivariate Descriptive Statistical Analysis: Correspondence Analysis and Related Techniques for Large Matrices
McNEIL • Interactive Data Analysis
MAINDONALD • Statistical Computation
MANN, SCHAFER and SINGPURWALLA • Methods for Statistical Analysis of Reliability and Life Data
MARTZ and WALLER • Bayesian Reliability Analysis
MIKÉ and STANLEY • Statistics in Medical Research: Methods and Issues with Applications in Cancer Research
MILLER • Beyond ANOVA, Basics of Applied Statistics
MILLER • Survival Analysis
MILLER, EFRON, BROWN, and MOSES • Biostatistics Casebook
MONTGOMERY and PECK • Introduction to Linear Regression Analysis
NELSON • Applied Life Data Analysis
OSBORNE • Finite Algorithms in Optimization and Data Analysis
OTNES and ENOCHSON • Applied Time Series Analysis: Volume I, Basic Techniques
OTNES and ENOCHSON • Digital Time Series Analysis
PANKRATZ • Forecasting with Univariate Box-Jenkins Models: Concepts and Cases
PIELOU • Interpretation of Ecological Data: A Primer on Classification and Ordination
POLLOCK • The Algebra of Econometrics
PRENTER • Splines and Variational Methods
RAO and MITRA • Generalized Inverse of Matrices and Its Applications
RIPLEY • Spatial Statistics
SCHUSS • Theory and Applications of Stochastic Differential Equations
SEAL • Survival Probabilities: The Goal of Risk Theory
SEARLE • Linear Models
SEARLE • Matrix Algebra Useful for Statistics
SPRINGER • The Algebra of Random Variables

(*continued from front*)